Methods for Real-Time Measurement of THMs and HAAs in Distribution Systems—Part 2

Awwa Research Foundation

About the Awwa Research Foundation

The Awwa Research Foundation (AwwaRF) is a member-supported, international, nonprofit organization that sponsors research to enable water utilities, public health agencies, and other professionals to provide safe and affordable drinking water to consumers.

The Foundation's mission is to advance the science of water to improve the quality of life. To achieve this mission, the Foundation sponsors studies on all aspects of drinking water, including supply and resources, treatment, monitoring and analysis, distribution, management, and health effects. Funding for research is provided primarily by subscription payments from approximately 1,000 utilities, consulting firms, and manufacturers in North America and abroad. Additional funding comes from collaborative partnerships with other national and international organizations, allowing for resources to be leveraged, expertise to be shared, and broad-based knowledge to be developed and disseminated. Government funding serves as a third source of research dollars.

From its headquarters in Denver, Colorado, the Foundation's staff directs and supports the efforts of more than 800 volunteers who serve on the board of trustees and various committees. These volunteers represent many facets of the water industry, and contribute their expertise to select and monitor research studies that benefit the entire drinking water community.

The results of research are disseminated through a number of channels, including reports, the Web site, conferences, and periodicals.

For subscribers, the Foundation serves as a cooperative program in which water suppliers unite to pool their resources. By applying Foundation research findings, these water suppliers can save substantial costs and stay on the leading edge of drinking water science and technology. Since its inception, AwwaRF has supplied the water community with more than $300 million in applied research.

More information about the Foundation and how to become a subscriber is available on the Web at **www.awwarf.org.**

Methods for Real-Time Measurement of THMs and HAAs in Distribution Systems—Part 2

Prepared by:
Gary L. Emmert, **Michael A. Brown**, **Paul S. Simone Jr.**, **Gija Geme**, and **Gang Cao**
The University of Memphis
Department of Chemistry
Rm 213 Smith Chemistry Building
Memphis, TN 38152

Jointly sponsored by:
Awwa Research Foundation
6666 West Quincy Avenue, Denver, CO 80235-3098

and

U.S. Environmental Protection Agency
Washington D.C.

Published by:

Awwa Research Foundation

American Water Works Association

IWA Publishing

DISCLAIMER

This study was jointly funded by the Awwa Research Foundation (AwwaRF) and the U.S. Environmental Protection Agency (USEPA) under Cooperative Agreement No. R-82940901. AwwaRF and USEPA assume no responsibility for the content of the research study reported in this publication or the opinions or statements of fact expressed in the report. The mention of trade names for commercial products does not represent or imply the approval or endorsement of AwwaRF or USEPA. This report is presented solely for informative purposes.

Copyright © 2007
by Awwa Research Foundation

All Rights Reserved

CONTENTS

TABLES

APPENDIX TABLES

FIGURES

APPENDIX FIGURES

FOREWORD

The Awwa Research Foundation is a nonprofit corporation that is dedicated to the implementation of a research effort to help utilities respond to regulatory requirements and traditional high-priority concerns of the industry. The research agenda is developed through a process of consultation with subscribers and drinking water professionals. Under the umbrella of a Strategic Research Plan, the Research Advisory Council priorities the suggested projects based upon current and future needs, applicability, and past work; the recommendations are forwarded to the Board of Trustees for final selection. The foundation also sponsors research projects through the unsolicited proposal process; the Collaborative Research, Research Application, and Tailored Collaboration program; and various joint research efforts with organization such as the U.S. Environmental Protection agency, the U.S. Bureau of Reclamation, and the Association of California Water Agencies.

This publication is a result of one these sponsored studies, and it is hoped that its findings will be applied in communities throughout the world. The following report serves not only a means of communicating the results of the water industry's centralized research program but also as a tool to enlist the further support of the nonmember utilities and individuals.

Projects are managed closely from their inception to the final report by the foundation's staff and large cadre of volunteers who willingly contribute their time and expertise. The foundation serves a planning and management function and awards contracts to other institutions such as water utilities, universities, and engineering firms. The funding for this research effort comes primarily from the Subscription Program, through which water utilities subscribe to the research program and make an annual payment proportionate to the volume of water they deliver and consultants and manufacturers subscribe based on their annual billings. The program offers a cost-effective and fair method for funding research in the public interest.

A broad spectrum of water supply issues is addressed by the foundation's research agenda: resources, treatment and operations, distribution and storage, water quality and analysis, toxicology, economics, and management. The ultimate purpose of the coordinated effort is to assist water suppliers to provide the highest possible quality of water economically and reliably. The true benefits are realized when the results are implemented at the utility level. The foundation's trustees are pleased to offer this publication as a contribution toward that end.

Walter J. Bishop
Chair, Board of Trustees
Awwa Research Foundation

Robert C. Renner, P.E.
Executive Director
Awwa Research Foundation

ACKNOWLEDGMENTS

The authors of this report are indebted to the following water utilities, organizations and individuals for their cooperation and participation in this project.

Memphis Light Gas and Water (MLGW), Water Quality Assurance Laboratory, Memphis Tennessee.

City of Houston, Public Works and Engineering Department, Water Production Branch, Houston, Texas.

The Department of Chemistry, College of Arts and Sciences and Office of the Vice-Provost, The University of Memphis, Memphis Tennessee

The authors would like to extend thanks to Alice Fulmer of AwwaRF who served as Project Manager for this study. Alice has provided excellent guidance for the authors and the Project Advisory Committee (PAC). She has done a superb job at keeping us on track and focused. The authors are indebted to the members of the PAC who provided exceptional advice during the review of this Report. Their thorough and thoughtful comments each reporting period have elevated our work beyond what we could have accomplished without their insightful feedback. The PAC members include: Bill B. Potter, Research Chemist at the USEPA, Cincinnati, OH; Claudia Wheeler, Metropolitan Water District of Salt Lake City & Sandy, Sandy, UT; and Kenneth Froese, Associate Professor of Public Health Sciences, Department of Public Health Sciences, Faculty of Medicine and Dentistry, University of Alberta.

We could not have done this project without the help of our colleagues in Memphis and Houston. Torrence Myers, Tom Winford, Glarretter Carter-Burks, and all of the Water Quality Laboratory at MLGW, are a constant resource for us at the University of Memphis. We are proud to call them our sister laboratory. We also thank Ying Wei and her staff at the City of Houston Water Quality Laboratory for their technical help and that fine Texas hospitality extended to us while we were on-site at Houston.

The authors would also like to extend thanks to Hugh Goldsmith of SRI Instruments, Inc., Don Olson and Woody Wolcott at Global FIA Inc., Janardhan Madabushi and Jim Whitford of VICI Inc., Einar Pontén of SeQuant Inc., Rob Buck and Larry Meeker of Waters Corporation, and Rick Jones, Tom Short and Ted Cook of Dionex.

EXECUTIVE SUMMARY

Chlorination is the most widely used water disinfection process in the United States. Chlorine, typically applied as gaseous chlorine, sodium hypochlorite, or calcium hypochlorite is a safe and economical water disinfectant and has been used with great success for over a century. Unfortunately, water chlorination results in the formation of halogenated disinfection by-products (DBPs). These DBPs include the trihalomethanes (THMs) and haloacetic acids (HAAs). Maximum contaminant levels (MCL) have been established by the USEPA for drinking water due to the possible adverse health effects from ingestion of THMs and HAAs. The current MCL for total THMs and total HAAs in drinking water are 0.080 mg/L and 0.060 mg/L, respectively.

It is possible that future regulations might lower the MCL for these DBPs, requiring plant operators in water utilities to develop new strategies aimed at minimizing total THMs and total HAAs concentrations in drinking water. They might also look toward optimizing their removal processes. To accomplish these goals, analytical methods are needed that will provide monitoring data in real-time or near-real time. The development of such measurement methods for on-line monitoring will improve understanding of variations in THM and HAA concentrations in treatment plants and drinking water distribution systems. The monitoring data will increase understanding of how peak THM and HAA formation events occur, and when they do not occur. This allows operators to adjust treatment processes accordingly at the plant and in the distribution system. Finally, the data that such monitoring provides can be used to develop better models of the drinking water distribution system.

RESEARCH OBJECTIVES

The overall goal of this research was to develop chemical analyzers and associated analytical methods for on-line monitoring of THM and HAA concentrations directly from drinking water distribution systems. There were several approaches that were explored for on-line monitoring of THMs including: (1) on-line monitoring purge and trap gas chromatography (OPTGC); (2) supported capillary membrane sampling-gas chromatography (SCMS-GC); (3) capillary membrane sampling-flow injection analysis (CMS-FIA); (4) capillary membrane sampling-gas chromatography (CMS-GC). The methods that were explored for monitoring HAAs included: (1) ion chromatography with membrane suppressed conductivity detection (IC-MSCD); (2) CMS-FIA; (3) post column-ion chromatography (PC-IC).

Preliminary studies were designed to evaluate the method detection limits (MDL), accuracy and precision of these analytical methods. For the methods that showed promise, optimization and linearity studies were performed. Additional experiments were designed to evaluate the utility of these methods to provide on-line monitoring data from water distribution systems. The proposed analytical methods were compared to appropriate EPA methods for THMs and HAAs. The methods were judged based on the following parameters: (1) the magnitude of the effect of total organic carbon (TOC) and ionic strength on the analytical signal; (2) the adaptability of the proposed method to real-time or near real-time measurements including the sample analysis rate; (3) the field portability of the instrumentation and method; (4) the user friendliness (or expertise level) of the instrumentation/method; and (5) the affordability of the instrumentation/method.

APPROACH

The candidate analytical methods were subjected to the following generalized format as they were developed, optimized and evaluated. A preliminary familiarization study was conducted to estimate the basic abilities of the method. For those methods that showed promise, a more detailed MDL, accuracy and precision study was done. For those methods that continued to appear promising after these initial studies, additional optimization studies were typically done. Preliminary linearity studies were carried out and the linear ranges were estimated. The analyzers were also evaluated with respect to their ruggedness by operating the promising methods in the presence of possible interferences in the laboratory and real-world situations. When possible, parameters were identified to establish control charts that will help to insure that the analyzer is operating within set limits in future studies. Ultimately the methods were subjected to validation studies by carrying out side-by-side comparisons to EPA methods in real world systems.

SUMMARY OF RESULTS

Several on-line monitoring methods were explored, along with a detailed report on construction, operation and maintenance of the promising analyzers. These methods included OPTGC, CMS-GC, CMS-FIA, and PC-IC. In each case, these methods were subjected to detailed comparison studies to EPA methods in the laboratory as well as diverse and challenging real world situations.

The OPTGC is a commercially available approach to purge and trap sampling that was developed in conjunction with SRI instruments and was applied to individual and total THM4 analysis. The OPTGC approach combined membrane sampling and purge and trap sampling to produce a relatively simple approach for on-line monitoring of THM4. The MDLs for each THM4 species using OPTGC are all less than 0.5 μg/L, with accuracies, measured as mean % recoveries ranging from 103 to 127%. The OPTGC also provides for relatively precise data exhibiting % RSD values less than 5%, typically ranging from 2.1-4.9%. The OPTGC was developed based on EPA methods, being similar to purge and trap methods such as EPA 502.2 and EPA 524.2. The OPTGC instrument is relatively inexpensive with an estimated price in the neighborhood of $15-20K. It can achieve sampling rates of 2-3 samples per hour depending on the specific sample matrix and it can be easily automated to report data in near-real time from remote locations via wired or wireless internet.

The CMS sampling device developed for FIA and GC can distinguish between volatile and non-volatile species. This device is simple and inexpensive to construct and can be made from commercially available components for less than $200.

The CMS device was interfaced to a GC to develop a new approach to on-line monitoring of THM4 that provides individual and total THM4 concentrations. The CMS-GC can be interfaced to an existing GC with an ECD, thus offers alternatives to laboratories that may already have such an instrument available. The CMS-GC method with ECD provides MDL values for the THM4 species in the range 0.3 to 0.5 μg/L, with accuracies ranging from 100 to 125% and precision estimates ranging from 3.5 to 8.6%. If a utility already has a GC with ECD, then this approach is very economical. Even if it is necessary to purchase a low end GC-ECD, the cost of this approach is still low, in the neighborhood of $8-12K. The CMS-GC can reasonably be expected to achieve sampling rates of 2-3 samples per hour and exhibits all the same reporting and communications options discussed with OPTGC.

A new approach to simplifying GC analysis, called GCOV, was detailed. GCOV is a practical starting point for miniaturizing analytical instrumentation useful for on-line monitoring. GCOV offers a low cost approach to analysis that is perhaps a bit more challenging to construct compared to CMS-GC. GCOV is still a relatively new approach, but may offer unique advantages, and uses a commercially available PDPID detector to detect THM4 species. It is possible to achieve sampling rates of 3-4 samples per hour with this device, but the operator skill level is slightly higher than other GC based methods. Preliminary studies gave MDL estimates in the range 1.1-1.6 μg/L with mean % recoveries ranging from 91-119% and % RSD values ranging from 8.6 to 16%.

The analytically important reaction between nicotinamide (NCA) and THM4 and HAA9 species was examined in detail. NCA reacts with THM4 and HAA9 species to form fluorescent products. Two analytical methods, CMS-FIA and PC-IC, were developed based on this chemistry that allowed for on-line monitoring in near-real time.

The CMS-FIA approach is, to the best of our knowledge, the only method that allows for selective on-line monitoring of total THM4 and total HAA9 in near-real time. This method is automated and allows for direct on-line monitoring of both total THM4 and total HAA9 concentrations directly from the drinking water distribution system. The MDL estimates for the CMS-FIA method are 1.1 μg/L for total THM4 and 1.6 μg/L for total HAA9. The CMS-FIA method gave accuracies of 111-115% and precision estimates of 2-3%. Perhaps the most important aspect to the CMS-FIA method is that it is based on florescence detection, which simplifies low level detection on-line. The operator skill needed to operate the CMS-FIA is relatively low compared to the other methods in this project. Correspondingly, the simplicity has an analytical cost in that CMS-FIA cannot be used to speciate THM4 and HAA9 components, but thanks to the CMS device it can distinguish between total THM4 and total HAA9. The CMS-FIA method can provide sampling rates up to four samples per hour. The fluorescence detector needed to construct the CMS-FIA can be a low end filter based unit that can greatly reduce expense compared to high end monochromator based detectors. The entire CMS-FIA analyzer could be constructed from commercially available components for ~$10K, or significantly cheaper if the laboratory already has a fluorescence detector available.

The PC-IC method allows for on-line monitoring of individual HAA5 species as well as total HAA5. The initial PC-IC analyzer was developed to quantify HAA5 and could not resolve the monohalogenated species. The method exhibits good selectivity for HAA5 species and offers a much simpler approach to sample collection; processing and analysis compared to common derivatization procedures that employ liquid-liquid extraction. The PC-IC method separates the HAA5 species using an anion exchange column followed by post column reaction with NCA and fluorescence detection. The MDL estimates using a state-of-the-art dual monochromator based fluorescence detector range from 1.2 to 2.1 μg/L with accuracies of 79.4 to 105% and precision estimates of 5.9-11%. This PC-IC method achieved sampling rates of 1 sample per hour and requires a higher level of operator skill and instrumentation compared to other methods.

Three on-line monitoring methods (CMS-GC, CMS-FIA, and PC-IC) were used in side-by-side studies with appropriate EPA methods in Memphis, TN and Houston, TX. A systematic approach for evaluating the data analytically and strategies for carrying out on-line monitoring studies were proposed. The analyzers and methods were revised based on limitations identified in real world monitoring studies. The specific analytical shortcomings that were identified were related mostly to chromatography and interference issues. It was proposed that prior to conducting on-line monitoring studies; analysts should conduct detailed premonitoring surveys using both standard EPA methods as well as the proposed on-line monitoring methods. Chromatographic issues were modified to allow for better resolution with interferents. Major advances were made with respect to the NCA based methods for HAAs as the methods were extended to analyze all HAA9 species and thereby eliminate four additional chemically similar species that would certainly interfere in measurements. A revised PC-IC method was developed that allows for measuring all HAA9 species. The MDL estimates range from 5-16 μg/L using an inexpensive filter based fluorescence detector and 0.6 to 10 μg/L using a state-of-the-art dual monochromator based fluorescence detector. The filter based PC-IC reports accuracies in the range 73.8 to 99.7% with precision estimates of 7-21 %, close to the MDLs of the method. The higher-end dual monochromator based PC-IC gave accuracies of 79-161 percent with precision estimates of 3.5 to 32%, again operating near the MDLs of the proposed method. Adding the additional four HAA species did result in a significant decrease in the sampling rate for the PC-IC method, from 1 to 0.5 samples per hour. However, the additional information provided by this revised method makes the decrease in sampling rate worthwhile.

Four revised chemical analyzers and associated methods (OPTGC, CMS-GC, CMS-FIA and PC-IC) were used in additional side-by-side studies with appropriate EPA methods in Memphis, TN and Houston, TX. A detailed statistical analysis of the results of the monitoring studies was done to compare the methods of choice. As long as one keeps in mind these analyzers are designed for treatment practice optimization studies rather than compliance monitoring, all four methods provided acceptable results when used in on-line monitoring studies in water distribution systems. It is possible to monitor individual THM4 and HAA9 concentrations using either OPTGC or CMS-GC for THM4 and PC-IC for HAA9. The CMS-FIA provides for a mass balance with the other methods as it can provide total THM4 and total HAA9. Used in concert, these three analyzers provide detailed on-line monitoring data directly from drinking water distribution systems over the concentration range ~1 to 100 μg/L with sampling rates of up to 3-4 samples per hour for all methods with the notable exception of PC-IC. For PC-IC, sampling rates are restricted to ~0.5-1 sample per hour with the current instrument.

A broader overview, a laboratory perspective, was presented that describes the implications of the results reported here, with specific challenges faced by relatively small water quality laboratories with limited resources. Particular attention was paid to keep methods economical and relatively easy to operate. Specific analytical issues with EPA 552.3 were identified and discussed, including most importantly an apparent positive bias. Broadly, this research provides a starting point for on-line monitoring with other species of interest such as emerging DBPs, pesticides, and other possible contaminants in drinking and wastewater. This is especially true of other VOCs and halogenated species, but also of compounds of interest to security monitoring. These methods also provide a fundamental chemical basis for the development of miniaturized analytical devices such as lab-on-a-chip technologies.

CONCLUSIONS AND RECOMMENDATIONS

We have proposed a practical low cost approach to on-line monitoring using well established analytical instrumentation that have been modified for on-line monitoring directly from water distribution systems. Three methods are recommended for monitoring THM4 and total THM4 concentrations: CMS-GC, OPTGC, CMS-FIA. Two methods are proposed for monitoring HAA9 concentrations: CMS-FIA and PC-IC. These recommended analyzers can be used in concert to get detailed concentration over time profiles with sampling rates up to 4 samples per hour. The combination of the CMS-FIA method with one of the selective methods for THM4 or HAA9, allows for a mass balance in these studies and agree within reason. Alternatively, the methods could be used alone and provide valuable data. The CMS-FIA is particularly attractive in that it provides data on both total THM4 and total HAA9 concentrations and could likely be subjected to further miniaturization for low cost near real-time chemical sensors that could be located throughout a distribution system. The analyzers were subjected to detailed studies in real world situations and were shown to provide reliable data when compared to EPA methods. Procedures are proposed to allow for statistical analysis of the absolute and relative errors associated with these on-line monitoring methods. Strategies for on-line monitoring were proposed and evaluated. Most notably, it is important to use established EPA procedures to conduct premonitoring surveys of the distribution system prior to to running on-line methods. The information obtained from the promonitoring surveys allows the analyst to understand the potential for interferences peculiar to the system. Armed with a detailed knowledge of the components of the distribution systems, the analyst can then use the on-line monitoring analyzers to obtain meaningful time dependent measurements directly from the distribution system. Studies of this type can provide important information for optimizing treatment practices. The use of multiple analyzers simultaneously throughout the distribution system could provide both spatial and temporal data. In particular, observations have been made related to a possible bias in the EPA 552.3 method for measuring HAA9 species. Sampling directly from the distribution system and using the PC-IC and CMS-FIA approach may provide a more accurate measure of the HAA9 concentrations than the EPA 552.3 since halogenation appears to continue in the sample vial after more than a two day holding time. This effect needs to be explored in more detail as it is a very economically important chemistry. Compliance with existing and developing regulations might hinge in some cases on how long the sample is held before analysis. The analyzers proposed here may provide a more accurate snapshot of the concentrations of THM4 and HAA9 species at a given time, especially with HAA9 species.

CHAPTER 1
OVERVIEW OF METHODS FOR ON-LINE MONITORING OF THM4 AND HAA5 SPECIES IN DRINKING WATER DISTRIBUTION SYSTEMS

OVERVIEW OF RESEARCH PROJECT

The overall goal of this research was to develop chemical analyzers and associated analytical methods for on-line monitoring of trihalomethane (THM) and haloacetic acid (HAA) concentrations in drinking water distribution systems. The THM methods that were developed, optimized and evaluated included: (1) an on-line monitoring purge and trap gas chromatograph (OPTGC) equipped with a dry electrolytic conductivity detector (DELCD); (2) a supported capillary membrane sampler (SCMS) interfaced to a GC with electron capture detector (ECD), or a pulsed discharge photoionization detector (PDPID), or an ion trap mass spectrometer (MS); (3) a capillary membrane sampling (CMS)-flow injection analysis (FIA) analyzer using the reaction of nicotinamide (NCA) with THM4 and HAA5 species with fluorescence detection of the products; (4) a CMS-GC analyzer with an ECD. The HAA methods that were developed, optimized and evaluated included: (1) an ion chromatography based method using commercially available membrane suppressed conductivity detection (IC-MSCD); (2) a CMS-FIA method for total HAA5 species (THAA5); (3) a method that used IC separation followed by post column reaction of HAA5 species with NCA, followed by detection of the fluorescent products (PC-IC).

Preliminary studies were designed to evaluate the method detection limits (MDL), accuracy and precision of these analytical methods. For the methods that showed promise, optimization and linearity studies were performed. Additional experiments were designed to evaluate the utility of these methods to provide on-line monitoring data from water distribution systems. The proposed analytical methods were compared to appropriate EPA methods for THM and HAA species. The methods were judged based on the following parameters: (1) the magnitude of the effect of total organic carbon (TOC) and ionic strength on the analytical signal; (2) the adaptability of the proposed method to real-time or near real-time measurements including the sample analysis rate; (3) the field portability of the instrumentation and method; (4) the user friendliness (or expertise level) of the instrumentation/method; and (5) the affordability of the instrumentation/method.

Background: THMs and HAAs in Drinking Water

Chlorination is the most widely used water disinfectant process in the United States. Chlorine, typically applied as gaseous chlorine, sodium hypochlorite, or calcium hypochlorite, is a safe and economical water disinfectant. Chlorination has been used with great success in the water industry for over a century. Unfortunately, water chlorination results in the formation of halogenated disinfection by-products (DBPs). These DBPs include THMs and HAAs. Concern over the possible health effects of THMs and HAAs has led to the legislation of maximum contaminant levels (MCL) in drinking water. The THMs include chloroform ($CHCl_3$), bromodichloromethane ($CHBrCl_2$), dibromochloromethane ($CHBr_2Cl$) and bromoform ($CHBr_3$). The current MCL for total THMs (TTHM4) in drinking water is 0.080 mg/L. The five HAAs that are commonly monitored include monochloroacetic acid (MCAA), dichloroacetic acid (DCAA), trichloroacetic acid (TCAA), monobromoacetic acid (MBAA) and dibromoacetic acid (DBAA).

These five HAAs are referred to collectively as HAA5. The current MCL for HAA5 is 0.060 mg/L for total concentration of HAA5 (THAA5).

Stage Two of the Disinfectant-Disinfection By-Product Rule (2003) may lower the MCL for TTHM4 and THAA5 in the near future, leading many plant operators to develop new strategies for minimizing THM4 and HAA5 concentrations in drinking water. They might also look toward optimizing THM4 and HAA5 removal processes. To accomplish these goals, analytical methods are needed that will provide reliable data on THM4 and HAA5 concentrations in real-time or near-real time. The development of such measurement methods for on-line monitoring will improve understanding of variations in THM4 and HAA5 concentrations in treatment plants and drinking water distribution systems. The monitoring data will increase understanding of how peak THM4 and HAA5 formation events occur (and when they do not occur). Real-time monitoring allows operators to adjust treatment processes accordingly at the plant and in the distribution system. Finally, the data that such monitoring provides can be used to develop better models of the water distribution system.

Review of Existing Analytical Methods for Monitoring THMs in Drinking Water

While THM4 and HAA5 species have in common that they are all DBPs of water chlorination, they are very different in their chemical structure and properties. The THM4 species are non-polar and volatile while the HAA5 species are polar and non-volatile. These differences have significant implications in their analytical chemistry applied to drinking water, particularly the sampling-sample preparation step. The concentrations of THM4 species can be determined using gas chromatography (GC). The major advantage of GC is its simplicity. GC is easy to use and automate and GC methods are well-established. GC is inexpensive and produces minimal waste. The analysis of the HAA5 species using GC is more complicated. To be analyzed by GC, HAA5 species must be converted to their corresponding methyl-esters (or some other volatile derivative) before they are volatile enough for routine GC analysis. This adds additional, typically manual, steps to the sample preparation process. Many of these steps are susceptible to transfer losses or carry-over of the analyte.

Much effort has been exerted to minimize these sample preparation steps required by GC. Liquid chromatographic (LC) techniques such as high performance liquid chromatography (HPLC), ion chromatography (IC) and HPLC-MS methods have been developed for HAA5 species. While to many, HPLC and IC analysis may seem more complex (meaning they require a higher level of analytical expertise to obtain reliable results), the reality is that these methods are becoming well established and routine. The cost of the technology, i.e., high-pressure pumps, injection valves and simple detectors, has dropped considerably. Some of the LC based methods for HAA5 species can be applied with minimal investment. IC is of particular interest because it can be readily applied to separation of HAA5 species as they are predominantly ionized at typical drinking water pH values.

Due to the demands of on-line monitoring, alternatives to GC and LC technologies such as spectroscopic methods of analysis are of particular interest. Spectroscopic detectors such as ultraviolet (UV), visible (Vis) and fluorescence (FL) have been used to detect THM4 and HAA5 species. Spectroscopic detection can be used in LC and IC methods, positioned at the end of the chromatography column ready to detect the separated species that elute. Additionally, this type of detection can be applied to FIA, sequential injection analysis (SIA) or miniaturized "lab-on-a-chip" measurements. In some instances, chromatographic methods and FIA methods can be coupled to obtain additional advantages (PC-IC).

The literature of analytical chemistry contains a wealth of information pertaining to the analysis of THM4 and HAA5 in drinking water and many details were summarized in the Phase One Report (Emmert et. al 2004b). The goal of Phase One was to collect data about sampling, sample preparation, and specific operation of the instruments that were used for measuring the concentrations of THM4 and HAA5 in drinking water. These methods were evaluated as to their adaptability for on-line real-time monitoring in drinking water distribution systems. The promising analytical methods included a wide array of methodologies that ranged from simple colorimetric methods to more advanced GC, GC-MS and LC based methods, and finally to less routine methods like capillary electrophoresis and LC-MS methods. The methods that are applied to monitoring THM4 and HAA5 are very different. These differences stem largely from the difference in chemical properties exhibited by THM4 and HAA5, specifically polarity and volatility. Based upon the review, there are some specific options that could be adapted for on-line monitoring of the concentrations of THM4 and HAA5. However, before methods are reviewed with respect to application to monitoring, it would be helpful to review the characteristics developed in cooperation with the Project Advisory Committee (PAC) to evaluate any proposed "Ideal Method."

Defining the Ideal Method

The focus of the project is to ultimately develop real-time or near real-time methods for monitoring THM4 and HAA5 in drinking water distribution systems. Ideally, it would be desirable to develop one analytical method that would be capable of monitoring both THM4 and HAA5 concentrations in the distribution system. Realistically, it is unlikely that this will be possible given the differences in chemical properties that exist between these two classes of DBPs. An immediate and perhaps more obtainable compromise would be to have two Ideal Methods, one for THM4 and one for HAA5. When dealing with complex samples such as drinking water, it has long been a goal of environmental analytical chemists to have available two methodologies, based upon different chemistries, for measuring concentrations in complex matrices as this allows for confirmation and quality assurance. Ideally, it would be desirable to have two analytical methods based on two different chemistries for monitoring THM4 and two different analytical methods based upon two different chemistries for monitoring HAA5.

From the standpoint of fundamental analytical chemistry, the method detection limit (MDL) is a very important parameter for determining the reliability and applicability of an analytical method. The MDL will be calculated using the EPA recommended procedure (Glaser and Forest 1981, USEPA 1996a). For this particular study, the MDLs will be considered acceptable if it falls in the range 0.5 to 5 μg/L. Two other parameters that are commonly used to compare analytical methods are accuracy and precision. For our purposes, the mean % recovery will be used to evaluate accuracy. Accuracy is defined as how close the value is to the true value. Precision, defined as the amount of scatter in the data, will be estimated using the % RSD. The accuracy and precision are calculated based on EPA recommendations (USEPA, 1996b). As a rule of thumb, the target goals of the mean % recovery will be 100% ± 5%. The % RSD will be deemed acceptable for monitoring studies if it is less than 10%.

Any analytical method developed for drinking water monitoring in distribution systems should not be adversely affected by the presence of natural organic matter. As a rule of thumb, it has been agreed that any method developed should provide acceptable results when TOC concentrations are in the range 1 to 20 mg/L. In other words, TOC should not be a significant interferent in the proposed analytical methods.

It is expected that the ideal analytical method or methods would provide data in real-time or near real-time. Real-time data collection typically implies that the analytical response will be instantaneous. Such instantaneous data collection, while certainly desirable, may be limited when measuring a class of compounds such as THM4 species and HAA5 species. For example, such methods might provide only TTHM4 or THAA5. While this information could be quite valuable, it is recognized that in many cases, it is desirable to be able to distinguish the different THM4 and HAA5 compounds. To achieve this goal, it is likely that some sort of chromatography or other separation technique will be needed. This means that the data collection rate will be limited by the timescale of the separation process, for example the chromatogram. With this in mind, in the ideal case it would be desirable to collect THM4 or HAA5 concentration data, and have it available for decision-making in the water treatment plant as soon as possible. For example, it would be desirable to have access to monitoring data every five to 20 minutes. A more reasonable, and perhaps obtainable, sampling rate might be at least one sample per hour. Thus the ideal method is expected to provide data as fast as possible, but at least one sample per hour.

An additional characteristic of the ideal method would be that it could be accomplished with portable instrumentation and specifically designed for monitoring in the field. With respect to drinking water monitoring, it is assumed that electrical power will be available in the field. So perhaps a more appropriate way of thinking about field sampling would be carried out at monitoring stations maintained throughout the system---or more specifically at different locations in the city. The ideal method would be able to be applied in such conditions.

It is expected that any methods and portable analyzers that are developed will be user friendly and that the skill level of the operator will likely not be at the typical level of the laboratory chemist with an associates or bachelor's degree—but significantly less specialized in analysis. An obvious path toward user friendliness would be for the method to be largely automated and have the ability to operate with minimal operator attention. The desire to have a simple automated method may seem to circumvent the well-established USEPA method for drinking water monitoring. This is not the case. Any portable automated methods that are developed should naturally be compared to the well-established methods that are proposed by the USEPA for drinking water monitoring. This means that the USEPA methods will be the standard by which the Ideal Method will be compared.

It is expected that the Ideal Method will be affordable—even inexpensive. Expense is a relative term. After all, what might be affordable for one utility might be impossible for another. It is desirable that an array of methods be developed that range from simple automated tests to more advanced techniques that may involve some sort of automated chromatography or electrophoresis. It is desired that the high end cost limit be no more than $20,000. It is also desirable that the analysis time be as short as possible and requires minimal calibration and maintenance. An additional attractive characteristic would be for the device to be able to interface to existing communication networks for early warning capabilities—perhaps through computer or cellular phone networks.

With these characteristics of the Ideal Method in mind, it should be noted that it is unlikely that any one analytical method will meet all these requirements. However, it may be possible to develop several methods that will provide all the desired characteristics through overlapping capabilities.

Generalized procedure used to evaluate the analytical methods

The candidate analytical methods were subjected to the following generalized format as they were developed, optimized and evaluated. A preliminary familiarization study was conducted to estimate the basic abilities of the method. For those methods that showed promise, more detailed MDL, accuracy and precision studies were done. For those methods that continued to appear promising after these initial studies, additional optimization studies were typically done. Preliminary linearity studies were carried out and the linear ranges were estimated. The analyzers were also evaluated with respect to their ruggedness by operating the promising methods in the presence of possible interferences in the laboratory and real-world situations. When possible, parameters were identified to establish control charts that would help to insure that the analyzer is operating within set limits in future studies. Ultimately the methods were subjected to validation studies by carrying out side-by-side comparisons to EPA methods in real world systems. While there was slight variation from method to method, a generalized description of the method development and evaluation procedures follows.

Familiarization Study

A preliminary familiarization study was done for each proposed method. The goal of this study was to establish the instrumental parameters and analytical conditions needed for the analyzer. Exploratory analyses were done using calibration standards. With the instrument stable, a preliminary calibration curve was prepared to evaluate the method. A check standard was analyzed. The percent recovery (% Recovery) and percent relative standard deviation (% RSD) were calculated. The analytical parameters were subjected to a preliminary optimization toward the desired goals of the study.

MDL, Accuracy and Precision Studies

Once the proposed method was subjected to a familiarization study, the optimized method was subjected to preliminary MDL, accuracy and precision studies. Each measurement was used to calculate the following statistics for each analyte (1) the average concentration (2) the standard deviation (3) percent relative standard deviation (4) the mean percent recovery and (5) the preliminary MDL.

The precision of the monitoring methods will be measured using % RSD on the consecutive injections of a check standard. The % RSD should be less than 10%. The MDL was calculated using the USEPA procedure (Glaser and Forest 1981, USEPA 1996a) and the precision and accuracy was calculated based on EPA recommendations (USEPA, 1996b).

The proposed research used the calibration plots to calibrate the methods. Stock standard solutions were prepared containing each analyte and were used to prepare the primary dilution standards. The primary dilution standard was analyzed and each component's retention time was identified for the particular method (when appropriate). It is the practice in the laboratory to construct calibration curves from a total of five primary dilution standards. Linear regression was used to determine the slope and intercept of the calibration curve. A check standard solution is typically analyzed seven consecutive times. From these studies, the MDL, the mean % recovery and the % RSD were calculated. The MDL was calculated by using Equation 1.1. The % recovery was calculated by using Equation 1.2.

$$MDL = t_{(n\text{-}1,\ 1\text{-alpha} = 0.98)} S \quad (1.1)$$

where MDL = Method Detection Limit

$t_{(n\text{-}1,\ 1\text{-alpha} = 0.98)}$ = Student's t value for 98% confidence level with (n-1) degrees of freedom

n = number of replicates

S = standard deviation of check standard analyses

$$\% REC = (B/A)*100 \quad (1.2)$$

where $\% REC$ = % recovery

A = known concentration of the check standard

B = measured concentration of check standard

The mean % recovery was calculated by taking the average of the % recovery data. Ideally, the mean % recovery should fall within the range 100 ± 5 %. The % RSD was calculated by dividing the standard deviation for the check standard solution concentration results by the average value. The % RSD should be less than 10%.

Instrument Optimization Study

Based upon the first two preliminary studies, the proposed analytical methods were selected for further study (if they show promise) or rejected based upon unsatisfactory performance results. Methods that were classified as showing promise were further optimized, if possible.

Instrument Linearity Study

Candidate methods were subjected to a detailed linearity study. The linear ranges were established for each method. Statistical tests, where possible, were used to establish the significance of linearity of the all methods (Taylor, 1987). The linearity of a method was tested using direct plots of transformed data to represent the relationship between the analytical signal and concentration of analyte. Residual plots were also used to evaluate the method linearity. A plot of the residuals (the difference between the observed analytical signal and calculated analytical signal of a calibration plot) versus the concentration of analyte should show no trend. Residual plots will be done with at least seven standard solutions distributed equally over the concentration range of the analyte, when practical. An F-Test was also used to test the significance of the linearity by using Equation 1.3. The standard deviation of the measurement at each point (s_w) based on multiple measurements of standards was measured. The values for s_w at several calibration points were pooled to obtain an estimate based on an increased number of degrees of freedom. The value of the variance (s_w^2) is compared to the variance of the points (s_b^2) about the fitted function to obtain the value.

$$F = (s_b^2)/(s_w^2) \quad (1.3)$$

where F = F-test value to determine significance of linearity
s_w^2 = value of variance
s_b^2 = variance of points about the function

The degrees of freedom for s_b^2 is equal to n-2 where n is the number of fitted points. The degrees of freedom for s_w^2 represents the pooled standard deviation estimate (s_w). If the value of F is significant, then s_b^2 will be larger than s_w^2 suggesting that the fit is not consistent with the scatter of the points. Coefficients of determination (r^2) were sometimes used to make judgement calls about linearity and the fit of calibration data. While the r^2 was collected and sometimes used in preliminary studies of linearity, the proposed procedure above was used in detailed linearity studies and provided the ultimate test for linearity over the use of r^2.

Establish Method/Analyzer Ruggedness

The proposed analytical methods used for on-line monitoring were subjected to preliminary tests designed to evaluate the ruggedness of the method and the analyzer. For example, interference studies were done with known and possible interferents. In addition, studies were done to evaluate the effect of TOC concentrations and ionic strength on the analytical signal. Preliminary studies were carried out in real world situations where environmental conditions may cause problems. Ideally, problems were identified, addressed and their magnitude evaluated. When possible, potential problems were identified and solved before monitoring studies were done. Extended on-line monitoring studies were done (described in more detail later), and used to further evaluate the ruggedness of the analyzer and analytical methods. Limitations of the proposed methods were identified and recommendations were made.

Method Validation Study and On-Line Monitoring Studies

For the purposes of this study, method validation involves direct, side-by-side comparison to appropriate EPA methods. In essence, the method was judged as to whether it could provide the desired analytical data for a given circumstance. The methods that were explored should not be confused with analytical methods designed for compliance monitoring (such as EPA methods). In order to validate a method, a set of firm requirements for the data must be set ahead of time. This was done earlier in describing the proposed Ideal Method. Broadly, the analytical methods that were developed should first meet the generalized criteria of the Ideal Method.

In addition to the controlled studies in the laboratory, candidate analytical methods were subjected to extended monitoring studies at the three different locations during the course of the project. The first on-line monitoring site (Memphis 1) was established at the University of Memphis, specifically at the researcher's laboratory. The second on-line monitoring site (Memphis 2) was established at the MLGW South Service Center. The third on-line monitoring site (Houston) was established at the City of Houston Department of Public Works (Houston, TX).

Candidate methods were evaluated in side-by-side comparison studies with an appropriate EPA Method. The CMS-GC-ECD method for THM4 species was compared directly to EPA Method 502.2 (USEPA 1995b) in Memphis 1 and Memphis 2. For Houston the CMS-GC –ECD results were compared to both EPA Method 502.2 and EPA Method 524.2 (USEPA 1995a). The CMS-FIA methods for total THM4 (TTHM4) and total HAA5 (THAA5) were compared to the results of both EPA Method 502.2 and EPA Method 552.3 (USEPA 2003a). The PC-IC method was compared directly to EPA Method 552.3. Additional, typically more limited studies were done with other methods in some instances and follow this general approach for comparison. The chemical analyzer was initialized, allowed to equilibrate and then calibration was performed. The analyzer was placed on-line after calibration and a sample was analyzed at a sampling rate of 1 sample per hour with the exception that approximately twice a day (every 12 hours) a check standard was analyzed. Samples were collected to coincide with the on-line monitoring analyzer and were analyzed by the appropriate EPA Method. Typically, EPA samples were collected each hour for the first 48 hours of a monitoring study, then every two hours until the end of the study. It was possible to do a direct comparison at each sampling time where both a candidate method result and an EPA Method result were measured.

To facilitate conducting a statistical comparison of the two analytical methods, the average concentration and its standard deviation over the extended monitoring study were calculated for each method. The assumption is made that the EPA Method is the "true value". With this assumption, it is possible to calculate the bias as the difference between the candidate method result and the EPA Method result (Equation 1.4).

$$\text{Bias} = (\text{Candidate Method Result} - \text{EPA Result}) \qquad (1.4)$$

The average bias and its standard deviation were calculated for the monitoring period. Assuming that the EPA results are the true value, the percent recovery from bias studies (% REC_{Bias}) was calculated for each sample by dividing the candidate method result by the EPA Method result and multiplying by 100. The standard deviation and range of the % REC_{Bias} was also calculated. In addition, the range of concentration and bias over the course of the on-line monitoring study were calculated. The results are typically presented graphically in two forms: as a plot of the concentration data as a function of time and second and as a plot of the bias as a function of time.

Student's t-test was used to identify differences between a candidate method and appropriate EPA Method by comparing the standard deviations with their results. This determined if these two methods give results that are statistically different or not. Student's t-test can be applied on the individual differences for results for each sample analyzed (Harris 2001). For the on-line monitoring studies it was assumed that each sample either collected on-line or off-line (grab sample) is effectively the same sample as long as they are collected within a very short time of each other (typically less than five minutes). The $t_{calculated}$ is calculated by first determining the difference (bias) of results of the two methods at each sampling time, denoted as d_i. The mean difference ($\bar{d}$) was calculated for the entire set of results for a particular analyte. The standard deviation (s_d) was calculated for these differences by using the following equation (Harris 1999). The standard deviation was calculated by using equation 1.5.

$$s_d = \sqrt{\frac{\sum (d_i - \bar{d})^2}{n-1}} \quad (1.5)$$

where s_d = standard deviation between two methods
d_i = bias of results for the two methods
$\bar{d}$ = mean difference for entire set of results for particular analyte
n = number of samples

This s_d value is then substituted into Equation 1.6 for the number of samples (n) analyzed for a particular study.

$$t_{\text{calculated}} = \frac{\bar{d}}{s_d}\sqrt{n} \quad (1.6)$$

where $t_{\text{calculated}}$ = calculated t-value
s_d = standard deviation between two methods
$\bar{d}$ = mean difference for entire set of results for particular analyte
n = number of samples

This value is then compared to a value from the t-table, listing t values for number of samples at different confidence levels. In practice, a spreadsheet (MS Excel) was used to calculate t_{critical} and $t_{\text{calculated}}$ for n. The t-test used in this instance was a two-tailed test at the 98% confidence level for n-1 degrees of freedom. For two methods *not* to be significantly different the $t_{\text{calculated}}$ must be less than the t_{critical} (Harris 2001).

OVERVIEW OF FINAL REPORT

This Report presents the results of studies aimed at developing on-line monitoring methods for THM4 and HAA5 species. Each Chapter (2-7) focuses on a particular method that was studied in depth. Chapter 2 presents the results of OPTGC studies for monitoring THM4 species. Chapter 3 outlined the results of SCMS-GC studies for THM4 species. Chapter 4 describes the CMS-FIA method using the reaction of NCA followed by FL detection to selectively monitor TTHM4 and THAA5 concentrations. Chapter 5 explores the use of CMS-GC-ECD for monitoring THM4 species. Chapter 6 presents the results of an evaluation of IC-MSCD for measuring HAA5 species. Chapter 7 combines the results of the CMS-FIA NCA-FL method with the separation of IC (called PC-IC) to selectively measure HAA5 species. Chapter 8 summarizes the results of the initial round of monitoring studies, identifies analytical issues with conducting online monitoring studies and proposes revised strategies for carrying out such studies. The HAA5 methods were expanded to include HAA9 species thereby eliminating four potential interferents. Chapter 9 presents these results of additional on-line monitoring studies and compares the methods to each other with respect to their results for THM4 and HAA9 species. Chapter 10 summarizes the accomplishments of the research along with recommendations for further studies. Appendices A-F provides details about the cost and construction of the three most promising analyzers developed in this study and the details of their operation in on-line sampling mode.

CHAPTER 2
ON-LINE MONITORING PURGE AND TRAP GAS CHROMATOGRAPHY WITH A DRY ELECTROLYTIC CONDUCTIVITY DETECTOR

INTRODUCTION

SRI Instruments has constructed for this project a prototype (but commercially available through special order) purge and trap (PT) GC (Figure 2.1). This instrument was derived from their USEPA Method 5030 (1995) compliant PT-GC and can be configured to sample directly from water distribution systems. This instrument is inexpensive, rugged, and has the capability of being fully automated. The gas extraction cell (GEC) is the integral component that makes on-line monitoring possible with the on-line monitoring purge and trap gas chromatograph (OPTGC) (Figure 2.2).

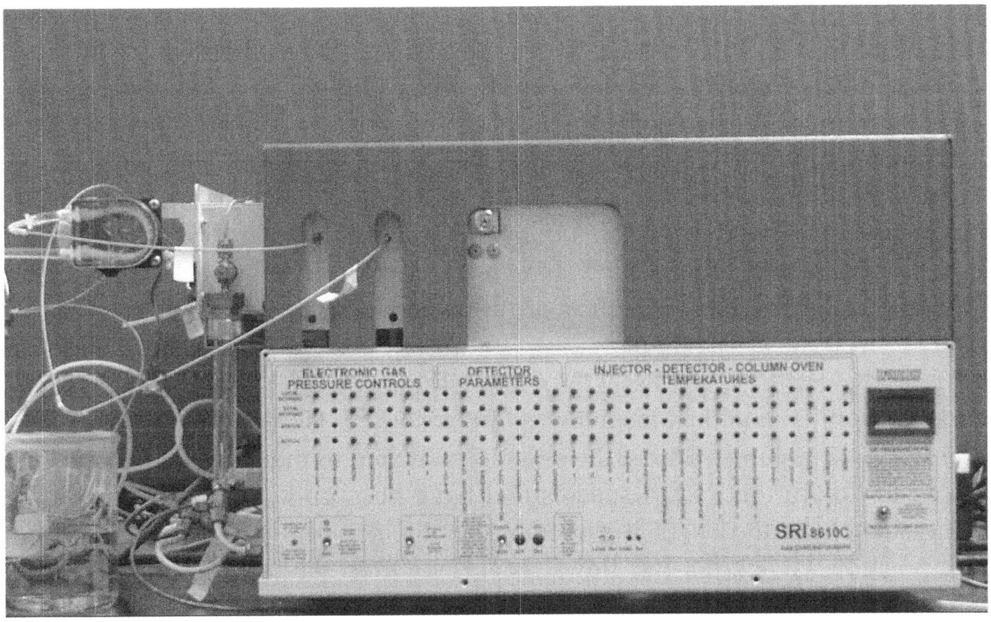

Figure 2.1 Picture of the SRI Instruments OPTGC-DELCD prototype

EXPERIMENTAL

All reagents, reagent water and glassware must be very clean to minimize sample contamination. All water used was distilled and then deionized by a Millipore Mill-Q Plus/Mill-RO®10 Plus Water Purification System to obtain a resistivity of at least 18.2 MΩcm and total organic carbon (TOC) concentrations of 10 μg/L or less (reagent grade water). Glassware and sample containers were meticulously cleaned with concentrated detergent and soaked for 24 hours and then rinsed with distilled water three times followed by reagent grade water several times. All chemicals were HPLC reagent grade or better and were purchased from Fisher Scientific Inc. Stock standard solutions for the THM4 species were purchased from Aldrich Chemical Company. Different aliquots (1.0 to 2.0 μL) of each stock standard solution were each added with a separate airtight 10 μL syringe (Hamilton-107) to a single 10 mL volumetric flask containing methanol to make the primary dilution standard solution. This primary dilution standard solution was then added in desired amounts (0.2 to 14.0 μL) for each preparation to a 100 mL volumetric flask containing reagent grade water to produce different concentrations of the calibration standard solutions or check standard solutions.

The OPTGC-DELCD Analyzer

The GEC selectively allows only volatile organics like the THM4 species to flow into traps, so that they can later be desorbed into the GC injection port. The GEC body is constructed of two, 3.0 cm long, hexagonal stainless steel (SS) fittings, and a 4.0 cm length of 1/4" SS tubing to join the fittings. The fittings each have a single 1/16" 0.75 mm i.d. SS tubing segment attached to them, which serves as the sparge gas (He) transfer (in/out) line. Inside the GEC, there is a length of silicone rubber tubing. The silicon rubber tubing has an exposed length of ~ 5.0 cm, an o.d. of 2.4 mm, i.d. of 1.6 mm, and wall thickness of 0.4 mm (Dow Corning). At the ends of each SS fitting a length of 1/16" 0.75 mm i.d. PEEK tubing from the "sample in" port and "sample out" port is inserted into silicone tubing. PEEK tubing is secured by two 1/16" PEEK fittings. The total length of the GEC is ~9.5 cm. The water sample is drawn into the GEC by means of a peristaltic pump with grab/batch samples or flowed through by water pressure in a typical distribution system for on-line samples.

The water sample flowing through the silicone rubber membrane is in constant contact with the inner wall of the membrane. The THM4 species in the water sample permeate the inner wall of the membrane, dissolve into the membrane, pass to the outer wall of the membrane, and desorb from the outer wall into the PT sparge gas stream. Once the THM4 species exit the membrane with the sparge gas (He) they are transported to a Valco ten-port heated injection valve in the LOAD position (Figure 2.2). In the LOAD position, the THM4 species are transported onto two traps (Trap 1 has TENAX™-GR packing and Trap 2 has no packing) kept at low temperature (~40°C) and thus the analytes are adsorbed. During this time the GC carrier gas (He) is directed to the column. Before the valve is switched to the INJECT position, the traps are rapidly heated to desorb the THM4 species that have been collected on the traps. The valve is switched immediately to the INJECT position after the pre-heating of the traps is complete. In the INJECT position, GC carrier gas flows the THMs from the traps into the GC injection port and to the column. At the same time, the GEC is being purged with both sparge gas and reagent water to minimize any carry-over effects. The entire process is fully automated.

The OPTGC is equipped with a modified version of an electrolytic conductivity detector (Hall)—called the dry electrolytic conductivity detector (DELCD). The DELCD is selective for halogenated organic compounds. It consists of a small ceramic tube that serves as the DELCD reactor that is heated to 1000°C. A platinum thermocouple inside the reactor measures the electrode temperature and a nichrome collector electrode measures the conductivity of the gases flowing through the detector from the GC column. Halogenated organic compounds such as THMs will react with oxygen in the make-up gas (compressed air) to form ClO_2 and BrO_2 that are detected. This differs from the traditional Hall detector that detects HCl and HBr. The Hall detector also requires a solvent electrolyte for detection of analytes in the gas phase. The DELCD has detection limits in the single μg/L concentration range.

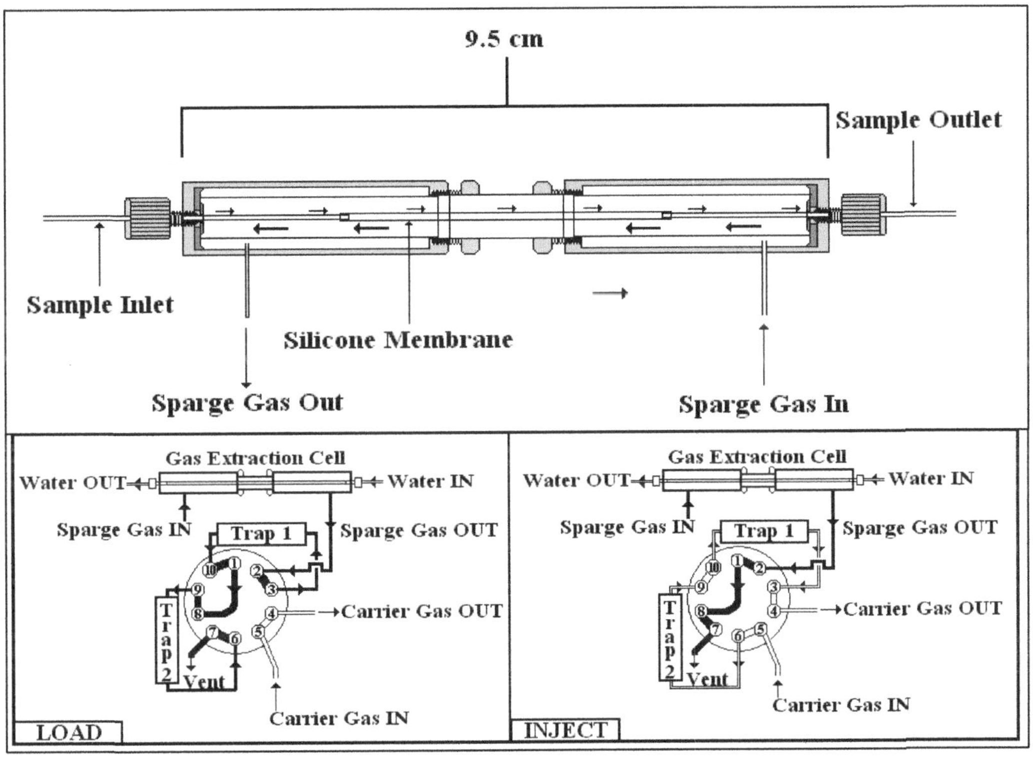

Figure 2.2 Schematic diagram of the GEC sampling device and valve configuration for the OPTGC-DELCD

RESULTS AND DISCUSSION

Optimization Studies of the OPTGC Method

There were three specific parameters that were optimized before proceeding with the OPTGC MDL, accuracy and precision studies. These parameters included the sample flow rate, sparge time and the sampling time. The sample flow rate is the flow rate of the drinking water sample to the membrane device used for extraction of THM4 and for collection on the traps. The sampling time refers to how long sample is flowed through the membrane before a sample is analyzed. The sparge time refers to how long the analytes are loaded onto the traps.

The sample flow rate was the first parameter to be optimized. Figure 2.3 shows the result of experiments aimed at determining the optimum sample flow rate. This figure is a plot of the analytical signal versus the sample flow rate. From the study, the optimum sample flow rate was chosen as 2.5 mL/min. While this may not appear to be the optimum sample flow rate, the choice is better understood given the following additional information. Certainly, it appears that the signal from $CHCl_3$ continues to increase as sample flow rate increases. However, $CHBrCl_2$ begins to decrease after 2.5 mL/min while the remaining two THM4 species stay relatively constant. But flow rates higher than 2.5 mL/min are difficult to constantly maintain with the available peristaltic pump, while a sample flow rate of 2.5 mL/min can be maintained at a relatively constant rate. Thus a sample flow rate of 2.5 mL/min was chosen as the optimum sample flow rate for OPTGC studies.

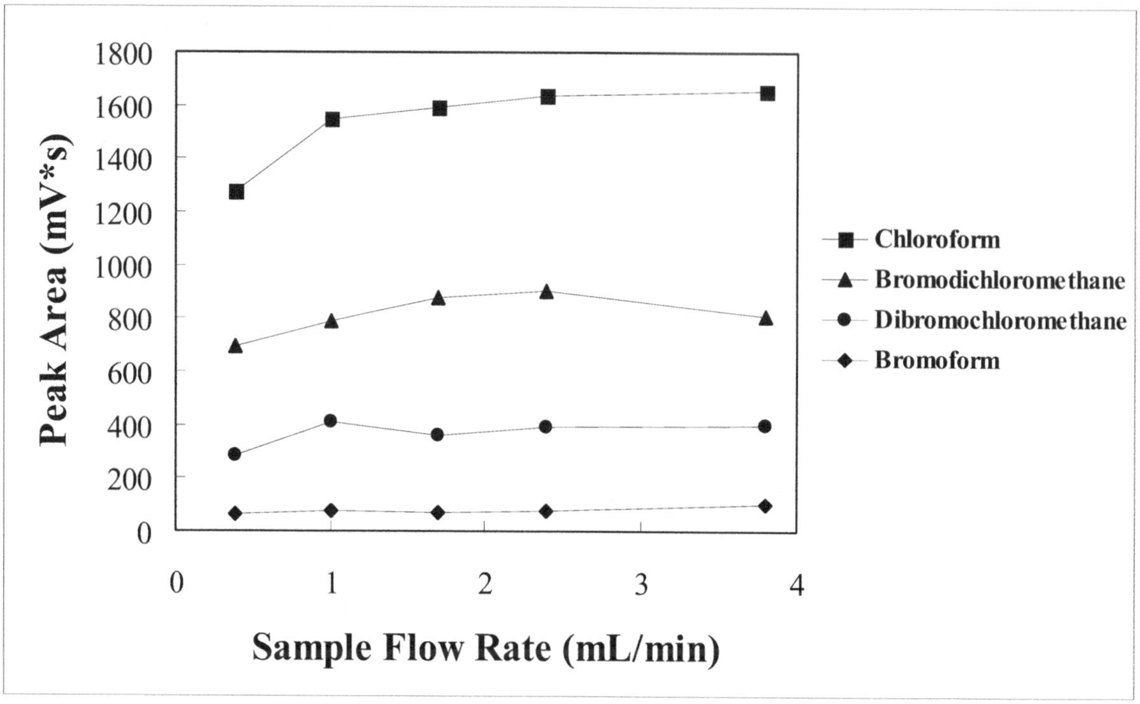

Figure 2.3 Optimization plot of sample flow rate for the OPTGC-DELCD

A plot of the analytical signal versus sampling time is shown in Figure 2.4. From the plot, it is clear that the analytical signal continues to increase for all THM4 species as sampling time is increased. However, to meet the target sampling rate of at least one sample per hour, a sampling time of 15 minutes was chosen. This sampling time gave the maximum analytical signal while still meeting the desired sample analysis rate once the separation and other parameters added into the analysis time.

A plot of the analytical signal versus the sparge time is shown in Figure 2.5. These results are very similar to the optimization of sampling time discussed previously in that the analytical signal continues to increase for all THM4 species. Similar to the previous situation, in a tradeoff between the desired MDL levels and the sample analysis rate, a sparge time of 0.9 minutes was chosen to be sufficient enough to achieve the desired MDL while still obtaining a sample analysis rate of one sample per hour.

Method Detection Limit, Accuracy and Precision Studies

With the optimized method, MDL, accuracy and precision studies were carried out (Glaser and Forest 1981; USEPA 1996a; and USEPA 1996b). Figure 2.6 shows five chromatograms for THM4 solutions over the concentration range of 1 to ~20 μg/L for each THM4 species. The results of MDL, accuracy and precision studies are shown in Table 2.1. Table 2.2 shows the instrument parameters used in the analysis.

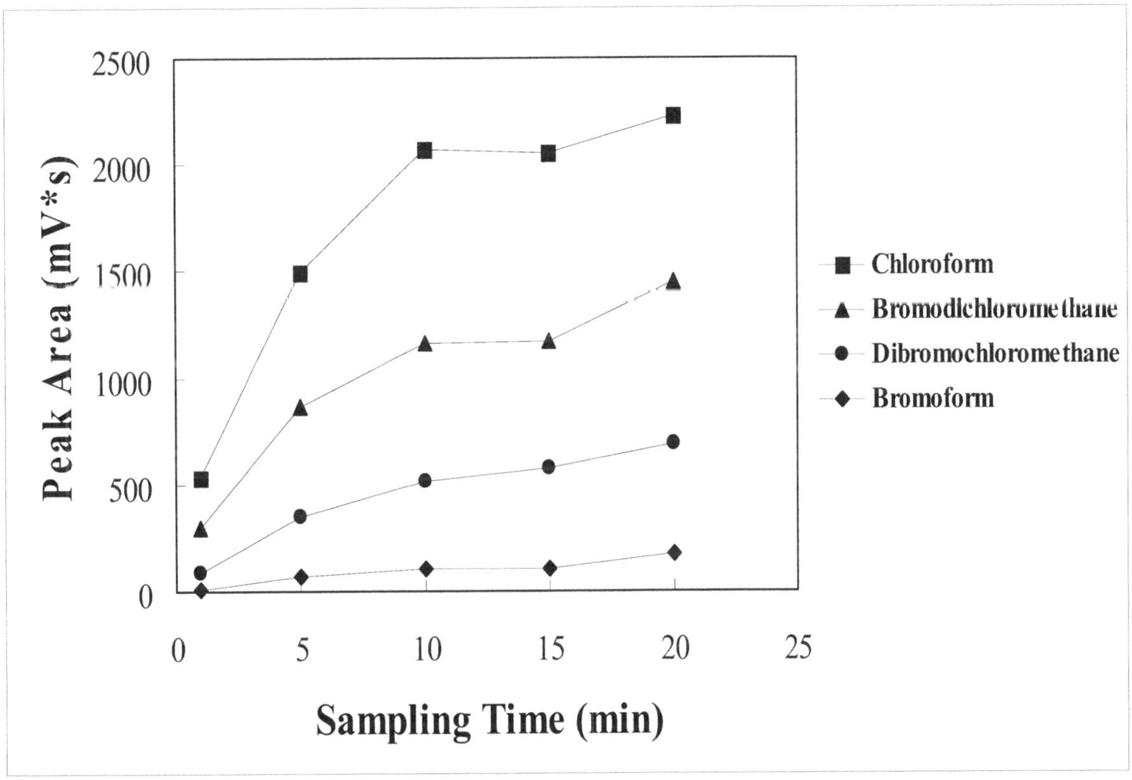

Figure 2.4 Optimization plot of sampling time for OPTGC-DELCD

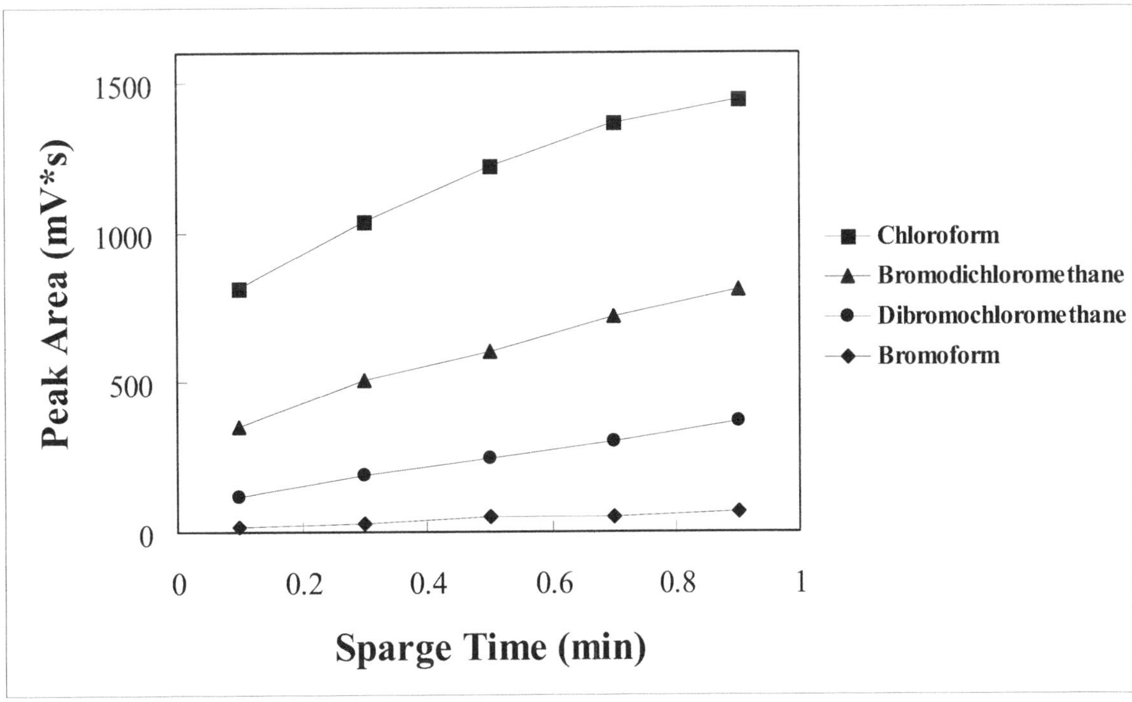

Figure 2.5 Optimization plot of sparge time for OPTGC-DELCD

Table 2.1
Results from MDL, accuracy and precision studies for the OPTGC-DELCD

THM4 Species	MDL (μg/L)[†]	Mean % Recovery	% RSD	$r^{2‡}$	Conc. Range (μg/L)*
$CHCl_3$	0.4	125	3.6	0.999	1-20
$CHBrCl_2$	0.3	115	2.3	0.999	1-20
$CHBr_2Cl$	0.3	127	2.1	0.998	1-20
$CHBr_3$	0.4	103	4.9	0.999	1-20

* Each standard was analyzed once and five standards were analyzed to construct the calibration curves used to make MDL, accuracy and precision estimates.

† The concentration of the check standard was ~3 μg/L for each THM and analyzed seven consecutive times.

‡ Estimates from using quadratic fits ($y=mx^2+bx+C$).

Table 2.2
Final optimized settings for the OPTGC-DELCD

Parameter	Setting
GC Column	Restek MXT-1 30 m x 0.53 mm i.d. x 5.0 μm f.t
Carrier Gas	Helium (99.999% purity) at 10.0 mL/min
Sparge Gas	Helium (99.999% purity) at 30.0 mL/min
Sampling Time	15.0 minutes
Sparge Time	0.90 minutes
Injector Temperature	175°C
DELCD Temperature	225°C
DELCD Reactor Value	260 (1000°C)
DELCD Gas	Extra Dry Air at 2 psi
Trap 1	Blank 40°C to 220°C
Trap 2	Tenax 40°C to 220°C
Valve Temperature	65°C
OPTGC Temperature	65°C
Initial Oven Temperature	50°C for 4.50 minutes
Temperature Ramp 1	50°C to 100°C at 20°C/min
Temperature Ramp 2	100°C to 200°C at 50°C/min
Final Oven Temperature	200°C for 2.00 minutes

These MDL estimates are much improved over preliminary studies that used linear fits and all are below the low-end of our target range of 0.5 to 5 μg/L. The mean % recoveries are improved as well and while they are still above 100% for three of the THM4 species, the % RSD values are all less than 5%. The concentrations of sample could be adjusted (corrected) based on the mean % recovery data without introducing more than 5% error. These improved results were made possible through the use of a non-linear fit of the calibration data. The calibration plots with the quadratic fit are shown in Figure 2.7. Such fits with the DELCD detector are to be expected (SRI Instruments 2002) and the determination coefficients (r^2) for the fits are very good.

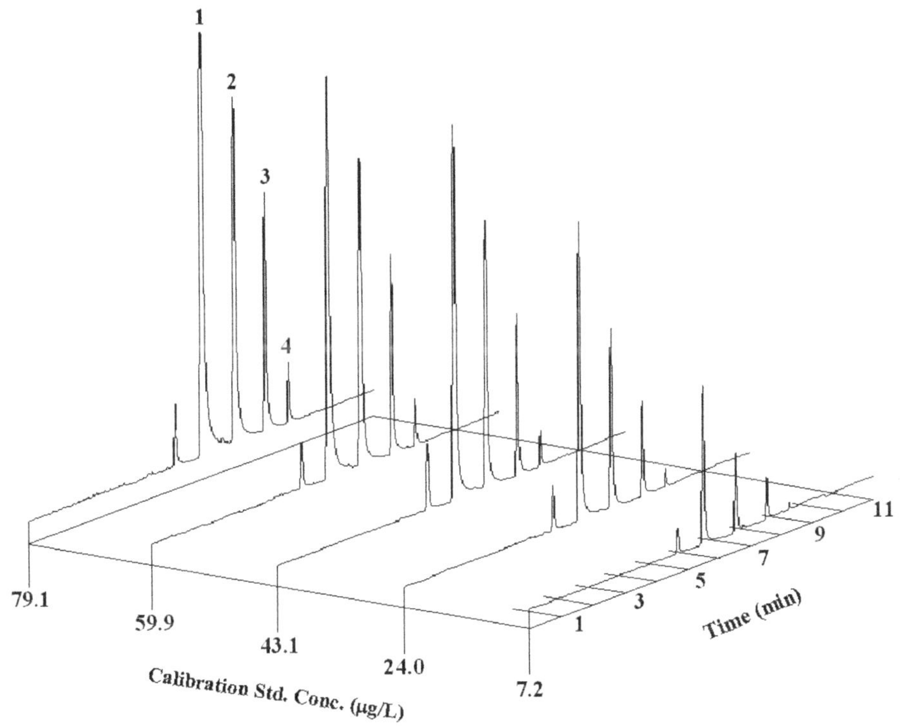

Figure 2.6 Chromatograms of five calibration standards of THM4 for the OPTGC-DELCD. The concentration of the THM4 standards ranged from ~7.2 to 79.1 µg/L. The peaks are identified as follows: (1) $CHCl_3$, (2) $CHBrCl_2$, (3) $CHBr_2Cl$, (4) CHBr3

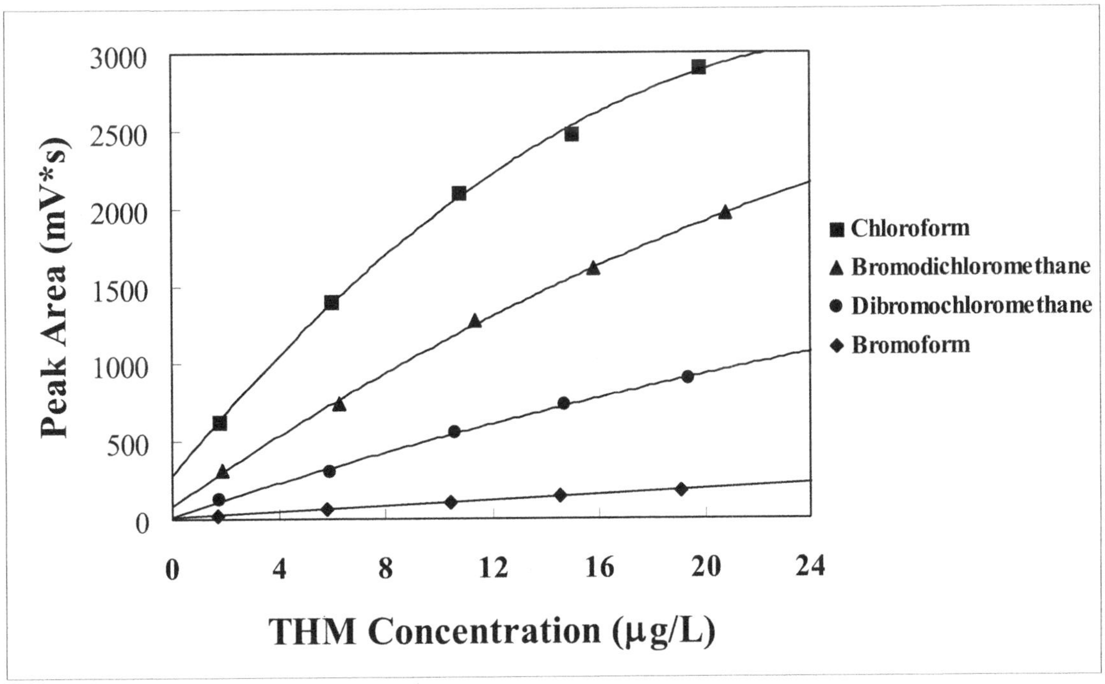

Figure 2.7 Quadratic fitting of calibration curves of THM4 species for OPTGC-DELCD

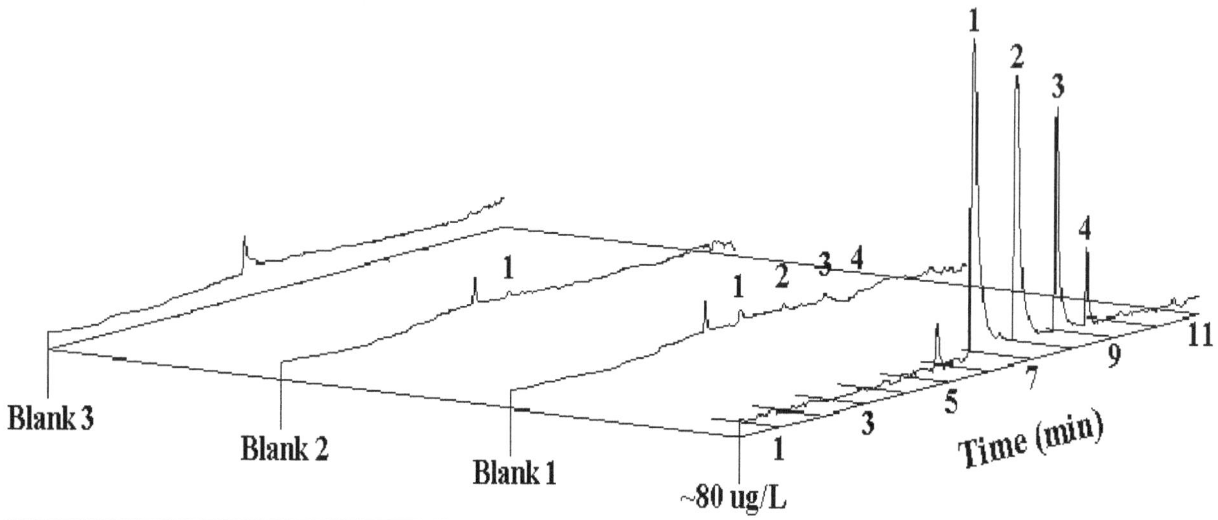

Figure 2.8 Results of revised carry-over experiments with the OPTGC-DELCD analyzer

Evaluating Sample Carry-over Effects

Sample carry-over from previous injections (Figure 2.8) may be a problem, so optimization of operating conditions was performed to minimize these issues. In preliminary studies, it was suspected that carry-over problems may not have been completely minimized as observed with the large and increasing mean % recoveries. In revised studies, Figure 2.8 shows sample injection approximately every 15 minutes. Sample carry-over has been minimized after 60 minutes. With the revised OPTGC method, a sample can be injected every 60 minutes without concern of carry-over issues.

Stability Issues with the DELCD

The successful operation of the DELCD (and thus the OPTGC) depends on maintaining a constant temperature of 1000°C at the location of the electrode (cathode/anode) (Figure 2.9). To reproducibly detect the oxidized species (ClO_2 and BrO_2) formed from THM4 species in the GC column effluent, the DELCD electrode must be centered in the DELCD body (or at least at a constant position). Unfortunately, the DELCD exhibits instability (poor reproducibility) after an extended periods of time that appear to be directly related to a "drooping" of the electrode caused by the immense heat within the DELCD body. Figure 2.10 is a picture of a typical electrode that bent after 13 hour of use. The electrode rod by itself is very sturdy, but the combination of heat and gravity eventually will cause it to bend closer and closer to the walls of the detector.

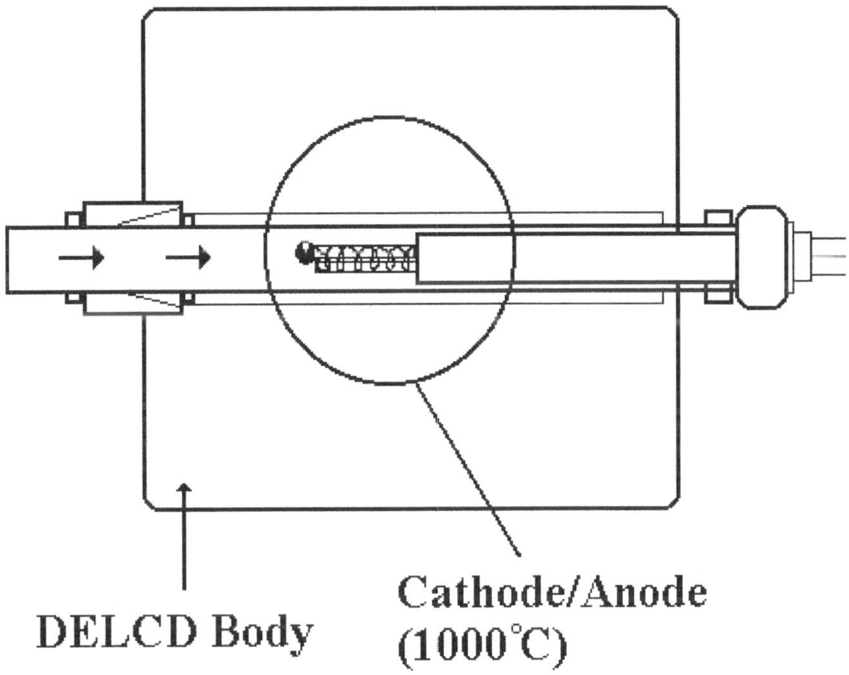

Figure 2.9 Internal diagram of the DELCD

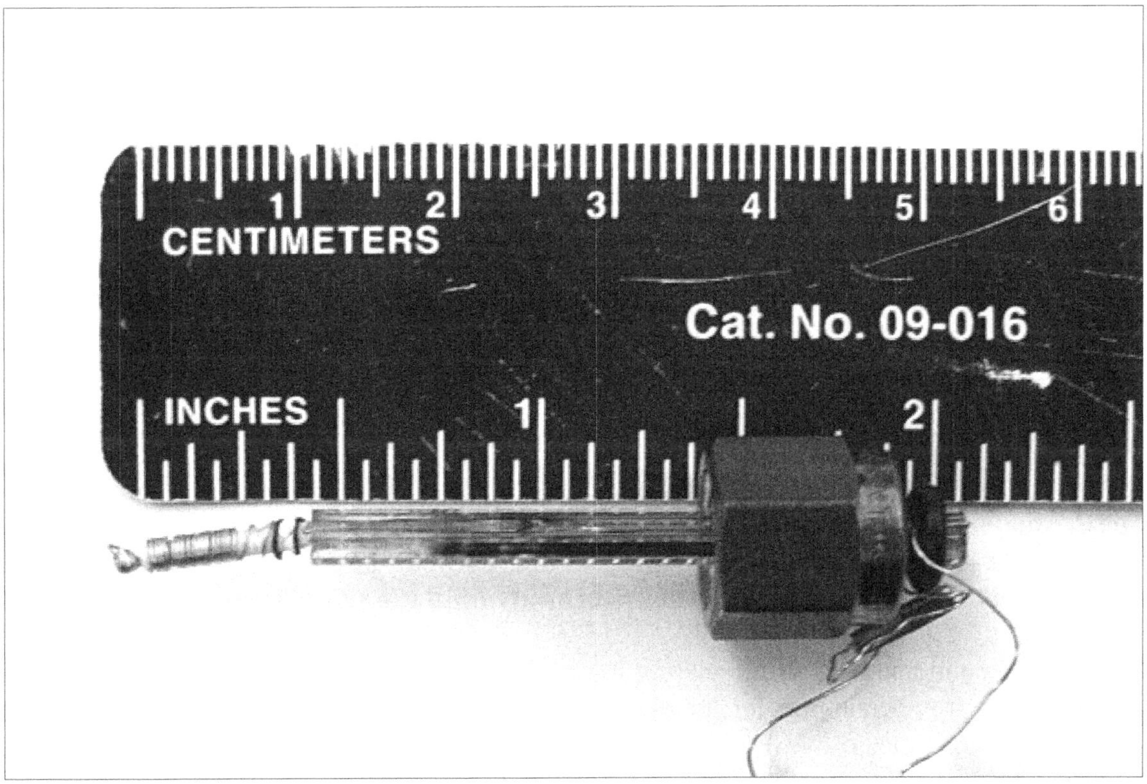

Figure 2.10 Picture of a DELCD electrode bent from extended usage

Figure 2.11 demonstrates the affect that this "drooping" has on analytical signal and peak shape by comparing a chromatogram recorded ~15 minutes after calibration and initial startup to a chromatogram recorded ~13 hours after startup. Most noticeable is the distinct change in peak shape that occurs and hampers quantification. One might think that after cooling, the electrode will bend back to its original position; however, after extended usage the electrode maintains a bent shape, which seems to get worse with each use. A possible solution to minimize this problem might be to mount the detector with the electrode aligned vertically rather than horizontally. This was attempted on the OPTGC, but the vertical configuration did not give improved performance. The chromatogram exhibited low sensitivity and poor peak shape for the THM4 components. This is likely because the additional fittings installed to allow for vertical mounting formed slightly cooler regions just prior to entering the DELCD. For optimum sensitivity, the species exiting the GC column need to be immediately directed towards the electrode and not allowed to cool.

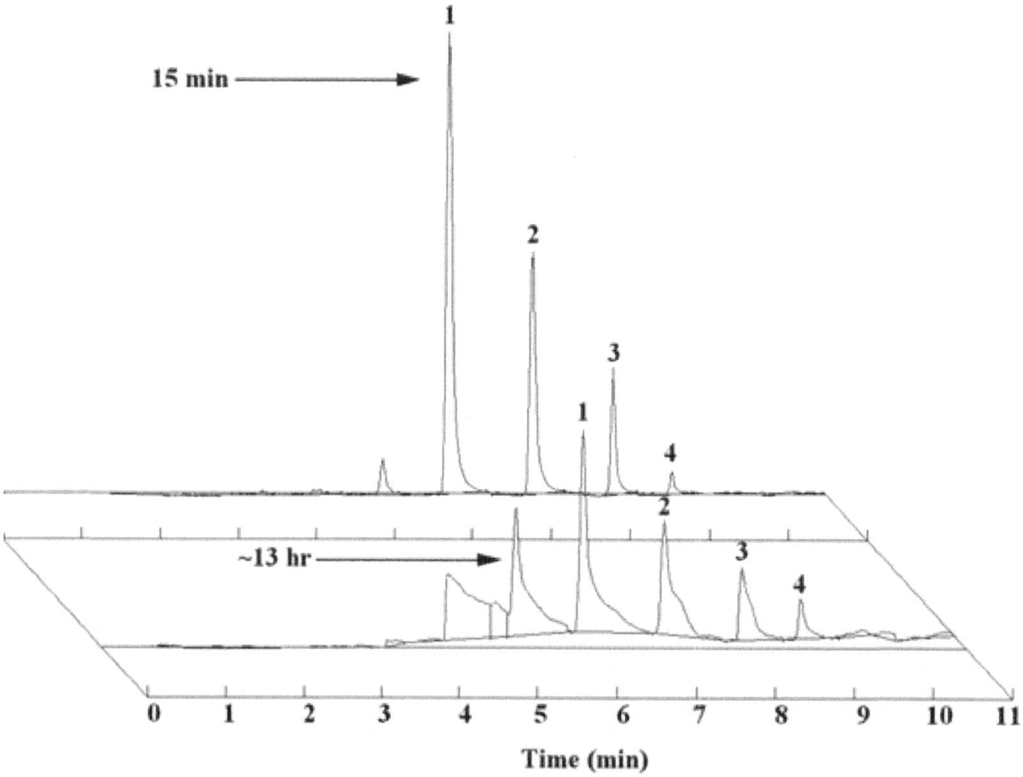

Figure 2.11 Results of detector stability studies with the OPTGC-DELCD. Note the change in peak shape and height after about 13 hours of on-line study. The change is thought to occur as a result of bending or drooping of the cathode on the DELCD. The drooping continues to occur through about 18 hours of study until the signal is completely unusable

The second approach at alleviating the problems with the OPTGC involved redesigning the DELCD to provide support for the electrode and minimize drooping over time. A modified assembly was constructed by SRI Instruments and assembled on site at the U of M laboratory. The modified electrode assembly (Figure 2.12) had a 24 gauge steel wire support inserted into the glass base of the electrode. The electrode rod was then secured to the support. To test this electrode an extended study (50 hr) was conducted. The modified electrode assembly was installed and the entire system was allowed to reach operating conditions. A THM4 standard (~9 μg/L each) was initially analyzed at hour "zero" and then periodically during the daytime hours of the study. The system was left on overnight. Figure 2.13 shows a plot of THM4 analytical signal versus time (hr) during the 50 hour stability study for each THM. It was observed that with this new electrode assembly the signal for all four THMs was relatively stable for ~ 9 hours; however, the analytical signal for each THM increased overnight ~ 14 hours. Operation time has a notably greater effect upon the more chlorinated species such as chloroform, which is to be expected, since the DELCD response is greater for more chlorinated species. Once the detector stabilized, revised MDL, accuracy, and precision studies were performed.

The calibration plots with a quadratic fit are shown in Figure 2.14. The results of MDL, accuracy and precision studies are shown in Table 2.3. The MDL estimates were again in the sub μg/L range. The mean % recoveries are similar to values previously reported as was the % RSD values, which were slightly improved. The results looked promising; however, once the system was shutdown following this calibration and brought back up the calibration was no longer reliable after 24 hours. This means that the OPTGC must be on for at least 24 hours before calibration and must stay on if a study is to be conducted following calibration. In summary, there are two points, or steps, to consider. First, the OPTGC must be operational for at least 24 hours prior to performing calibration studies and subsequently on-line monitoring studies. Second, if for any reason the OPTGC must be taken off-line and allowed to cool to ambient temperature, the first step must be repeated.

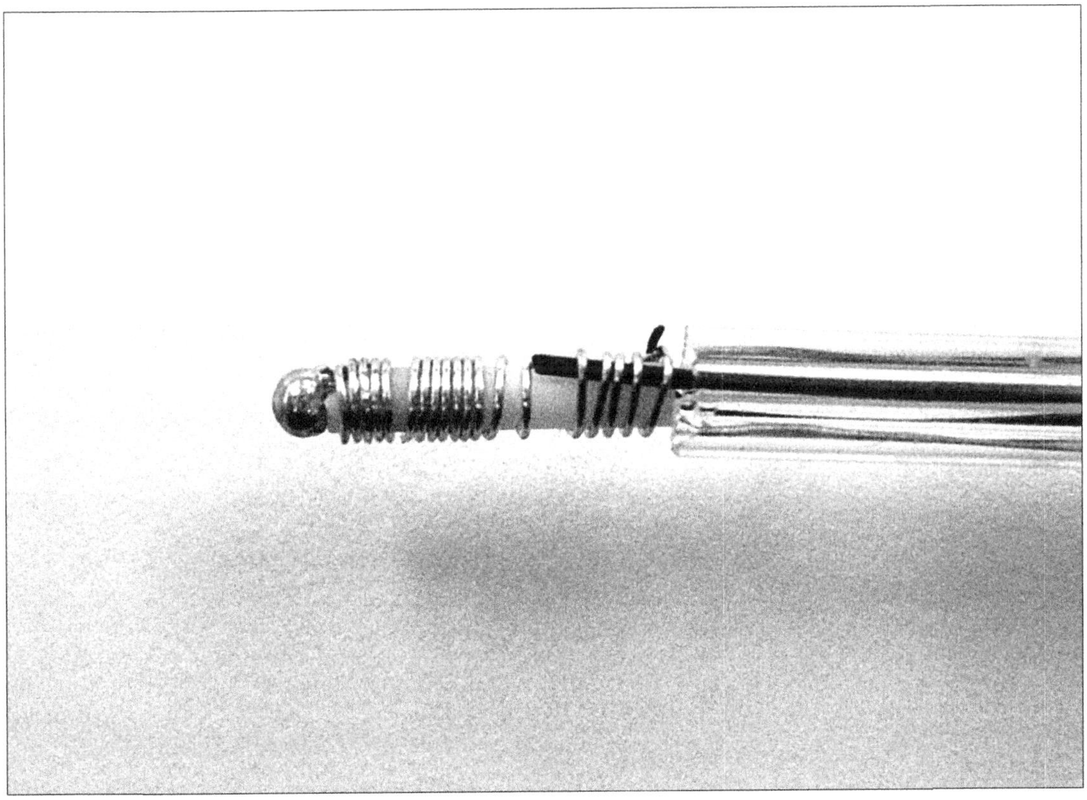

Figure 2.12 Picture of modified DELCD electrode (black rod is the steel support for the white electrode rod)

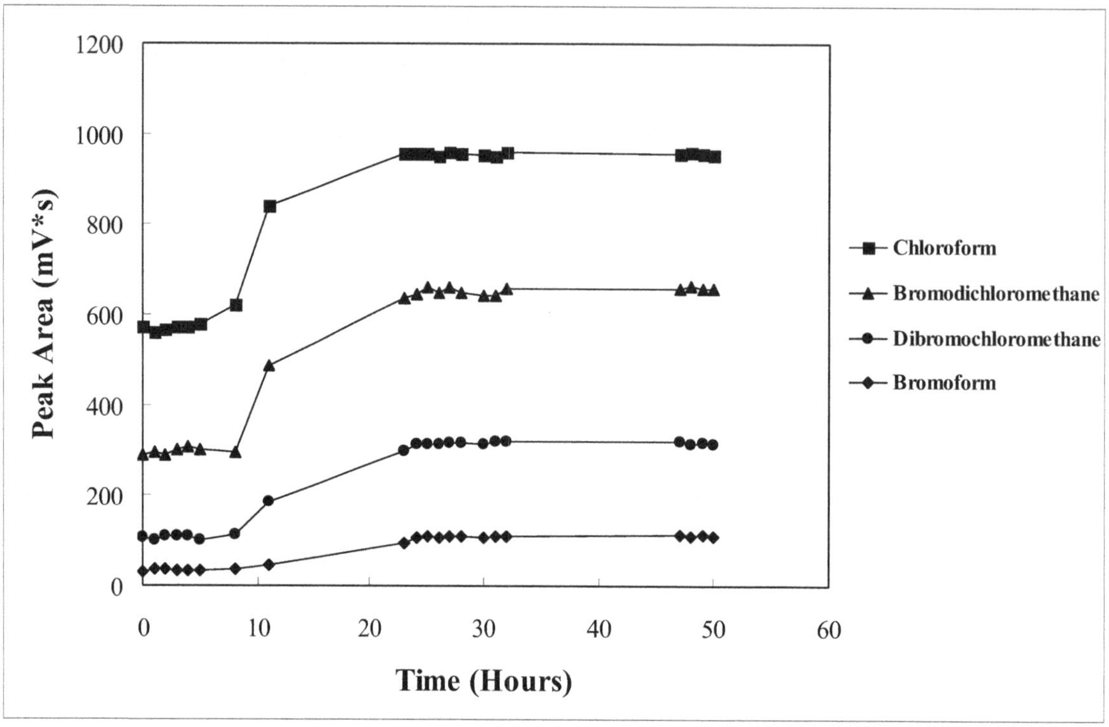

Figure 2.13 Fifty hour stability study for the modified DELCD electrode assembly

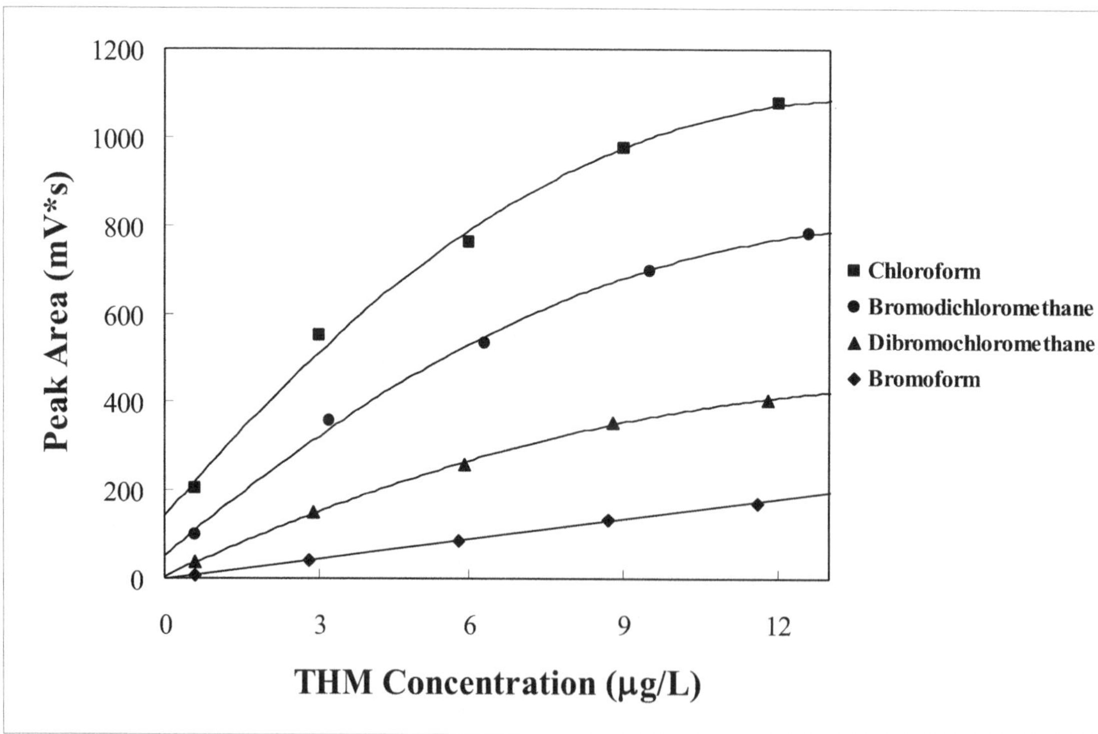

Figure 2.14 Quadratic fitting of the calibration curves of THM4 species for OPTGC-DELCD using the modified electrode

Table 2.3
Results from revised MDL, accuracy and precision studies for the OPTGC-DELCD

THM4 Species	MDL (μg/L)†	Mean % Recovery	% RSD	$r^{2‡}$	Conc. Range (μg/L)*
$CHCl_3$	0.2	128	2.1	0.994	1-12
$CHBrCl_2$	0.2	121	2.2	0.998	1-12
$CHBr_2Cl$	0.1	116	1.2	0.999	1-12
$CHBr_3$	0.2	110	2.8	0.999	1-12

* Each standard was analyzed once and five standards were analyzed to construct the calibration curves used to make MDL, accuracy and precision estimates.

† The concentration of the check standard was ~1.8 μg/L for each THM and analyzed six consecutive times

‡ Estimates from using quadratic fits ($y=mx^2+bx+C$).

CONCLUSIONS AND RECOMMENDATIONS FOR THE OPTGC ANALYZER

There are several reasons why there was interest in OPTGC method: the instrument can be purchased as a special order from SRI Instruments, it is fully automated and offers all the same advantages of other methodologies that use PeakSimple software. The preliminary estimates of MDL values were good, even though the method suffered from accuracy, precision and non-linearity problems. Most of these problems were resolved using a non-linear curve fit. When extended monitoring studies were done, stability issues arose. The instability was thought to arise from the bending of an electrode in the DELCD during extended term monitoring studies. The bending (or drooping) of the electrode is not a problem during normal operation, but becomes apparent through a change in chromatogram peak shape after approximately 13 to 18 hours. Two revisions of the OPTGC DELCD were evaluated. One revision was the remounting of the DELCD, so that the electrode was vertical. The other involved providing a support for the electrode. The remounting of the electrode vertically leads to additional issues with analytical signal. The configuration with the supported electrode provided improved results after an extended initial operating period of approximately 24 hours. After this initial period, the OPTGC was stabilized and gave acceptable results. Each time the OPTGC is powered down (with heated zones allowed to cool to ambient temperature), it has to be re-stabilized for ~24 hours before being calibrated and placed on-line.

These stability issues stem largely from the DELCD. The researchers believe the OPT sampling system offers an attractive on-line sampling approach for several reasons. It is commercially available, easy to interface to drinking water distribution systems and fully automated. If the OPTGC were fitted with a detector that was more stable, such as an ECD, it is expected that this method would provide outstanding results. Initially the OPTGC was not used for the first round of monitoring studies. This was largely due to the time delay for the OPTGC to reaching a stable during monitoring studies as well as logistical reasons. However, the OPTGC-DELCD in regards to the stability study suggests that it would be able to perform an extended on-line study if additional evaluation studies were conducted. This analyzer was not chosen to participate in initial monitoring studies, but was employed for later for revised monitoring studies presented in Chapter 9.

Recommendation 2.1: The OPTGC-DELCD is recommended for on-line monitoring of THM4 species in drinking water if it is shown to stable for a preliminary study. While the OPTGC offers many attractive features, the instability of the DELCD limits its use for on-line monitoring over extended periods of time.

Recommendation 2.2: It is suggested that the OPTGC might provide excellent results if another detector was used. For example, the ECD would provide a much more stable GC detector that would be expected to work well with the automated purge-and-trap system, thus the OPTGC warrants further study.

Recommendation 2.3: A more robust peristaltic pump (continuous duty-medical) could increase the flow rate of sample through the GEC and possibly increase analytical response. And minimize any variations in the flow rate through the GEC that might result from pressure changes in the distribution system.

Recommendation 2.4: It is also desirable to develop a way to accurately and precisely introduce one or more internal standards into the sample stream. Internal standardization could account for changes in the instrument response that might occur during the extended monitoring studies.

CHAPTER 3
SUPPORTED CAPILLARY MEMBRANE SAMPLING-GAS CHROMATOGRAPHY

INTRODUCTION

The supported capillary membrane sampling (SCMS) probe has been used in the laboratory as a sampling device for GC analysis. In this section, three different approaches are described that use the SCMS probe for THM4 analysis. In particular, the bench-top SCMS-GC with an electron capture detector (ECD) is discussed as well as the development of a more portable supported capillary membrane sampling-gas chromatography with electron capture detection (SCMS-GC-ECD) analyzer specifically designed for on-line monitoring of THM4 species. An additional approach, called SCMS-GC on a Valve (SCMS-GCOV) that uses a pulsed discharge photoionization detector (PDPID) is presented. Studies are also presented that interfaced the SCMS probe to a GC with an ion trap mass spectrometer. With each, the focus is on THM4 analysis.

THE SCMS-GC-ECD APPROACH TO THM4 MONITORING

Duty (2000), Liao (2001), Cao (2004), and Emmert et al. (2004a) have used supported capillary membrane sampling-gas chromatography (SCMS-GC) and bench top GCs to measure THM4 concentrations in water. The SCMS probe (Figure 3.1) is comprised of a silicone rubber capillary membrane tubing wound around a metal body that is approximately five inches in length. The membrane is connected to inlet and outlet fittings. The portion of the SCMS probe containing the wrapped membrane is placed directly in the water to be analyzed. The SCMS-probe can be assembled in a calibration configuration or in a fast-loop/flow cell (FL/FC) configuration, thus allowing for on-line monitoring. In the calibration configuration (Figure 3.2), the SCMS probe is immersed in a water sample contained in the calibration vessel and connected to the SCMS carrier gas (N_2). The outer wall of the membrane is in direct contact with the water sample containing THM4 species.

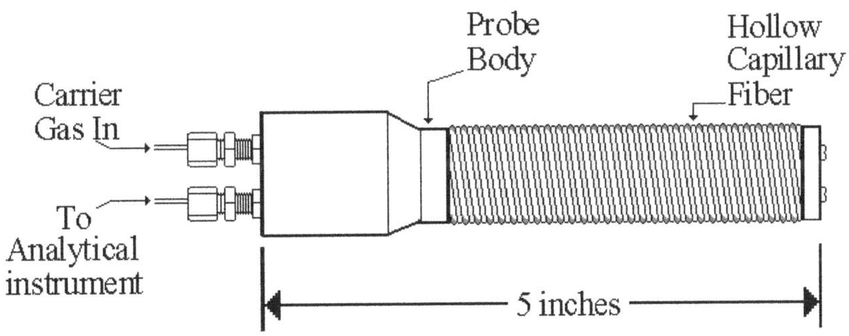

Figure 3.1 SCMS probe diagram (adapted from Global FIA, 1999)

The THM4 species permeate the outer wall of the capillary membrane and move through the capillary membrane to the inner wall where they desorb into the flowing SCMS carrier stream of N_2. The SCMS carrier gas transports the THM4 species to a six-port injection valve fitted with a sample loop. The injection valve is connected to an analytical instrument. When the injection valve is in the LOAD position, the mixture of carrier gas/THM4 species mixture passes through a sample loop and is vented to waste. When the injection valve is switched to the INJECT position, a volume of sample is presented to the analytical instrument.

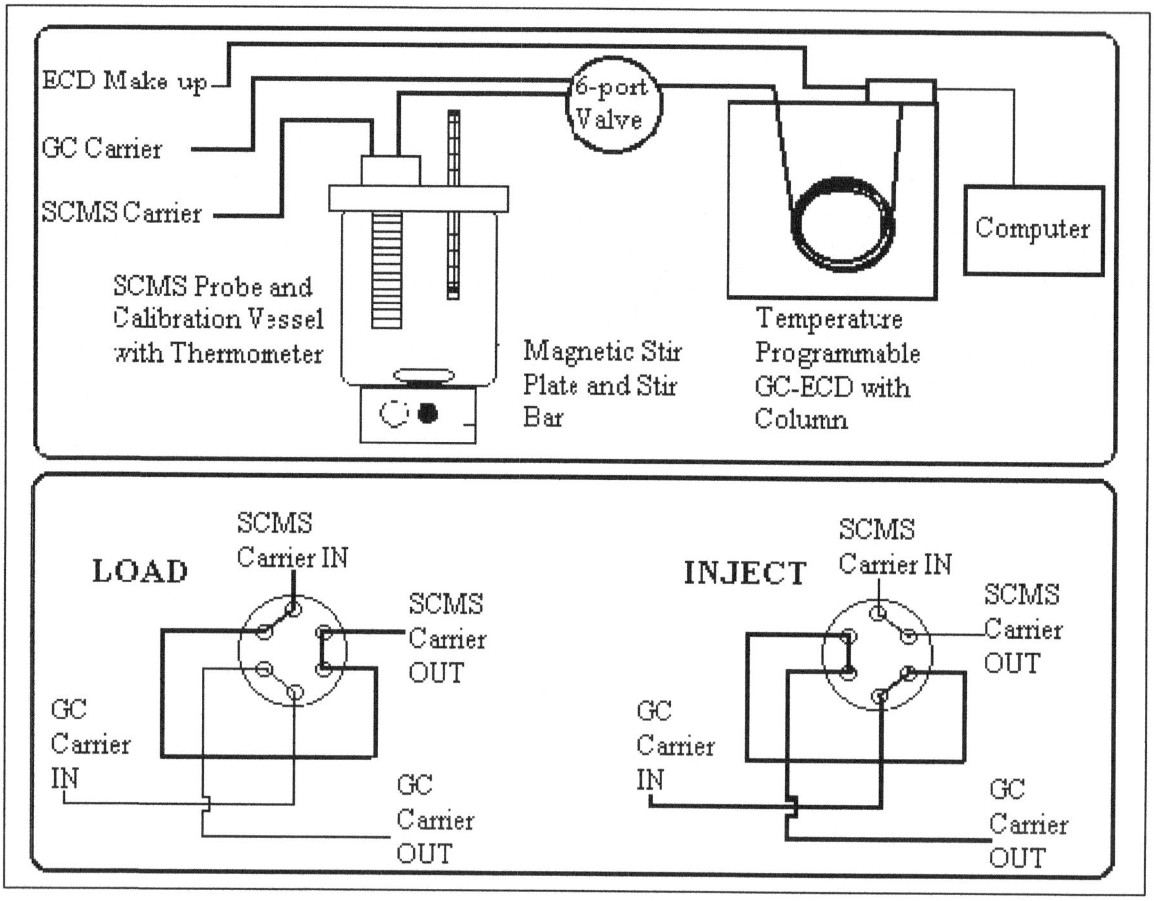

Figure 3.2 The top schematic shows the flow diagram of the bench-top SCMS-GC-ECD. The bottom drawing shows the configuration of the two-position six-port injection valve when in the LOAD and INJECT positions. The top sample port is position number "1" and the ports are numbered 1 to 6 going clockwise (Source: Emmert et al. 2004a)

The effectiveness of the permeation process depends on: (1) mass transport of the THM4 species to the membrane outer surface; (2) the solubility of the THM4 species in the membrane material; (3) the vapor pressures of the THM4 species involved; and (4) the sample temperature (Global FIA 1999). The mass transport rate and vapor pressures of the analytes are strongly affected by the temperature (Global FIA 1999). This means that the SCMS probe must be calibrated to achieve quantitative results. This is done by adjusting the SCMS carrier flow rate to an optimum point where the internal flow velocity becomes low enough to allow the analyte to reach vapor-liquid equilibrium with the sample solution. Here, it is possible to use Henry's Law (Nicholson, Maguire, and Bursill 1984) and Antoine's Equation (Dean 1985), shown as Equation 3.1, to correct for temperature changes in the sample matrix. Henry's Law predicts that the solubility of a gas is directly proportional to the partial pressure exerted by the gas.

$$p_i = K_H X_i \quad (3.1)$$

where p_i = the partial pressure of the compound in the gas phase (atm)
X_i = the concentration (mole/m^3)
K_H = the Henry's law constant (m^3 atm/ mole)

K_H depends upon the temperature and can be empirically calculated for $CHCl_3$ as described by Nicholson and coworkers (1984). A similar approach could be applied for other compounds. Antoine's Equation (Dean 1985) describes the partial pressure of a pure compound over its liquid form as a function of temperature. The partial pressure can be calculated at a particular temperature using Antoine's Equation and substituted into the Henry's Law equation. Since the Henry's Law constant for the compound of interest can be found in literature (Global FIA 1999; Dean 1985), the mole fraction of the analyte can be calculated.

In this section, the construction of a bench-top SCMS-GC-ECD is presented. In addition, calibration methods consistent with USEPA recommendations (Glaser and Forest 1981; USEPA 1996a; and USEPA 1996b) for measuring the concentrations of THM4 in water are described. The effects of experimental parameters such as flow rate of the SCMS probe carrier gas; sample temperature, sample ionic strength and transfer line temperature on the system performance are investigated. MDL, accuracy and precision studies are presented for measuring THM4 concentrations in water. The results of the studies provide the foundation for the construction and operation of the portable SCMS-GC discussed in the next section.

Experimental

Unless otherwise noted, all water used was distilled and deionized using a Barnstead/NANO pure water purification system. The resulting water was of the highest quality and did not have any measurable organic or inorganic impurities. All glassware and sample vessels were scrupulously cleaned. All chemicals were reagent grade or better. The THM4 standards were purchased from Aldrich Chemical Company. A stock standard solution of a THM4 mixture was prepared in 2 mL volumetric flasks containing methanol by addition of aliquots of each compound. Appropriate amounts of the stock standard solution were spiked into 2 L reagent water to give various concentrations of the working standard solutions.

Figure 3.1 shows the typical flow diagrams for constructing a typical SCMS-GC. The SCMS probe was constructed from a metal body with helical shaped grooves along the body. A 35.5 inch length of silicone membrane (Cat#: SRT-020-032) was wrapped along the grooves. The exposed length of the membrane in the sampling system was about 30 inches.

The SCMS probe carrier gas flowed through the hollow membrane while the water sample was in contact with the outside of the membrane. When the probe was immersed into the water sample, the THM4 species permeated the outer wall of the membrane, dissolved into the membrane material, passed through the inner wall and then desorbed from the inner wall into the SCMS probe carrier gas that entered the SCMS-probe inlet fitting through a transfer line (Valco, TNl130) from a flow controller (Brooks, Model 8744). The SCMS probe carrier gas transported the THM4 from inner wall of the silicone membrane through the outlet fitting connected to a transfer line (Valco, TNl130) that flowed to a two position air actuated six-port valve (Valco, Serial No. I×0143). The valve was fitted with a 100 μL sample loop. When the six-port valve was switched to the INJECT position via a digital valve interface (DVI: Valco, 110VAC), 100 μL of the SCMS probe carrier gas was loaded onto the GC column. The injection time interval (the time from switching from the INJECT position to LOAD position) was set at 30 seconds. The volume of carrier gas injected was about 1.56 mL containing 100 μL analyte. In order to minimize condensation of the THM4 on the inside the transfer lines, a heating cord (Fisher, Cat#: 11-463-99) was wrapped around them and they were covered with aluminum foil. The temperature of the heating cord was controlled via a transformer (Matheson Scientific, Cat#: 21530-10).

The optimized bench-top system presented here uses a Varian 3380 GC with a ^{63}Ni ECD. A VB-624 (30 m × 0.32 mm i.d. × 1.80 μm f.t.) capillary column was used for separation. The SCMS-probe carrier gas was N_2 at 6 mL/min. The GC carrier gas and the ECD make-up gas were N_2 at 13 mL/min and 25 mL/min, respectively. The injector temperature was 175°C. The detector temperature was 300°C. The oven temperature program was as follows: the initial temperature was held at 35°C for one minute, increased to 75°C at 20°C/min, increased to 200°C at 50°C/min and held for three minutes. The total chromatographic run time was 8.5 minutes. The data acquisition software was Varian Star® Chromatography Workstation version 6.00.

Results and Discussion

SCMS Probe Carrier Gas Flow Rate Optimization

The flow rate of carrier gas significantly influences the performance of the SCMS probe. Sufficiently low internal flow rate can allow the THM4 species to reach true vapor-liquid equilibrium with the sample solutions. The optimum flow rate for the SCMS probe was experimentally determined for the SCMS-GC. The flow rate of the SCMS probe carrier gas was initially set to 20.0 mL/min. The analytical signal was measured. The SCMS probe carrier flow rate was reduced in 2.0 mL/min increments, and the analytical signal was recorded. This procedure was repeated until the SCMS probe carrier flow rate was lowered to 3.0 mL/min. The optimum flow rate was determined to be the point where the response no longer increased with decreased flow rate. For the prototype instrument, the optimum flow rate for the SCMS probe carrier gas was determined to be 6.0 mL/min for $CHCl_3$, $CHBrCl_2$ and $CHBr_2Cl$. The response of $CHBr_3$ did not change appreciably while decreasing the flow rate from 15.0 to 3.0 ml/min. Therefore, 6.0 mL/min was selected as the optimum flow rate for the analysis of THM4 species.

Effect of Transfer Line Temperature

The temperature of the SCMS probe to GC transfer line affected both peak area and the peak shape. Figure 3.3 shows two chromatograms obtained under the same conditions varying only the transfer line temperature. Optimization studies were conducted over from 38 to 144°C and showed that the peak area of each THM reached the maximum when the transfer line temperature was 65°C. This is consistent with the observations of Allonier and coworkers (2000) who tested the effect of PT transfer line temperature on determination of THM4 ranging from 100 to 200°C. Their results showed that at 200°C, the efficiencies for all THM4 were decreased, which led them to select a transfer line temperature of 100°C. Similarly, these studies confirm that increasing the temperature does not necessarily increase sample sensitivity. Ultimately, this means optimization studies are needed to determine the optimum temperature of the transfer line.

MDL, Accuracy and Precision Studies

MDL, accuracy and precision studies were conducted using the bench-top SCMS-GC instrument. THM4 standards were prepared and calibration plots for the SCMS-GC-ECD method were constructed by plotting the peak area for each THM component as a function of concentration. The MDL for each THM component was calculated using the USEPA recommended protocol (Glaser and Forest 1981; USEPA 1996a). The precision and accuracy were calculated based on EPA recommendations (USEPA 1996b). The results of these studies are presented in Table 3.1.

Table 3.1
Results from preliminary MDL, accuracy and precision studies for the optimized SCMS-GC-ECD

THM4 Species	MDL (μg/L)†	Mean % Recovery	% RSD	r^2	Conc. Range (μg/L)*
$CHCl_3$	0.6	132	4.1	0.986	1.2-21.5
$CHBrCl_2$	0.9	95	5.3	0.992	1.2-22.1
$CHBr_2Cl$	0.3	95	3.9	0.995	1.2-22.5
$CHBr_3$	0.4	117	3.4	0.998	1.7-32.0

* Each standard was analyzed five times five standards were analyzed to construct the calibration curves used to make MDL, accuracy and precision estimates.

† The concentration of the check standard was ~2.5 μg/L for each THM and analyzed seven consecutive times.

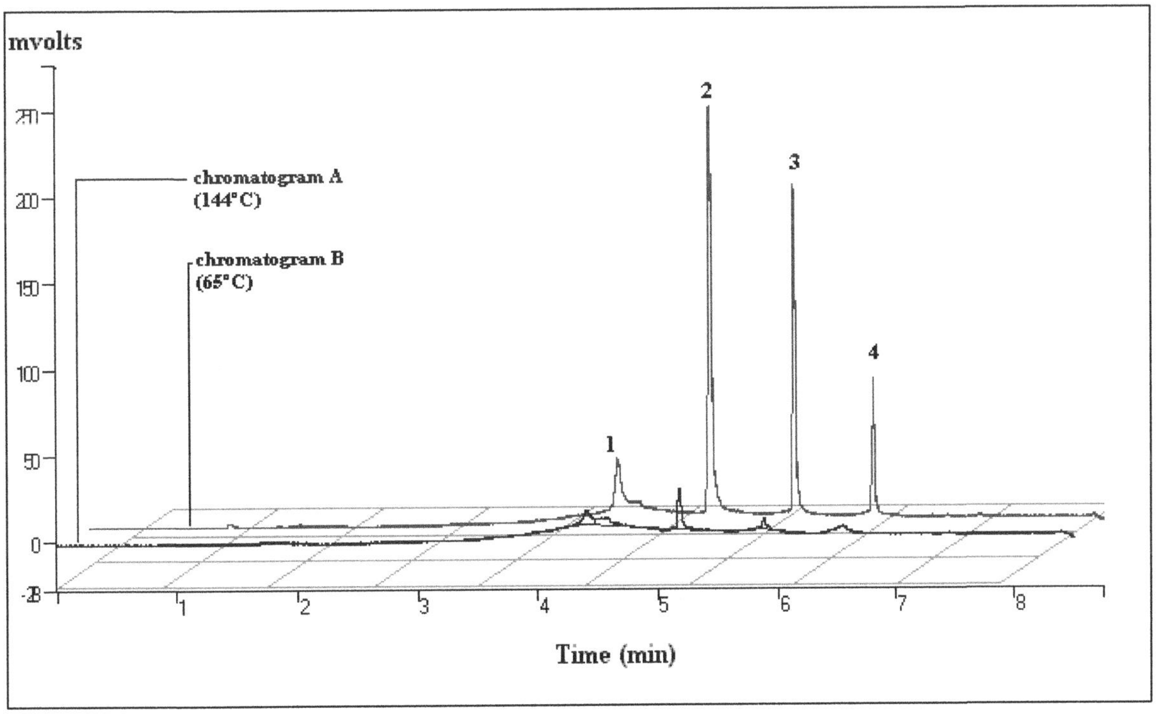

Figure 3.3 Bench-top SCMS-GC-ECD chromatogram of THM4. 1. $CHCl_3$ (8.86 μg/L); 2. $CHBrCl_2$ (15.88 μg/L); 3. $CHBr_2Cl$ (17.39μg/L); 4. $CHBr_3$ (25.79μg/L) (Source: Emmert et al. 2004a)

With the bench-top SCMS-GC-ECD instrument, more detailed experiments were carried out to evaluate the behavior of the analytical signal as a function of time once the SCMS-probe was immersed in the calibration vessel. The general trend was for the analytical signal to increase over time, reach a plateau where the analytical signal for each THM species is relatively constant and then begin to decrease. This time interval for reaching this plateau was somewhat different for each THM species. Specifically, the plateau time intervals for CHBr Cl_2 and $CHBr_3$ were approximately 30 minutes. The plateau time intervals for $CHCl_3$ and $CHBr_2Cl$ were 40 and 90 minutes, respectively. However, as long as sample injection is reproducible, it is not necessary to wait for the analytical signal to plateau. With automated and very reproducible sample injection, it is possible to sample accurately and precisely on the shoulder of the signal plot as a function of time. This latter approach was used for all work with the portable SCMS-GC and applied to on-line monitoring studies.

With the bench-top SCMS-GC, all calculations were done using the peak area as the response of analytical signal. The MDL value for each THM species was less than 1 μg/L. The MDL for TTHM4 was calculated by adding the MDL values for each component and it was 2.2 μg/L. The mean % recoveries ranged from 95 to 132% for the THM4 species, but were very reproducible as exhibited by the % RSD values that were less than 6% for each THM species. The method showed good sensitivity as measured by the slope of the calibration curves. The % error on the slope of each calibration curve was less than 6% for each THM4 species. The % error on the slope of the calibration curves for the THM4 species averaged 4%. The r^2 values for the calibration curves used in the MDL, accuracy and precision studies showed good linearity, but more detailed experiments were conducted to further explore the linear range. These results illustrate that the SCMS-GC approach shows promise for monitoring the concentrations of THM4 in water distribution systems.

Linearity

The linear range of the SCMS-GC-ECD method for THM4 was evaluated by plotting the analytical signal (peak area) as a function of concentration. Table 3.1 shows the concentration range studied and the associated r^2 values for calibration curves with each THM species. The results show acceptable linearity over the concentrations ranges typically monitored in drinking water. Plots of the residuals did not show any significant trends. Evaluation of the linearity using the more rigorous F-test did suggest that the calibration curves were non-linear.

Effect of ionic strength

Standard solutions of THM4 were prepared in solutions of differing ionic strength. Sodium sulfate was used to vary the ionic strength of the solution. As expected, increasing the ionic strength of the THM standards increased the analytical signal from the "salting out" effect. More specifically, the addition of sodium sulfate changed the physical properties of the analytes, such as the vapor and partial pressure, solubility, thermal conductivity, surface tension and etc., all of which, if occur, may lead to the variation of the vapor-liquid equilibrium (Zygmunt 1996; Wang, Simmons, and Deininger 1995). Over the ionic strength range of 0 to 3.21 M, the analytical signal increase for $CHCl_3$, $CHBrCl_2$, $CHBr_2Cl$ and $CHBr_3$ was 13 %, 20 %, 21 %, and 17 % signal increase per 1 M increase in ionic strength, respectively.

At first glance this seems to be a large difference. However closer examination shows that in typical drinking water, this difference may not be as pronounced as it seems. Consider a 5 μg/L standard solution of $CHCl_3$: From a typical calibration curve, the predicted peak area for this standard would be 77289.5 μV*s. The ionic strength calculated for Memphis drinking water is ~0.002 M. Using the correction factor for $CHCl_3$ given above, the expected increase in peak area would be about 0.26 %. For a given peak area of 77289.5 μV*s for a standard in Memphis drinking water would give a peak area of 77490.5 μV*s. A 5 μg/L standard in Memphis drinking water would give a calculated concentration of about 5.02 μg/L less than a 0.5 % difference. Similar calculations can be done for the remaining THM4 with similar results. At typical conditions, the ionic strength effect is expected to be relatively low from water to water. In practice, analysts can evaluate the magnitude by spiking known concentrations into raw drinking water matrix and observing the relative analytical signals in laboratory water and drinking water samples.

Effect of solution temperature

Increasing the temperature of the sample increases the vapor pressure of the THM4 species and increases the analytical signal. The temperature coefficient of response for THM4 species was investigated over the temperature range 18 to 30°C. The results showed that the average temperature increase over the range of 20 to 30 °C for $CHCl_3$, $CHBrCl_2$, $CHBr_2Cl$ and $CHBr_3$ was 3%, 4%, 3%, and 4% signal increase per °C, respectively. These results are in good agreement with Blanchard and Hardy (1984) who used a silicone polycarbonate membrane for sampling volatile organic compounds. Their results showed that over the temperature range 0.4 to 29.2°C, the effect of temperature on the permeation rate constant varied from 2.60 to 4.00% change per °C. The effect of temperature on $CHCl_3$ was 2.80% change per °C. Temperature should be monitored during calibration and sampling but under normal circumstances, it is not be expected to change significantly from calibration to calibration. However if it did change, a simple correction factor of about 3 % increase in analytical signal/°C can be applied to correct the results.

Conclusions of Benchtop SCMS-GC

A bench-top SCMS-GC-ECD instrument was constructed and used to measure the concentrations of THM4 in water. The effects of experimental parameters, such as flow rate of carrier gas, water temperature, ionic strength of solutions and transfer line temperature were investigated. The results of MDL, accuracy and precision studies establish that the method shows promise for monitoring the concentrations of THM4 in drinking water distribution systems. The method requires that the temperature of the water be monitored, but the results can be readily corrected to take temperature variability into account. Changing the ionic strength of the sample will affect the analytical signal. Preliminary results indicate that this effect is not large comparing laboratory and Memphis drinking water. Under the optimal conditions, MDL estimates were less than 1 μg/L with good accuracy, precision and linearity. The SCMS-GC approach offers several advantages. The components are commercially available and easily connected to commercially available GC instruments. Solvent extraction, pre-concentration and sample handling are virtually eliminated. The SCMS probe was used to sample directly in water distribution systems or other industrial processes. The SCMS probe is rugged and relatively inexpensive compared to other sampling devices. It can be easily automated via a simple six-port sample injection valve. Additional studies will be discussed in next section on the construction of a portable SCMS-GC analyzer specifically adapted for on-line monitoring of THM4 species.

THE PORTABLE SCMS-GC-ECD

Overview of Instrument

The research for the first SCMS-GC instruments was conducted using a bench top GC and a 2 L calibration vessel. There have been studies done for the SCMS-GC-ECD using a 1 L calibration vessel configuration. Even though the results reported are acceptable, it was suspected that these calibration configurations did not truly represent the conditions experienced by the SCMS probe inside the FL/FC and would not be satisfactorily applied to on-line monitoring. With this in mind, a smaller calibration vessel was constructed to better approximate the conditions inside the FL/FC. The new calibration vessel was made using standard glass tubing, which was blown to produce a single closed end vessel with a 2.5 cm diameter and 14 cm length. These dimensions allow the SCMS to fit tightly inside the vessel, while allowing for a magnetic stirrer to be placed beneath. Due to the displacement of water by the SCMS, the vessel can only accept approximately 50 mL of water sample to create a zero headspace environment. Hopefully, this configuration would better mimic the FL/FC. Additionally, other modifications were made to the SCMS-GC. Figure 3.4 shows the current portable SCMS-GC analyzer with the miniaturized calibration vessel configuration.

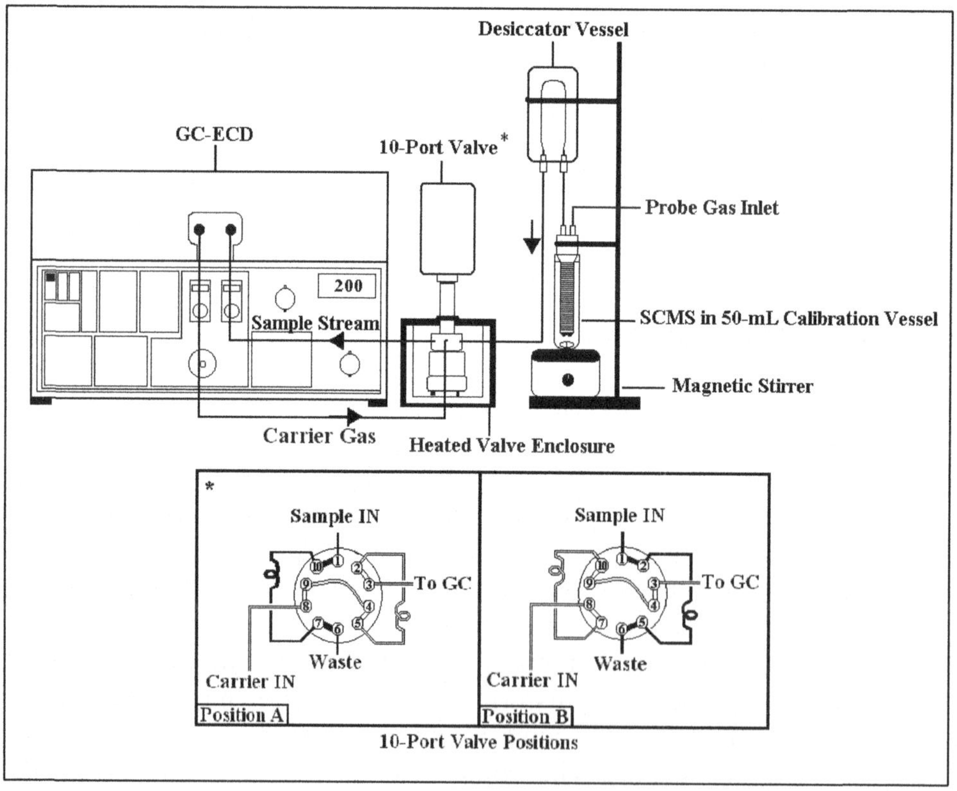

Figure 3.4 Schematic diagram of the portable SCMS-GC-ECD with miniaturized calibration vessel

Summary of MDL, Accuracy and Precision Studies

Typical calibration curves for the SCMS-GC-ECD method are shown in Figure 3.5. MDL, accuracy and precision studies were carried out using the USEPA recommended protocol (Glaser and Forest 1981; USEPA 1996a; and USEPA 1996b). In summary, these results were in agreement with previous results reported with a typical laboratory based instrument versus a portable GC. The detailed MDL, accuracy and precision results are shown in Table 3.2 for the portable SCMS-GC-ECD. The MDL values for each of the THM4 species range from 1.9 to 4.2 µg/L. The mean % recoveries ranged from 115 to 214% with % RSD from 4.9 to 8.3%.

Table 3.2
Results from preliminary MDL, accuracy and precision studies for the portable SCMS-GC-ECD

THM4 Species	MDL (µg/L)	Mean % Recovery	% RSD	R^2	n_{std}*	n_{chk}† (29.2 µg/L)	Conc.Range (µg/L)
$CHCl_3$	2.2	115	4.9	0.970	4	3	3.3-16.6
$CHBrCl_2$	4.0	128	8.3	0.987	4	3	4.3-21.4
$CHBr_2Cl$	1.9	163	5.1	0.967	4	5	5.4-27.2
$CHBr_3$	4.2	214	8.3	0.977	3	4	6.4-32.2

* Each standard was injected three or four times as indicated, and each calibration curve was based on five calibration standards.

† The check standard (29.2 µg/L TTHM4) was analyzed 3, 4 or 5 times as indicated.

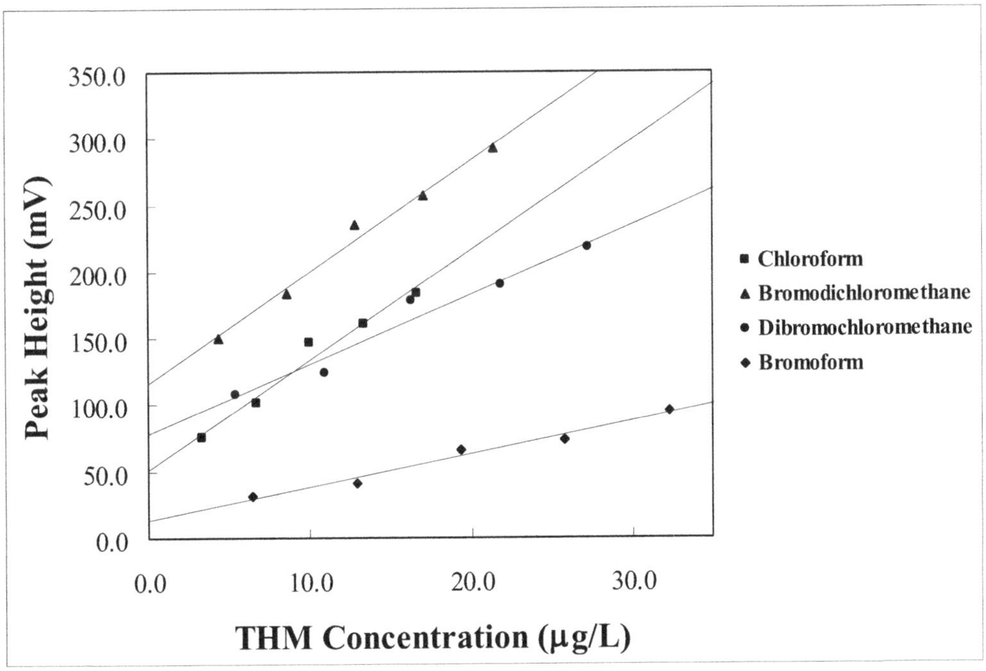

Figure 3.5 Typical calibration plots from the preliminary MDL, accuracy and precision studies using the portable prototype SCMS-GC-ECD for THM4 analysis

Over the course of the study, there were electronic problems with the detector board on the prototype's GC. Later studies involved a newer SRI GC resulting in a "second generation" portable SCMS-GC-ECD analyzer (Figure 3.6). All other SCMS-GC-ECD studies were done with this "second generation" instrument. This instrument was used to carry out revised MDL, accuracy and precisions studies as well as some preliminary monitoring studies in Memphis drinking water. The calibration curves for the portable SCMS-GC-ECD method are shown in Figure 3.7 and revised MDL, accuracy and precision studies were carried out using the USEPA recommened protocol (Table 3.3). The MDL values for each of the THM4 species range from 0.6 to 1.2 μg/L. The mean % recoveries ranged from 104 to 120% with % RSD from 5.5 to 10%.

Table 3.3.
Results from revised MDL, accuracy and precision studies for the portable SCMS-GC-ECD analyzer.

THM4 Species	MDL (μg/L)†	Mean % Recovery	% RSD	r^2	Conc Range (μg/L)*
$CHCl_3$	1.0	120	8.4	0.999	1 – 20
$CHBrCl_2$	1.2	115	10	0.995	1 – 20
$CHBr_2Cl$	0.6	112	5.5	0.996	1 – 20
$CHBr_3$	0.6	104	6.6	0.988	1 – 20

* Each standard was injected one time. Each calibration curve was based on five calibration standards.
† The check standard (~3 μg/L of each THM) was analyzed seven consecutive times.

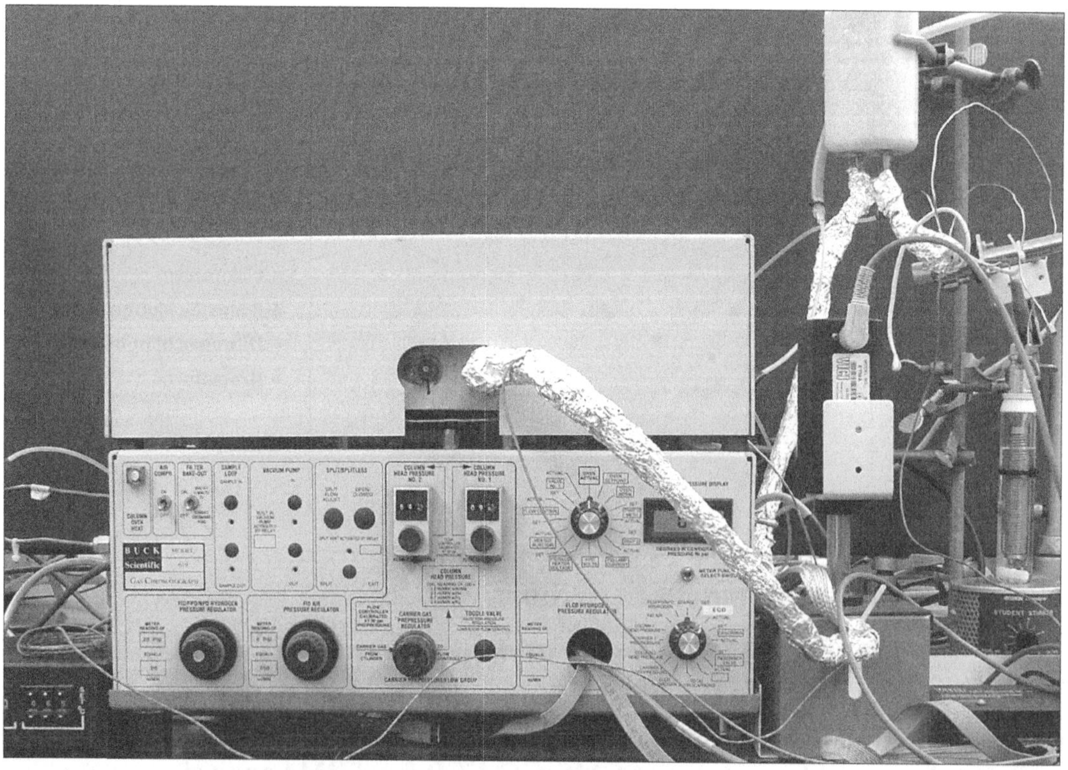

Figure 3.6 Picture of the portable SCMS-GC-ECD analyzer

Evaluating Sample Carry-over Effects

As discussed with the OPTGC analyzer (Chapter 2), sample carry-over from previous injections might be a problem, so care must be taken to minimize such problems. Figure 3.8 shows the results of these carry-over studies. The figure shows sample injection of ~ 80 µg/L of THM4 followed by blank injections every 15 minutes. The study showed that sample carry-over has been minimized (illustrated by the lack of THM4 peaks) after 60 minutes. Thus using the portable SCMS-GC-ECD will allow for one sample to be injected every 60 minutes with minimal carry-over issues.

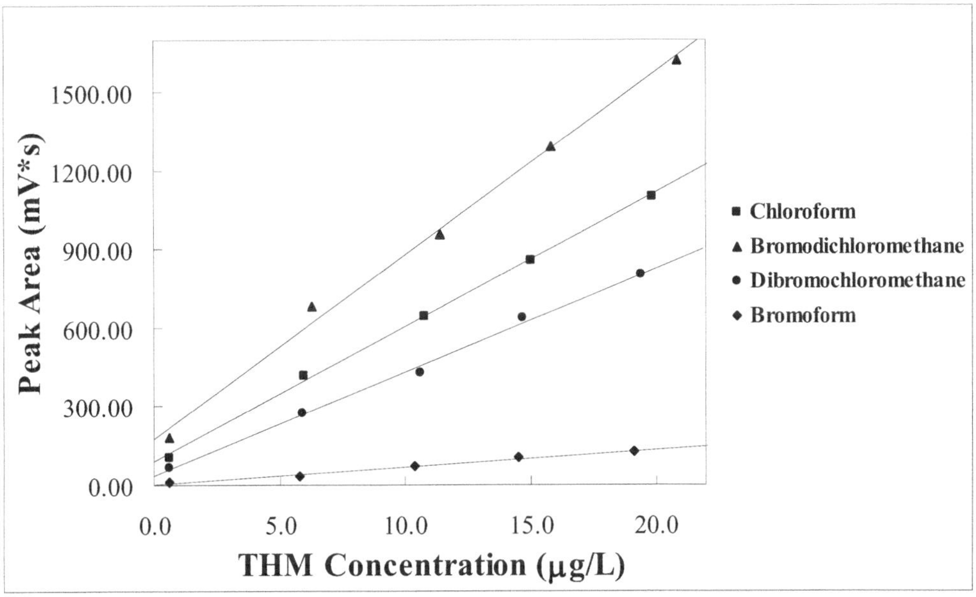

Figure 3.7 Calibration plots from the preliminary MDL, accuracy and precision studies using the portable SCMS-GC-ECD for THM4 analysis

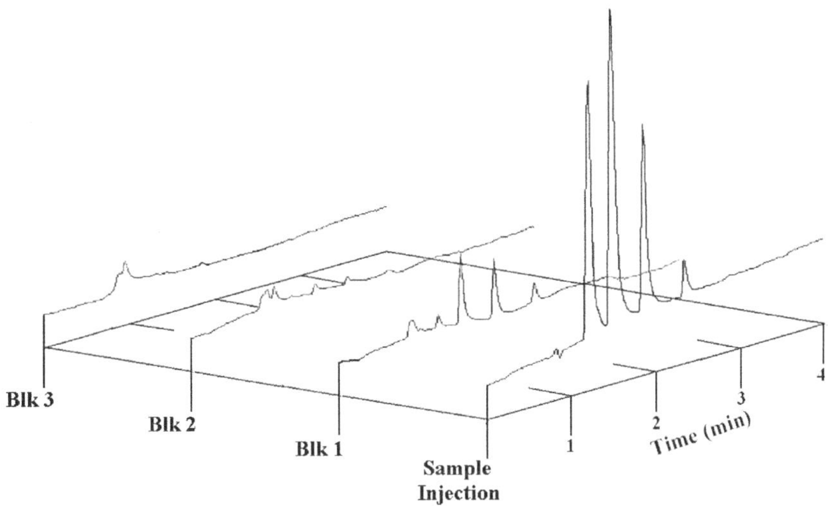

Figure 3.8 Results of carry-over studies conducted with the portable SCMS-GC-ECD

On-line Monitoring with SCMS-GC

The SCMS-GC approach is a very promising technique for on-line monitoring of THM4 concentrations in drinking water distribution systems. As presented here, the instrument requires two gas cylinders, and thus use of the word portable to describe the technique is used with reserve. The SCMS probe is commercially available as an adapter that allows for on-line monitoring via a FL/FC configuration. Pre-concentration of analyte is not required. Sample losses or contamination are largely reduced. The SCMS method eliminates the need for extraction solvents. The cost of an SCMS probe and its maintenance is significantly less than other comparable sample introduction devices such as PT. The SCMS-GC is easier to automate depending only on controlling a two-position injection valve and oven temperature.

Adapting the Portable SCMS-GC-ECD toward On-Line Monitoring Studies

The FL/FC has been constructed for on-line monitoring studies. A schematic diagram of this configuration is shown in Figure 3.9. This configuration allows the SCMS-GC to be applied to on-line monitoring directly in the drinking water distribution system. The basic FL/FC device is commercially available and was purchased from Global FIA. Researchers at the U of M lab have made modifications to the basic FL/FC device in order to adapt it to drinking water monitoring. The FL/FC has a chamber for insertion of an SCMS probe. The cylindrically shaped SCMS probe chamber has a diameter of 2.0 cm and a length of 14.2 cm. These dimensions are very similar to the dimensions of the miniaturized calibration vessel described earlier in this report. The volume of the probe chamber is ~ 75 mL when it contains the probe. Table 3.4 lists the typical settings used for the SCMS-GC-ECD operated in FL/FC configuration.

Initial studies with the FL/FC identified two issues. The first issue was that an unusually high flow rate (~3 L/min) was necessary to completely fill the probe chamber and prevent accumulation of trapped air pockets or bubbles. The high flow rates required were undesirable because they would be extremely wasteful for studies that would be carried out continuously. The high flow rates also caused severe turbulence in the tubing and limited the contact time for efficient permeation of THM4 across the probe membrane. It was decided to restrict the flow rate through the FL/FC to the EPA recommended value for sampling of 450-500 mL/min. To do this, it was necessary to add two external adapters (purchased from Swagelok®) to the FL/FC. These adapters lowered the flow rate to 500 mL/min by controlled restriction at the effluent end of the flow cell. The second issue was the adaptation of the FL/FC to the drinking water tap. Tygon® tubing was used to connect the FL/FC to the water tap. It should be noted that 316 SS tubing can be used in the final optimized configuration, but adds considerable expense. Tygon® tubing is more inexpensive, readily available, and can be more readily adapted to various types of water taps. It is also easier to cut into sections and replace if necessary. Figure 3.9 shows that the SCMS probe portion of the set-up has been adapted to the FL/FC, but the rest of the SCMS-GC configuration has remained unchanged. The point is that it is now possible to switch back and fourth between the calibration vessel configuration and the FL/FC configuration in about five minutes. A 1.0 m length of 2.5 cm o.d. Tygon® tubing was used to transport water from the water tap to the first adapter. The first adapter is connected to the FL/FC using a ~300 cm length of 2.5 cm o.d. Tygon® tubing. The water effluent exits the flow cell through another piece of Tygon® tubing with a second Swagelok® adapter identical to the first.

Table 3.4

Typical settings for the portable SCMS-GC-ECD operated in the FL/FC configuration

Parameter	Setting
GC Column	Valco VB-5, 15.0m x 0.53 mm x 1.0 µm f.t.
Carrier Gas	Helium (99.999% purity) at 5.0 mL/min
SCMS Carrier Gas	Nitrogen at 6.0 mL/min
ECD Make-up Gas	Nitrogen at 35.0 mL/min
Transfer Line Temperature	65 °C
Valve Enclosure Temperature	65 °C
ECD Temperature	300 °C
Initial GC Oven Temperature	60 °C for 2.20 minutes
Flow Cell Flow Rate	Between 450-500 mL/min
Temperature Ramp	60 °C to 220 °C at 60°C/min
Final GC Oven Temperature	220°C for 4.33 minutes
Total Chromatographic Time	9.20 min
Return to Initial Conditions	7.00 min
Total Run Time	16.20 min

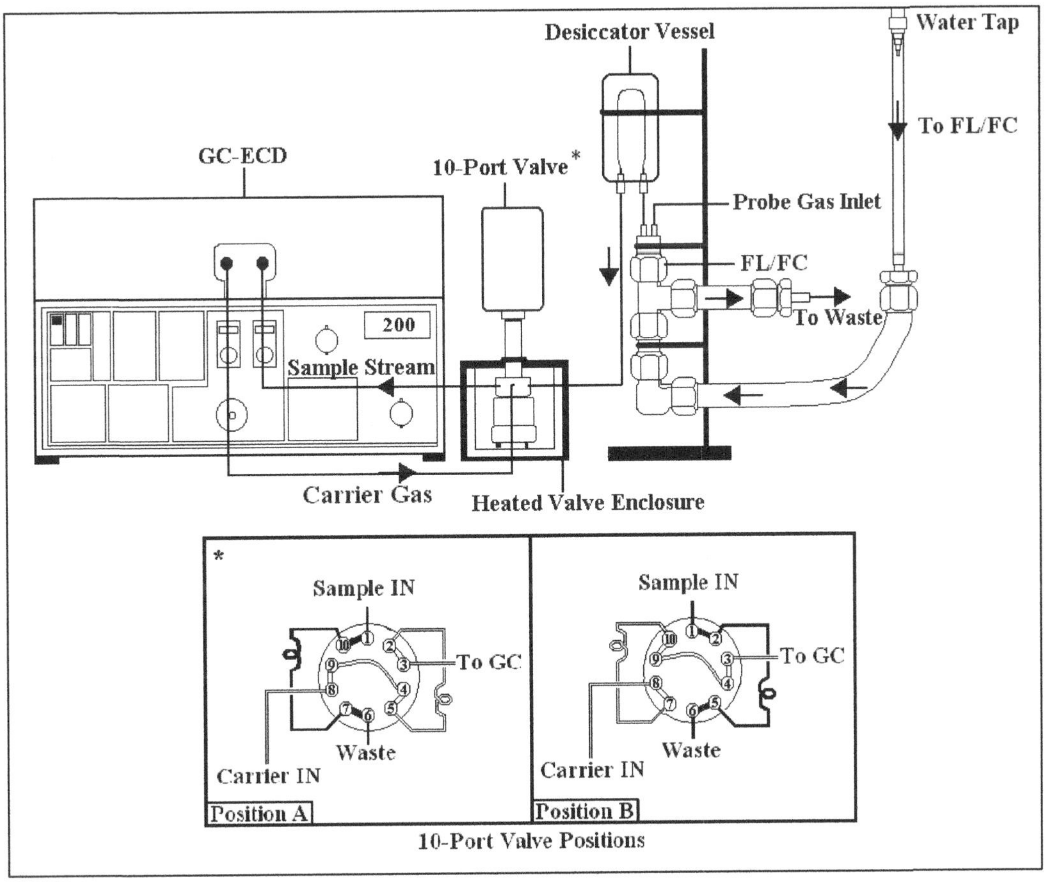

Figure 3.9 Schematic diagram of the SCMS-GC-ECD in the fast loop/flow cell configuration for on-line monitoring of THM4 in drinking water distribution systems

The SCMS-GC can be fully automated and can send data to a FTP server if desired. The latest version of PeakSimple has included fail safe mechanisms such as relay shutdowns for the valve to prevent instrument damage if the power goes out; data warnings if THM concentrations exceed a specified amount; data back up for any given amount of time and executable files which could alert an instrument operator of any potential problems with the system. These features are crucial to developing a reliable system for data collection and analysis. Data quality is of equal importance when considering the operation of an instrument.

On-Line Monitoring Study at the University of Memphis Sampling Site

The SCMS-GC-ECD was used to conduct a ~six day monitoring study (Sept. 24-30, 2004) at the University of Memphis (Memphis 1) sampling site. The SCMS-GC-ECD was interfaced to the drinking water distribution system using the FL/FC adaptor described earlier. The concentrations of each THM4 species was monitored at a rate of ~1 sample/hour (though the analyzer was typically taken off-line each 12 hours to run a check standard). The results for each THM4 species were summed and are reported as TTHM4. The results of these studies along with details used in the analysis of the data are presented in Table 3.5.

Table 3.5
Memphis 1 monitoring study: monitoring THM4 concentrations using SCMS-GC-ECD

	$CHCl_3$		$CHBrCl_2$		$CHBr_2Cl$		$CHBr_3$		TTHM4	
Statistic/*Method*	*EPA*	*SCMS*	*EPA*	*SCMS*	*EPA*	*SCMS*	*EPA*	*SCMS*	*EPA*	*SCMS*
N	48	48	48	48	48	48	48	48	48	48
Avg. Conc	1.1	1.2	1.2	1.3	1.5	1.9	2.2	0.8	5.9	5.2
Std. Dev Conc.	0.1	0.2	0.1	0.2	0.1	0.3	0.1	0.2	0.2	0.8
Conc. Range	1.0 to 1.4	0.8 to 1.6	1.1 to 1.4	0.8 to 1.9	1.3 to 1.6	1.3 to 3.1	2.2 to 2.3	0.3 to 1.3	5.6 to 6.4	3.5 to 7.3
Avg. Bias	---	0.1	---	0.1	---	0.5	---	–1.5	---	–0.8
Std. Dev Bias	---	0.2	---	0.2	---	0.3	---	0.2	---	0.7
Bias Range	---	–0.3 to 0.5	---	–0.3 to 0.6	---	–0.1 to 1.6	---	–1.9 to –0.9	---	–2.2 to 1.0
% REC_{Bias}	---	108	---	109	---	133	---	35	---	87
s.d. % REC_{Bias}	---	19	---	17	---	20	---	10	---	11
% REC_{Bias} Range	---	70.8 to 147	---	73.4 to 149	---	95.2 to 202	---	14.8 to 59.6	---	61.4 to 115
t-test*		Yes		Yes		Yes		Yes		Yes
$t_{critical}$	---	2.408	---	2.408	---	2.408	---	2.408	---	2.408
$t_{calculated}$		2.701		3.774		11.218		43.327		8.147

* The $t_{critical}$ and $t_{calculated}$ values listed were calculated for the N listed using a two-tailed t-test with 98% confidence. Typically a $t_{calculated}$ value should be less than the $t_{critical}$. Yes or No as indicated means if statistically different.

A plot of the concentration of $CHCl_3$ versus time is shown in Figure 3.10. For 48 direct comparisons, the concentration ranges from 1.0 - 1.4 μg/L for EPA 502.2 and 0.8 - 1.6 μg/L for the SCMS-GC. The average concentration was 1.1 ± 0.1 μg/L (EPA 502.2) and 1.2 ± 0.2 μg/L (SCMS-GC-ECD). Figure 3.11 shows a plot of the bias as a function of time for the SCMS-GC analyzer compared to EPA 502.2. The bias, a measure of the absolute error of the candidate method, ranges from –0.3 - 0.5 μg/L with an average of 0.1 ± 0.2 μg/L. The % REC_{Bias} ranged from 70.8 - 147 % with an average of 108 ± 19 %.

Similar results were observed for $CHBrCl_2$ (Figure 3.12). The results of 48 direct comparisons, gave concentration ranges from 1.1 - 1.4 μg/L (EPA 502.2) and 0.8 - 1.9 μg/L (SCMS-GC). The average concentration was 1.2 ± 0.1 μg/L (EPA 502.2) and 1.3 ± 0.2 μg/L (SCMS-GC). The bias (Figure 3.13) ranged from –0.3 - 0.6 μg/L (averaging 0.1 ± 0.2 μg/L) and % REC_{Bias} ranged from 73.4 - 149 % with an averaging 109 ± 17 %.

For $CHBr_2Cl$ and $CHBr_3$, the results begin to show more deviation. For $CHBr_2Cl$ (Figure 3.14), the concentrations ranged from 1.3 - 1.6 μg/L (EPA 502.2) and 1.3 - 3.1 μg/L (SCMS-GC). The average concentration was 1.5 ± 0.1 μg/L (EPA 502.2) and 1.9 ± 0.3 μg/L (SCMS-GC). The bias (Figure 3.15) ranged from –0.1 - 1.6 μg/L (averaging 0.5 ± 0.3 μg/L) and % REC_{Bias} ranged from 95.2 - 202 % (averaging 133 ± 20 %). For $CHBr_3$ (Figure 3.16), the concentrations ranged from 2.2 - 2.3 μg/L (EPA 502.2) and 0.3 - 1.3 μg/L (SCMS-GC). The average concentration was 2.2 ± 0.1 μg/L (EPA 502.2) and 0.8 ± 0.2 μg/L (SCMS-GC). The bias (Figure 3.17) ranged from –1.9 to –0.9 μg/L (averaging –1.5 ± 0.2 μg/L) and % REC_{Bias} ranged from 14.8 - 59.6 % (averaging 35 ± 10 %).

Overall, the results for TTHM4 were as follows: the concentrations (Figure 3.18) ranged from 5.6 - 6.4 μg/L (EPA 502.2) and 3.5 - 7.3 μg/L (SCMS-GC). The average concentration was 5.9 ± 0.2 μg/L (EPA 502.2) and 5.2 ± 0.8 μg/L (SCMS-GC). The bias (Figure 3.19) ranged from –2.2 - 1.0 μg/L (averaging –0.8 ± 0.7 μg/L) and % REC_{Bias} range from 61.4 - 115 % (averaging 87 ±11 %).

A statistical analysis was performed on the data. A $t_{critical}$ for the 48 comparisons was estimated to be 2.408. For $CHCl_3$ $t_{calculated}$ was 2.701, which means that there is a statistical difference between the methods for $CHCl_3$. The $t_{calculated}$ for $CHBrCl_2$ is slightly larger than $CHCl_3$ at 3.774; therefore a statistical difference between the two methods is also evident for $CHBrCl_2$. For $CHBr_2Cl$ and $CHBr_3$, the $t_{calculated}$ values were 11.218 and 43.327, respectively, signifying statistical difference for these species. These values correlate with the bias and % REC_{Bias} results in that the differences of the methods for these species. For TTHM4, the $t_{calculated}$ was also larger than $t_{critical}$ with a value of 8.147. Overall, statistical agreement is better for the more chlorinated species than with the more brominated species.

Qualitatively, the agreement between the two methods appears promising for the $CHCl_3$ and $CHBrCl_2$, but shows deviation for $CHBr_2Cl$ and $CHBr_3$. The concentrations of THM4 species are all at relatively low levels (1-10 μg/L) compared to those experienced by most drinking water utilities. These measurement tax the MDL estimates of the SCMS-GC-ECD method and are likely some of the lowest concentrations that would likely be encountered. These concentrations range from ~8 - 80 times lower that the MCL for TTHM4 or individual THM4 species. Despite the differences between the individual species that exist, the results for TTHM4 appear to compare favorably.

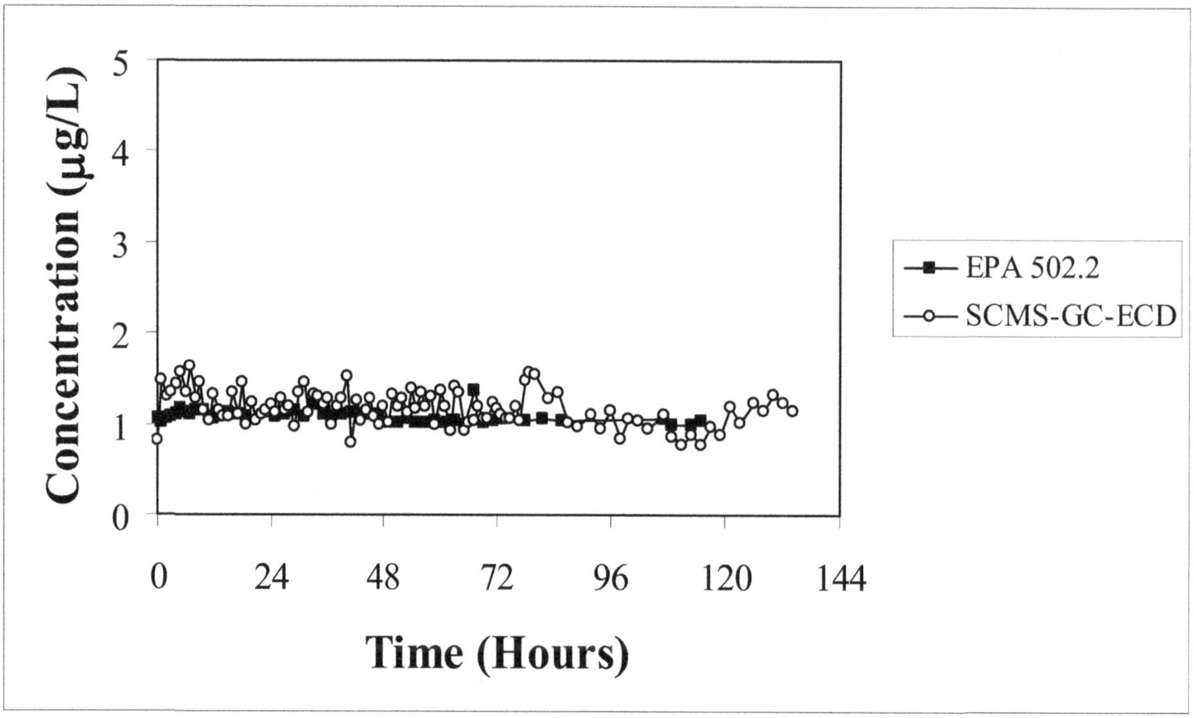

Figure 3.10 Memphis 1 concentration plot of $CHCl_3$ for EPA 502.2 and SCMS-GC-ECD

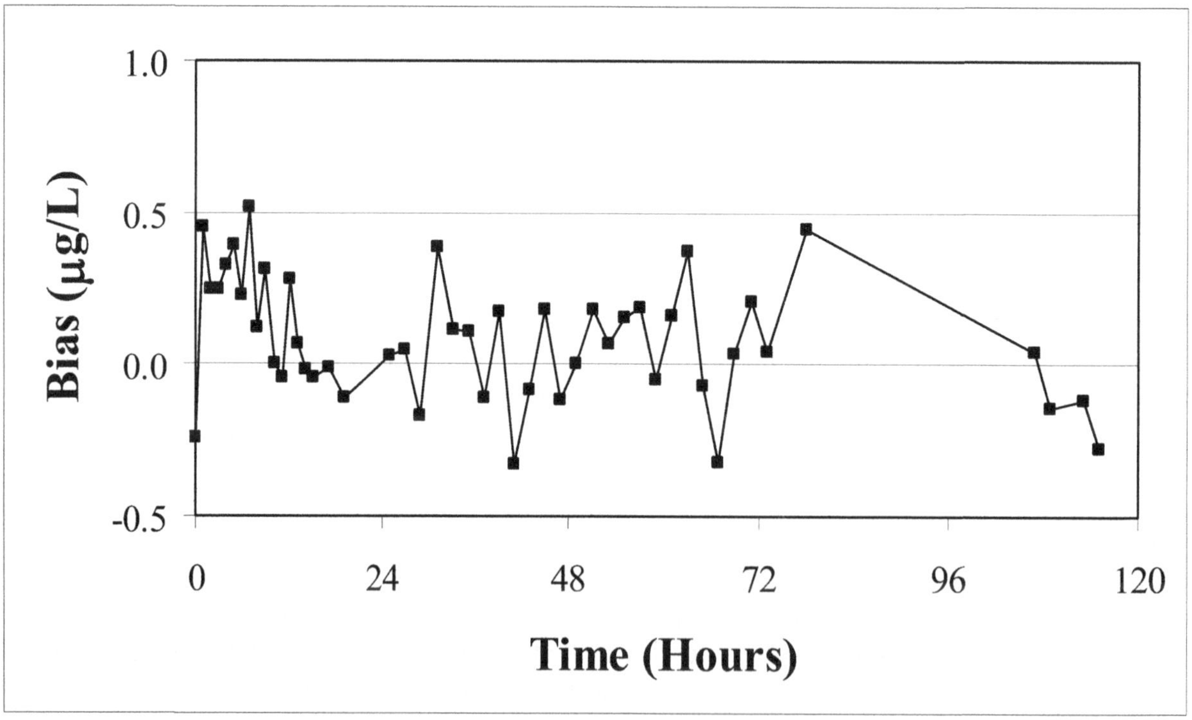

Figure 3.11 Memphis 1 bias plot of $CHCl_3$ comparing SCMS-GC-ECD to EPA 502.2

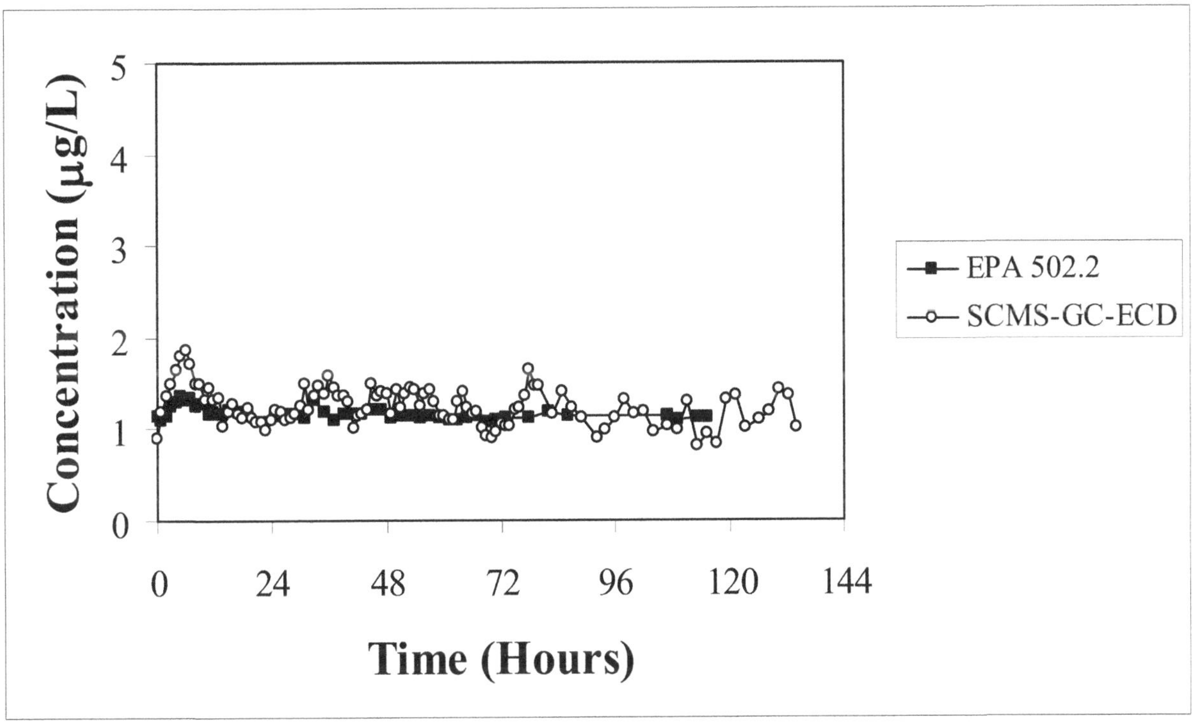

Figure 3.12 Memphis 1 concentration plot of $CHBrCl_2$ for EPA 502.2 and SCMS-GC-ECD

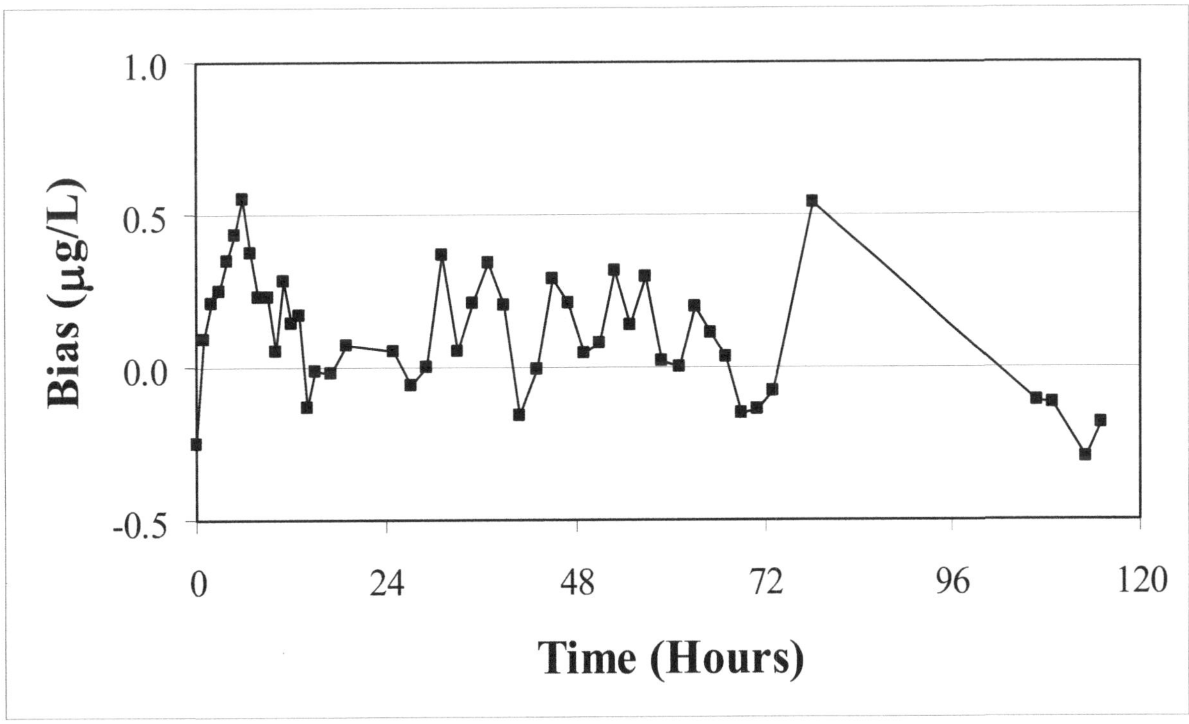

Figure 3.13 Memphis 1 bias plot of $CHBrCl_2$ comparing SCMS-GC-ECD to EPA 502.2

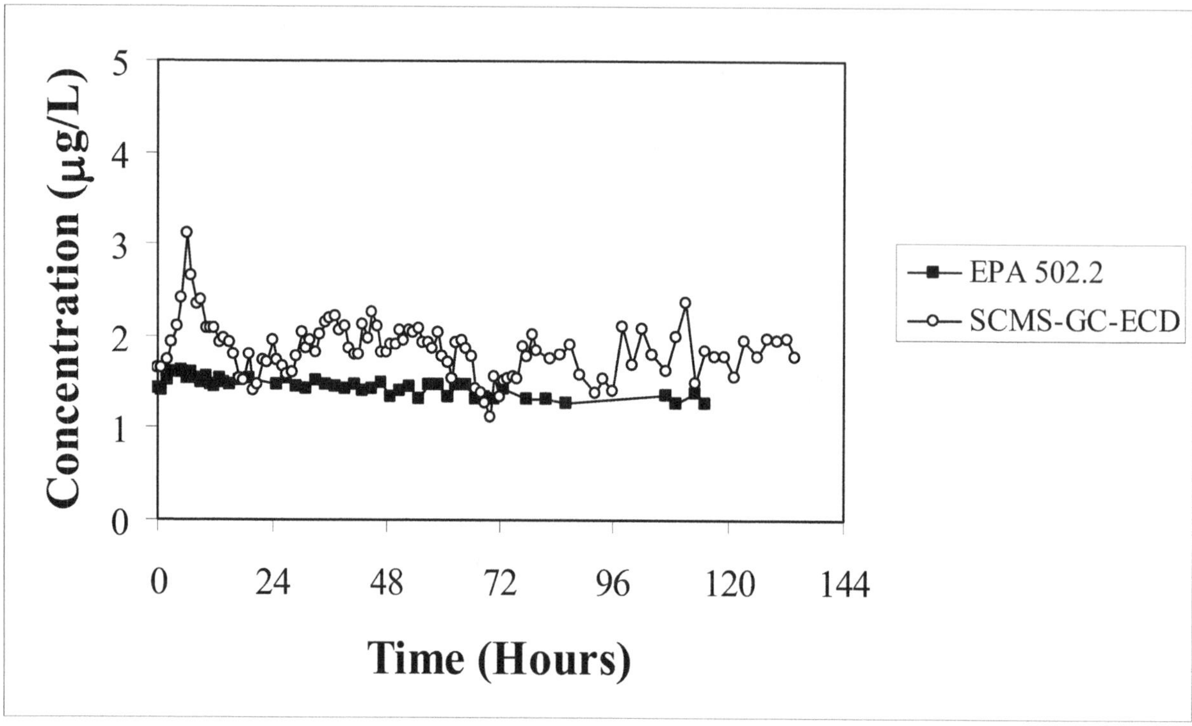

Figure 3.14 Memphis 1 concentration plot of $CHBr_2Cl$ for EPA 502.2 and SCMS-GC-ECD

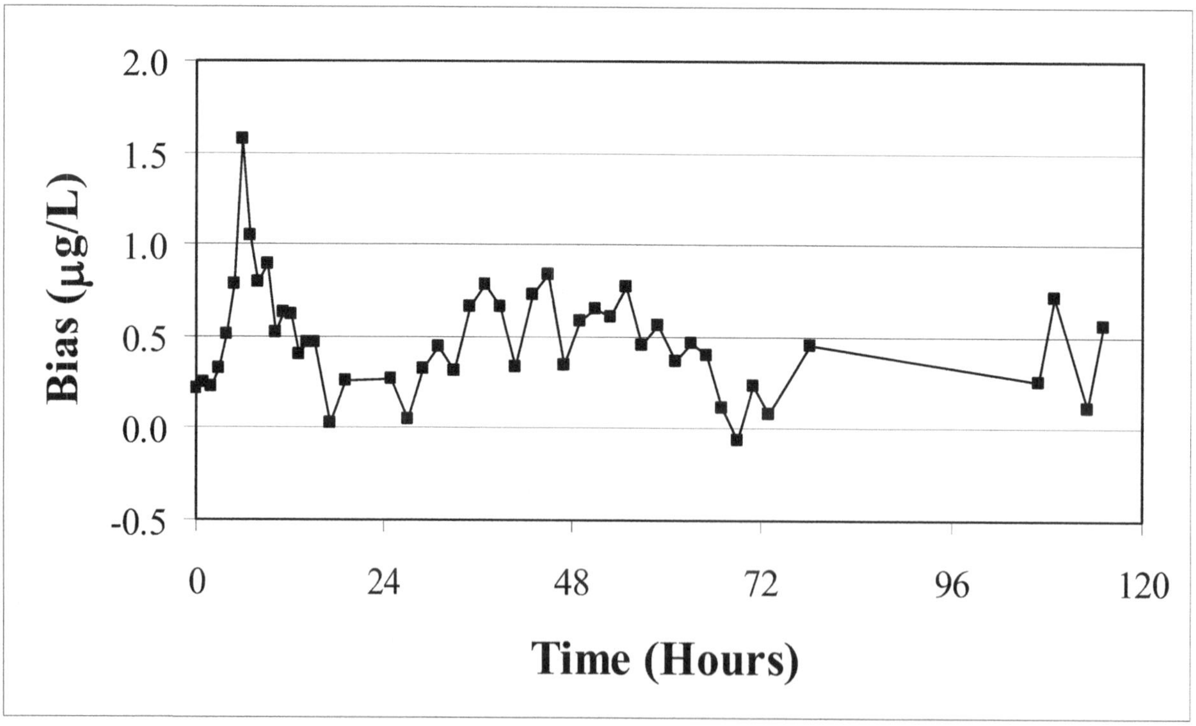

Figure 3.15 Memphis 1 bias plot of $CHBr_2Cl$ comparing SCMS-GC-ECD to EPA 502.2

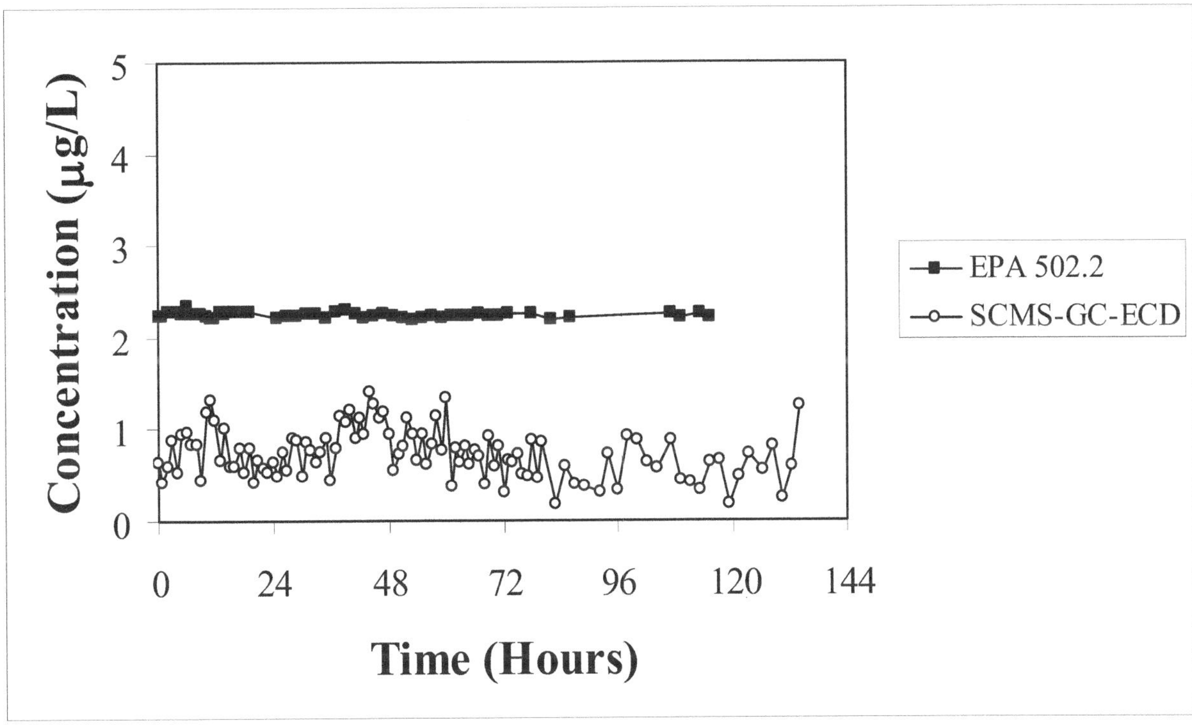

Figure 3.16 Memphis 1 concentration plot of $CHBr_3$ for EPA 502.2 and SCMS-GC-ECD

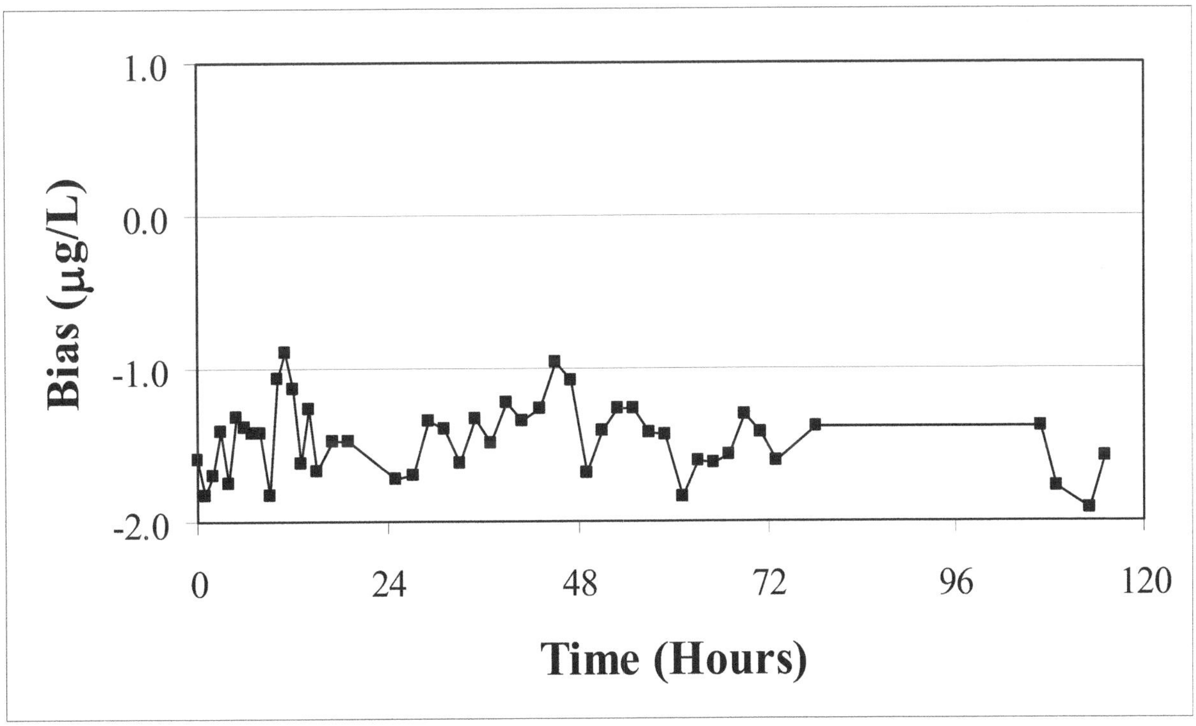

Figure 3.17 Memphis 1 bias plot of $CHBr_3$ comparing SCMS-GC-ECD to EPA 502.2

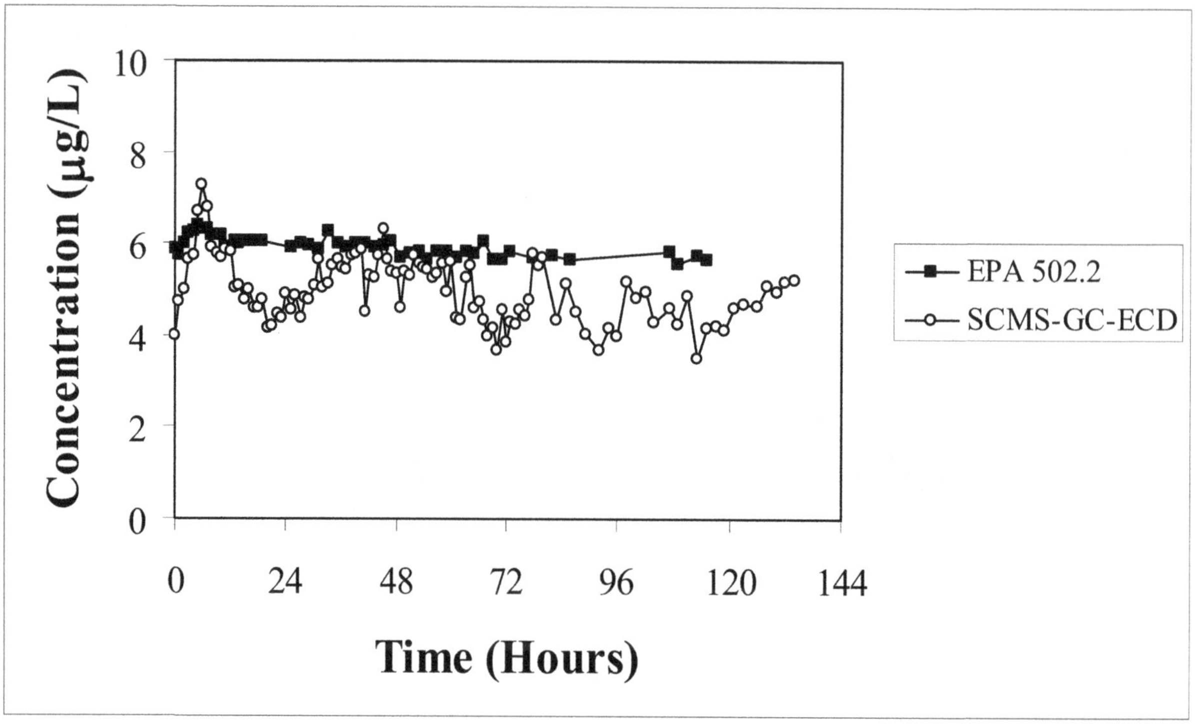

Figure 3.18 Memphis 1 concentration plot of TTHM4 for EPA 502.2 and SCMS-GC-ECD

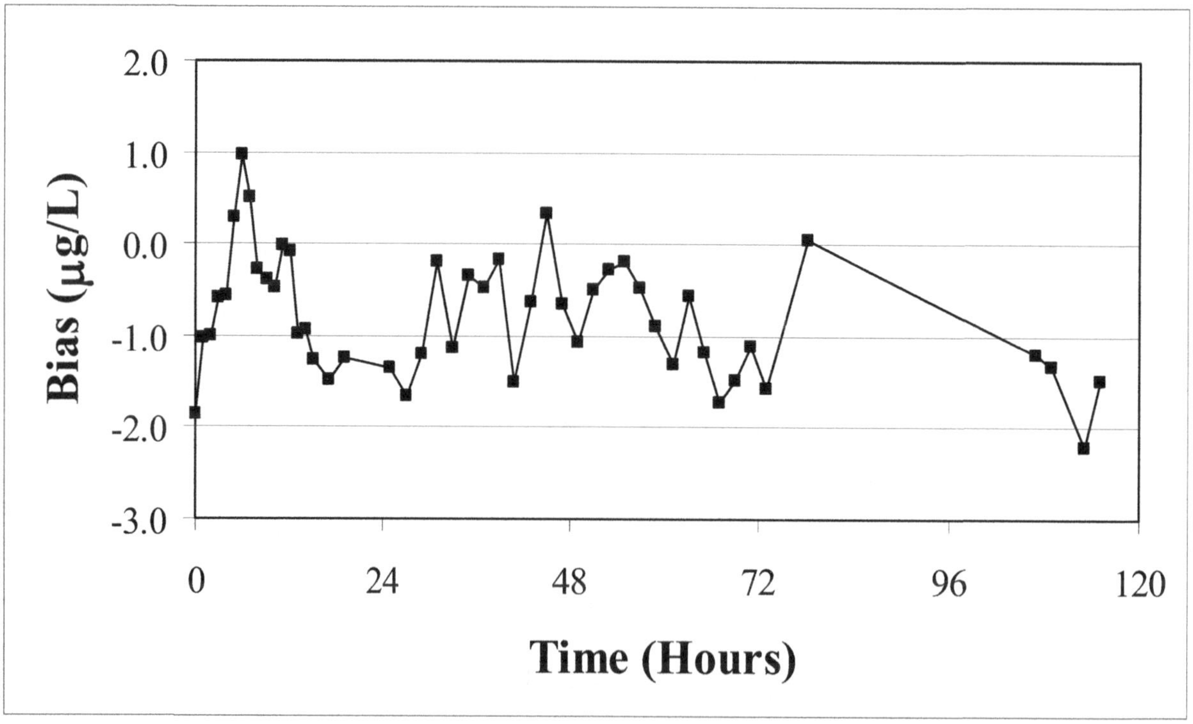

Figure 3.19 Memphis 1 bias plot of TTHM4 comparing SCMS-GC-ECD to EPA 502.2

Conclusions and Recommendations for the Portable SCMS-GC-ECD

The SCMS-GC-ECD offers many of characteristics desired for an on-line monitoring method. The SCMS probe can be interfaced to a portable GC-ECD with minimal expense. The SCMS-GC-ECD can be calibrated using a calibration vessel, or sample from the drinking water distribution systems using the FL/FC configuration. It offers a sampling rate of ~1 sample/ hour with minimal sample carry-over and is fully automated following calibration. The SCMS-GC, as configured, uses PeakSimple version 3.32 to communicate data to remote servers as long as a wired or wireless internet is available. It can be constructed with commercially available components, but to be fair, the skill level required of the analyst is slightly higher that imagined for an ideal analyzer. Most of the medium to high skill level would be required in constructing the analyzer. Fortunately, once constructed, the SCMS-GC-ECD is largely automated and thus the necessary skill level for operation relaxes somewhat.

The MDL, accuracy and precision studies, though not ideal, are promising. The MDL values are in the range 0.5 to 5 μg/L. The mean % recovery results are slightly outside the desired range of 100 ± 5%. (closer to ± 20%) but the % RSD values are within the ideal range of ±10%. Logistically, the researchers were not able to us both SCMS-GC and CMS-GC (Chapter 5) at the same time due to limited GC availability. The SCMS-GC-ECD presented in this Chapter and the CMS-GC-ECD (Chapter 5) use the same GC-ECD. For this reason only, the SCMS-GC was not used for more extensive monitoring studies at Memphis 2 and Houston. Compared to the CMS-GC, the FL/FC sampling device uses much more water (~500-1000 mL/min as opposed to ~2.6 mL/min), but more importantly, it is harder to maintain a constant flow rate through the FL/FC than the CMS device. Moreover, despite limited progress during the research project, it is still a desire of the researchers to develop a manner to introduce internal standards directly into the sampling line while doing on-line monitoring. It is much more likely that this can be done accurately and precisely with the CMS device than the SCMS probe.

Recommendation 3.1: The SCMS probe can be used for on-line monitoring; however, a redesign of the FL/FC is suggested. The current FL/FC must be changed, allowing a peristaltic pump to flow sample through. It is also suggested that this revised FL/FC be constructed of glass or SS. Factors such as temperature, pH, conductivity, and flow rate changes can change calibration parameters. In real world situations, it may be difficult to isolate and investigate these effects. The necessity of a separate calibration vessel needs to be eliminated, and further, it is desirable to integrate the calibration vessel and flow cell so that they are the same device.

Recommendation 3.2: Automated internal standard addition needs to be further explored for all on-line methods, but would be particularly attractive for SCMS-GC. Both calibration protocols (calibration curve and internal standardization) should be used when possible during on-line monitoring studies.

Recommendation 3.3: The SCMS-GC-ECD method needs to be further explored in more demanding sample matrices. It is desirable to use the method, not only for THM4 species, but other compounds of interest to drinking water professionals—for example, other common volatile compounds found in drinking water.

Recommendation 3.4: The sampling rate of the SCMS-GC-ECD method needs to be improved.

THE SCMS-GCOV APPROACH TO THM4 MONITORING

Introduction

The SCMS-GC approach has also been used as a sample introduction method for the "gas chromatography on a valve" approach (GCOV) that was developed in this research. The detector used for GCOV is the pulsed discharge photoionization detector (PDPID). The PDPID has been described in the Phase One Report (Emmert et al. 2004b). The PDPID uses a stable, low-power, pulsed discharge in He as the ionization source. Compounds eluting from the GC column are ionized by photons from the discharge zone. Resulting electrons are focused toward the collector electrode by the two bias electrodes. The principal mode of ionization is photoionization by radiation arising from the transition of diatomic helium (He_2) to the dissociative ground state (energy range from the He_2 is 13.5 eV to 17.7 eV). The PDPID can be operated in several modes to achieve nearly universal detection. For example, in helium ionization mode (PDHID), the detector is essentially non-destructive and highly sensitive. The response to organic compounds is linear over five orders of magnitude with detection limits in the low or sub picogram range. The PDHID mode is the mode that was used in this research. Earlier studies were conducted with a Model D-2 PDPID detector. This detector can be operated in several different modes. More recently, a newer D-4 Model PDPID detector was applied to THM4 monitoring (Emmert et al. 2006).

In this section, the construction, operation and evaluation of a prototype miniaturized version of the SCMS-GC with PDPID that uses the GCOV approach is discussed. The prototype SCMS-GCOV was used to conduct MDL, accuracy and precision studies. The GCOV method was then compared to EPA 502.2 for measuring THM4 concentrations in Memphis drinking water.

Constructing the Prototype Miniaturized SCMS-GCOV Analyzer

Unlike the original SCMS-GC-PDECD method mentioned earlier the SCMS-GCOV uses a Model D-4 PDPID that can only be operated in the HID mode. A two valve enclosure (HVE2, VICI) was used as the GC oven for the prototype as it provided enough space to accommodate both the PDPID (D-4-I, VICI), GC column and injection valve. A support was constructed to mount the PDPID to the two valve enclosure. A ~ 1 mm thick aluminum sheet (3.0 cm x 36.0 cm) was cut and four holes (diameter ~ 4 mm) were drilled. These holes were made for sheet metal screws (length ~1.0 cm, diameter < 4 mm) used to attach the PDPID to the standard bracket (SABB, VICI) and to the support. A pencil was used to trace the locations of the holes and lines shown in Figure 3.20. The exact locations of the holes for the standard bracket are given; however, the location of the holes for the detector may differ for different models of the PDPID. The detector used for this prototype required that the holes be placed 0.8 cm up from the dotted line and 3.9 cm over from the solid black line. The final shape of the support was prepared by bending the sheet on the dotted lines and cutting on the solid black lines (Figure 3.20).

The PDPID kit included screws to secure the standard bracket to the enclosure. After the standard bracket was secured, the support was attached to the bracket by using two screws, two spacers, two washers, and two nuts. The PDPID was then lowered into the support and through the top of the enclosure and secured using appropriate screws and nuts. Finally, insulation was placed between the detector and enclosure.

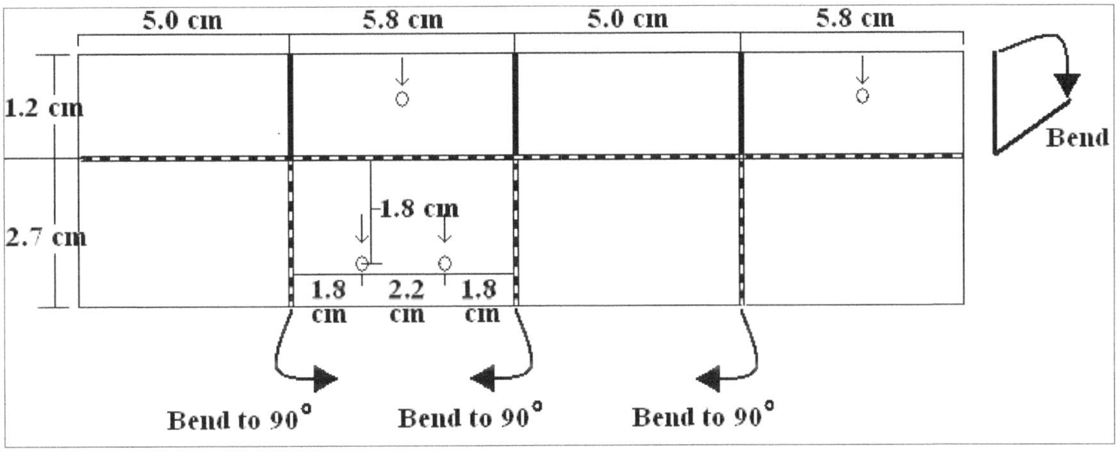

Figure 3.20 Diagram for constructing an aluminum support to mount the model D-4 PDHID on a two valve enclosure for SCMS-GCOV (Source: Emmert et al. 2006)

A picture of the SCMS-GCOV is shown in Figure 3.21 and a diagram of the assembled prototype SCMS-GCOV is shown in Figure 3.22. An eight-port air actuated injection valve (A4C8WT, VICI) was mounted on the two valve enclosure. Two 250 μL stainless steel (SS) sample loops (SL250CW, VICI) were attached to Ports 8, 4 and 6, 2 (gold plated ferrules were substituted for the SS ferrules). Lengths of SS tubing (1.6 mm o.d., 0.75 mm i.d.) were used to make sample transfer, vent, and GC carrier gas transfer lines. These lines were then connected to the specified valve ports (Figure 3.22). The heater block for the valve enclosure was attached to the valve head as described in the PDPID instrument manual. All operation procedures for the Instrument Temperature Controller (ITC10399, VICI), which controls the valve heater, were followed from the manual. A Digital Valve Interface (DVI, VICI) was attached to a nitrogen tank (80 psi) and then to an actuator on the valve.

The SCMS sampling probe was assembled as described in a manner similar to the bench-scale prototype SCMS-GC PDPID described in the Phase One Report (2004b). N_2 was used as a SCMS probe carrier gas. A flow controller (FC22AS1S3, VICI) was attached to a high-purity SS regulator (Airgas, 30 psi) that was attached to a N_2 cylinder in order to maintain a constant flow of 6.0 mL/min through the membrane. Two sections of SS tubing, three 1/16" internal fittings and ferrules, and one external fitting and ferrule were needed to complete this connection. A Restek MXT®-1 (Silcosteel® 20 m × 0.53 mm i.d. × 5.00 μm f.t.) coiled to a 3.5" diameter was installed to Port 1 of the injection valve. A 5 cm length of SS tubing, a SS 1/16" internal nut, a 1/16", a 1/16" external nut, two gold ferrules, a Swagelok SS 1/8" fitting with graphite ferrule, and a Swagelok SS 1/8" to 1/16" reducing union was used to connect the column to the injection valve. A commercial fused silica adaptor is available (FS11.58-5, VICI) for a megabore column that could replace these materials; however, the laboratory made configuration was readily available. A tee was installed as described in the operating manual to supply both the PDPID and column with carrier gas (He). A micrometering valve (ZBNV1LF-D, VICI) was attached to the tee to control the carrier gas flow rate to the valve. Before installation to the PDPID, the column was conditioned for several hours at 220 °C inside the valve enclosure. The carrier gas flow rate through the column was set at 5.0 mL/min. The PDPID temperature was held at 320°C during this time. After this conditioning, the column was inserted ~11.0 cm into the PDPID column adapter.

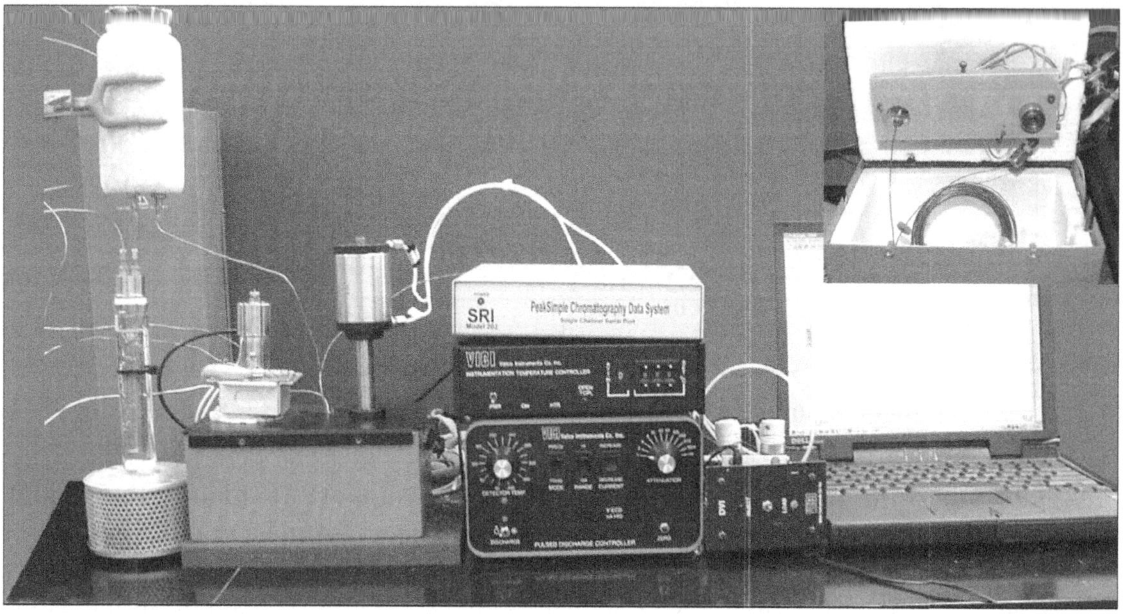

Figure 3.21 Picture of the SCMS-GCOV with PDHID. (Inset; Upper Right) Picture showing inside valve enclosure with valve, GC column and PDPID detector connection

The start-up procedures for the PDPID are described in the operation manual. A high purity regulator with SS diaphragms (Airgas) was attached to a 30 mL/min restrictor connected to the PDPID gas inlet. Details of the assembly and operation of the D-4 PDPID are found in the operating manual. After startup and conditioning the detector at 320°C for approximately 12 hours, the background current dropped to the ideal range of ~1.0 to 3.0 nA.

A single channel data acquisition system (SRI Instruments Model 203) was used to collect the non-attenuated signal output from the PDPID control module. All data was collected using PeakSimple software (v3.21). The eight-port sample injection valve was automated by connecting a digital valve interface (DVI, VICI) to a relay on the data acquisition board.

Because this prototype can only operate isothermally no connections were made with the temperature programming of the data system. The column was typically conditioned through an overnight bake-out (300°C) after each day. The detector was also left to bake at 320 °C at the end of each day. With previous research the transfer line that transported the THM4 to valve was heated to ~65 °C to prevent condensation on the SS tubing walls. On the prototype this transfer line was much shorter than with previous instruments, and it was placed above the PDPID. The heat from the PDPID was sufficient enough to prevent any noticeable condensation. The final parameters and settings for the SCMS-GCOV were as follows: The carrier gas was He (99.9999% purity) at 5.0 mL/min. The SCMS probe carrier gas was N_2 at 6.0 mL/min. The detector make-up gas was He (99.9999% purity) at 30.0 mL/min. The heated valve enclosure temperature (oven temperature) was 95 °C. The detector temperature was 320 °C. The total chromatographic time was 5 min.

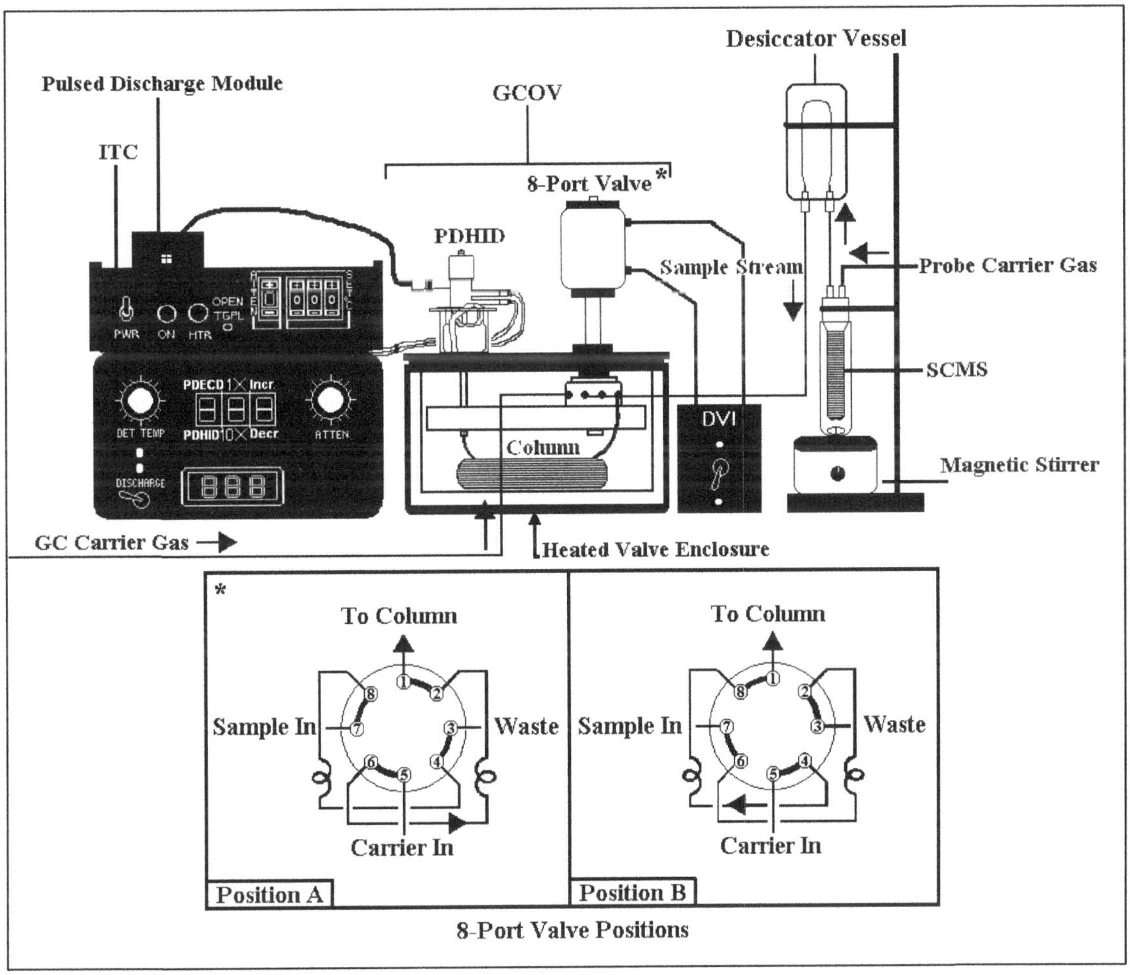

Figure 3.22 Schematic diagram of the SCMS-GCOV with PDHID (Source: Emmert et al. 2006)

Results Using the Prototype SCMS-GCOV with PDHID

With the SCMS-GCOV-PDHID, the calibration vessel used for this prototype was considerably smaller (50 mL) than the one used with the SCMS-GC-PDECD (2 L), thus allowing for faster permeation of the THM4 species through the membrane. Four calibration standards were analyzed for each calibration curve followed by five consecutive analyses of the check standard. The results of these studies are summarized in Table 3.6. Typical SCMS-GCOV-PDPID chromatograms are shown in Figure 3.23 and calibration plots are shown in Figure 3.24. The MDL values for the THM4 species ranged from 1.1 to 1.6 μg/L. The mean % recoveries averaged 101%. The % RSD values that averaged 13%.

Table 3.6
Results from MDL, accuracy and precision studies for the SCMS-GCOV analyzer

THM4 species	MDL (µg/L)†	Mean % Recovery	% RSD	r^2	Conc Range (µg/L)*
$CHCl_3$	1.6	91	16	0.984	1.8-15.0
$CHBrCl_2$	1.4	94	13	0.989	1.9-15.4
$CHBr_2Cl$	1.6	99	14	0.988	1.8-14.7
$CHBr_3$	1.1	119	8.6	0.976	1.7-14.5

* Each standard was injected one time. Each calibration curve was based on four calibration standards.
† The check standard (~12 µg/L TTHM4) was analyzed five consecutive times.

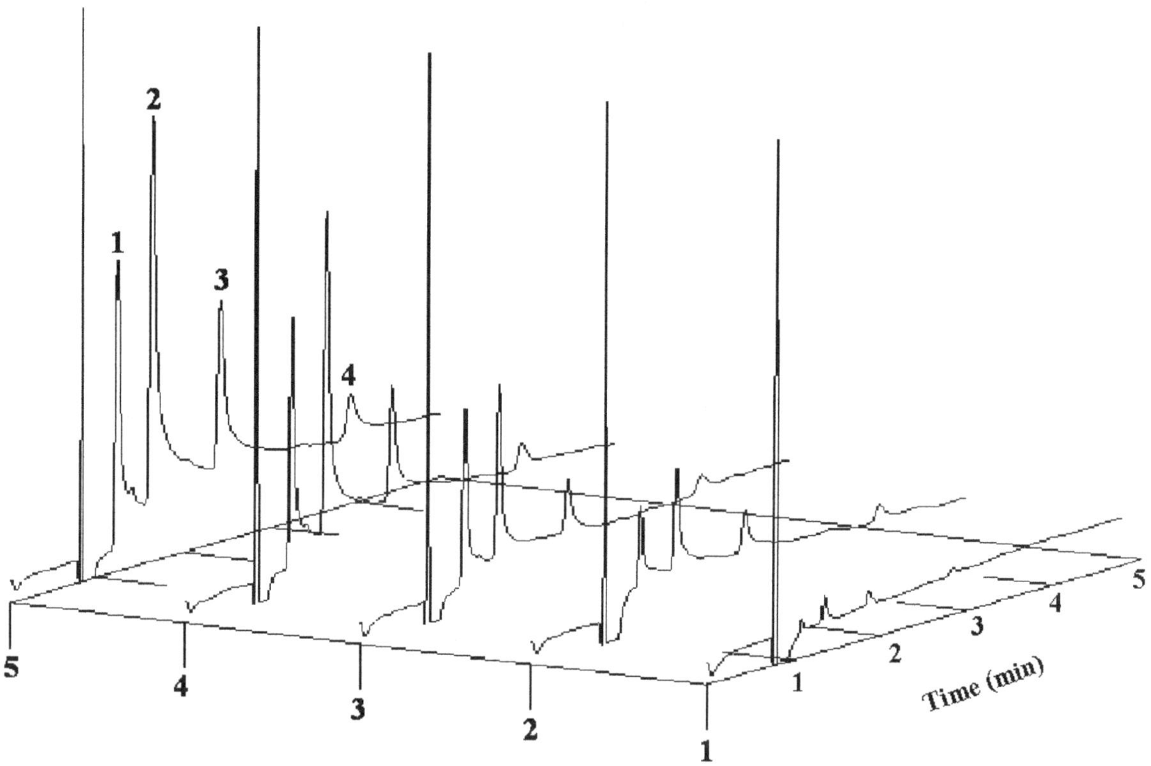

Figure 3.23 Typical chromatograms for THM4 using the "GC on a Valve" Prototype: (1) Memphis Water Sample #1, (2) Check Standard ~ 3.0 µg/L of each THM4, (3) Calibration Std. #2 ~ 6.0 µg/L each, (4) Calibration Std. #3 ~ 11.0 µg/L each, and (5) Calibration Std. #4 ~ 15.0 µg/L each (Source: Emmert et al. 2006)

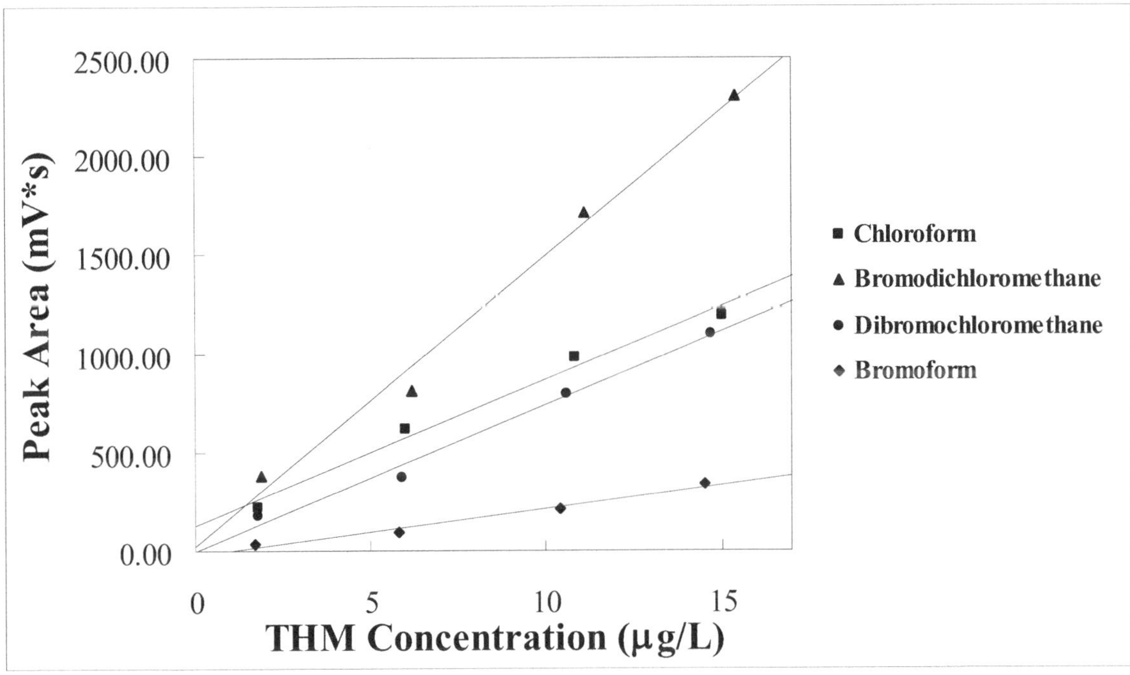

Figure 3.24 Typical calibration plots for the SCMS-GCOV with PDHID detector

On-Line Monitoring Studies Using SCMS-GCOV

A study was performed where five samples were analyzed using the GCOV and EPA 502.2. EPA Method 502.2 was taken to be the "true value" for the concentration of each THM4 species. Figure 3.23 shows five chromatograms from the SCMS-GCOV analyzer (four standards and a chromatogram for Memphis drinking water).

A summary of the study's results is presented in Figures 3.25 and 3.26 showing a plot of the raw data (concentration of THM4 species versus time) from using SCMS-GCOV and EPA Method 502.2. The average percent difference for $CHCl_3$, $CHBrCl_2$, $CHBr_2Cl$ and $CHBr_3$ was 24, 9, 17, and 12%, respectively. The average percent difference for TTHM4 was 6%. These values seem high, but it should be noted that the values of THM concentrations were in the range 1.0 to 4.0 µg/L. At these low levels, the percent difference will be expected to be high. Viewed from another angle, Figure 3.27 shows the plot of the difference between the results for EPA Method 502.2 and the SCMS-GCOV analyzer. This plot is commonly called a bias plot. The average difference between the two methods for $CHCl_3$, $CHBrCl_2$, $CHBr_2Cl$ and $CHBr_3$ was 0.14, 0.07, 11, and 0.20 µg/L, respectively. The average difference for the THM4 species was 0.24 µg/L.

The GCOV approach is feasible. However, it was not expected to be used as a monitoring tool. The technique might ultimately be useful; however, the PDHID requires a relatively high level of analyst skill, and some problems with high background current with this detector have been experienced. These are likely due to carry-over issues; therefore, this limits the use of the detector. When preliminary monitoring studies were done, the detector background current would eventually rise to unacceptable levels. This problem might be circumvented if temperature programming of the oven were possible. Unfortunately, the present state-of-the-art does not allow for this. Thus no extended monitoring studies were conducted beyond those presented.

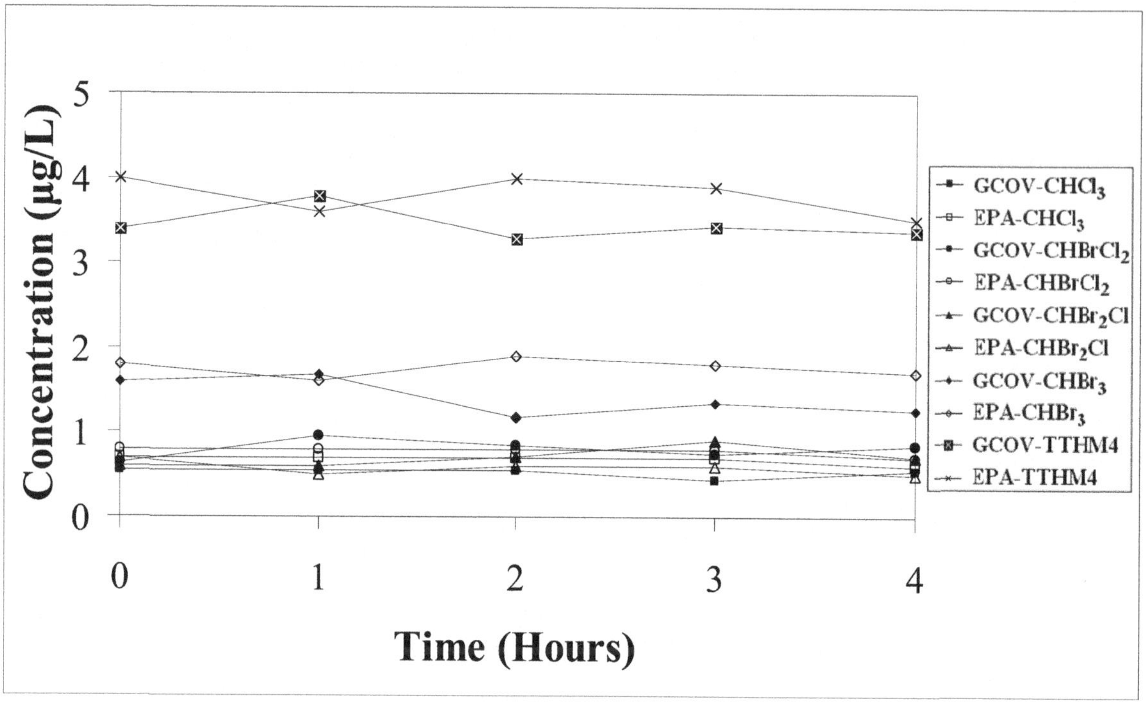

Figure 3.25 Memphis 1 concentration plots of THM4 species for SCMS-GCOV-PDHID and EPA 502.2. Full scale 0 to 5.0 µg/L (Source: Emmert et al. 2006)

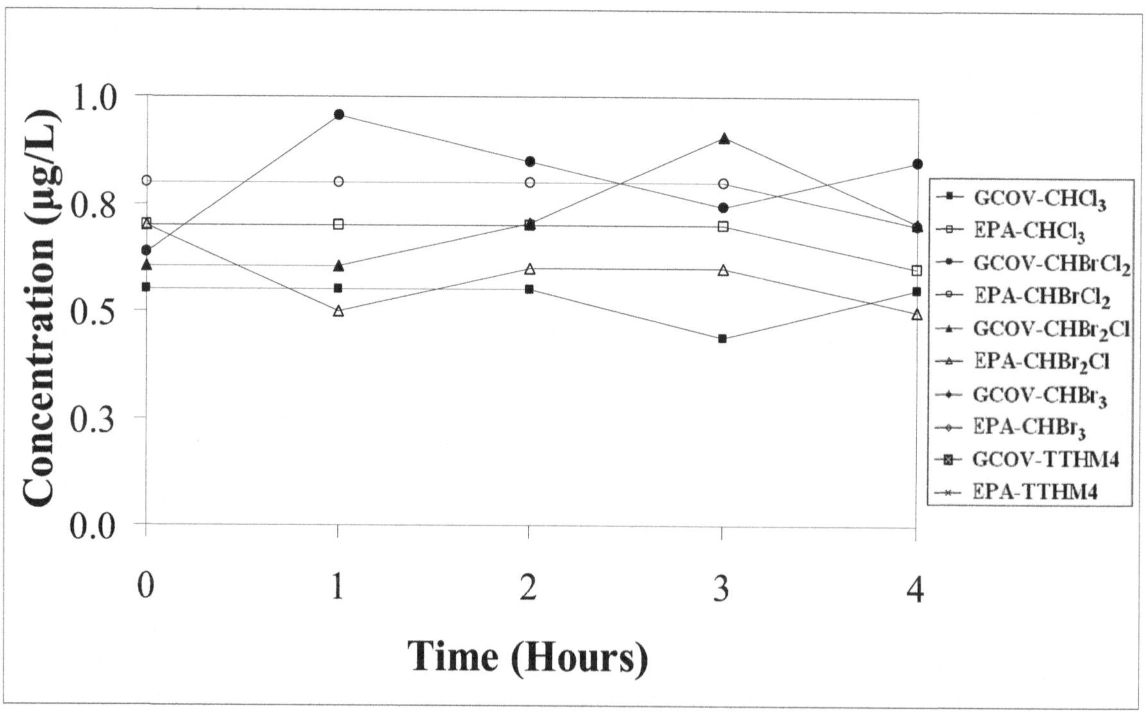

Figure 3.26 Memphis 1 concentration plots of THM4 species for SCMS-GCOV-PDHID and EPA 502.2. Enlarged scale 0 to 1.0 µg/L

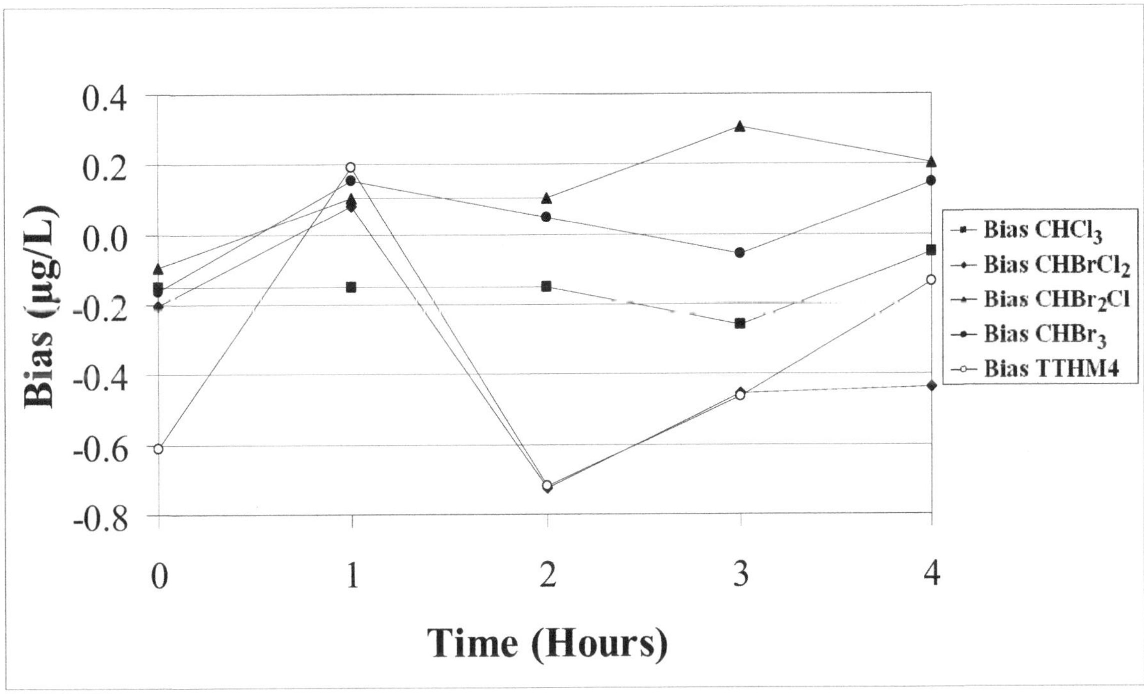

Figure 3.27 Memphis 1 bias plots of THM4 species comparing SCMS-GCOV-PDHID to EPA 502.2 (Source: Emmert et al. 2006)

Summary of SCMS-GCOV Studies

These results demonstrate the feasibility of the GCOV approach. The SCMS-GCOV is a simple system constructed with commercially available components and provides promising results for on-line THM4 monitoring. Certainly the mean % recovery and % RSD are not ideal, but they can possibly be improved. The MDLs are promising as well. Considering the relatively low cost of the GCOV approach (less than $10K), this approach warranted further study.

Recommendation 3.5: Temperature programming is necessary on the heated valve enclosure to decrease contamination of the system during repeat analysis, and improve resolution of analytes in the presence of more complicated sample matrices.

Recommendation 3.6: The electronics used to control detector and heated valve enclosure needs to be further miniaturized to fully exploit the advantages of GCOV, and make it very portable. Overall, it should be possible to further reduce the size of the GCOV system.

Recommendation 3.7: A revised SCMS-GCOV would benefit from the same calibration advances as described for the SCMS-GC-ECD and should also be applied to other common volatile organic compounds found in drinking water.

THE SCMS-GC-MS APPROACH TO THM4 MONITORING

Introduction

Mass spectrometry provides a powerful tool for identification, quantitation and confirmation of THM4 in drinking water. A goal of this research is to develop a confirmatory method for THM4 in drinking water using the supported capillary membrane sampler-gas chromatography with mass spectrometry (SCMS-GC-MS) system. The idea behind this method is that during monitoring studies, if the identity of a chemical species that interferes is unknown, the SCMS-GC-MS could be used to confirm the identity of the species. This research has focused on interfacing the SCMS-GC to an ion trap MS operating in either full scan (FS) mode or selected ion monitoring (SIM) mode. The internal standard method was used as the calibration method. MDL, accuracy and precision studies for the SCMS-GC-MS were carried out. For SCMS-GC-MS, internal standard calibration was accomplished using fluorobenzene (C_6H_5F).

Experimental

Figure 3.28 shows the typical flow diagrams for constructing the SCMS-GC-MS. The flow of the carrier gas (N_2) through the SCMS probe was the same as in the SCMS-GC-ECD system described earlier. After flowing through the SCMS probe, the probe carrier gas transported the THM4 through a transfer line connected to a Nafion-Drierite drying apparatus. The THM4 continue to a two position air actuated six-port valve, fitted with a 250 μL sample loop (Valco). The valve was switched to the injection position using a digital valve interface (DVI); 250 μL of the SCMS probe carrier gas was loaded onto the GC column. The injection time interval was set at 30 seconds and, the volume of carrier gas injected was approximately 0.5 mL containing 250 μL analyte.

GC-MS analyses were performed on a Varian GC 3380 with a Saturn 2000 ion trap MS that was operated in the electron impact mode with automatic gain control (AGC). The FS and SIM modes were used for this study. The data acquisition software was Varian Saturn® GC/MS Workstation Version 5.52. Table 3.7 displays the GC-MS operation parameters used in the study.

The identification of THM4 species was based on their retention times and mass spectra. Under the above optimal conditions in Table 3.7, the retention times of each THM4 species was 3.2 min ($CHCl_3$), 5.0 min ($CHBrCl_2$), 8.3 min ($CHBr_2Cl$), 10.1 min ($CHBr_3$); the retention time of C_6H_5F was 4.0 min. The total analysis time was 14.5 min. Figure 3.29 and 3.30 show the GC-MS chromatograms obtained at the same concentration level of THM4 in the SIM mode and the FS mode, respectively. In this study, MDL, accuracy and precision studies were conducted using the SCMS-GC-MS instrument. The MDL for each THM4 species was calculated using EPA recommended protocol (Glaser and Forest 1981; USEPA 1996a). The accuracy and precision were calculated based on EPA recommendations (USEPA 1996b). These results are shown in Table 3.8.

The internal standard method was also used as a calibration method for the SCMS-GC-MS. Calibration plots were constructed by plotting the ratio of peak area of each THM4 compound to that of internal standard from the chromatogram as a function of concentration ratio of each THM4 compound to the internal standard. Figures 3.31 and 3.32 show the calibration curves for the quantitative analysis of THM4 using C_6H_5F as an internal standard in the SIM mode and the FS mode.

Table 3.7
Typical settings for the SCMS-GC-MS

Parameter	Setting
GC Column	VF-5MS (30m × 0.25mm × 0.25μm)
Carrier Gas	Helium (ultra pure carrier gas grade) at 1.0 mL/min
Injector Temperature	175°C
Split Ratio	1:15
Initial Oven Temperature	30°C for 7 minutes
Temperature Ramp	30°C to 180°C at 20°C/min
Transfer Line Temperature	280°C
Trap Temperature	200°C
Acquisition Mode	Selective Ion Monitoring (SIM)
	83($CHCl_3$, $CHBrCl_2$)
	96 (C_6H_5F, internal standard)
	129 ($CHBr_2Cl$)
	173 ($CHBr_3$)
	Full Scan (FS)
EMV	1650 V (SIM), 1800 V (FS)
Emission Current	90 μA (SIM), 75 μA (FS)

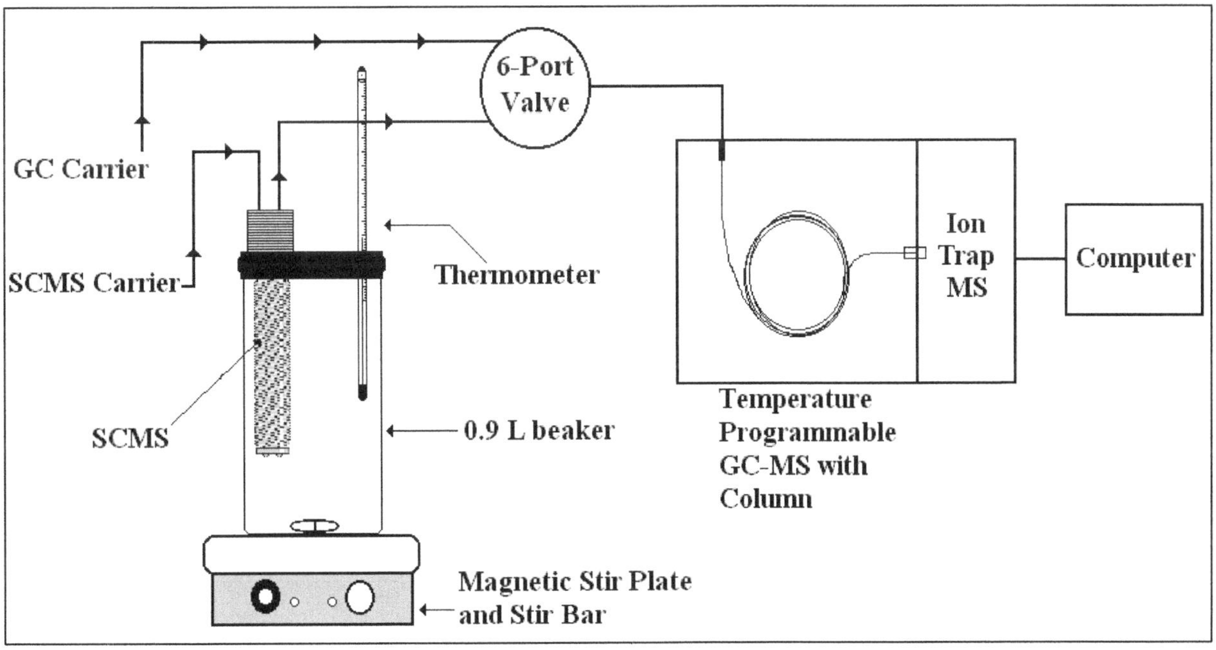

Figure 3.28 The flow diagram of the SCMS-GC-MS

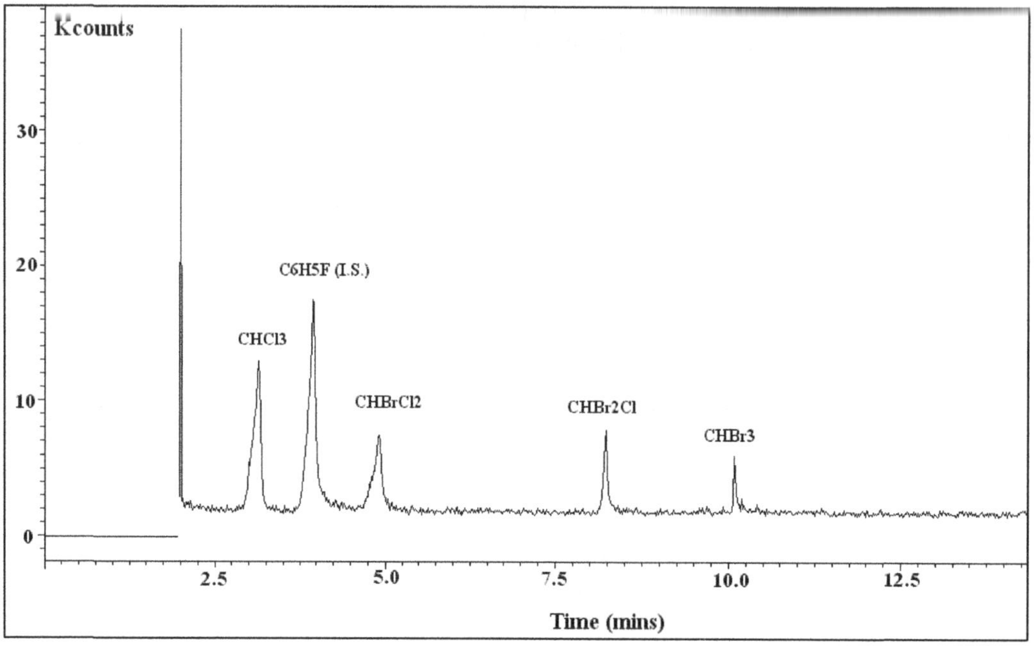

Figure 3.29 SCMS-GC-MS chromatogram (SIM) of THM4 species: The concentrations of THM4 were: (1) $CHCl_3$ 14.92 μg/L; (2) $CHBrCl_2$ 19.25 μg/L; (3) $CHBr_2Cl$ 24.51 μg/L; (4) $CHBr_3$ 28.94 μg/L

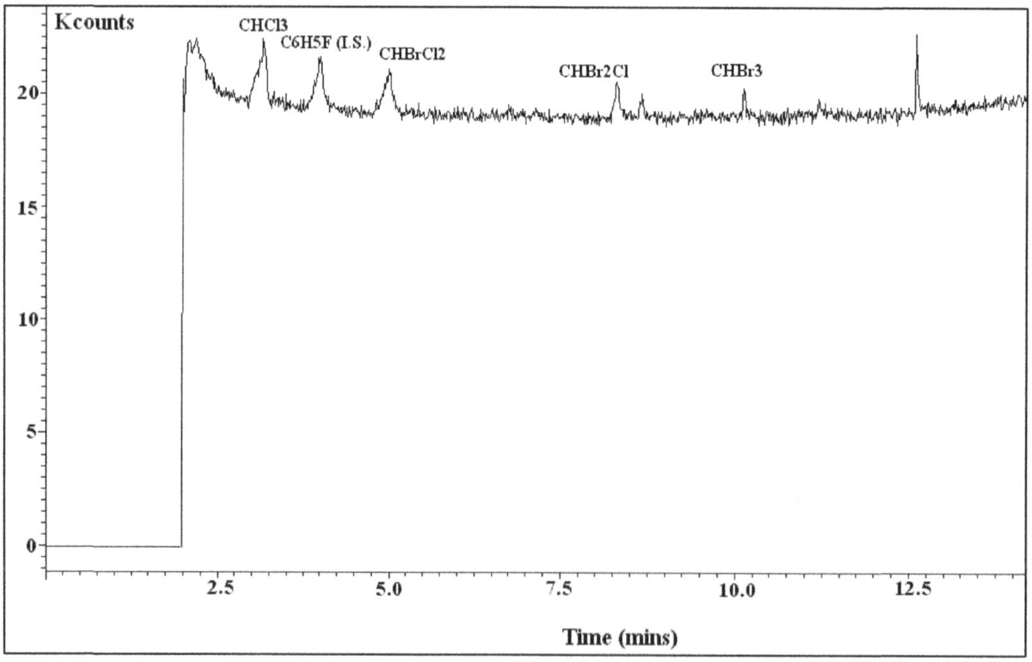

Figure 3.30 SCMS-GC-MS chromatogram (FS) of THM4 species: The Concentrations of THM4 were: (1) $CHCl_3$ 11.51 μg/L; (2) $CHBrCl_2$ 15.16 μg/L; (3) $CHBr_2Cl$ 19.30 μg/L; (4) $CHBr_3$ 22.79 μg/L

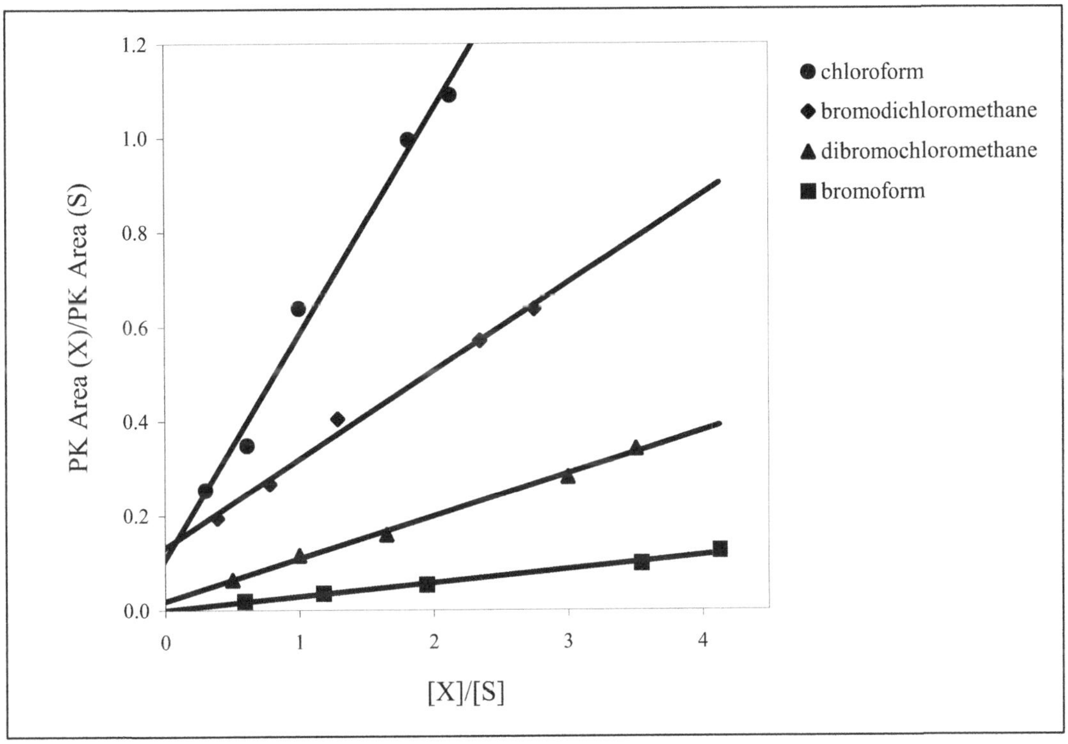

Figure 3.31 SCMS-GC-MS calibration curve for the quantitative analysis of THM4 using fluorobenzene as an internal standard (SIM)

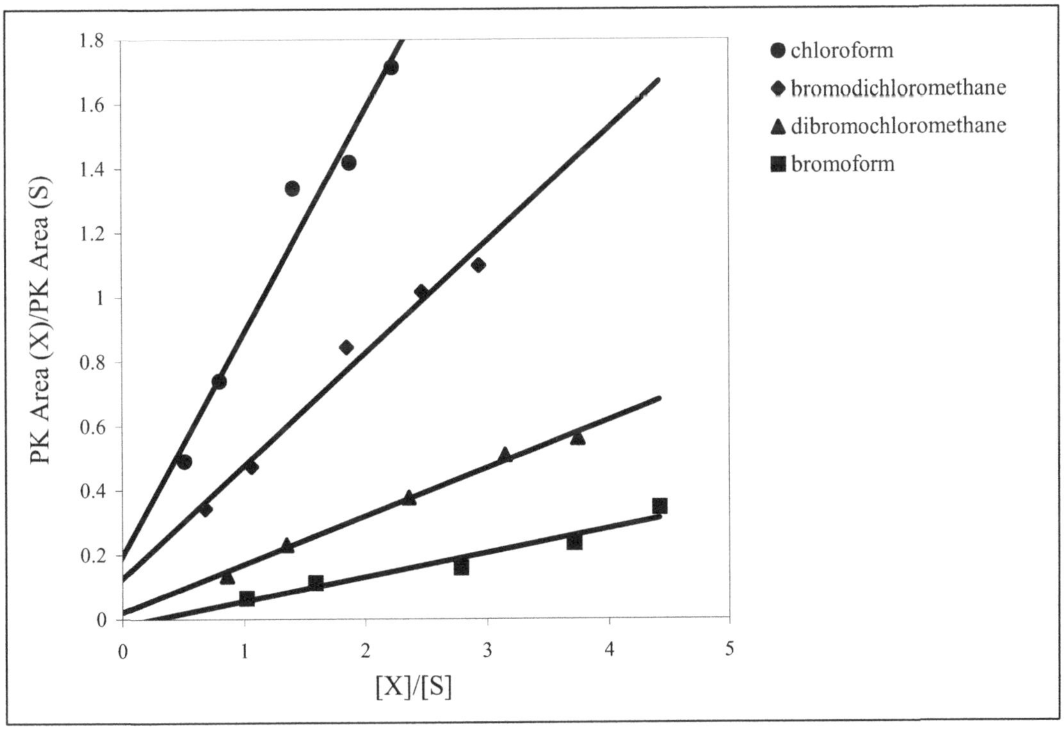

Figure 3.32 SCMS-GC-MS calibration curve for the quantitative analysis of THM4 using fluorobenzene as an internal standard (FS)

A "check standard" was analyzed four or five consecutive times depending on the individual THM4. The ratio of peak area between THM4 and the internal standard was calculated using the peak area of each compound and the internal standard obtained from the chromatogram. Using the calculated peak area ratios, the calibration curve equation, and the known concentration of the internal standard, the calculated concentration of each THM4 compound was obtained for each analysis. The average and the standard deviation of four or five calculated concentrations were calculated. The MDL was calculated by the standard deviation multiplied by the associated t-value at the 98% confidence level for four (t=4.541) or five analyses (t=3.747).

Results and Discussion

Table 3.8 shows the results of calibration MDL, accuracy and precision studies on the SCMS-GC-MS instrument in two modes. In the SIM mode, the MDL value for each THM4 species was on the order of single μg/L. The MDL for THM4 was calculated by summing the MDL values for each component, which yields a calculated value 6.74 μg/L. This value is far below the current MCL (80 μg/L). The mean % recoveries ranged from 98.2 to 109% for the three THM4 species with the %RSD of less than 7%, except $CHBrCl_2$ at about 14%. The % error on the slope of each calibration curves for $CHCl_3$, $CHBrCl_2$, $CHBr_2Cl$ and $CHBr_3$ was 6.5, 5.5, 3.6 and 5.1 %. The average % error on the slope of calibration curves for the four THM4 species averaged 5.2 %.

Table 3.8
Results from MDL, accuracy and precision studies for the SCMS-GC-MS operated in the SIM and FS modes

Mode /THM4	Calib. Range (μg/L)	MDL (μg/L)	Mean% Recovery	% RSD	Slope (response factor)	% Error on Slope	r^2
SIM mode							
$CHCl_3$	2.8-15.3	0.4	104	2.2	0.4716±0.0309	6.5%	0.988
$CHBrCl_2$	3.6-19.7	3.2	98.2	14.1	0.1872±0.0103	5.5%	0.991
$CHBr_2Cl$	4.4-24.3	2.1	102	6.8	0.0901±0.0032	3.6%	0.996
$CHBr_3$	7.0-38.2	1.1	109	2.7	0.029±0.0015	5.1%	0.992
FS mode							
$CHCl_3$	1.2-21.5	0.9	80.7	4.8	0.6917±0.0782	11.3%	0.963
$CHBrCl_2$	1.2-22.1	2.2	70.8	10.9	0.3489±0.0304	8.7%	0.978
$CHBr_2Cl$	1.2-22.5	2.2	125	4.0	0.1493±0.0076	5.1%	0.992
$CHBr_3$	1.7-32.0	4.1	103	7.6	0.0752±0.0109	14.5%	0.941

In the FS mode, the MDL value for each THM4 species was on the order of single μg/L. The MDL for THM4 was 9.42 μg/L, which was far below the MCL (80 μg/L). The mean % recoveries ranged from 70.8 to 125% for the four THM4 species with the %RSD of less than 11%. The % error on the slope of each calibration curve for $CHCl_3$, $CHBrCl_2$, $CHBr_2Cl$ and $CHBr_3$ was 11.3, 8.7, 5.1 and 14.5 %. The average % error on the slope of all four calibration curves for THM4 averaged 9.9%. Comparing the results from the SIM mode and the FS mode, it is obvious that the data improved in the SIM mode in terms of the accuracy estimated using mean % recovery. The MDL values for $CHCl_3$, $CHBr_2Cl$ and $CHBr_3$ in the SIM mode were lower than those in the FS mode. In addition, the SIM mode provided better precision for $CHCl_3$ and $CHBr_3$.

Summary of SCMS-GC-MS Methods

The SCMS-GC-MS approach provides all the advantages and disadvantages of the MS detector. The chief advantage is that the MS approach can be used to provide for confirmation of chemical identity. The SCMS-GC-MS approach would be especially useful as a confirmation method in the event there were more complex sample matrix issues. However when examined in terms of portability (compared to other methods in this study), it is not as portable. It is also true that the cost of the MS alone will make the SCMS-GC-MS too expensive to meet the goals of this study. There are advantages and disadvantages in both SIM mode and FS mode. The SIM mode provides maximum sensitivity to the monitored compounds of known spectral characteristics, but loses analytical information of the non-monitored unknown compounds in the sample. Therefore, the SIM mode is useful if the focus of the experiment is only to monitor and detect the analytes of interest in a sample. On the other hand, the FS mode gives the entire spectra of all the ions reaching the detector but offers poor sensitivity for each compound. As a result, the FS mode is generally used if the goal of the analysis is to discover and identify the unknown compounds. In this study, the SIM mode is more appropriate than the FS mode in that the research is aimed at detecting the specific compounds such as THM4.

Recommendation 3.8: It is likely that the SCMS-GC-MS results could be further optimized and the sensitivity could be improved with SIM using a different temperature programming, EMV setting, emission current, and trap conditions.

Recommendation 3.9: The SCMS-GC-ECD need to be applied to real-world monitoring studies using modified calibration methods as well as a modified FL/FC as described earlier for the other SCMS-GC methods.

Recommendation 3.10: The SCMS-GC-MS method might be particularly useful in complex drinking water samples with a very demanding matrix. Other compounds could be quantified including common volatile compounds found in drinking water.

Recommendation 3.11: The SCMS-GC-MS is recommended primarily as a tool to confirm the chemical identities of possible DBPs in demanding sample matrices when the need arises. This sort of study should be conducted prior to carrying out extended monitoring studies when possible.

CONCLUSIONS FOR THE SCMS-GC ANALYZERS

The SCMS sampling probe is a versatile sampling device for GC. It has been used successfully with several GC detectors. In this Chapter, the SCMS probe has been interfaced with the ECD, the PDPID and an ion trap MS for the analysis of THM4 species. The SCMS-GC-

ECD analyzer seems particularly attractive for on-line monitoring of THM4 concentrations, because the THM4 species readily permeate the probe's silicone rubber membrane. A sample injection valve can be used to introduce the THM4 species directly into a GC for separation, detection and quantification, and of the three SCMS-GC detectors in this study, the ECD seems the most promising with the goals of the study in mind. The later versions of the SCMS-GC-ECD are relatively portable (though it does require carrier and make-up gas cylinders). It is relatively inexpensive and constructed with commercially available components. Preliminary studies indicate that it should be able to sample directly from drinking water distribution systems in near-real time. The sampling rate for the SCMS-GC-ECD method meets the goal of at least one sample per hour. The MDL estimates fall in the approximate range 0.5-5 μg/L. The mean percent recoveries fall slightly outside the desired range of 100 ± 5% at 104 to 120%. The precision (% RSD) is within the desired value of less than 10%., with the actual range being 5.5-10%.

In certain situations, the additional analyzers presented here might also be useful. For example, the SCMS-GCOV approach offers a simplified and more portable analyzer that could be constructed for less expense. Though it does exhibit poorer MDL, accuracy and precision estimates and at present only allows for isothermal GC separation. The SCMS-GC-MS approach costs considerably more than either of the other approaches, but offers the ability to confirm chemical identity of the analytes. Of the three approaches, the SCMS-GC-ECD was the only one subjected to extended on-line monitoring studies, and for logistical reasons, this was done only at the Memphis 1 sampling site. Ultimately, when compared to the CMS-GC-ECD method of Chapter 5, the researchers chose to use the CMS-GC-ECD as the method of choice.

CHAPTER 4
CAPILLARY MEMBRANE SAMPLING-FLOW INJECTION ANALYSIS USING THE REACTION OF NCA WITH THM4 AND HAA5 SPECIES

INTRODUCTION

Early in the research project, a simple method was explored that was based upon the reaction between NCA and the THM4 and HAA5 species. These DBPs would react with NCA under basic conditions to produce a fluorescent product. Since NCA is similar to pyridine, it is thought that the mechanism is similar to the Fujiwara colorimetric reaction discussed in the Phase One Report (Emmert et. al 2004b). Reckhow and Pierce (1992) described the Fujiwara chemistry in detail. Nitrogenated polyenes were formed when dihalogenated or trihalogenated species reacted with pyridine. N-alkylated pyridines were formed when a polyhalogenated compound underwent nucleophilic substitution by pyridine. This reaction was first order in pyridine and alkyl halide. N-alkylated pyridines then underwent alkaline ring cleavage and a hydroxylated intermediate was formed by attack at the pyridinium-2 position. Hydrolysis of this compound yielded the salt of the amidine with an absorption maximum at 420 nm. If the pyridine concentration was high, an alkylpyridinium intermediate was formed. Hydrolysis of this compound yielded another sodium salt of amidine with an absorption maximum at 530 nm. Each of the absorbing products would slowly decompose to form glutaconaldehyde with an absorption maximum at 370 nm (Figure 4.1).

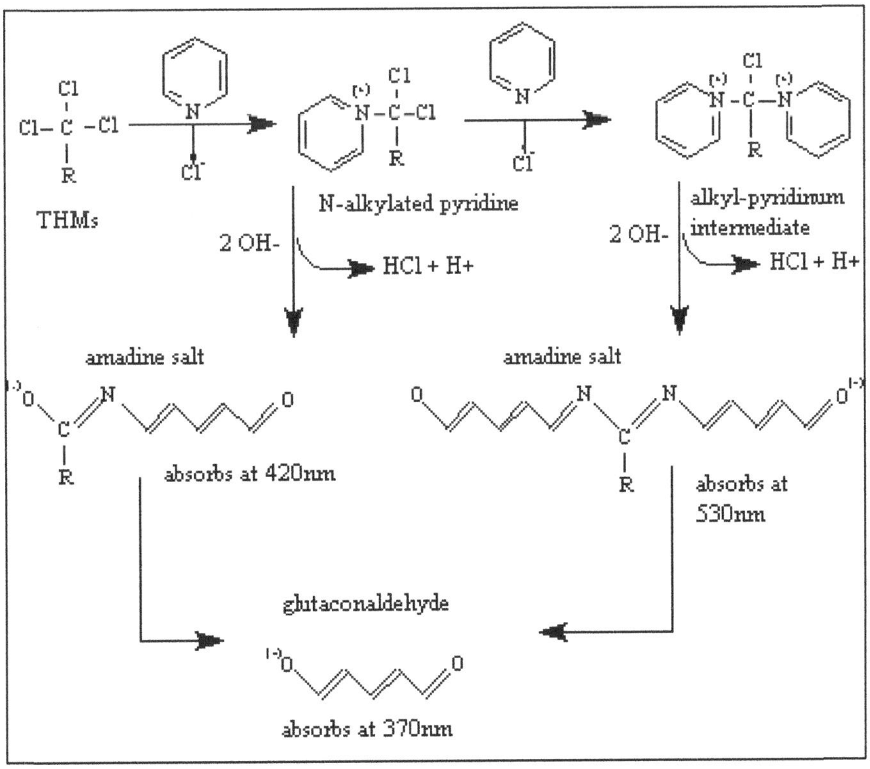

Figure 4.1 The classical pathway for the Fujiwara reaction (Adapted from Reckhow and Pierce (1992)

The main problem with using pyridine chemistries for analytical methodologies is that they are not sensitive. The molar absorptivities of the products are simply not large enough for analysis at sub-µg/L concentration levels. In our studies, it should be noted that NCA is a pyridine derivative and only difference is that it has a formamide moiety meta to the N in the aromatic ring. Preliminary results showed that the NCA-DBP reaction produces fluorescence which allows for sub-µg/L analyses of DBPs. We believe the formamide moiety affects the chemistry (perhaps by preventing the ring from opening) and makes the products of the NCA-DBP reaction just rigid enough to enhance fluorescence. This may be why fluorescence occurs with NCA and not with pyridine—as fluorescence is typically enhanced by a more rigid structure. If the ring does not open, then the molecule undergoes fluorescence rather than dissipating the energy through rotations (as might be the case with pyridine). More detailed studies of this chemistry are currently underway, yet it can still be used from an Edisonian view to develop a method for detecting THM4 and HAA5.

In this Chapter, a capillary membrane sampling (CMS) flow injection analysis (FIA) method is used to distinguish between TTHM4 and THAA5 concentrations in drinking water (Geme et al. 2005). Building on the work of Aoki and Kawakami (1989), both THM4 and HAA5 were analyzed by the CMS-FIA method. The THM4 and HAA5 species react with NCA in basic solution to form a fluorescent product (NCA-FL). A practical analyzer (CMS-FIA NCA-FL) that combines this NCA-FL chemistry with CMS-FIA has been developed for on-line monitoring of TTHM4 and THAA5 concentrations in drinking water capillary membrane sampling flow injection analysis with nicotinamide-fluorescence (Figure 4.2). The results of MDL, accuracy and precision studies are reported. Interference, selectivity, carry-over and linearity studies are reported as well as the effect of temperature and ionic strength changes. Drinking water samples were analyzed by each proposed method and the results were compared to USEPA Method 502.2 (1995b) and USEPA Method 552.3 (2003a).

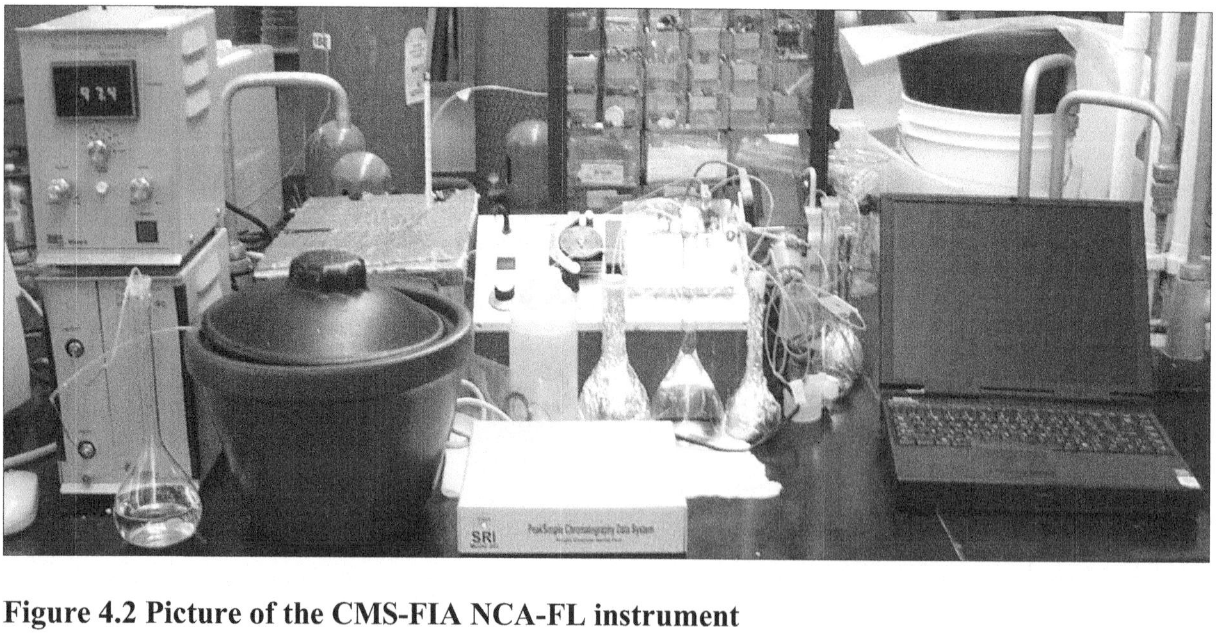

Figure 4.2 Picture of the CMS-FIA NCA-FL instrument

EXPERIMENTAL

Unless otherwise stated, all water used was distilled and then deionized using a Millipore Milli-Q/ Milli RO® 10 Plus Water Purification System. The resulting water was of the highest quality and did not have any measurable organic or inorganic impurities. All chemicals were reagent grade or better. The THM4, HAA5, NCA, carbon tetrachloride (CCl_4), dibromomethane (CH_2Br_2), 1,3-dichloroacetone ($C_3H_4Cl_2O$), trichloroacetonitrile (Cl_3CCN) and tribromoacetaldehyde (Br_3CCHO) standards were purchased from Aldrich Chemical Company. Sodium hydroxide (NaOH) was purchased from Fisher Scientific Inc. Standard stock solutions of the individual THM4 at a concentration of 1000 mg/L were prepared by dissolving each THM in methanol. Appropriate amounts of the stock standard solution were added to a 100 mL volumetric flask and diluted to volume using reagent water to give various concentrations of the working standard. Standard stock solutions of the individual HAA5 species were prepared by dissolving each HAA in reagent water, and the working standards were prepared in the same way as THM4. A 30% NCA solution was prepared using reagent water immediately prior to use. 0.2M NaOH was prepared by dissolving NaOH pellets in reagent water.

The Flow Injection Analyzer

The FIA analyzer used in this study was a modified FIALab 2000 flow injection analyzer. The stock, manually operated six-port sample injection valve was replaced with a microelectrically actuated Valco ten-port sample injection valve. The ten-port valve was fully automated using PeakSimple software and a single channel serial port data acquisition system (Model 203, SRI Instruments) that also collected the data from a Fluorichrom fluorescence detector (Model 430020-02, Varian).

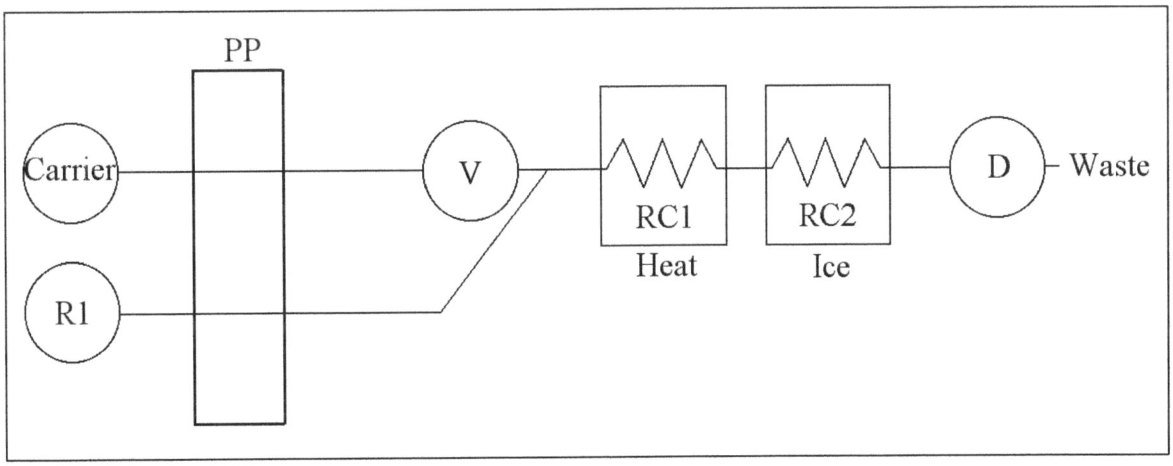

Figure 4.3 Schematic Diagram of CMS-FIA NCA-FL instrument for TTHM4 and THAA5. Carrier: NaOH stream; R1: NCA reagent stream; PP: peristaltic pump; RC1: reaction coil in water bath (97°C); RC2: reaction coil; D: fluorescence detector

The four-channel flow injection manifold (Figure 4.3) had the sample continuously flowing through a sample injection loop (60 μL) at 0.25 mL/min on the ten-port injection valve. Sample was then injected (30 seconds injection time) into a carrier stream (0.2M NaOH: flow rate = 0.05 mL/min). This carrier stream was merged with a reagent stream of 30% NCA: flow rate = 0.08 mL/min. The reagents were mixed using a heated reaction coil (RC1) held at 97°C. The reaction coil was a knitted open tubular reaction coil (o.d. 1.6 mm, i.d. 0.50 mm, length 2500 mm, SeQuant). The THM4 or HAA5 species reacted with the NCA forming a fluorescent product which was detected. In theory, the flow injection gradient is monitored as a function of time, and the height of the gradient peak is proportional to the analyte concentration. In reality, both THM4 and HAA5 reacted; therefore, the FIA method alone could not distinguish between these chemical species; therefore, it was fundamentally a non-selective method.

Capillary Membrane Sampling Device

The schematic diagram for the CMS device is shown in Figure 4.4. A 94 cm length of silicone rubber membrane tubing was placed inside Tefzel® tubing and then was coiled. The two ends of the membrane were directed through two separate tee unions (Swagelok®) and adapted to a 1/16″ to 1/8″ int/ext union (Valco) using two SS tubing sections (o.d. 0.5 mm, length 13 mm, Global FIA). The "sample in" and "sample out" connections consisted of 1/16″ FEP tubing adapted to fittings (Swagelok®) using two 1/16″ to 1/8″ int/ext unions (Valco).

Additional pieces of Tefzel® tubing were used to adapt these connections to the tees. 1/16″ fittings (Valco) were used to secure the FEP tubing to the unions (Valco). The CMS was used to sample TTHM4 or THAA5 species, depending on how the CMS was connected to the FIA analyzer.

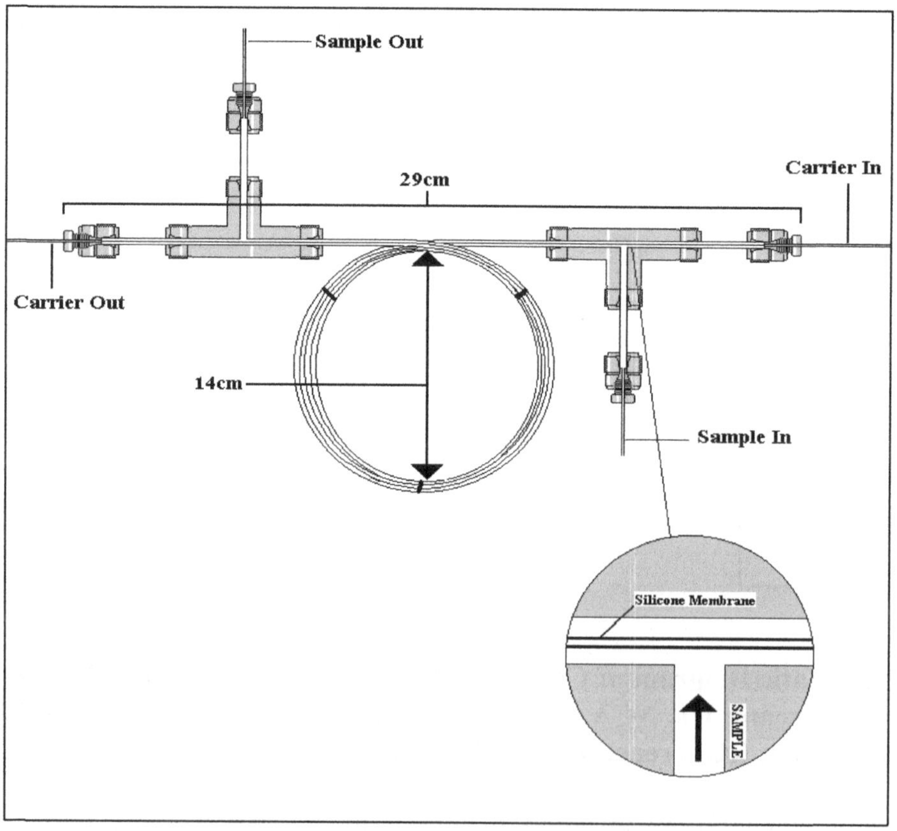

Figure 4.4 Diagram of the prototype CMS device (Source: Geme et al. 2005)

In the method for TTHM4, the CMS was connected to the FIA analyzer as shown (Figure 4.5, top left). In the LOAD position (Figure 4.5, bottom left), a carrier stream consisting of only reagent water passed through the interior of the entire length of the silicone membrane tubing and was sent out to Port 1 on the ten-port microelectrically actuated valve. The carrier stream transported THM4 that permeated the membrane into a sample loop connected to Ports 10 and 3, then out of Port 2 to a waste container. Simultaneously, a drinking water sample flowed through the first tee union and around the entire membrane present in the length of Tefzel® tubing exiting to waste. In the INJECT position (Figure 4.5, bottom right), the volume in the sample loop flowed from Port 8 to RC1 where it was mixed with the 30% NCA solution at 97°C. After exiting RC1, the reaction mixture flowed through RC2 where it was cooled down to minimize air bubble formation and maximize fluorescent intensity (Aoki and Kawakami 1989) and then entered the flow-cell of the fluorescent detector. Concurrently, reagent water (through Ports 1 and 2) and sample were pumped continuously to waste.

An alternative configuration can be used that enhances selectivity for THAA5 over TTHM4 by modifying the flow path from the CMS sampler. With the configuration for TTHM4, the stream inside the capillary membrane flowed to the sample injection valve. This stream was enriched in THM4 species as they demonstrated a higher permeability rate through the capillary membrane. Thus selectivity was enhanced for THM4 species. With the alternate configuration (Figure 4.5., top right), the stream on the outside of the capillary membrane flowed to the injection valve. This stream was "effectively enriched" in HAA5 species as the THM4 species were transported to the inner stream as in the THM4 method. It should be noted that there was really no enrichment occurring, but more appropriately, interfering THM4 species were being removed. The result was selectivity enhancement in this stream for HAA5 species over interfering THM4 species.

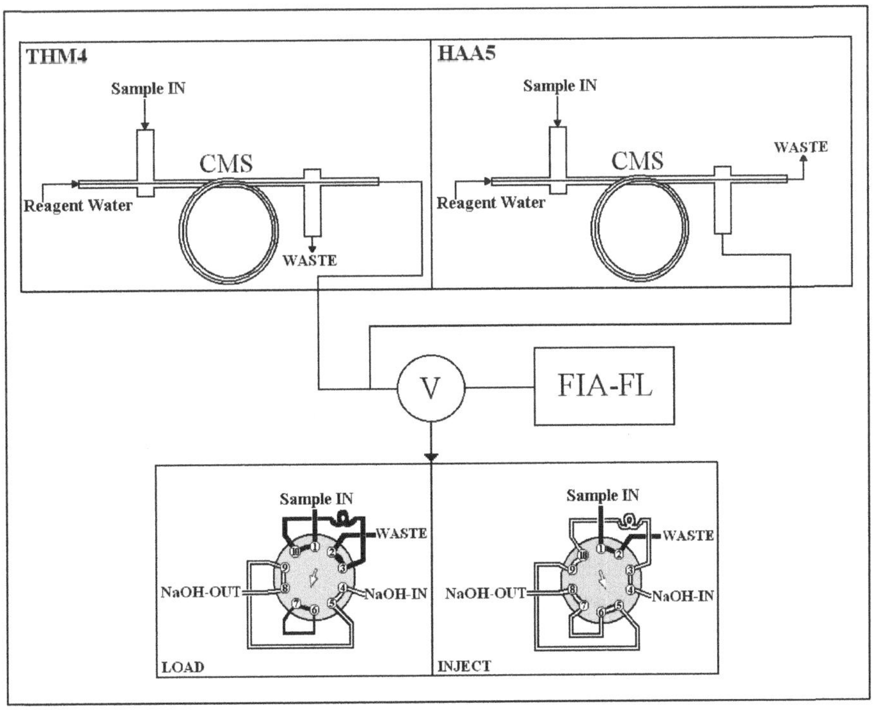

Figure 4.5 Schematic diagram for the CMS–FIA NCA-FL configuration for monitoring TTHM4 and THAA5 in water distribution systems (Source: Geme et al. 2005)

RESULTS AND DISCUSSION

MDL, Accuracy and Precision Studies

Standards were prepared for the individual THM4 and HAA5 species as well as for TTHM4 and THAA5 and were analyzed using the appropriate THM4 or HAA5 CMS device configuration. Figure 4.8 shows FIA gradients for TTHM4 and THAA5 solutions at the concentration of 50 μg/L each. Calibration plots were prepared by plotting the peak height of the FIA gradient for each THM4 and HAA5 species or for TTHM4 or THAA5 as a function of concentration (Figure 4.6 and 4.7). MDLs for individual and TTHM4 and THAA5 components were calculated using the recommended USEPA protocol (Glaser and Forest 1981; USEPA 1996a). The precision and accuracy were calculated based on USEPA recommendations (USEPA 1996b). The results of these studies are shown in Table 4.1.

The MDL, accuracy and precision results showed promise for both THM4 and HAA5 on-line monitoring studies. For THM4, the MDL values ranged from 1.7 to 9.4 μg/L with an average MDL value of 3.8 μg/L. Mean % recoveries ranged from 110 to 142% with an average of 121%. The % RSD values ranged from 2.9 to 14% with an average value of 6.4%. For HAA5, the MDL values ranged from 2.0 to 5.9 μg/L with an average MDL value of 2.9 μg/L. Mean % recoveries ranged from 104 to 122% with an average of 113%. The % RSD values range from 3.7 to 11% with an average value of 6.0%.

Table 4.1
Results from MDL, accuracy and precision studies for the CMS-FIA NCA-FL method for THM4 and HAA5 species measured separately

Analyte Species	MDL (μg/L)†	Mean % Recovery	% RSD	$r^{2‡*}$
$CHCl_3$	2.1	110	4.1	0.981
$CHBrCl_2$	4.8	115	8.8	0.989
$CHBr_2Cl$	9.4	142	14	0.976
$CHBr_3$	1.7	124	2.9	0.986
TTHM4	1.1	115	2.1	0.985
MCAA	5.9	112	11	0.996
DCAA	2.8	122	4.9	0.987
TCAA	2.2	118	4.0	0.978
MBAA	3.1	104	6.3	0.993
DBAA	2.0	111	3.7	0.981
THAA5	1.6	111	3.2	0.974

* Six standards were analyzed for THM4. Five standards were analyzed for HAA5. Each calibration curve was based on five calibration standards.

† The check standard concentration was 25 μg/L.

‡ The concentration ranged from 10 to 100 μg/L.

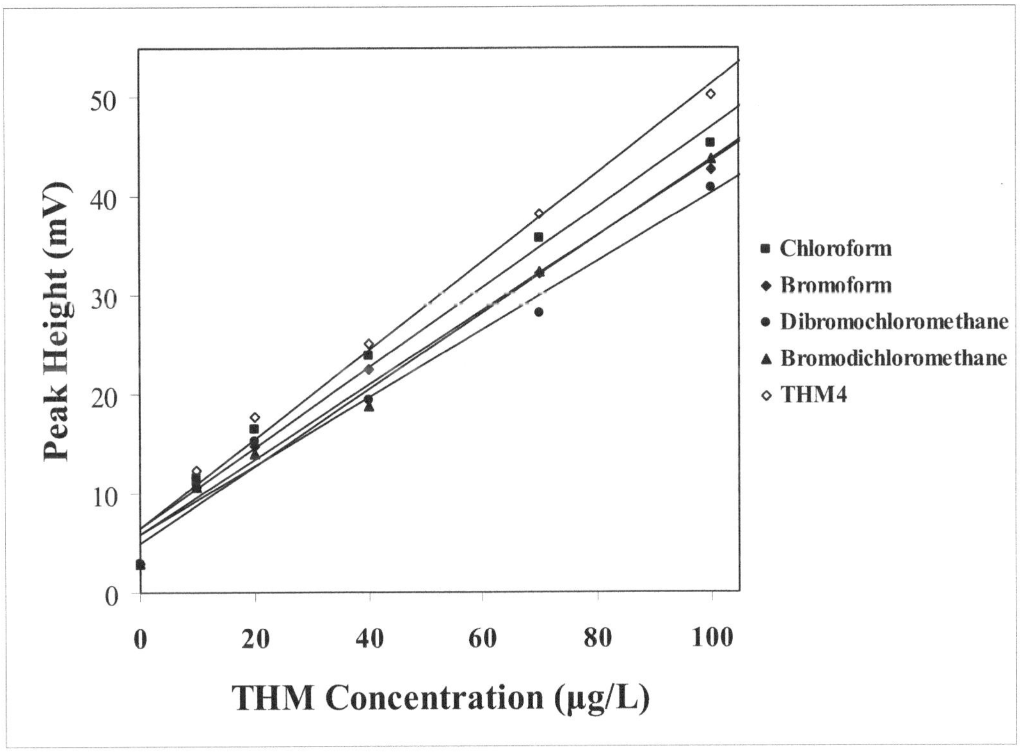

Figure 4.6 CMS-FIA NCA-FL calibration plots of individual and TTHM4 species

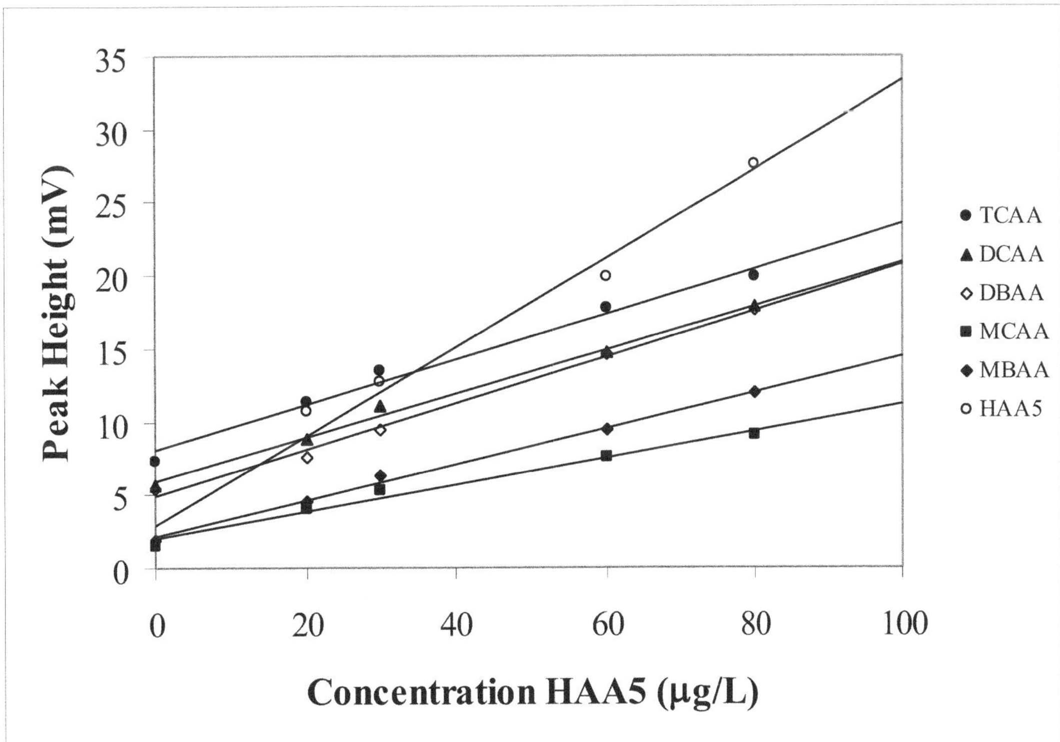

Figure 4.7 CMS-FIA NCA-FL calibration plots of individual and THAA5 species

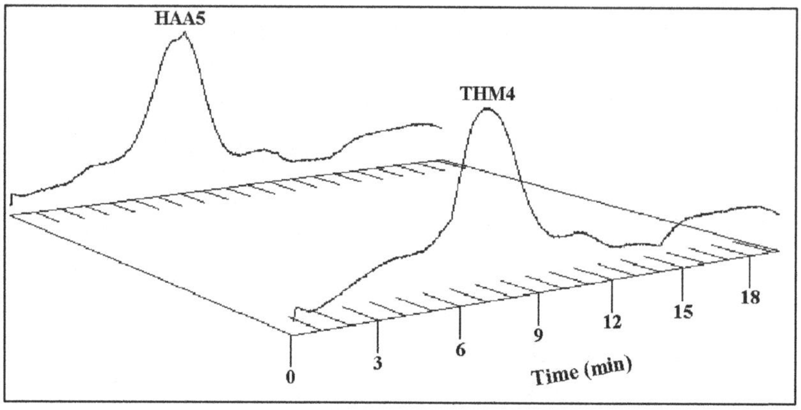

Figure 4.8 Typical CMS-FIA NCA-FL gradients of TTHM4 and THAA5 species

Linearity Studies

The methods that were used to calculate the MDL, accuracy and precision estimates were assumed upon a linear model and provided acceptable results for both TTHM4 and THAA5. Additional experiments were designed to more closely evaluate the linear range of the CMS-FIA NCA-FL methods over the concentration range of 10 to 200 μg/L (Figure 4.9). With each, the linear range was evaluated by plotting the analytical signal (peak height) as a function of analyte concentration. The coefficients of determination for the calibration plots were generally >0.98. A more rigorous evaluation of linearity using an F-test at the 95% confidence level (Walpole and Mayers 1989) showed that the calibration plots were generally non-linear over this concentration range. At concentrations above 100 μg/L, the deviation from linearity became more pronounced as the calibration curves reached a plateau. This could be because the loading capacity of the CMS membrane has been exceeded. Once the concentration of THM4 species exceeds the membrane capacity, the amount of THM4 species crossing the membrane is limited and analytical signal will plateau. The HAA5 nonlinearity is attributed to the FL detector.

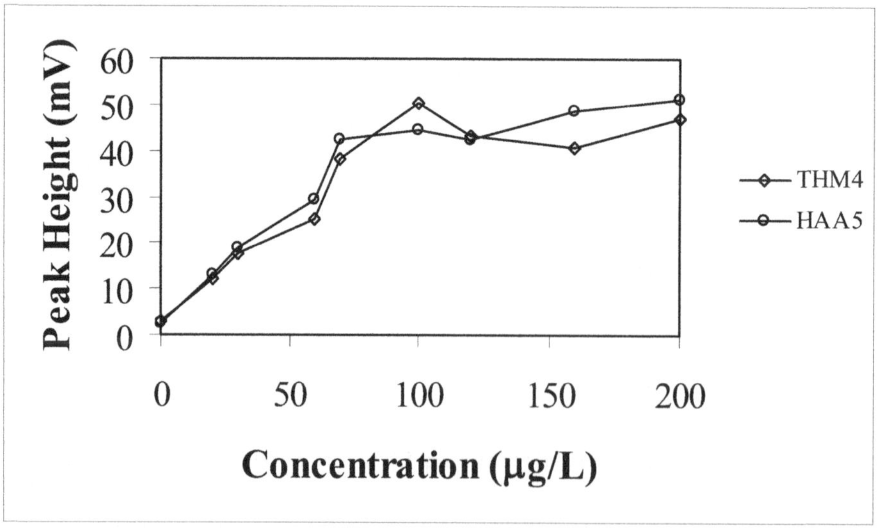

Figure 4.9 Results of the linearity study of TTHM4 and THAA5 species using the CMS-FIA NCA-FL method

Interference and Selectivity Studies

One of the main concerns over using the CMS-FIA method is a potential problem associated with interferences. The chemistry is relatively non-selective. This is because the same chemistry can be applied to TTHM4 and THAA5. The idea behind CMS was that it would be possible to achieve some selectivity because the THM4 species would more readily permeate a silicone rubber membrane. These results showed a slight selectivity enhancement for THM4 over all the species studied (Table 4.2). For example, when a 100 μg/L solution of HAA5 was analyzed it appeared as about 30 μg/L THM4. This means that the magnitude of HAA5 interference was on the order of 30% or in other words, the method exhibited a selectivity factor of 3.5 to 1 for THM4 over HAA5 in the THM4 configuration. Similar studies that were carried out with (CCl_4; 100 μg/L) gave less than 1% interference and the calculated selectivity factor averaged 830. For (CH_2Br_2; 100 μg/L), the magnitude of the interference averaged 32% with a calculated selectivity factor of 3.0. Additional studies were carried out to evaluate the magnitude of the interference of haloacetones, haloacetonitriles and haloacetoaldehydes on measurements of THM4 species. Three representative species (one for each class) were used and included $C_3H_4Cl_2O$, Cl_3CCN and Br_3CCHO. No significant interferences (<0.5%) were observed with these species during THM4 measurements.

Table 4.2
Results of preliminary interference studies for the CMS-FIA NCA-FL method for TTHM4

Interferent Species	Conc of Interferent (μg/L)	Peak Height (mV)	Conc. Expressed as $CHCl_3$	Magnitude of Interference	Selectivity Factor for THM4
HAA5	100	15.569	30	30	3
	100	15.200	29	29	4
AVG	100	15.385	30	30	4
CCl_4	100	4.772	0.03	< 1	3333
	100	4.834	0.20	< 1	500
AVG	100	4.803	0.12	< 1	833
CH_2Br_2	100	16.380	32	32	3
	100	16.179	32	32	3
AVG	100	16.280	32	32	3
$C_3H_4Cl_2O$	100	5.529	0.5	< 1	200
	100	5.726	0.6	< 1	167
AVG	100	5.628	0.55	< 1	184
Cl_3CCN	100	5.148	0.3	< 1	333
	100	4.926	0.2	< 1	500
AVG	100	5.037	0.25	< 1	417
Br_3CCHO	100	5.642	0.5	< 1	200
	100	5.804	0.6	< 1	167
AVG	100	5.723	0.55	< 1	184

Likewise, studies were carried out to evaluate the selectivity of the alternative configuration for THAA5 (Table 4.3). For TTHM4, a 100 μg/L solution appeared on average as 5.0 μg/L HAA5. This means that the magnitude of the interference was on the order of 5% resulting in a selectivity factor of 20 to 1 for THAA5 over TTHM4. CCl_4 interference averaged 8% leading to a selectivity factor of about 12 while CH_2Br_2 exhibited an interference of ~2% with selectivity factor of ~67. As with the THM4 study above, additional studies were done to evaluate the interference of haloacetones, haloacetonitriles and haloacetaldehydes. The same three species used in the THM4 studies were used. For $C_3H_4Cl_2O$, Cl_3CCN and Br_3CCHO, the interference was significant at 44.0, 42.2 and 53.0%, respectively. This means that if these species are present in comparable concentrations to the HAA5 species, they could present a significant interference. Yet, if the concentrations of these possible interferents are low (<1 μg/L) versus the THAA5 (~20 to 60 μg/L), then these interferences are manageable. We recommend that preliminary screening be done using USEPA methods to have some knowledge of the matrix to be monitored *a priori* if such species are suspected to be in comparable or high concentrations.

Table 4.3
Results of preliminary interference studies for the CMS-FIA NCA-FL method for THAA5

Interferent Species	Conc of Interferent (μg/L)	Peak Height (mV)	Conc. Expressed as HAA5	Magnitude of Interference	Selectivity Factor for HAA5
$CHCl_3$	100	5.886	5.2	5.2	19
	100	5.710	4.7	4.7	21
AVG	100	5.798	5.0	5.0	20
CCl_4	100	6.920	7.9	7.9	13
	100	7.142	8.5	8.5	12
AVG	100	7.031	8.2	8.2	13
CH_2Br_2	100	4.284	1.0	<1	104
	100	4.689	2.0	2.0	50
AVG	100	4.487	1.5	1.5	67
$C_3H_4Cl_2O$	100	24.954	38	38	3
	100	20.310	50	50	1
AVG	100	22.632	44	44	2
Cl_3CCN	100	22.005	43	43	2
	100	21.758	41	41	2
AVG	100	21.882	42	42	2
Br_3CCHO	100	26.910	55	55	2
	100	26.190	53	53	2
AVG	100	26.550	54	54	2

An additional concern when analyzing drinking water samples is the effect of total TOC concentrations. Preliminary studies were carried out in the laboratory to evaluate the effect of TOC interference over the TOC concentration range of ~0 to 25 mg/L. Standard solutions of THM4 and HAA5 were prepared in solutions of differing TOC concentrations. Anhydrous potassium biphthalate was used as the standard TOC compound as it is commonly used as a standard for calibration of common TOC measurements (APHA, AWWA, WEF 1992).

Increasing the TOC concentration in the analyte matrix resulted in an increase in the analytical signal for both THM4 and HAA5. The results of the TTHM4 analysis showed an average increase in analytical signal of 0.5% for each 1.0 mg/L increase in TOC concentration. Similar results were observed for HAA5 except the average increase in analytical signal was 0.4% per 1 mg/L increase in TOC concentration. Decreasing the concentration of TOC resulted in lowering the analytical signal by the same magnitude. In practice, the TOC concentration would not be expected to vary greatly from analysis to analysis. If matrix effects were demonstrated to be problematic in analyzing real samples, then standards could be prepared with TOC concentrations matching the system to be monitored.

Effects of Temperature and Ionic Strength on Analytical Signal

Temperature effects were investigated over the range of 2 to 30°C. The average increase in analytical signal over the range of 2 to 30°C for TTHM4 and THAA5 was 4% increase/°C and 1% increase/°C, respectively. These results were in good agreement with Blanchard and Hardy (1984) and Emmert et al. (2004a) who studied hollow capillary samplers. Temperature should be monitored during calibration and sampling; but under normal circumstances, it is not expected to change significantly from calibration to calibration. If changes did occur, a simple correction factor can be applied to take this into account.

The effect of changing the ionic strength of the sample on the analytical signal was also investigated. Standard mixtures of TTHM4 and THAA5 were prepared in solutions of differing ionic strength from ~0 – 3.0 M (this equates to ~0 to 1.0 M Na_2SO_4). Increasing the ionic strength of the TTHM4 standards increased the analytical signal from the "salting out" effect, while increasing the ionic strength of THAA5 standards exhibited a markedly smaller effect. Over the range of these experiments, the average increase caused from ionic strength for TTHM4 and THAA5 was 27 and 2% signal increase per 1M increase in ionic strength, respectively. Initially, this seems like a large increase for TTHM4. Consider a 10 μg/L standard solution of TTHM4, the predicted peak height from a typical calibration curve would be 9.040 mV. The ionic strength calculated for Memphis drinking water is ~0.002M. Using the correction factor for THM4 given above, the expected increase in peak height would be about 0.05%. This means that instead of giving a peak height of 9.040 mV, the standard in Memphis drinking water would give a peak height of 9.045 mV. A 10 μg/L standard in Memphis drinking water would give a calculated concentration of 10.01 μg/L less than a 0.1 % difference. Similar calculations can be done for HAA5 standards. A 10 μg/L standard in Memphis drinking water would give less than a 0.05% difference. At typical conditions, the effect of changes in ionic strength is expected to be relatively low from water to water. In practice, analysts can evaluate the magnitude by spiking known concentrations into the raw drinking water matrix and observing the relative analytical signals in laboratory water and drinking water samples.

Spiking and Standard Addition Studies

The method of standard addition was used to evaluate whether the CMS-FIA method suffered from matrix issues when applied to real drinking water samples. Studies were done in two different water samples: an Arkansas (AR) sample and a California (CA) sample. The AR sample was representative of a relatively simple matrix (ground water source). The CA sample was an example of a more complicated sample matrix with relatively high bromide ion concentrations (surface water source).

Standard Addition Studies for TTHM4 Concentrations

Each sample was analyzed using two different approaches to calibration, a calibration curve and standard addition. The standard addition curve was based on analysis of four different spike concentrations plus the original sample. For the AR sample, the results for TTHM4 with the calibration curve and standard addition methods were 62 and 61 μg/L, respectively. The difference between the two results (the absolute error, if one assumes that the standard addition method is the "true value") was 1 μg/L and the percent relative error was 2%. In the same order, the results for the CA sample were 69 and 80 μg/L. The absolute error for the CA results was 11 μg/L and the percent relative error was -16%.

It is possible that a slight, but correctable, matrix effect may present itself as one moves from relatively clean sample matrices (ground water) to more complex matrices (surface water). One would expect that as the matrix became more complex, the THM4 species would be "salted out" and the calibration curve might overestimate the predicted THM4 concentrations. For the AR sample, the results did show slightly higher calculated concentrations from the calibration curve (TTHM4 concentrations ~60 μg/L). For the CA sample, the results appeared to show the opposite effect. Since the method was at the upper limit of its linear range, the concentrations were under-predicted. In the standard addition plots, this had the effect of lowering the regression line at the upper end of the THM4 concentration range leading to a more negative x-intercept and thus a higher predicted concentration using standard addition.

Standard Addition Studies for THAA5 Concentrations

For the AR sample, the results for THAA5 with the calibration curve and standard addition methods were 58 and 47 μg/L, respectively. The absolute error was 11 μg/L and the percent relative error was 19%. In the same order, the results for the CA sample were 36 and 40 μg/L. The absolute error for the CA results was 4 μg/L and the percent relative error was -11%.

With the HAA5 species, it was expected that a slight positive bias might be observed, since not all THM4 species would permeate the CMS membrane. As one moves from simple to more complex sample matrices, one might expect this effect to become less pronounced as the THM4 species are salted out and permeate the membrane at a higher rate. Comparing the AR to the CA samples, this trend was observed. The AR samples showed that the calibration curve result was slightly higher than that predicted by standard addition. With the CA samples, the calibration curve gave slightly lower results. Again, a slight, but quantifiable and correctable, matrix effect was observed in the AR and CA samples.

Comparing the CMS-FIA NCA-FL Methods to USEPA Methods

Comparing CMS-FIA NCA-FL Method to USEPA Method 502.2

Comparison studies were carried out between the CMS-FIA NCA-FL method and USEPA Method 502.2 (Table 2.4.). In this study, the results for the individual THM4 species from USEPA Method 502.2 (1995b) were summed to obtain TTHM4. In the study, grab samples were collected from locations in Arkansas, California, Louisiana, Maryland, Wisconsin and seven different locations in Tennessee. The two TN 2 samples are not duplicates, but were collected at different times of the year.

The results from analysis of the real world samples were promising. For example, taking the USEPA Method 502.2 THM4 concentration to be the "true value," it was possible to calculate the percent recovery, the magnitude of the difference between the CMS-FIA NCA-FL method and USEPA Method 502.2 and the % difference between the two values. The average % recoveries from all samples ranged from 71 to 329% and averaged 114%, including three samples with results that were considered low concentrations (TTHM4 ≤ 10 μg/L). This seems to be in poor agreement until one considers that the three samples were ≤ 10 μg/L skew the results and arguably make them appear worse than they really are. Neglecting these three samples (simply for the sake of argument), the % recoveries ranged from 86 to 99% with an average of 93%. The absolute value of the difference between the two methods (considering all data) ranged from 1 to 9 μg/L with an average magnitude of 4 μg/L. When the same three samples are neglected, the range remained from 1 to 9 μg/L and the average was 5 μg/L. The % difference ranged from 1 to 229% and averaged 29% when all samples were considered. Neglecting the three samples, the % difference ranged from 1 to 14% with an average of 7%. Student's t-test (Harris 2001) was used to evaluate if the results from the two methods were statistically different at the 95% confidence level. The results of this statistical comparison indicated that the two different methods did not give statistically different results.

Table 4.4
Comparison of the CMS-FIA NCA-FL method for TTHM4 to USEPA Method 502.2

Analytical Method	**Results of Analysis of THM4 using EPA 502.2 P&T GC with PID and Hall Detector**					**CMS-FIA NCA-FL**
Location	**$CHCl_3$**	**$CHBrCl_2$**	**$CHBr_2Cl$**	**$CHBr_3$**	**TTHM4***	**Total THM4**
AR 1	67	9.0	1.8	n.d.[†]	**78**	**72**
CA1	31	20	19	3.0	**73**	**69**
LA 1	65	13	1.5	n.d.	**80**	**74**
MD 1	78	18	2.8	1.8	**101**	**92**
TN 1	66	7.3	0.7	n.d.	**74**	**70**
TN 2 (1)	74	10	0.7	n.d.	**85**	**84**
TN 2 (2)	29	4.9	1.7	n.d.	**36**	**31**
TN 3	4.3	3.0	2.4	n.d.	**10**	**7.1**
TN 4	36	4.8	1.7	n.d.	**43**	**40**
TN 5	n.d.	0.3	1.1	n.d.	**1.4**	**4.6**
TN 6	66	9.2	1.1	n.d.	**76**	**73**
TN 7	36	8.1	2.4	n.d.	**47**	**43**
WI 1	3.4	2.0	1.7	n.d.	**7.1**	**11**

* TTHM4 was calculated as the sum of the concentrations of the individual THM species measured using USEPA Method 502.2.

† n.d. – not detected in collected water samples.

Comparing CMS-FIA NCA-FL Method to USEPA Method 552.3

Comparison studies were carried out between the CMS-FIA NCA-FL method and USEPA Method 552.3 (2003a). The results of these studies are shown in Table 4.5. In this study, the individual results for the individual HAA5 species from USEPA Method 552.3 were summed to obtain the result for THAA5. The study was done in drinking water samples collected from the same locations as described in the THM4 study. In each case, the results showed good agreement and the same trends were observed in the sample set.

As with the THM4 study, taking the USEPA Method 552.3 concentration to be the "real value", it was possible to calculate the percent recovery, the magnitude of the difference between the CMS-FIA method and USEPA Method 552.3 and the % difference between the two values. Considering all samples, the average % recoveries ranged from 81 to 280% and averaged 125%. If the USEPA Method 552.3 results that were less than 10 µg/L are neglected as before, the average % recovery ranges from 81 to 135% with an average of 111%. The absolute value of the difference between the two methods (considering all data) ranged from 2 to 14 µg/L with an average magnitude of difference of 6 µg/L whether the two samples were neglected or not. The % difference ranged from 2 to 180% and averaged 32% when all samples were considered. Neglecting the two low level samples, the results ranged from 2 to 35% with an average of 19%. The same calibration issues discussed earlier with the THM4 method would apply here as well. Student's t-test (Harris, 2001) was used to determine if the results from the two methods were statistically different at the 95% confidence level. The results of this statistical comparison indicated that the results using the two different methods were not statically different.

Table 4.5
Comparison of the CMS-FIA NCA-FL method for THAA5 to USEPA Method 552.3

Analytical Method	**Results of Analysis of HAA5 using EPA 552.3 Microextraction, Derivatization with analysis using GC-ECD**						**CMS-FIA NCA-FL**
Location	**MCAA**	**DCAA**	**TCAA**	**MBAA**	**DBAA**	**THAA5***	**Total HAA5**
AR 1	0.5	7.7	22	1.1	0.3	**32**	**30**
CA 1	0.3	15	10	n.d†	5.0	**30**	**36**
LA 1	n.d.	19	20	n.d.	0.9	**40**	**54**
MD 1	1.9	15	22	1.2	3.3	**43**	**45**
TN 1	n.d.	12	18	n.d.	1.0	**31**	**40**
TN 2 (1)	6.8	20	23	n.d.	0.9	**51**	**57**
TN 2 (2)	1.1	20	24	0.1	0.5	**45**	**44**
TN 3	1.0	4.7	4.4	0.1	1.4	**12**	**16**
TN 4	1.6	20	21	0.1	0.9	**43**	**35**
TN 5	n.d.	1.4	n.d.	n.d.	1.2	**3**	**4**
TN 6	n.d.	15	18	n.d.	1.0	**34**	**44**
TN 7	1.4	21	23	0.7	0.8	**47**	**38**
WI 1	n.d.	2.5	2.3	n.d.	n.d.	**5**	**14**

* THAA5 was calculated as the sum of the concentrations of the individual HAA species.
† n.d. – not detected in collected water samples.

CMS-FIA NCA-FL Monitoring Studies

Description of On-Line Monitoring of THM4 and HAA5 Using Combined CMS-FIA

In the CMS-FIA method for on-line monitoring of TTHM4 and THAA5 concentrations in a water distribution system, the CMS device was connected to a ten-port microelectrically actuated valve and a FIA analyzer as shown in Figure 4.10. With the ten-port valve in position (A) (Figure 4.10), an acceptor stream of only reagent water flowed through the entire interior of the silicone membrane tubing and was sent out to Port 8 then out of Port 9 to waste. A carrier stream of just NaOH solution was flowed through sample loop 1 (60 μL) continuously from Port 1 to Port 10, through the sample loop, to Port 7 and out to the manifold from Port 6. Simultaneously, a drinking water sample flowed through the first tee union of the CMS and around the entire exterior wall of the membrane inside a Tefzel® tubing shell. HAA5 sample flowed into sample loop 2 (60 μL) connected to Ports 2 and 5 through Port 4 and then out Port 3 to waste container. In Position (B) (Figure 4.10), THM4 sample was loaded onto the sample loop connected to Ports 7 and 10 through Port 8, then out Port 9 to waste. The HAA5 sample present within sample loop 2 was carried by the NaOH solution from Port 1 through Ports 2 to 5, then out to Port 6 to the RC1 where it was mixed with the 30% NCA solution and heated at 97°C in

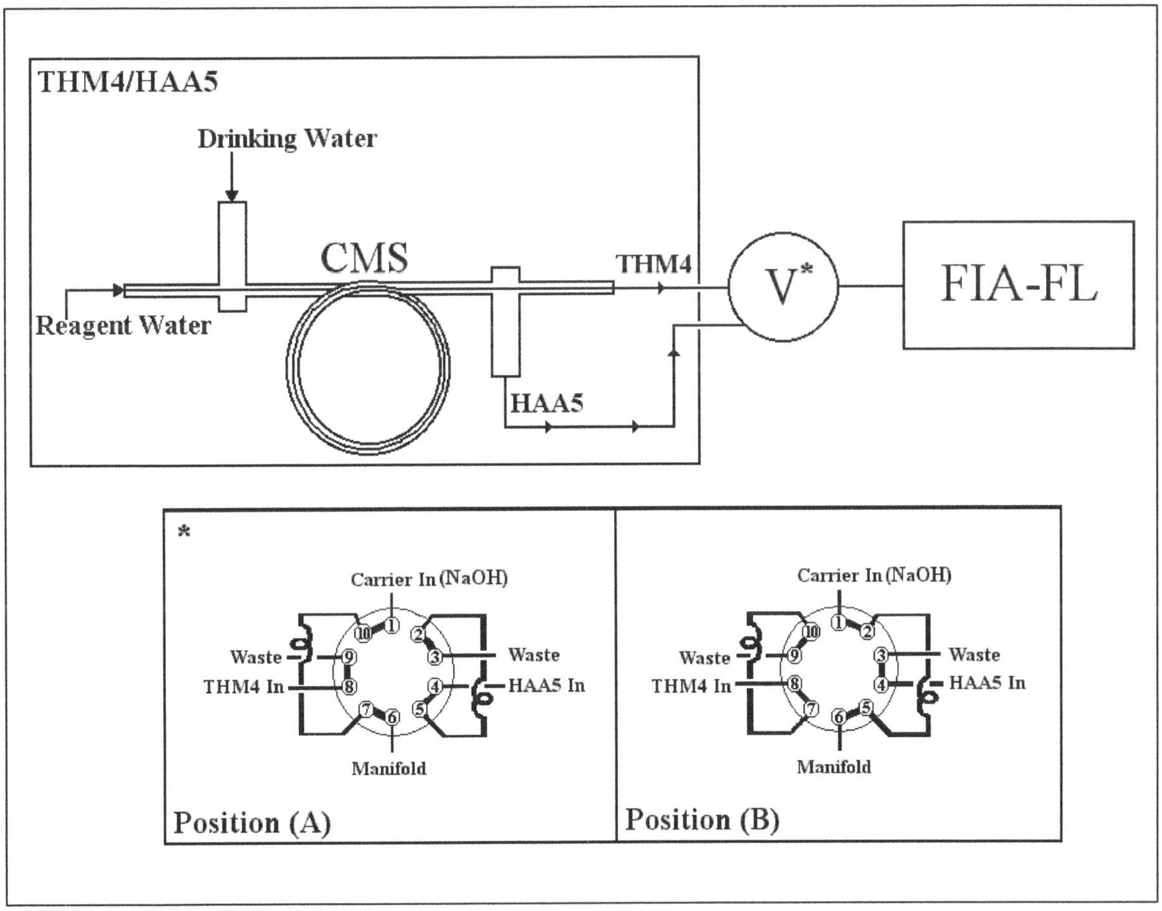

Figure 4.10 Schematic diagram of the CMS-FIA NCA-FL in the on-line sampling configuration

the heat bath. After exiting RC1, the reaction mixture flowed through RC2 where it was cooled in an ice bath to maximize fluorescent intensity and minimize air bubble formation, and then entered the flow-cell of the fluorescent detector. The valve was held in Position (B) for 6.0 minutes and then was switched back to the Position (A). In Position (A), sample loop 1 filled with THM4 enriched reagent water was back flushed from Port 1 through Port 10 and 7, and through Port 6 to the manifold, where it is mixed, heated, and cooled just as the HAA5 sample previously was. The use of a ten-port injection valve with two sample loops has allowed for the analysis of THM4 and HAA5 species simultaneously without changing the configuration of the CMS-FIA system.

Evaluating Sample Carry-Over Effects

A study was conducted to evaluate possible carry-over issues from one analysis time to the next (Figure 4.11). A THM4 and HAA5 standard of 100 µg/L concentration each was analyzed once followed by several "blank" analyses of reagent water. Blank injections were done every 45 minutes for three hours. After 45 minutes, there was a small carry-over with the THM4 and HAA5 species. However, after 60 minutes the CMS device did recover and it was concluded that one sample could be measured per hour without carry-over effects. It is possible that, if the membrane length was increased beyond the 120 cm length of the present device, the recovery time would be expected to increase. For the current studies, to achieve the goal of one sample per hour, the decision was made to use membrane length of 94 cm and to inject one sample each hour. The membrane length could likely be increased and optimized further.

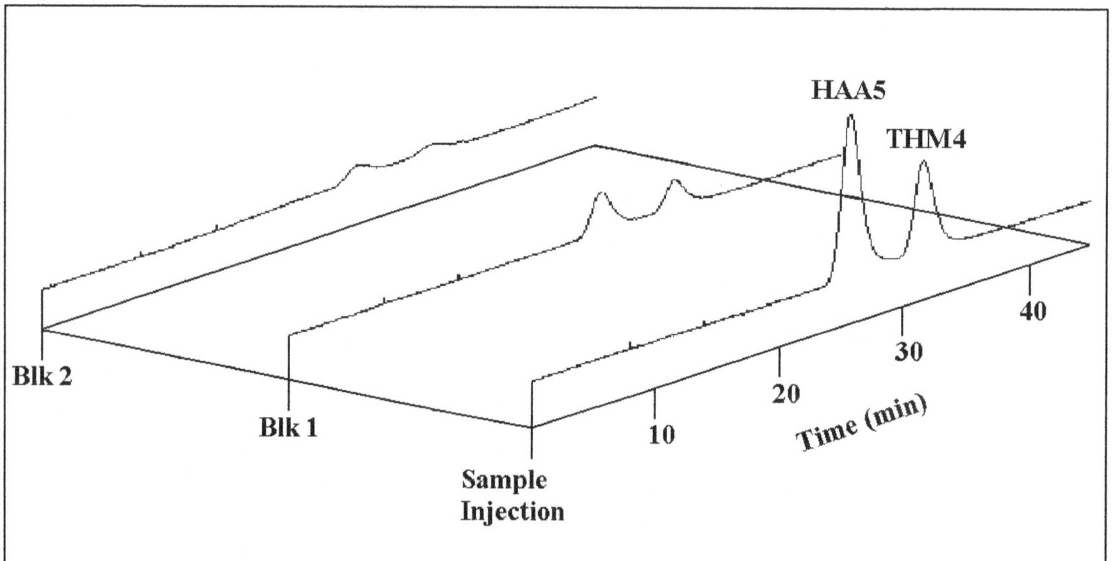

Figure 4.11 The results of carry-over experiments for CMS-FIA NCA-FL TTHM4 and THAA5 – 100 µg/L each

On-Line Monitoring Studies for TTHM4 using CMS-FIA NCA-FL

The first on-line monitoring study for CMS-FIA was conducted from December 16 to 23, 2004 at the University of Memphis sampling site (Memphis 1). The second on-line monitoring study for CMS-FIA was conducted from May 19 to 27, 2005 at the MLGW South Service Center. The third on-line study was conducted from June 21 to 28, 2005 in Houston, TX. In each study, the CMS-FIA was interfaced to the drinking water distribution system as described earlier and in more detail in the Appendix A. The concentration of TTHM4 species was monitored at a rate of 1 sample/ hour (though the analyzer was typically taken off-line every 12 hours to run a check standard). The results of these studies along with details used in the analysis of the data are presented in Table 4.6. At the same time as these studies were carried out, simultaneous studies were done using CMS-GC-ECD (Chapter 5) at Memphis 1, Memphis 2, and Houston.

Table 4.6
Memphis 1, Memphis 2 and Houston (502.2) and Houston (524.2) monitoring studies: monitoring TTHM4 concentrations using CMS-FIA NCA-FL

	Memphis #1		**Memphis #2**		**Houston (502.2)**		**Houston (524.2)**	
Statistic/*Method*	*EPA*	*CMS-FIA*	*EPA*	*CMS-FIA*	*EPA*	*CMS-FIA*	*EPA*	*CMS-FIA*
N	114	114	99	99	35	35	24	24
Avg. Conc.	2.2	3.0	13.5	10.9	50.0	61.5	22.9	63.5
Std. Dev Conc.	0.9	1.3	3.5	4.3	4.0	15.8	4.2	13.1
Conc. Range	1.1 to 4.0	1.1 to 5.5	8.5 to 26.2	1.1 to 24.2	41.2 to 56.6	41.3 to 97.4	18.1 to 37.4	44.7 to 84.0
Avg. Bias	---	0.7	---	2.6	---	12	---	40.5
Std. Dev Bias	---	1.6	---	5.2	---	16	---	13.9
Bias Range	---	-2.4 to 4.1	---	-11.5 to 16.4	---	-9.9 to 43.2	---	15.5 to 62.0
% REC_{Bias}	---	151	---	85	---	124	---	289
s.d. % REC_{Bias}	---	97	---	39	---	34	---	73
% REC_{Bias} Range	---	3.7 to 447	---	4.3 to 191	---	81 to 200	---	141 to 421
t-test* / $t_{critical}$ / $t_{calculated}$	---	Yes / 2.360 / 5.578	---	Yes / 2.365 / 4.986	---	Yes / 2.441 / 4.183	---	Yes / 2.500 / 14.239

* The $t_{critical}$ and $t_{calculated}$ values listed were calculated for the N listed using a two-tailed t-test with 98% confidence. Typically a $t_{calculated}$ value should be less than the $t_{critical}$. Yes or No means if statistically different.

A plot of the concentration of TTHM4 versus time is shown in Figure 4.12 for Memphis 1 study. For 114 direct comparisons, the concentration for TTHM4 ranged from 1.1 - 4.0 μg/L for EPA 502.2 and from 1.1 - 5.5 μg/L for the CMS-FIA method. The average TTHM4 concentration from EPA 502.2 was 2.2 ± 0.9 μg/L and from CMS-FIA was 3.0 ± 1.3 μg/L. Figure 4.13 shows a plot of the bias as a function of time for the CMS-FIA analyzer compared directly to EPA 502.2. The bias, a measure of the absolute error of the candidate method, ranged from –2.4 - 4.1 μg/L (averaging 0.7 ± 1.6 μg/L) and % REC_{Bias} ranges from 3.7 - 447 % (averaging 151 ± 97%).

A plot of the concentration of TTHM4 versus time is shown in Figure 4.14 for Memphis 2 study. For 99 direct comparisons, the concentration for TTHM4 ranged from 8.5 - 26.2 μg/L (EPA 502.2) and from 1.1 - 24.2 μg/L (CMS-FIA). The average TTHM4 concentration was 13.5 ± 3.5 μg/L (EPA 502.2) and 10.9 ± 1.3 μg/L (CMS-FIA). Figure 4.15 shows a plot of the bias as a function of time for the CMS-FIA analyzer compared directly to EPA 502.2. The bias ranged from –11.5 - 16.4 μg/L with an average of 2.6 ± 5.2 μg/L. The % REC_{Bias} ranges from 4.3 - 191% with an average of 84.7 ± 38.7%.

For the Houston study, the CMS-FIA method for TTHM4 was compared to both EPA 502.2 and EPA 524.2. A plot of the concentration of TTHM4 versus time is shown in Figure 4.16 for Houston (502.2) study. For EPA 502.2 there were 35 direct comparisons with concentration ranges varying from 41.2 - 56.6 μg/L (EPA 502.2) and 41.3 - 97.4 μg/L (CMS-FIA). The average concentration was 50.0 ± 4.0 μg/L (EPA 502.2) and 61.5 ± 15.8 μg/L (CMS-FIA). Figure 4.17 shows a plot of the bias as a function of time for the CMS-FIA analyzer compared directly to EPA 502.2. The bias ranged from -9.9 – 43.2 μg/L and averaged 12 ± 16 μg/L. The % REC_{Bias} varies from 81 - 200% with an average of 124 ± 34%.

For the Houston 524.2 comparison, a plot of the concentration of TTHM4 versus time is shown in Figure 4.18. For EPA 524.2 there were 24 direct comparisons with concentration ranges varying from 18.1 - 37.4 μg/L (EPA 524.2) and 44.7 - 84.0 μg/L (CMS-FIA). The average concentration was 22.9 ± 4.2 μg/L (EPA 524.2) and 63.5 ± 13.1 μg/L (CMS-FIA). The bias plot for Houston (524.2) study is shown in Figure 4.19. The bias ranged from 15.5 to 62.0 μg/L and averaged 40.5 ± 13.9 μg/L. The % REC_{Bias} varies from 141 - 421% with an average of 289 ± 73%.

Statistically, For the Memphis 1 study, the $t_{critical}$ for the 114 comparisons for TTHM4 is estimated to be 2.360. The $t_{calculated}$ is 5.578, which means that there is a statistical difference between the methods. In the Memphis 2 study for 99 comparisons for TTHM4, the $t_{critical}$ is 2.365 and the $t_{calculated}$ is 4.986; therefore, there is a statistical difference between the two methods. With the Houston study (502.2) for 35 comparisons for TTHM4, the $t_{critical}$ is 2.441 and $t_{calculated}$ is 4.183, which also means the two methods are statistically different. With the Houston study (524.2) for 24 comparisons the $t_{critical}$ is 2.500 and $t_{calculated}$ is 14.239, the two methods are not statistically the same.

Qualitatively, the results do show some agreement, but this agreement seems to worsen as the sample matrix become more complicated. As discussed earlier, the three sampling sites were selected as they provide three different levels of complexity with respect to sample matrix. The sample matrix increases in complexity in the order Memphis 1, Memphis 2 and Houston as does the positive bias. It is not unexpected that the CMS-FIA NCA-FL method for TTHM4 should exhibit a positive bias. For example, as the sample matrix becomes more complicated, there are two practical concerns. It could be that some of the compounds present in the more complex matrix permeate the CMS membrane and might chemically react with NCA and fluoresce—thus increasing the analytical signal. These compounds might interfere in the CMS-FIA method, but would not be expected to exhibit interference in EPA 502.2 where speciation is possible. It is also possible that the more complex Houston matrix might physically increase the concentration of analyte over that of the calibration standards via processes such as the "salting out" effect or similar phenomena. Both of these factors could lead to positive bias.

Consider that there are several compounds that might interfere in the CMS-FIA method. For example, other halogenated species such as halonitriles, haloaldehydes and haloketones were all shown to interfere as might iodinated analogs of THM4 species, if present. This is why, in retrospect, the researchers suggest that prior to carrying out a monitoring study; pre-monitoring surveys of the compounds present in the matrix should be done. It is also possible that disinfectant species such as hypochlorous acid (HOCl), hypochlorite ion (OCl^-) and chloramines might interfere, but more work needs to be done to quantify this effect. It is also possible that increasing the sample matrix complexity might lead to increases in several parameters. For example, ionic strength, TOC concentrations and temperature have all been shown to increase the analytical signal. These effects have been quantified. There are other factors such as alkalinity, pH, and possibly other factors that need to be further investigated.

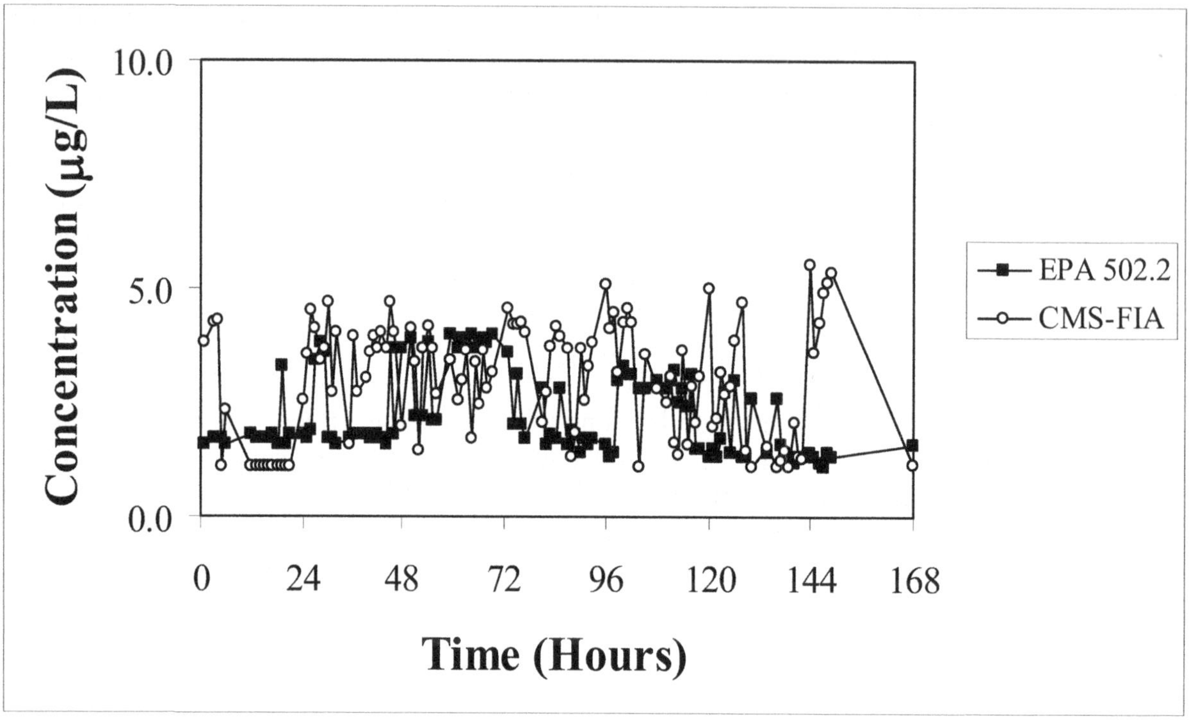

Figure 4.12 Memphis 1 concentration plot of TTHM4 for EPA 502.2 and CMS-FIA NCA-FL

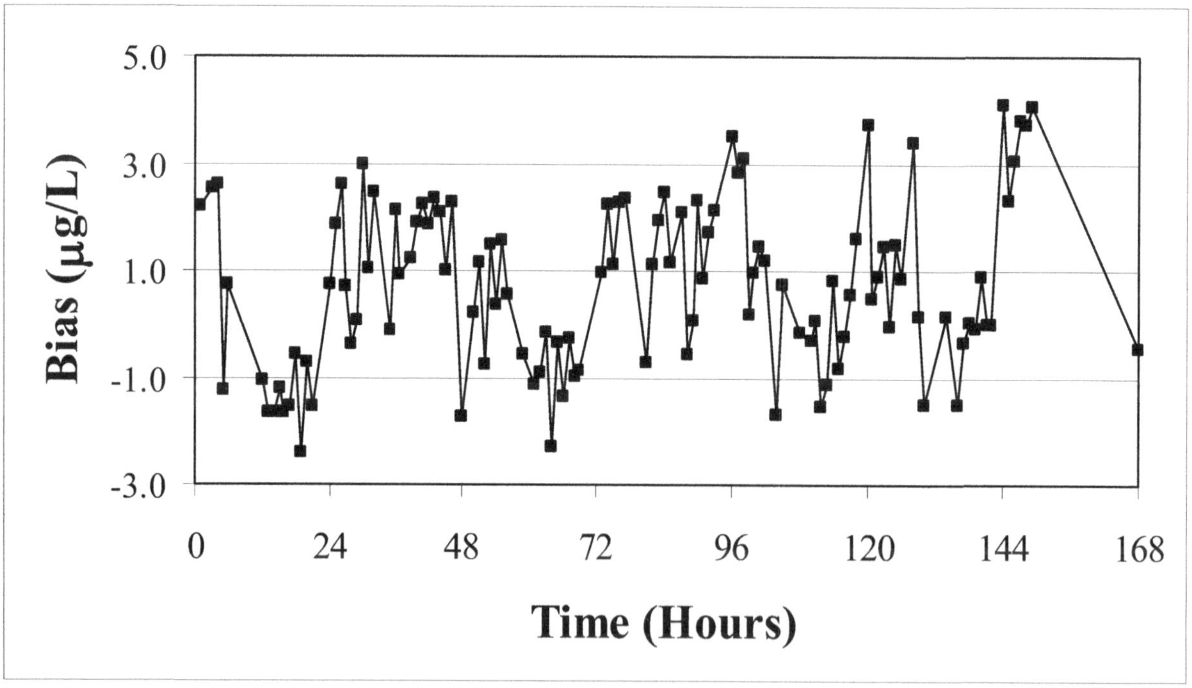

Figure 4.13 Memphis 1 bias plot of TTHM4 comparing CMS-FIA NCA-FL to EPA 502.2

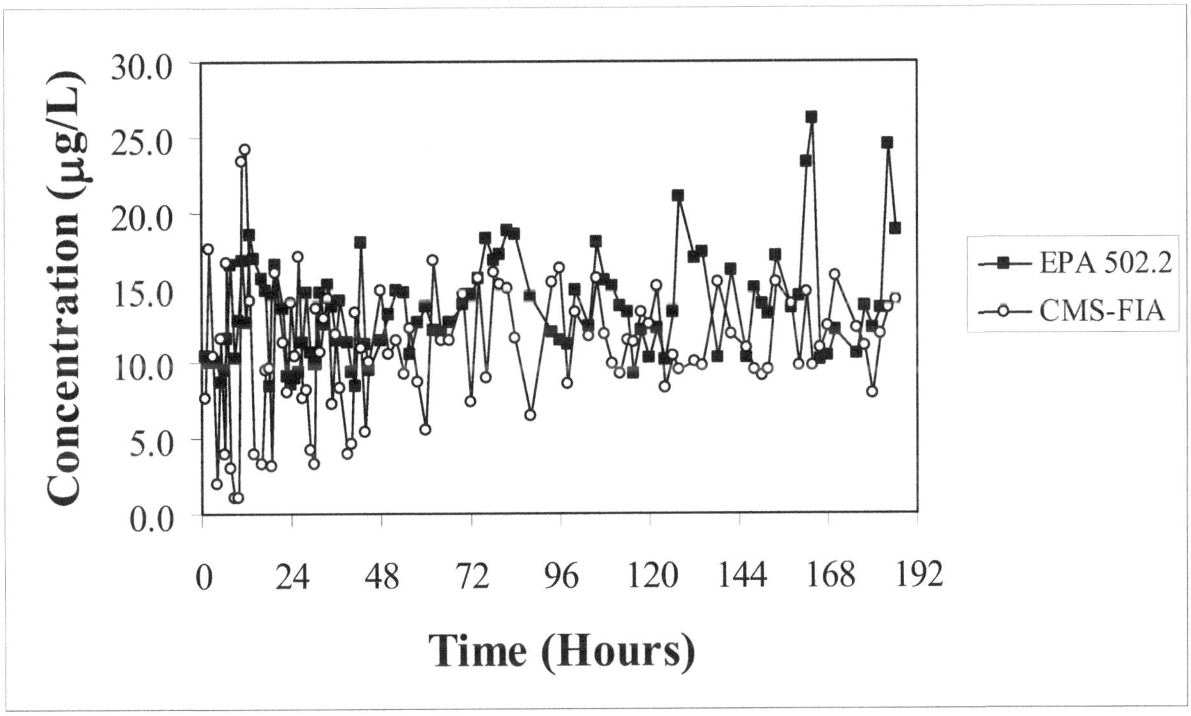

Figure 4.14 Memphis 2 concentration plot of TTHM4 for EPA 502.2 and CMS-FIA NCA-FL

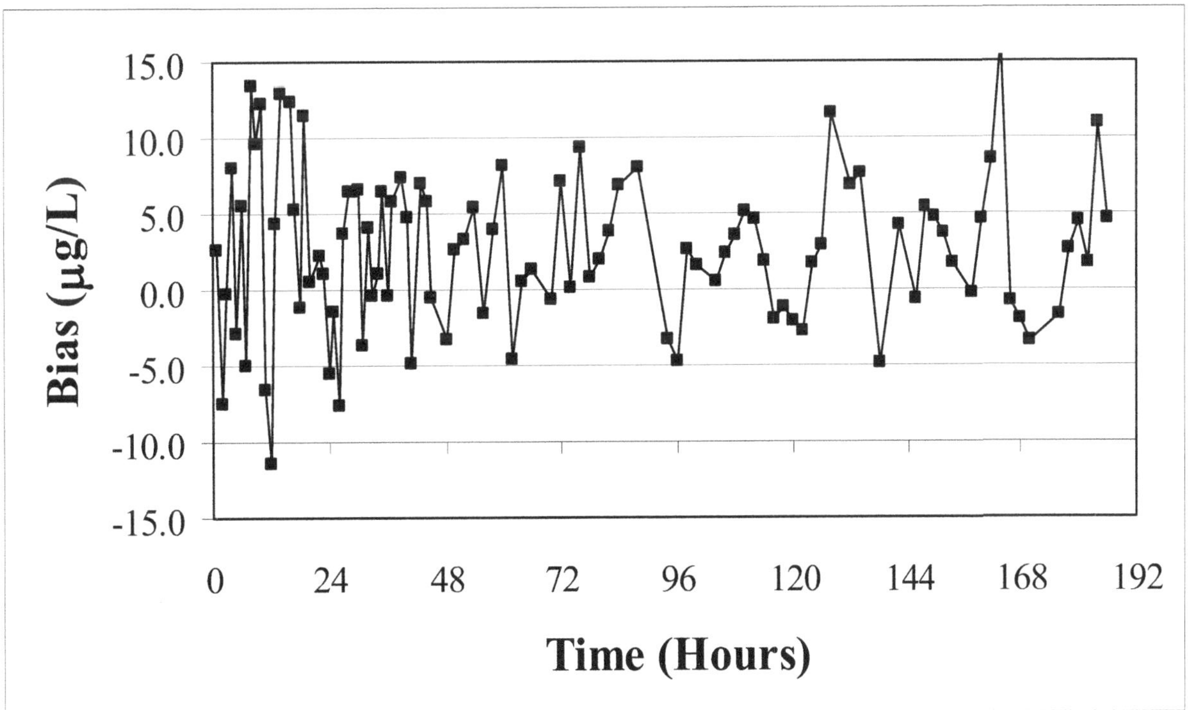

Figure 4.15 Memphis 2 bias plot of TTHM4 comparing CMS-FIA NCA-FL to EPA 502.2

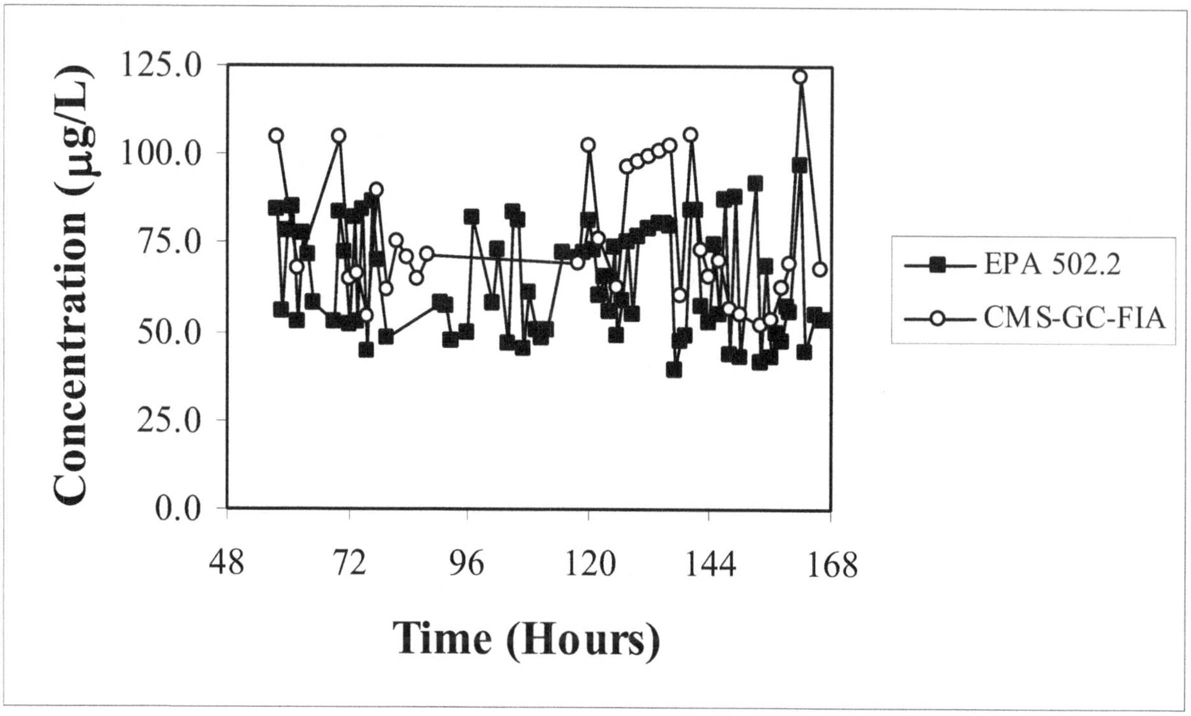

Figure 4.16 Houston concentration plot of TTHM4 for EPA 502.2 and CMS-FIA NCA-FL

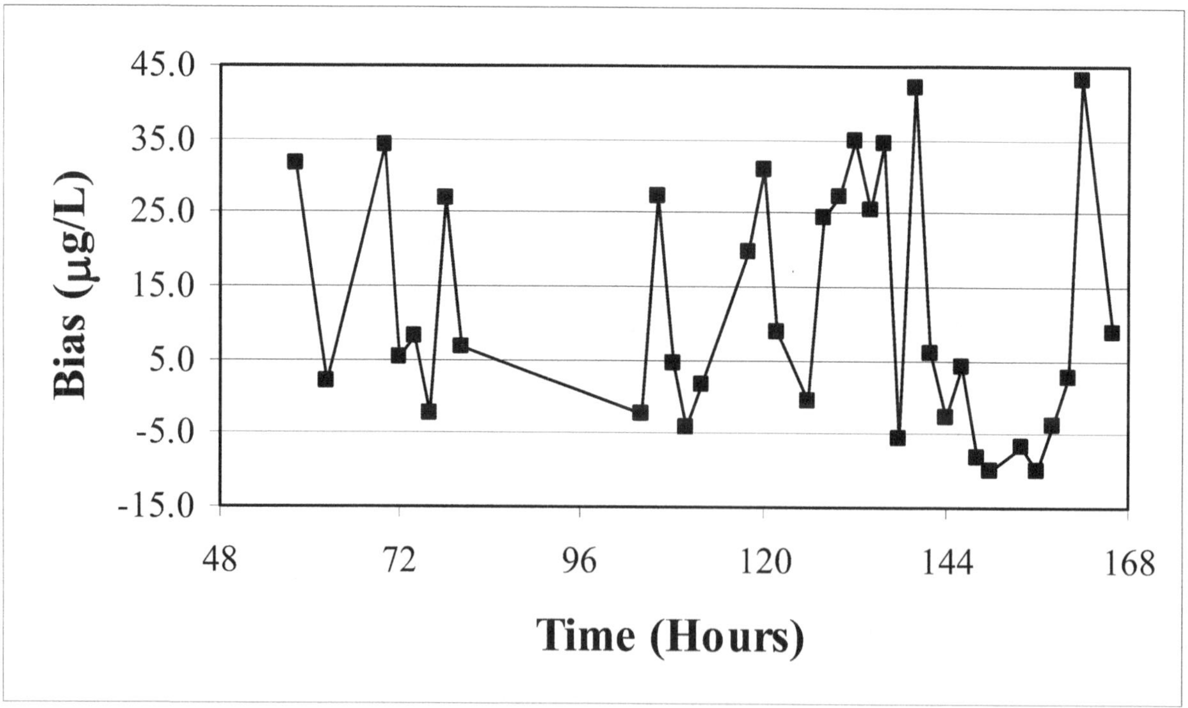

Figure 4.17 Houston bias plot of TTHM4 comparing CMS-FIA NCA-FL to EPA 502.2

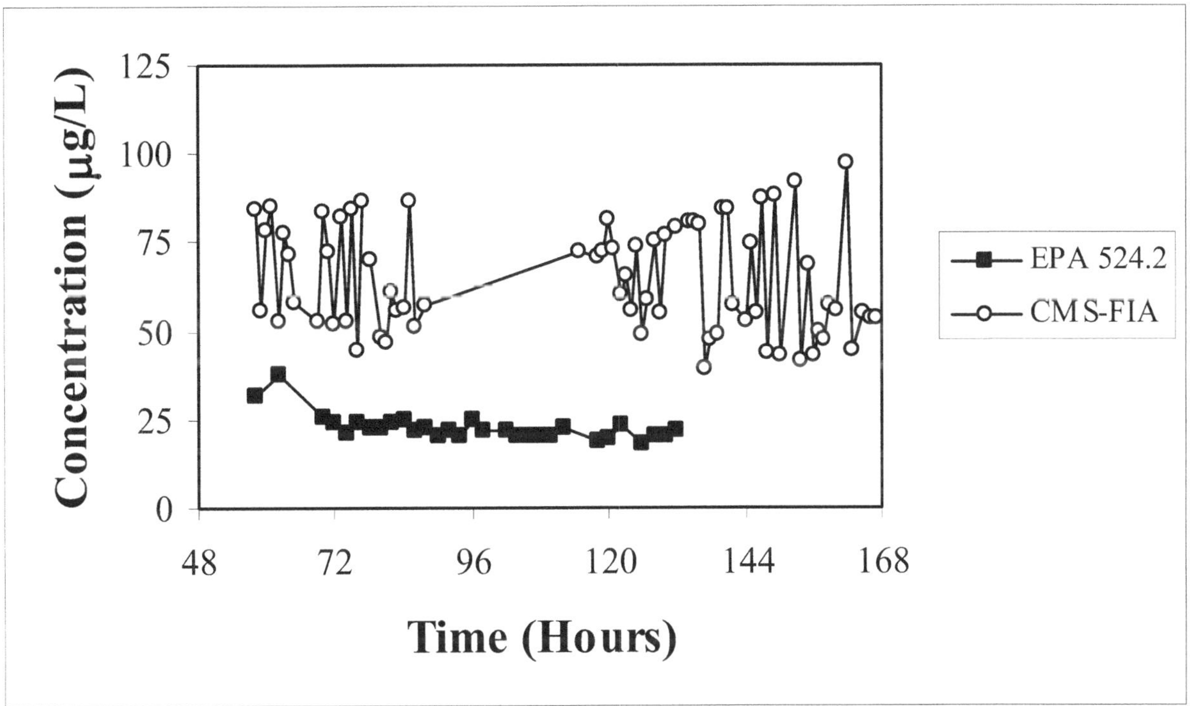

Figure 4.18 Houston concentration plot of TTHM4 for EPA 524.2 and CMS-FIA NCA-FL

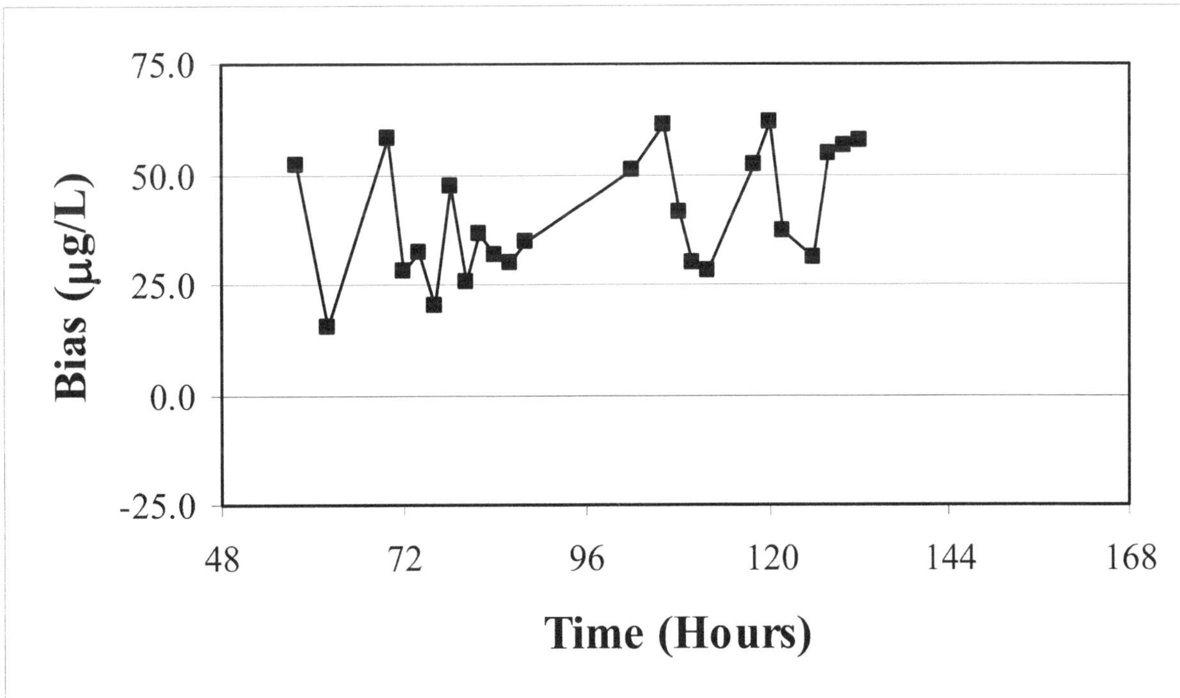

Figure 4.19 Houston bias plot of TTHM4 comparing CMS-FIA NCA-FL to EPA 524.2

Increasing the ionic strength has been shown to increase the analytical signal. The % increase in analytical signal per 1M increase in ionic strength was measured as 27%. The ionic strengths for Memphis 1, Memphis 2 and Houston were 0.0012 M, 0.0019 M and 0.0052, respectively. The expected increase in concentration for the three sampling sites are <0.001 μg/L (Memphis 1), ~0.001 μg/L (Memphis 2), and 0.016 μg/L (Houston). Since the average bias estimates for Memphis 1, Memphis 2, and Houston were +0.7 ± 1.6 μg/L, +2.6 ± 5.2 μg/L and 11.6 ± 16.4 μg/L, respectively, the contributions toward positive bias from ionic strength increase is not considered significantly large.

The concentration of TOC was shown to effect analytical signal. The TOC concentrations in Memphis 1, Memphis 2 and Houston were 0.4 mg/L, 0.7 mg/L and 3.9 mg/L, respectively. The effect of increasing TOC concentration was measured as ~+0.5% per mg/L TOC increase. For Memphis 1, Memphis 2, and Houston, this means one would predict a <0.01, 0.01 and 0.23 mg/L increase in TTHM4 concentrations. As noted before, the average bias estimates for Memphis 1, Memphis 2, and Houston were +0.7 ± 1.6 μg/L, +2.6 ± 5.2 μg/L and 11.6 ± 16.4 μg/L, respectively, and thus the contributions toward positive bias from TOC increase is not considered significant.

It is possible that alkalinity and pH might have effects on analytical signal. Unfortunately, these factors have not been evaluated. Alkalinity is the buffering capacity of water. The alkalinity of Memphis 1, Memphis 2, and Houston averages 41, 68 and 80 mg/L, respectively. The pH of Memphis water is ~7.3 while the pH of Houston water is ~8.0. The fluorescence intensity of the NCA reaction is strongly related to pH, occurring at more basic pH values (pH 10-12). It is possible that as the water sample becomes more buffered (more resistant to pH changes), more mixing time is needed in the FIA analyzer to achieve optimum FL intensity. Again, more work needs to be done to explore these issues.

It is possible that the detector response varies during the course of a monitoring study. To evaluate this variability, a check standard is run two times daily during a study. This application of the percent recovery is referred to here as % $REC_{Monitor.}$ This was done by taking the temperature corrected data, and using the % $REC_{Monitor}$ for each check standard, rather than the mean % recovery from the calibration curve (Org. CMS). The results of the corrections are presented in Table 4.7. The TTHM4 results from CMS-FIA had the % $REC_{Monitor}$ correction was applied and the results compared to EPA 502.2 as before (% REC Corr. CMS). The results from EPA 502.2 ranged from 41.2 - 56.6 μg/L (averaging 50.0 ± 4.0 μg/L). The results for the % REC_{Bias} (Figure 4.20) for the CMS-FIA TTHM4 concentration ranged from 47.0 - 122 (averaging 72 ± 19 μg/L). The bias (Figure 4.21) with EPA 502.2 ranged from –2.8 - 68.1 μg/L (averaging 22 ± 20 μg/L) with % REC_{Bias} ranging from 94.8 - 233 % (averaging 146 ± 41 %). Statistically, the $t_{critical}$ for the 35 comparisons was 2.441. For % $REC_{Monitor}$ corrected values for TTHM4 $t_{calculated}$ was 6.702 meaning that the methods are still statistically different. This indicates that the % $REC_{Monitor}$ correction factor does not account for the differences between the two methods.

Qualitatively, however, the average bias has doubled after the % $REC_{Monitor}$ correction factor is applied and the % REC_{Bias} has increased. (Figures 4.20 to 4.21). The difference between the CMS-FIA method and EPA 502.2, at this point may be due to either other halogenated species (such as haloaldehydes, haloketones halonitriles, iodinated THMs and disinfectant species). It is possible that alkalinity and pH differences might contribute some of the bias. Detector drift does not seem to be a problem as indicated by the % REC $_{Monitor}$ correction giving poorer results. More detailed interference studies need to be done for the CMS-FIA methods.

Using the CMS-FIA methods, researchers should conduct preliminary surveys and evaluate the interference effects of those compounds present in the sample.

Table 4.7
Houston monitoring study (502.2): comparison of % $REC_{Monitor}$ corrected (%REC Corr. CMS) and original data (Org. CMS) for CMS-FIA NCA-FL method for TTHM4 to USEPA Method 502.2

		CMS-FIA TTHM4	
Statistic/*Method*	*EPA 502.2*	*Org. CMS*	*%REC Corr. CMS*
N	35	35	35
Avg. Conc	50.0	61.5	72
Std. Dev Conc.	4.0	15.8	19
Conc. Range	41.2 to 56.6	41.3 to 97.4	47.0 to 122.3
Avg. Bias	---	11.6	22
Std. Dev Bias	---	16.4	20
Bias Range		-9.9 to 43.2	-2.77 to 68.1
% REC_{Bias}	---	124	146
s.d. % REC_{Bias}	---	33.9	41
% REC_{Bias} Range	---	81.2 to 200	94.8 to 233
t-test* $t_{critical}$ $t_{calculated}$	---	Yes 2.441 4.183	Yes 2.441 6.702

* The $t_{critical}$ and $t_{calculated}$ values listed were calculated for the N listed using a two-tailed t-test with 98% confidence. Typically a $t_{calculated}$ value should be less than the $t_{critical}$. Yes or No means if statistically different.

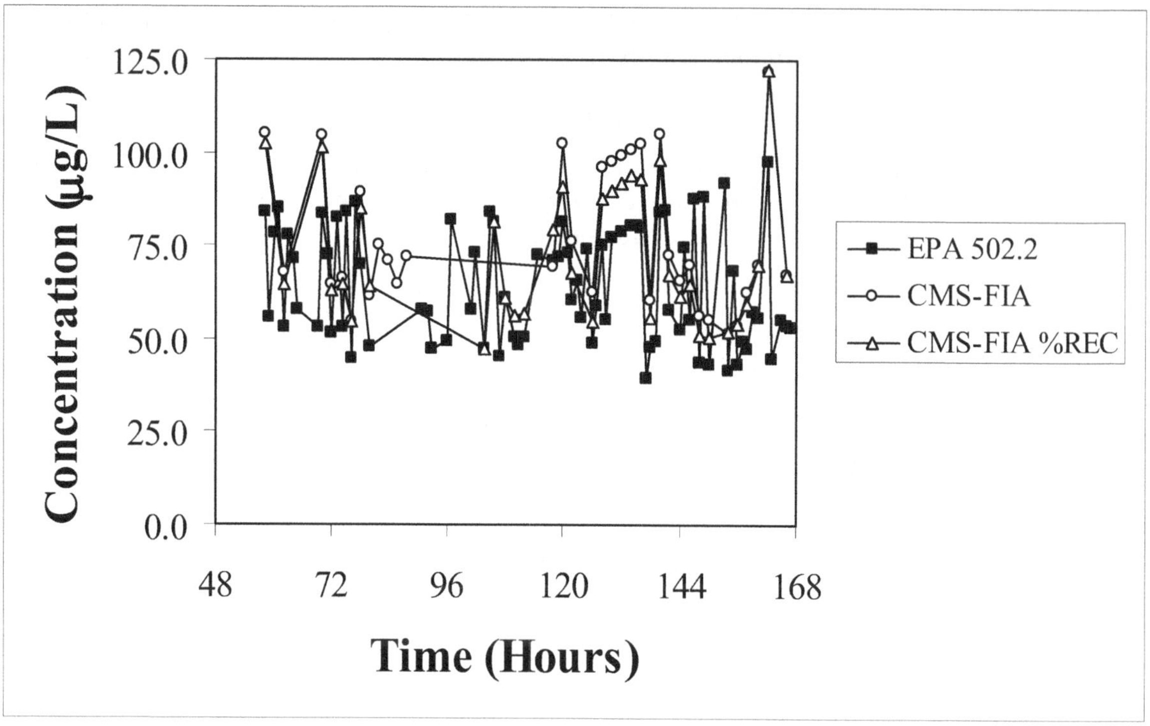

Figure 4.20 Houston concentration plot of TTHM4 for EPA 502.2, CMS-FIA NCA-FL and CMS-FIA NCA-FL % $REC_{Monitor}$ correction

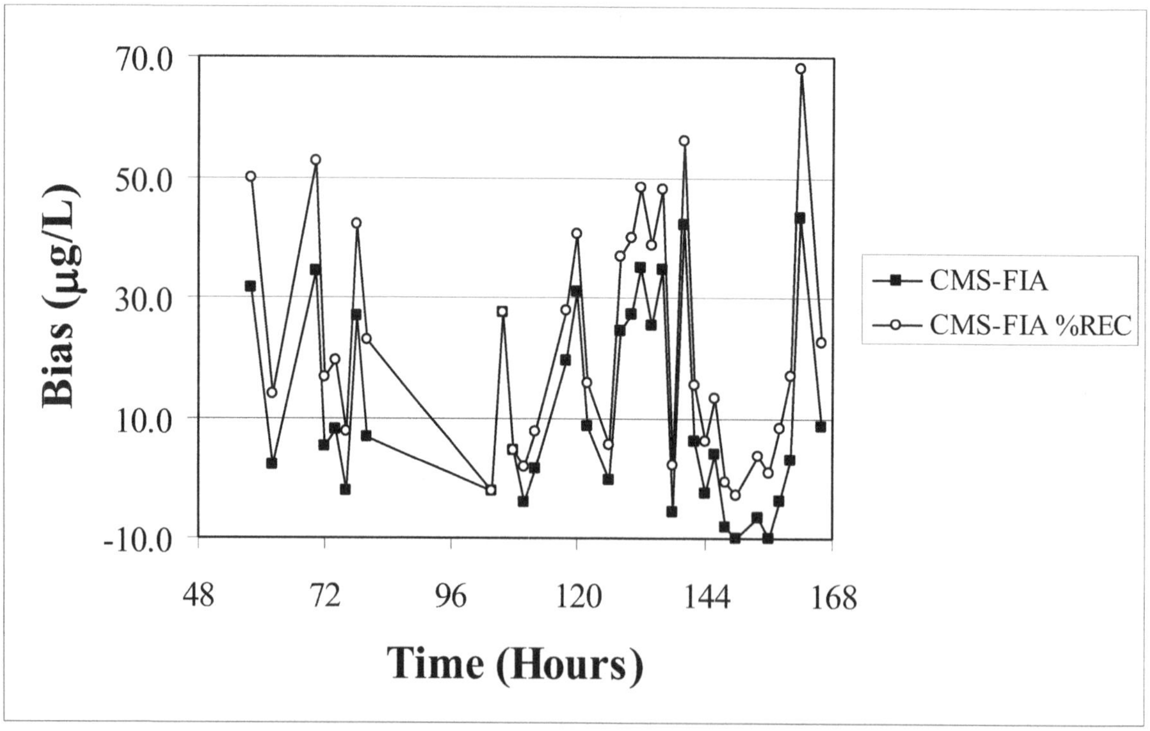

Figure 4.21 Houston bias plots of TTHM4 comparing CMS-FIA NCA-FL and CMS-FIA NCA-FL % $REC_{Monitor}$ correction to EPA 502.2

On-Line Monitoring Studies for THAA5 using CMS-FIA NCA-FL

These studies were carried out at the same time as the TTHM4 studies just presented. The first on-line monitoring study for CMS-FIA was conducted from December 16 to 23, 2004 (Memphis 1), May 19 to 27, 2005 (Memphis 2), and June 21-28, 2005 (Houston). In each study, the CMS-FIA was interfaced to the drinking water distribution system as described earlier and in the Appendix A. The concentration of THAA5 species was monitored at a rate of 1 sample/hour (though the analyzer was typically taken off-line every 12 hours to run a check standard). In both the TTHM4 and THAA5 CMS-FIA Houston data, there is a large gap between hours 80 and 100. During this time, the researchers took the instrument offline to recalibrate, as the flow rate for calibration was ~0.5 mL/min, and the on-line sampling was ~2.4 mL/min. The instrument was recalibrated at a standard flow rate of ~2.4 mL/min. After this recalibration, the instrument ran without failing, and was taken offline to run check standards. The results of these studies along with details used in the analysis of the data are presented in Table 4.8.

Table 4.8
Memphis 1, Memphis 2 and Houston monitoring studies: monitoring THAA5 concentrations using CMS-FIA NCA-FL

	Memphis #1		**Memphis #2**		**Houston (552.3)**	
Statistic/*Method*	*EPA*	*CMS-FIA*	*EPA*	*CMS-FIA*	*EPA*	*CMS-FIA*
N	60	60	105	105	34	34
Avg. Conc	1.1	4.1	1.8	4.6	44.9	58.8
Std. Dev Conc.	1.5	1.7	0.9	2.1	6.4	11.4
Conc. Range	0.4 to 11.2	1.6 to 8.1	0.2 to 5.0	1.6 to 9.2	34.0 to 61.6	43.2 to 93.1
Avg. Bias	---	3.1	---	2.8	---	14
Std. Dev Bias	---	2.4	---	2.3	---	13
Bias Range	---	-9.6 to 7.4	---	-2.0 to 9.0	---	-12.9 to 51.8
% REC$_{Bias}$	---	574	---	373	---	133
s.d. % REC$_{Bias}$	---	349	---	494	---	30
% REC$_{Bias}$ Range	---	14 to 1576	---	47 to 4579	---	79 to 225
t-test* $t_{critical}$ $t_{calculated}$	---	Yes 2.391 9.818	---	Yes 2.363 12.839	---	Yes 2.444 6.125

* The $t_{critical}$ and $t_{calculated}$ values listed were calculated for the N listed using a two-tailed t-test with 98% confidence. Typically a $t_{calculated}$ value should be less than the $t_{critical}$. Yes or No means if statistically different.

A plot of the concentration of THAA5 versus time is shown in Figure 4.22 for the Memphis 1 study. For 60 direct comparisons, the concentration for THAA5 ranged from 0.4 - 11.2 µg/L for EPA 552.3 and from 1.6 - 8.1 µg/L for CMS-FIA. The average THAA5 concentration for EPA 552.3 was 1.1 ± 1.5 µg/L and 4.1 ± 1.7 µg/L for CMS-FIA. Figure 4.23 shows the bias plot of the data comparing the CMS-FIA analyzer directly to EPA 552.3. The bias ranged from –9.6 - 7.4 µg/L with an average of 3.1 ± 2.4 µg/L. The % REC_{Bias} ranged from 14.3 – 1576 % with and average value of 574 ± 349 %.

The results for the Memphis 2 study (Figure 4.24) include 105 direct comparisons. The concentration of THAA5 ranged from 0.2 – 5.0 µg/L (EPA 552.3) and from 1.6 – 9.2 (CMS-FIA). The average THAA5 concentration for EPA 552.3 was 1.8 ± 0.9 and for the CMS-FIA was 4.6 ± 2.1 µg/L. The bias over the course of the monitoring study (Figure 4.25) ranged from –2.0 to 9.0 µg/L when the CMS-FIA analyzer was compared directly to EPA 552.3. The average bias was 2.8 ± 2.3 µg/L. The % REC_{Bias} ranged from 47 – 4579 % with an average of 373 ± 494 %.

The results of the Houston monitoring study (Figure 4.26) show 34 direct comparisons where the THAA5 concentration ranged from 34.0 - 61.6 µg/L (EPA 552.3) and from 43.2 – 93.1 µg/L (CMS-FIA). The average THAA5 concentration was 44.9 ± 6.4 µg/L (EPA 552.3) and 58.8 ± 11.4 µg/L (CMS-FIA). The plot of the bias as a function of time (Figure 4.27) compares the CMS-FIA analyzer to EPA 552.3. The bias ranged from -12.9 - 51.8 µg/L with an average of 14 ± 13 µg/L. The % REC_{Bias} ranges from 79 – 225 % with an average of 133 ± 30 %.

Statistically, the Memphis 1 study, the CMS-FIA results differ from EPA 552.3 results ($t_{critical}$ = 2.391 versus $t_{calculated}$ = 9.818). In the Memphis 2 study, the two method gave different results ($t_{critical}$ = 2.363 versus $t_{calculated}$ = 11.415). The same is true of the result from Houston where the result differ ($t_{critical}$ = 2.444 versus $t_{calculated}$ = 6.125).

Qualitatively, the THAA5 results for the Memphis studies were promising with an average bias of less than 5 µg/L; however, the average bias for the Houston study increased substantially. The effects of ionic strength and TOC concentration were considered. Without going into the detail as in the THM4 studies, similar calculations predict that at most ~2% increase in concentration could be the result of increasing TOC concentrations in Houston. For all other cases, interference calculations were considered negligible (<0.01%). It is likely that any other positive bias in considering the Houston data was due to additional chemical species. For example, bromochloroacetic acid (BCAA) was detected at significant concentrations in Houston water. It is likely that this species, as well as other halogenated species could contribute to a positive bias. As with the TTHM4 method, additional studies need to be done to consider possible interferences. Additional work needs to be done to evaluate the effects of alkalinity and pH. These possibilities need to be explored further.

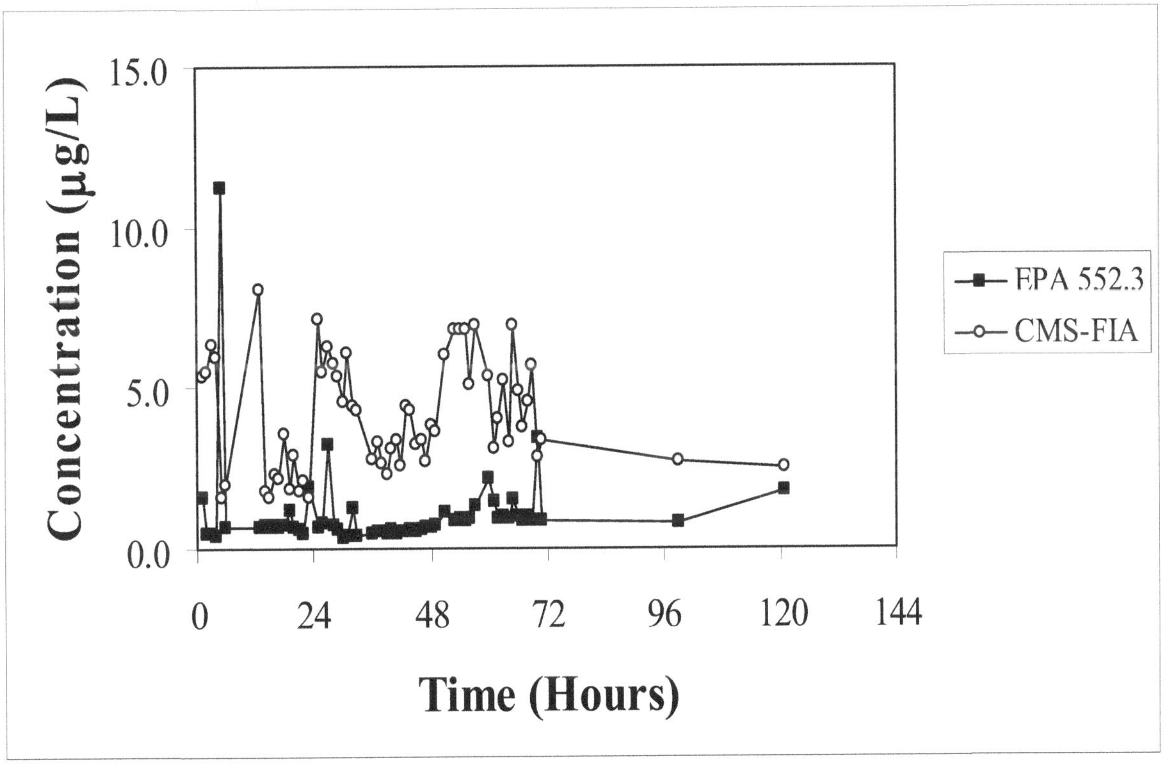

Figure 4.22 Memphis 1 concentration plot of THAA5 for EPA 552.3 and CMS-FIA NCA-FL

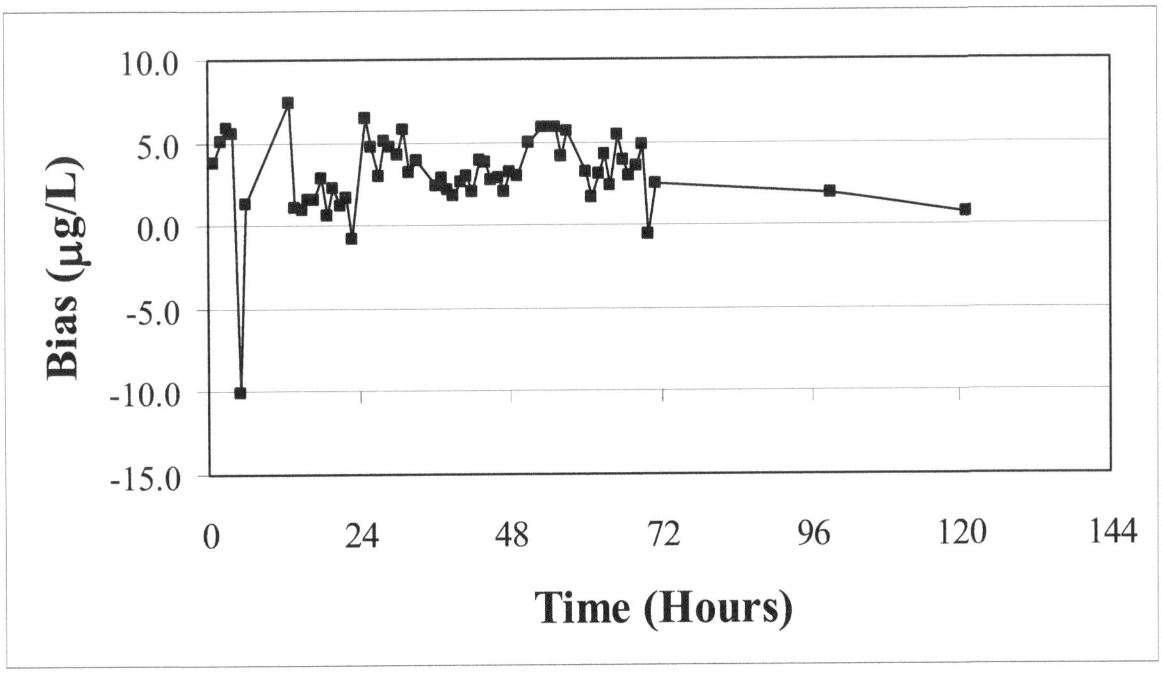

Figure 4.23 Memphis 1 bias plot of THAA5 comparing CMS-FIA NCA-FL to EPA 552.3

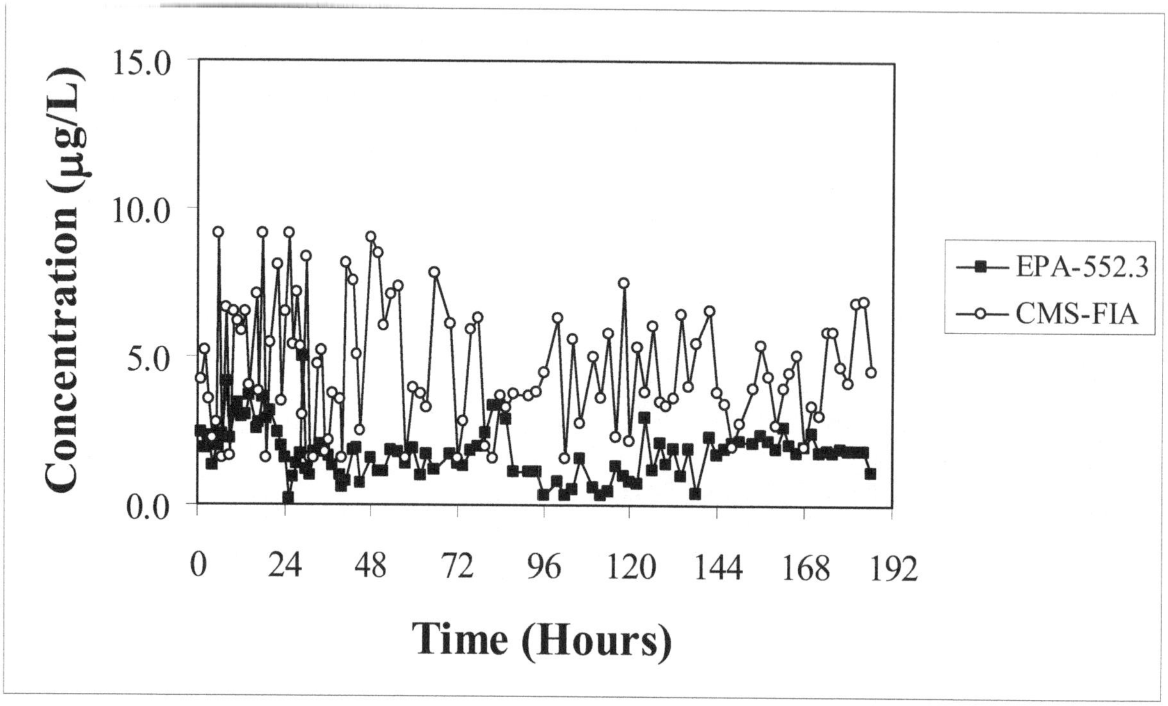

Figure 4.24 Memphis 2 concentration plot of THAA5 for EPA 552.3 and CMS-FIA NCA-FL

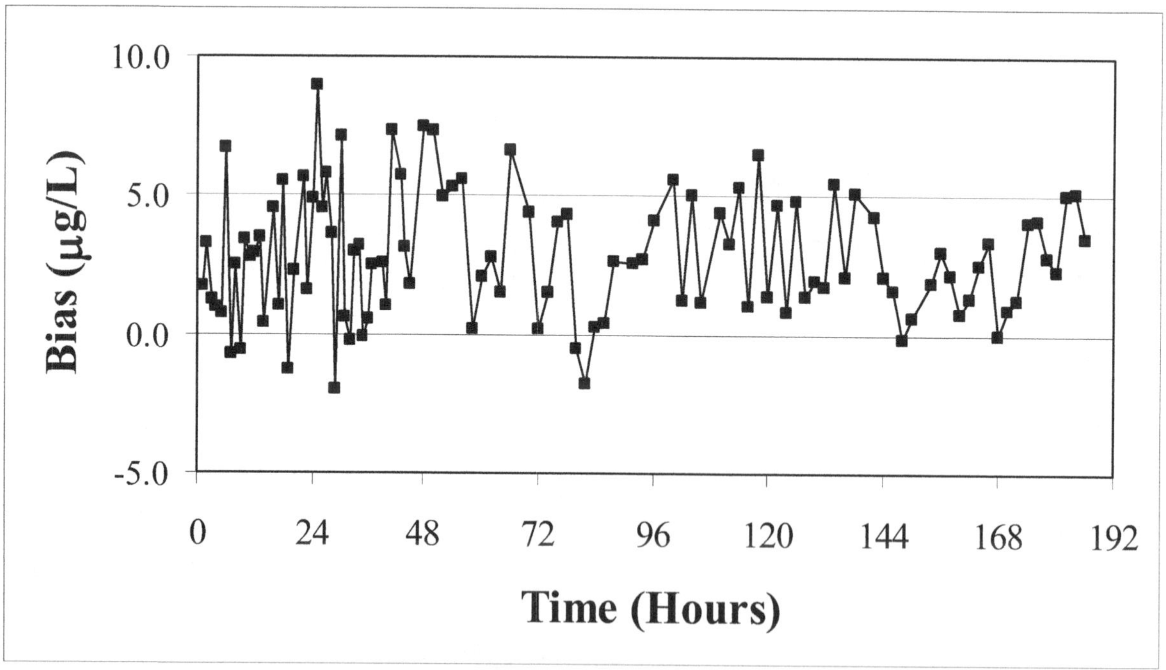

Figure 4.25 Memphis 2 bias plot of THAA5 comparing CMS-FIA NCA-FL to EPA 552.3

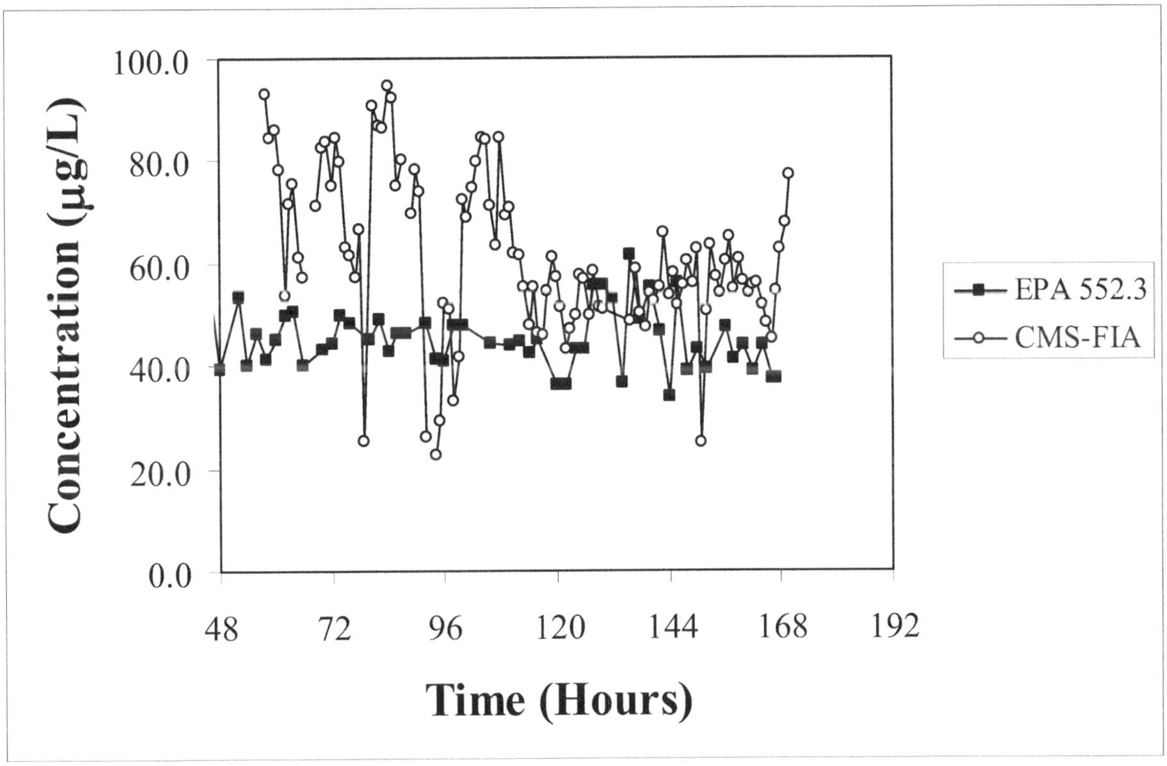

Figure 4.26 Houston concentration plot of THAA5 for EPA 552.3 and CMS-FIA NCA-FL

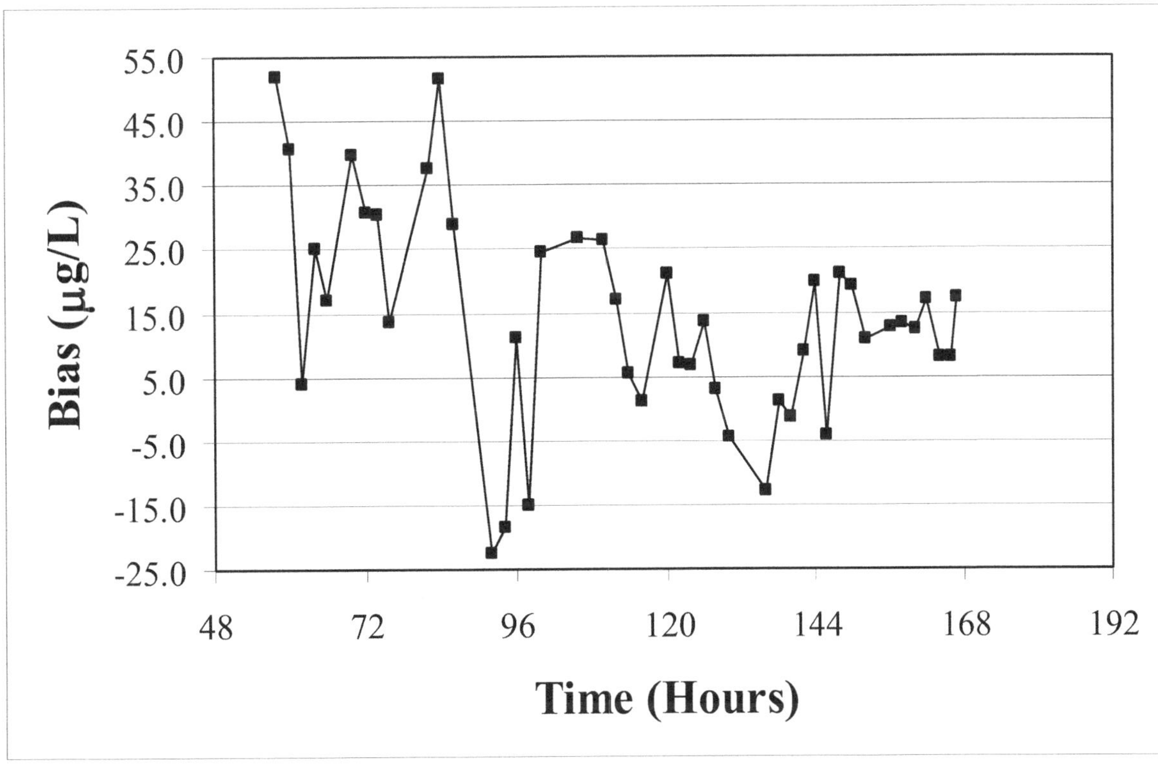

Figure 4.27 Houston bias plot of THAA5 comparing CMS-FIA NCA-FL to EPA 552.3

Results presented in Chapter 7, will demonstrate that all HAA9 species will react with NCA to form fluorescent products. It is also possible that other species such as iodated HAA species, if present might react with NCA as well. The presence of BCAA in Houston samples was confirmed using a 10.0 µg/L standard of BCAA analyzed by EPA 552.3. The concentration of BCAA was estimated using the single standard to get a rough approximation of the BCAA concentrations in samples. This estimated concentration was then added to THAA5 results to obtain the estimated "THAA6" concentration (referred to here as "BCAA Corr" results). Thus HAA6 for this research includes MCAA, DCAA, TCAA, MBAA, DBAA, and BCAA. The THAA6 concentration was then compared to the CMS-FIA THAA concentration with respect to the BCAA corrected bias, % REC_{Bias} and Student's t-test (Table 4.9).

Table 4.9
Houston monitoring study (552.3): comparison of % $REC_{Monitor}$, BCAA and % $REC_{Monitor}$ with BCAA corrected data for CMS-FIA NCA-FL method for THAA5 to USEPA Method 552.3

	CMS THAA BCAA Corr.		CMS THAA %$REC_{Monitor}$ Corr		CMS THAA BCAA+ % $REC_{Monitor}$	
Statistic/*Method*	*THAA6*	*CMS-FIA*	*THAA5*	*CMS-FIA*	*THAA6*	*CMS-FIA*
N	34	34	34	34	34	34
Avg. Conc	56.4	58.8	44.9	53.6	56.4	53.6
Std. Dev Conc.	7.8	11.4	6.4	16.2	7.8	16.2
Conc. Range	38.0 to 71.8	43.2 to 93.1	34.0 to 61.6	35.2 to 101.6	38.0 to 71.8	35.2 to 101.6
Avg. Bias	---	2.4	---	8.6	---	-2.8
Std. Dev Bias	---	15.0	---	17.2	---	19.2
Bias Range	---	-20.9 to 48.3	---	-19.9 to 60.3	---	-28.0 to 56.8
% REC_{Bias}	---	107	---	121	---	98
s.d. % REC_{Bias}	---	29	---	39	---	36
% REC_{Bias} Range	---	70 to 208	---	68 to 246	---	60 to 227
t-test* $t_{critical}$ $t_{calculated}$	---	No 2.445 0.934	---	Yes 2.445 2.920	---	No 2.445 0.844

* The $t_{critical}$ and $t_{calculated}$ values listed were calculated for the N listed using a two-tailed t-test with 98% confidence. Typically a $t_{calculated}$ value should be less than the $t_{critical}$. Yes or No means if statistically different.

Interestingly, when the concentration of BCAA is taken into account, the results of the CMS-FIA and EPA 552.3 are in statistical agreement ($t_{critical}$ = 2.245 versus $t_{calculated}$ = 0.934). This indicates that BCAA is likely a significant (influential) interferent that leads to primary difference between the two methods. Moreover, when BCAA is taken into account, the remaining estimates (average concentration, concentration range, bias etc) show better agreement. For example, the average bias drops from 14.2 ± 16.7 µg/L to 2.4 ± 15.0 µg/L. The average % REC_{Bias} changes from 88 ± 202% to 107 ± 29%. This very crude correction seems to account for much of the disagreement originally seen between the two methods.

Another possible correction (as discussed with the TTHM4 studies), is to take into account possible drift in the detector during the course of a study. When the % REC $_{Monitor}$ correction factor is applied with TTHM4, the results seemed to be in worse agreement. For the THAA5 studies, the temperature corrected data was subjected to % $REC_{Monitor}$ correction. In short, the % $REC_{Monitor}$ correction did not statistically agree with the EPA 552.3 results, ($t_{critical}$ = 2.245 versus $t_{calculated}$ = 2.920), but the average bias dropped from 14.2 ± 16.7 µg/L to 8.6 ± 17.2 µg/L and the % REC_{Bias} changed from 88 ± 202 % to 121 ± 39 %. This work is presented in more detail in Table 4.9.

When both BCAA correction and % $REC_{Monitor}$ correction are both applied, the results from CMS-FIA and EPA 552.3 remain in statistical agreement ($t_{critical}$ = 2.245 versus $t_{calculated}$ = 0.844)). With both correction factors are applied, the average bias drops from 14.2 ± 16.7 µg/L to –2.8 ± 19.2 µg/L. The average % REC_{Bias} estimate changes from 88 ± 202 % to 98 ± 36 %.

Thus, in summary, when BCAA concentrations are taken into account, the results from CMS-FIA and EPA 552.3 for the Houston study show improved results. When % $REC_{Monitor}$ correction is applied the result improve, but are still statistically different. When both corrections are applied, the results are also in statistical agreement. These corrected results are shown in Figures 4.28-4.31 and are compared to the original results.

As with the TTHM4 studies, it is possible that the THAA5 results might also be affected by other factors. For example, as discussed earlier, there are apparently significant differences in alkalinity and pH in going from Memphis 1 to Houston water. Further work needs to be done to evaluate these possible interference issues.

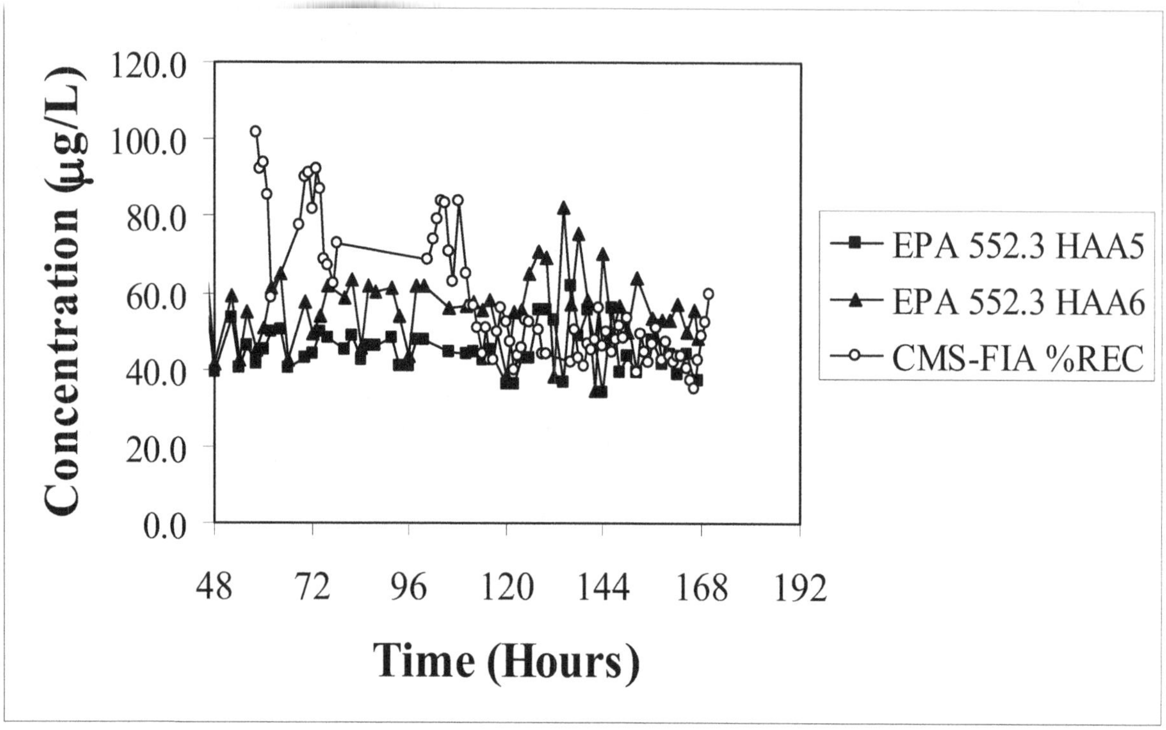

Figure 4.28 Houston concentration plot of EPA 552.3 THAA5, EPA 552.3 THAA6 and CMS-FIA NCL-FL THAA5 % $REC_{Monitor}$ correction

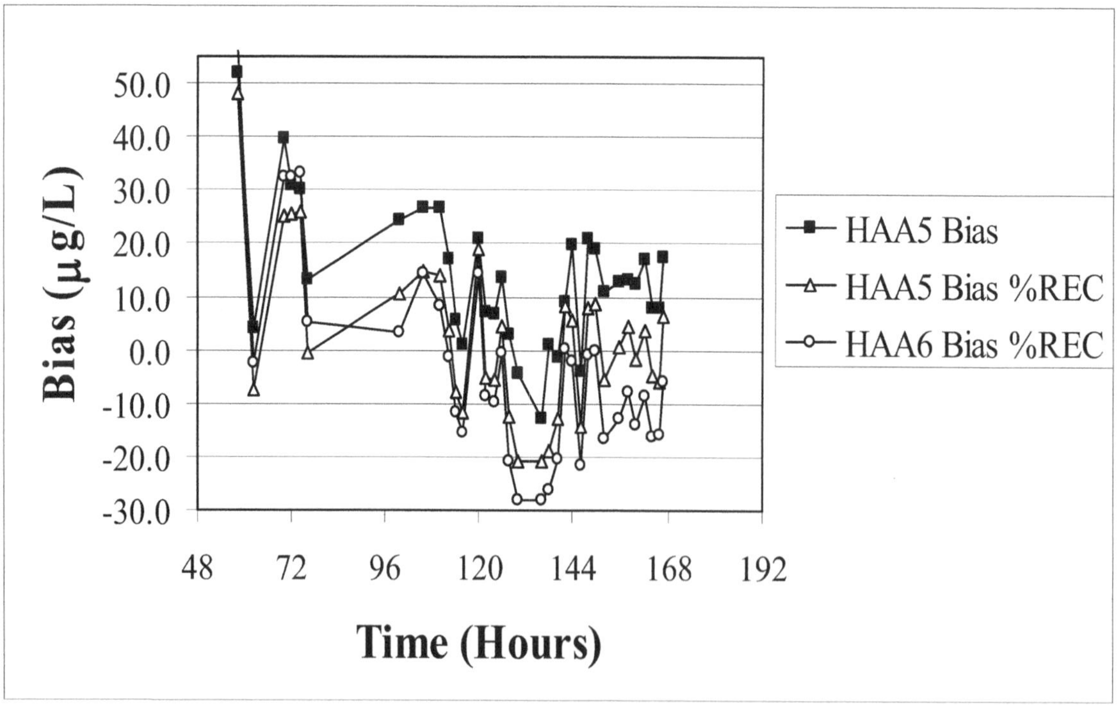

Figure 4.29 Houston bias plot comparing CMS-FIA NCA-FL to EPA 552.3 THAA5, and CMS-FIA NCA-FL % $REC_{Monitor}$ correction to EPA 552.3 THAA5 and THAA6

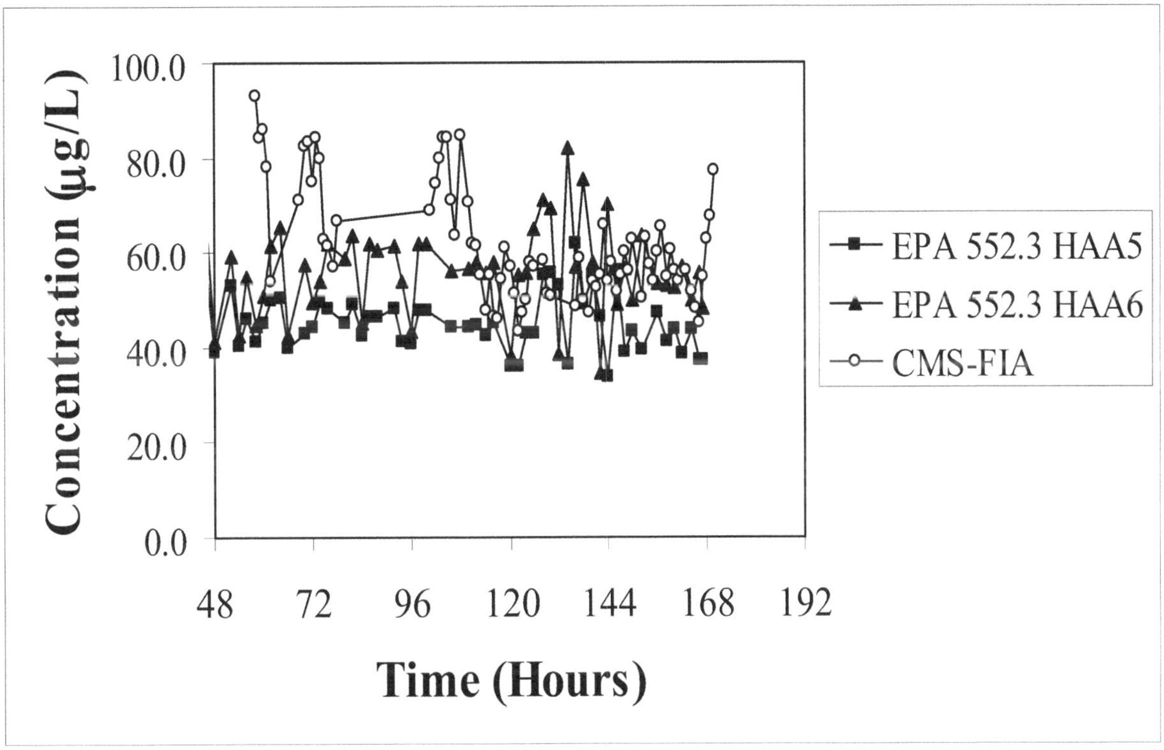

Figure 4.30. Houston concentration plot of EPA 552.3 THAA5, EPA 552.3 THAA6 and CMS-FIA NCL-FL THAA5

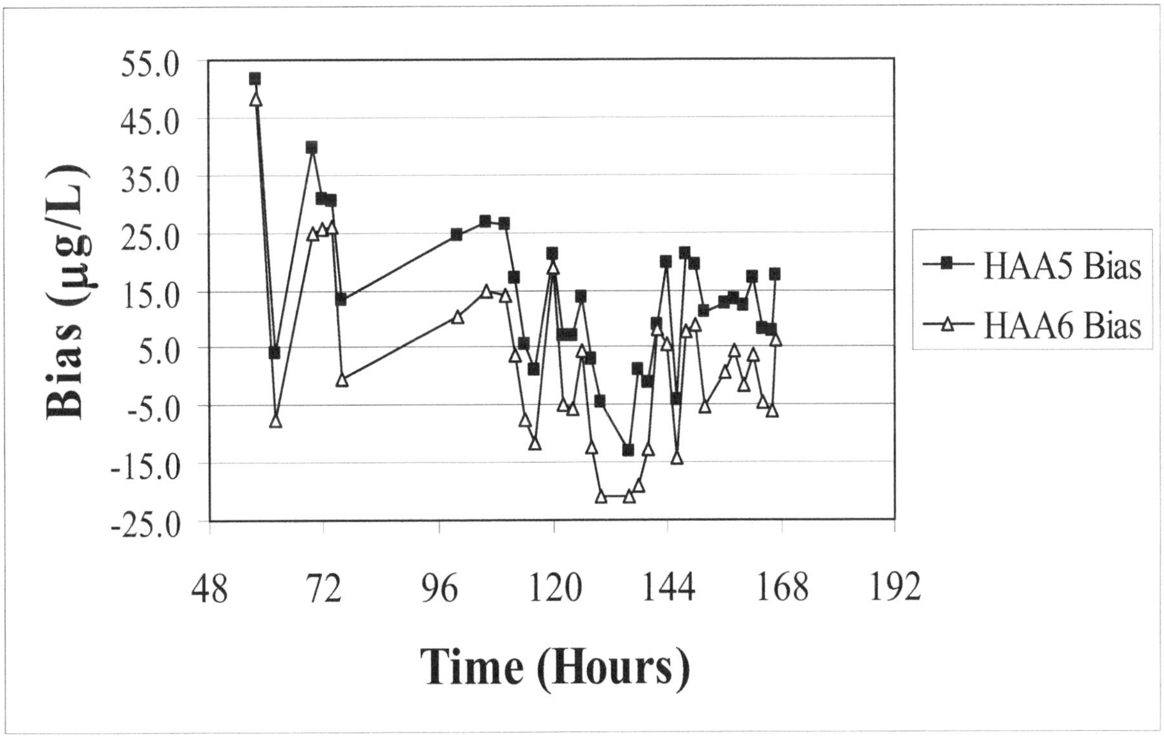

Figure 4.31. Houston bias plot comparing CMS-FIA NCA-FL THAA5 to EPA 552.3 THAA5 and EPA 552.3 THAA6

CONCLUSIONS AND RECOMMENDATIONS FOR THE CMS-FIA ANALYZER

The CMS-FIA NCA-FL method shows promise as a tool for on-line monitoring of TTHM4 and THAA5 concentrations in drinking water distribution systems. The MDL, mean % recoveries and % RSD estimates ranged from 1.1 to 1.6 μg/L, 111 to 115%, and 2.1 to 3.2%, respectively. Sample throughput was relatively high at 2 samples/hr. The analyzer was portable and user friendly. The operator skill level was relatively low as the method was largely automated.

However, it should be noted that the method appears to be especially vulnerable to interferences. This is apparent when one compares the Memphis 1 and Houston results and applies to both TTHM4 and THAA5 monitoring. With the CMS-FIA method, the importance of pre-monitoring survey studies is especially crucial, as this method does not provide a way to separate analyte from interfering species as is the case with methods that use chromatography. Of the methods that are recommended for further study (CMS-FIA, CMS-GC and PC-IC) only CMS-FIA does not use chromatography. This means that any species that can react with NCA will interfere. CMS-FIA is the simplest of the monitoring methods proposed. This work suggests that the simpler the method for monitoring, the more it is desirable to use more complex methods that can separate and speciate analyte and possible interferences in pre-monitoring surveys.

Regardless of the problems experienced by the CMS-FIA method, it is the only one of the proposed methods that can currently monitor both THM4 and HAA5 concentrations. With some minor modifications, it is likely that this method might be improved and adapted for HAA9 monitoring as well. As long as possible interferences are quantified, the CMS-FIA method can provide useful monitoring results. Additonal work needs to be done to improve the CMS-FIA method beyond these preliminary studies. Once the chemistries of interferences are firmly established, it might be possible to miniaturize the CMS-FIA method to the "lab-on-a-chip" scale. Such miniaturization might radically lower the cost and improve portability of such an analyzer.

While certainly not meant to replace the USEPA methods, the proposed methods compared favorably with USEPA Methods 502.2 and USEPA Method 552.3. Applying the method in more complex situations increased the difference compared to EPA methods. However, if all the characteristics of the CMS-FIA method are considered when applied in appropriate situations, such as when operators are trying to optimize treatment processes, these methods can provide on-line monitoring data in a simple and inexpensive manner directly from the distribution system.

Recommendation 4.1: The CMS-FIA NCA-FL method is recommended for further study, despite the problems shown with interference issues. These issues are fundamentally linked to the non speciation of this method but may be minimized with appropriate (and detailed) pre-monitoring surveys using more complex methods such as EPA 502.2, 524.4 and 552.3. All of these chromatographies might allow for possible (and likely) interferents to be detected and quantified. The CMS-FIA method should be applied to the problem of HAA9 analysis.

Recommendation 4.2: If possible, alternative variations of the CMS-FIA method should be explored. For example, higher reagent flow rates through both the CMS and FIA system should be explored to improve sample throughput (sampling rate). One might be able to inject multiple samples instead of injecting once each hour. Stopped-flow studies should be explored to increase analytical signal and thus lower (improve) MDL estimates.

Recommendation 4.3: If possible, the CMS-FIA needs to be modified to allow for on-line addition of standard to drinking water sample as it is taken into the analyzer. Such spiking or standard addition studies would allow for "on-line spiking" experiments so that the contribution of matrix effects could be better evaluated (and corrected). One particular advantage that lab-on-a-chip approaches could bring to the method is that arrays of perhaps three separate micromanifolds could be fashioned on a simple chip, each with an independent, relatively inexpensive, LED based FL detector. With such a system, it would be possible to generate standard addition using the graphical approach on the fly in real-time and thereby minimizing matrix effects.

Recommendation 4.4: More detailed interference studies need to be carried out with disinfectant species, other chlorinated compounds, as well as alkalinity and pH effects. These effects need to be firmly quantified with there contribution to TTHM4 and THAA9 analysis.

CHAPTER 5
CAPILLARY MEMBRANE SAMPLING-GAS CHROMATOGRAPHY

INTRODUCTION

In Chapter 4, a capillary membrane sampler (CMS) was used as a sampling device in FIA methods for TTHM4 and THAA5. Recently, the CMS device has been adapted for use as a sampler for GC (Brown and Emmert 2006). This CMS-GC can be used to monitor the concentrations of THM4 species in drinking water. In a broader sense, CMS-GC demonstrates the attractive characteristics of three methods previously presented in this research. As previously mentioned, the CMS device is very similar in design to the one used with the FIA studies. The membrane is the same material (silicone rubber) as used in the SCMS sampling probe, but the CMS device lacks the supporting metal body of the SCMS probe. Compared to SCMS probe, the CMS device is easier to interface to the water distribution system for on-line sampling. In fact, interfacing the CMS-GC to the distribution system is similar to OPTGC described in Chapter 2. A picture of the CMS-GC is shown in Figure 5.1. The CMS device is the white coil in the upper right hand side of the picture. The basic design of the CMS as applied to FIA is described in Chapter 4. With the CMS-GC configuration, the CMS carrier gas is connected to the silicone rubber tubing positioned inside the Tefzel® tubing. The THM4 standard or drinking water is connected to the CMS device, so that it flows around the silicone rubber tubing inside the Tefzel® tubing. The THM4 species will permeate the silicone rubber membrane into the CMS carrier gas, and be transported to a ten-port sample injection valve where a precise volume is loaded into the GC-ECD separation and detection of the THM4 species occurs. As with the SCMS-GC method presented in Chapter 3, the CMS-GC can be operated in two basic configurations: the calibration configuration and the on-line monitoring configuration.

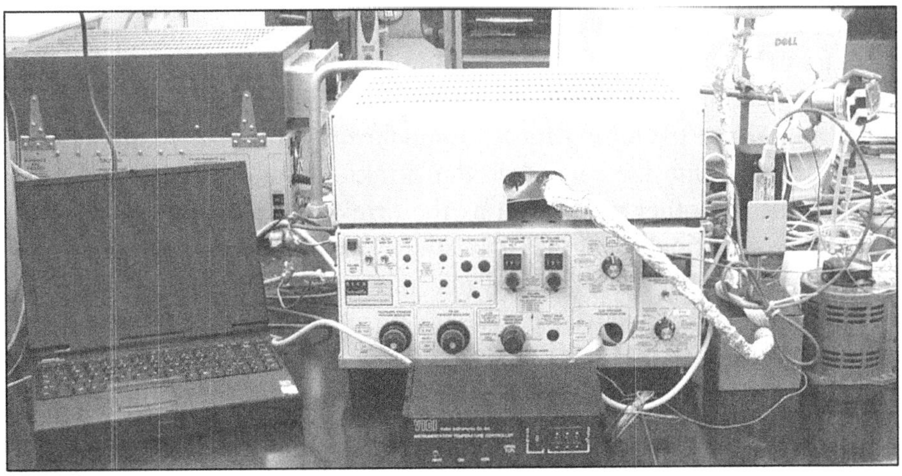

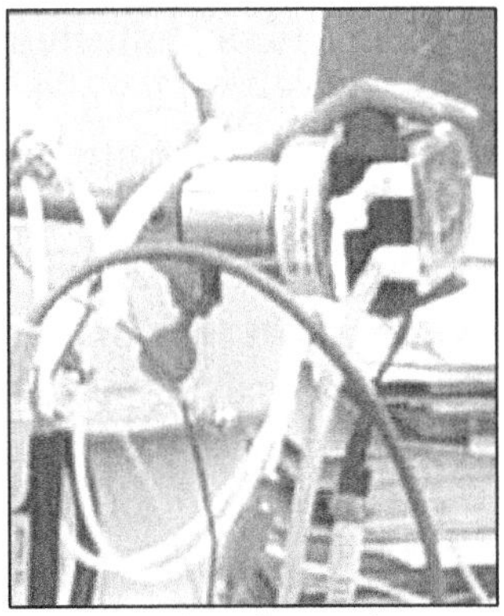

CMS Device

Figure 5.1 The CMS-GC with ECD used for monitoring THM4 concentrations and an enlarged picture of the CMS device used for CMS-GC-ECD

The objective of this study was to develop and evaluate a portable capillary membrane sampler-gas chromatograph with electron capture detection (CMS-GC-ECD) for the analysis of THM4 in drinking water. The construction of the CMS sampling device as well as the details of using it for GC is discussed. The results of optimization, MDL, accuracy and precision studies are presented. The results of side-by-side on-line monitoring studies comparing the CMS GC-ECD method to EPA Method 502.2 (1995b) and EPA Method 524.2 (1995a) are also presented.

EXPERIMENTAL

All reagents, reagent water and glassware must be very clean to minimize sample contamination. All water used was distilled and then deionized by a Millipore Mill-Q Plus/Mill-RO® 10 Plus Water Purification System to obtain a resistivity of at least 18 MΩcm and TOC concentrations of 10 μg/L or less. Glassware and sample containers were meticulously cleaned with concentrated detergent soaked for 24 hours and then rinsed with distilled water three times followed by reagent grade water several times. All chemicals were HPLC reagent grade or better and were purchased from Fisher Scientific Inc. Stock standard solutions for the THM4 were purchased from Aldrich Chemical Company. Different aliquots (1.0 to 2.0 μL) of each stock standard solution were added with a separate airtight 10 μL syringe (Hamilton-107) to a single 10 mL volumetric flask containing methanol to make the primary dilution standard solution. This primary dilution standard solution was then added in desired amounts (0.2 to 6.6 μL) to a 100 mL volumetric flask containing reagent grade water to produce different concentrations of the calibration standard solutions or check standard solutions.

Capillary Membrane Sampler and Sample Introduction

The CMS device used in this study was constructed in-lab (Figure 5.2), but similar devices are commercially available (Global FIA, Fox Harbor, WA) and could be ordered. The lab-built CMS device consisted of a length of silicone rubber tubing membrane. The silicone rubber (Global FIA part # SRT-020-032) has the following dimensions: i.d. 0.51 mm, o.d. 0.94 mm wall thickness 0.21 mm. There is similar tubing is available from VWR Scientific (part # - 60985-702 manufacturer Dow Corning). The silicone rubber tubing was inserted into a length of 1/8" Tefzel® tubing. Two SS tees (VICI-TCEF211T) with appropriate fittings and ferrules were attached at opposite ends of the 1/8" tubing. The ends of the silicone tubing were carefully pulled through each tee and attached to 1/16" FEP tubing by SS hypodermic tubing segments (Global FIA- SS-TUB) and 1/16" SS fittings. One tee served as the drinking water sample outlet and CMS carrier gas inlet section while the other tee served as the drinking water sample inlet and the CMS carrier gas outlet section. Additional lengths of 1/16" FEP tubing were also required to connect the CMS to a peristaltic pump or drinking water tap and waste outlet. The flow controller (Brooks, Model 8744) for the CMS carrier gas was connected to the CMS using a 1/8" to 1/16" internal reducing union (VICI-EZRU21).

The drinking water to be sampled is either pumped through the Tefzel® tubing using a peristaltic pump or flows through the device when connected to the water distribution system. The THM4 species in the drinking water sample permeated the silicone membrane's outer wall and moved through to the inner wall. The CMS carrier gas (N_2) flowed through the interior of the silicone membrane. Compounds that have permeated through to the membrane's inner wall were transported through a desiccator vessel that used Nafion® tubing (NT-060, Global FIA) and desiccant to remove excess water vapor (Duty 2000). Once through the desiccator vessel, the analytes flowed to an electrically actuated ten-port/two position sample injection valve (VICI-ET4C10UWE, HVEB, and ITC10399). The ten-port/two position injection valve was fitted with two 100 μL sample loops (VICI- SL100UW) for alternate loading of sample. Valve automation was accomplished using a relay closure on the GC. When the valve was in Position A, the mixture of carrier gas and analyte molecules filled sample loop 1 and flowed to waste. During the same time, sample loop 2 was constantly being purged with GC carrier gas. When the valve was switched to the Position B, the volume of sample present in sample loop 1 was loaded onto the GC also column (Figure 5.3). The transfer lines connecting the CMS to the desiccator vessel, to the valve configuration and valve configuration to the GC injection port were both heated to 65°C (Duty 2000). A GC with a ^{63}Ni ECD was used to separate and detect the THM4 species. This GC contained a built in data acquisition board. The data acquisition software used was PeakSimple (SRI Instruments Inc.) version 3.20, which also allowed for full automation of the injection valve and GC oven.

There are two sampling configurations for the CMS-GC method. The calibration configuration (Figure 5.3A) was used for calibration studies, optimization studies, and analyzing grab samples. The on-line configuration (Figure 5.3B) differs only in that a liquid flow controller connected to the typical water tap was substituted for the peristaltic pump. The sample was flowed through the CMS at approximately the same sample flow rate used for calibrating the system by using the process line's sufficient pressure.

Optimized CMS-GC-ECD Operating Conditions

Separation and detection of the THM4 species was accomplished using a Buck Scientific 610 GC equipped with a VICI Ni-63 ECD. A Valco VB-5 (15 m × 0.53 mm i.d x 1.00 μm f.t.) capillary column was used for separation. The GC carrier gas was 99.999% He at 5.00 mL/min and the detector gas was 99.999% N_2 at 40.0 mL/min. The detector temperature was 300°C. The transfer lines and valve enclosure were kept at 65°C. The CMS length was 120 cm. Sampling time was 15.0 min at 2.6 mL/min where the CMS volume was ~2.34 mL. The oven temperature programming was as follows: the initial temperature was held at 65°C for 1 min, increased to 225°C at ~30°C/min max rate (set as 60°C/min in software), and held at 225°C for 3.33 min. The total chromatographic run time was 7.00 min. The total time for analysis was 15.00 min.

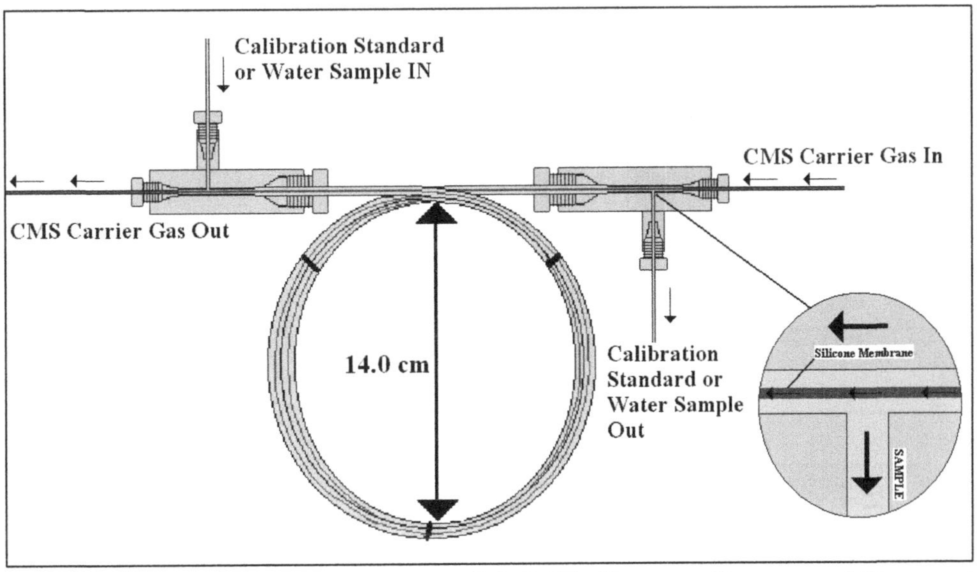

Figure 5.2 Diagram of CMS as adapted for GC-ECD (Source: Brown and Emmert 2006)

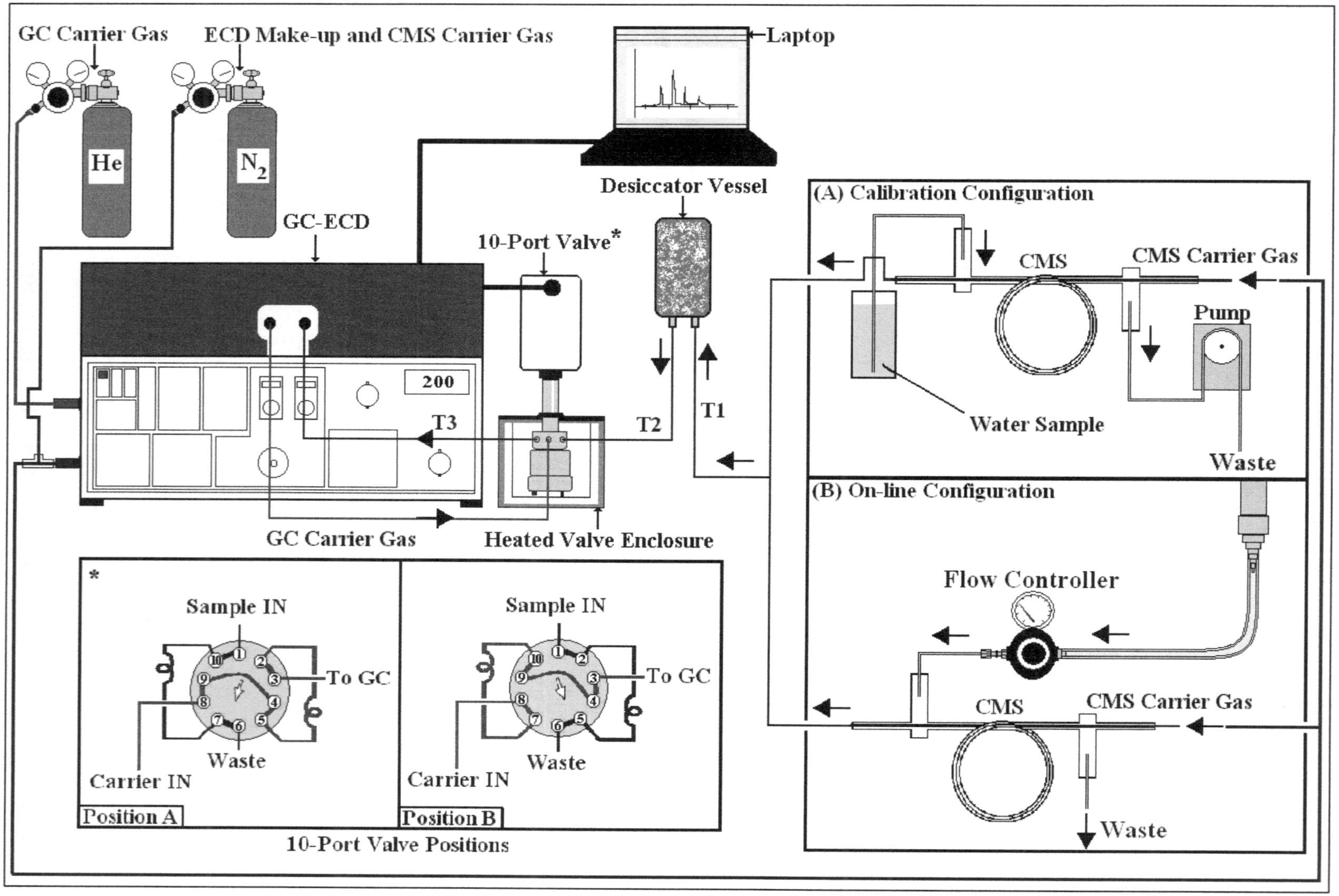

Figure 5.3 Schematic diagram of the portable CMS-GC-ECD operated in either (A) Calibration configuration or the (B) On-line Configuration, *Ten-port valve: Position one is the top port, and the ports are numbered consecutively going clockwise (Source: Brown and Emmert 2006)

With Memphis drinking water (groundwater source), the sample matrix was relatively clean. With this in mind, the researchers aimed for chromatography separation times approaching those of "fast GC analysis" with a 5 m column, 15 minutes. For Memphis drinking water, the researchers could accomplish these shorter separation times. In more complex systems, the chromatography may need to be modified.

RESULTS AND DISCUSSION

CMS Optimization Studies

Several parameters were varied to optimize and evaluate the performance of the CMS-GC-ECD analyzer. These included the length of the CMS sampling device, the flow rate of the sample and CMS carrier gas, the sampling time before injection, the time required before injection of another sample (carry-over studies), the effect on analytical signal of sample temperature, ionic strength, and TOC concentrations.

Effect of CMS Length on Analytical Signal

The optimum length of the silicone rubber tubing used in the CMS device for GC was investigated (Figure 5.4). CMS devices were constructed with silicone rubber membrane lengths of 15 cm, 30 cm, 60 cm and 120 cm. The concentrations of THM4 species were held constant and the analytical signal was recorded as a function of membrane length using the different CMS devices. For $CHCl_3$ and $CHBrCl_2$, the analytical signal continually increased as the length of the membrane increased. $CHBr_2Cl$ and $CHBr_3$ appeared to reach a maximum at ~30 cm membrane length and then slightly declined. The increase in the analytical signal for $CHCl_3$ and $CHBrCl_2$ was greater than the loss of analytical signal for $CHBr_2Cl$ and $CHBr_3$, thus the decision was made to use the 120 cm long CMS device. This CMS device with a 120 cm capillary membrane was used for the remaining optimization studies. At the present time it is not possible to prepare a CMS device with a membrane significantly longer than 120 cm, simply put it has not been possible to thread a membrane longer than 120 cm through the Tefzel® tubing without tearing it or having difficulty pulling it through. These results suggest that it is possible to increase the analytical signal by making device with longer membrane (larger surface area). However, it is likely that sample carry-over from previous injections might take longer to reduce to acceptable levels. Carry-over issues are described in more detail shortly.

CMS Carrier Gas and Sample Flow Rate Studies

Optimization studies were also conducted for the CMS carrier gas flow rate. These studies were similar to previous studies done with the SCMS probe (Duty 2000; Liao 2001; Emmert et al. 2004a). The optimum flow rate for the CMS carrier gas was similarly observed to be 6.0 mL/min. Unlike the SCMS probe, the CMS device requires that the flow rate of the sample throughout the device is optimized. This was done by evaluating the analytical signal as the sample flow rate was changed over the range of 0.4 to 3.2 mL/min. A standard that contained THM4 species was analyzed at each flow rate. The results of these studies showed that the analytical signal increased as the flow rate of the sample through the CMS device was increased (Figure 5.5). Larger flow rates should lead to improved efficiency. Practically, a flow rate between 2.2 to 2.6 mL/min was chosen as the optimum flow rate of sample throughout the CMS device. This choice was made because it can be easily maintained during routine operation using the available peristaltic pump. A higher flow rate (> 3.2 mL/min) might provide even better performance, but equipment was not available to consistently achieve these higher flow rates for routine analysis.

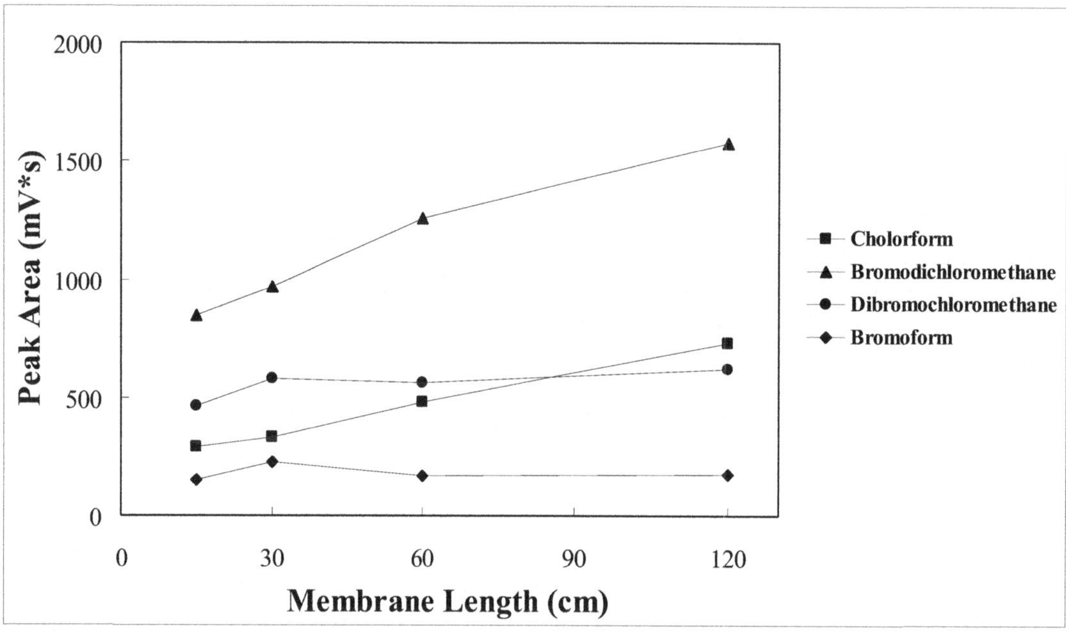

Figure 5.4 Results of membrane length optimization studies for the CMS adapted for GC-ECD (Source: Brown and Emmert 2006)

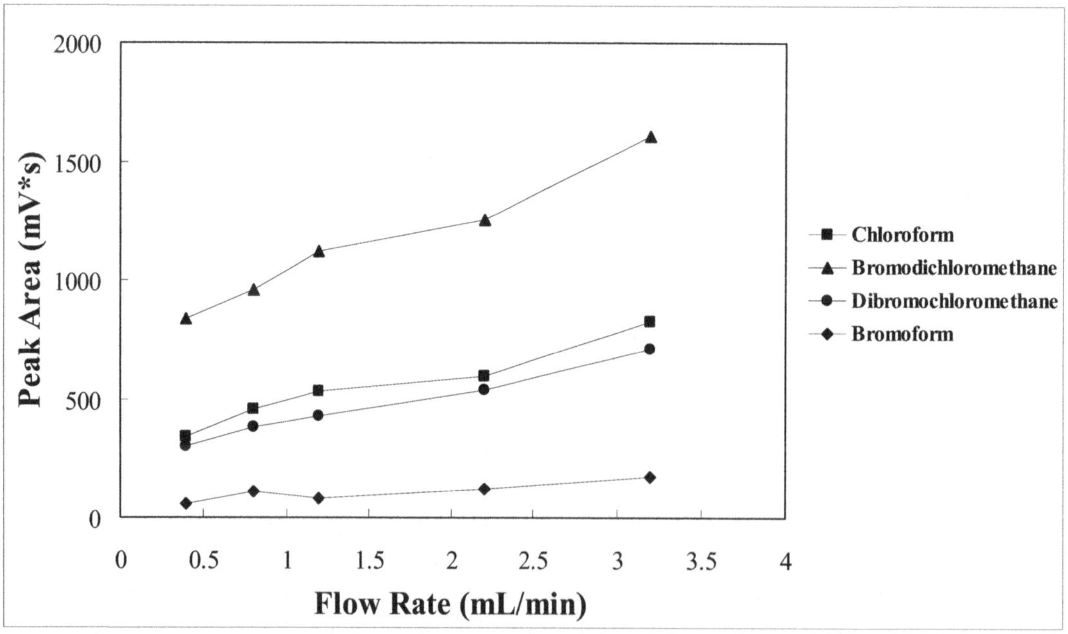

Figure 5.5 Optimization of sample flow rate through the CMS device adapted for GC-ECD (Source: Brown and Emmert 2006)

Sampling Time Optimization

With the optimum sample flow rate selected, the optimum sampling time for the CMS device was investigated. The sampling time is the time that a sample is flowed through the CMS before loading a sample onto the GC column. Different sampling times were evaluated over the range of 1.0 to 45.0 minutes. A standard containing the same concentration of THM4 species was analyzed at each different sampling time. The results from this study showed that for three THM4 species, the analytical signal reached a plateau at a sampling time of 15 minutes. In an effort to balance optimization with the desired goals of minimized carry-over and an analysis rate of one sample per hour, a sampling time of 15 minutes was chosen as the optimum sampling time. However, for $CHBrCl_2$ the analytical signal should be higher if the sampling time was increased. Ultimately, the optimum sampling time of 15 minutes was chosen as the best overall sampling time.

Evaluating Sample Carry-over Effects

Using the previously described optimum conditions, a study was conducted to evaluate possible carry-over issues from one analysis to the next (Figure 5.6). A THM4 standard was analyzed once followed by several "blank" analyses of reagent water. Blank injections were done every 15 minutes for 60 minutes. After 60 minutes, there was no detectable carry-over with any of the THM4 species. This time is termed the recovery time of the CMS device. Based on these

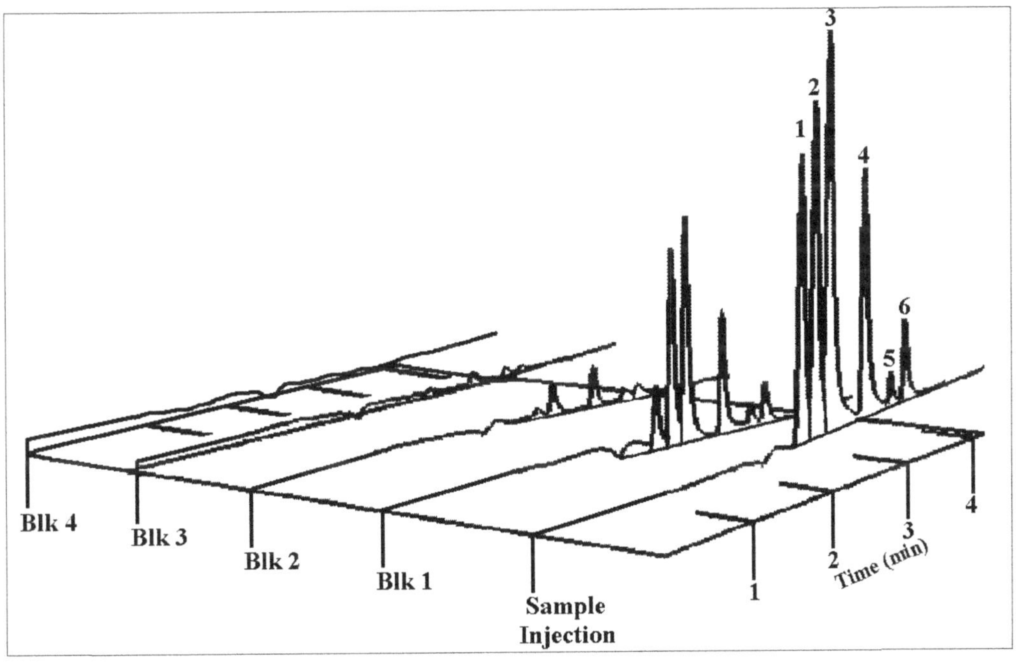

Figure 5.6 CMS-GC-ECD results from carry-over experiments for THM4 (and two possible internal standards). After four blank analyses, no detectable THM4 species are observed. Components: (1) $CHCl_3$ 9.8 μg/L, (2) CCl_4 3.8 μg/L (3) $CHBrCl_2$ 20.8 μg/L, (4) $CHBr_2Cl$ 19.4 μg/L, (5) $C_3H_6Br_2$ 11.6 μg/L, (6) $CHBr_3$ 19.1 μg/L (Source: Brown and Emmert 2006)

studies, it was determined that one sample could be measured per hour without carry-over effects. As mentioned earlier, if the membrane length was increased beyond the 120 cm length of the present device, the recovery time would be expected to increase. For the current studies, to achieve the goal of one sample per hour, the decision was made to use membrane length of 120 cm and to inject one sample each hour.

Effect of Ionic Strength on Analytical Signal

Standard solutions of THM4 species were prepared in solutions of differing ionic strength using sodium sulfate. Increasing the ionic strength is expected to increase the analytical signal via the "salting out" effect (Emmert et al. 2004a). The results showed that over the ionic strength range of ~ 0 (reagent water) to 3M, the average effects of ionic strength of solutions on the signal for $CHCl_3$, $CHCl_2Br$, $CHClBr_2$ and $CHBr_3$ was 10.7, 20.4, 24.9, and 31.6% signal increase per 1M increase in ionic strength, respectively.

Initially, this seems to be a large difference. However, in typical drinking water, this difference may not be as large as it appears. For example, consider a 5 μg/L standard solution of $CHCl_3$: from a typical calibration curve, the predicted peak area for this standard would be 407.8 mV*s. The ionic strength calculated for Memphis drinking water is ~0.002M. Using the correction factor for $CHCl_3$ given above, the expected increase in peak area would be about 0.214%. This means that instead of giving a peak area if 407.8 mV*s, the standard in Memphis drinking water would give a peak area of 408.7 mV*s. A 5 μg/L standard in Memphis drinking water would give a calculated concentration of 5.02 μg/L, less than a 0.5% difference. Similar calculations were done with the remaining THM4 species with like results. At typical conditions, the effect of changes in ionic strength is expected to be relatively low from sample to sample. In practice, analysts can evaluate the magnitude by spiking known concentrations into a raw drinking water matrix and observe the relative analytical signals in laboratory water and drinking water samples. These results are in very good agreement with other researchers who used similar sampling devices (Duty 2000; Emmert et.al 2004a; Liao 2001; Geme et al. 2005).

Effect of Water Temperature on Analytical Signal

The effect of sample temperature on the analytical signal was measured using the optimized configuration for the CMS-GC-ECD. Certainly, changes in the temperature of the sample will change the analytical signal due the equilibrium vapor pressure increase. Similar effects have been observed with the SCMS sampling probe. For quantitative analysis, changes in the temperature of the sample, if they occur during extended on-line monitoring studies must be taken into account. In other words, a correction factor needs to be determined to correct the results if such changes in temperature occur. A TTHM4 standard was measured repeatedly at different temperatures. The results showed that the temperature coefficients in the temperature range of ~0°C to 35°C for $CHCl_3$, $CHCl_2Br$, $CHClBr_2$ and $CHBr_3$ were 1.9%, 3.0%, 3.0%, and 3.9% analytical signal increase per °C, respectively. The results were similar to those of the SCMS probe which showed a ~ 1.6 to 4% increase in analytical signal per ^oC increase in temperature over the temperature range ~ 0 to 35^oC (Emmert et al. 2004a and Geme et al. 2005). This means if the sample temperature differs from calibration conditions over this range, a correction factor can be applied to on-line monitoring studies, which was the case in this research. Blanchard and Hardy, (1984), Emmert et al. (2004a), and Geme et al. (2005) all used similar devices and correction factors for monitoring THM4 species and other volatile organic compounds. These results support that real-time temperature monitoring is crucial to correct for the effects that temperature has on the permeation across the silicone rubber membrane.

Total Organic Carbon Interference Study

Experiments were designed to better understand the possible interferences of TOC using anhydrous potassium biphthalate (the standard used for TOC analysis) to simulate the effect of changes in TOC concentrations. A stock standard solution of potassium biphthalate was prepared (1.00 mL of solution = 1.00 mg) and added in varying amounts to reagent water in 100 mL volumetric flasks and diluted to the mark to prepare different TOC solutions in the concentration range of 0 to 20 mg/L (APHA, AWWA, WEF 1992). THM4 species were then added to each solution and the analytical signal measured. The results of these experiments showed that the simulated TOC concentrations have little effect on the analytical signal of THM4 species, even at the exaggerated concentrations of 20 mg/L. Since most drinking water has TOC concentrations in the 2 to 5 mg/L range, the effect is expected to be even less pronounced. This is a preliminary study in a simulated TOC matrix. Complete evaluation of TOC effects need to be done in real world studies. However, the preliminary results seem to indicate that changes in TOC concentrations are not a large problem. Studies mentioned later in this Chapter investigated the CMS-GC method in real world samples with relatively high TOC concentrations.

Linearity, MDL, Accuracy, and Precision Studies

Studies were done to evaluate the linear range of the CMS-GC-ECD method for measuring THM4 species in water. MDL, accuracy, and precision estimates were measured for the CMS-GC-ECD method. The MDL, mean percent (%) recovery and percent relative standard deviations were calculated based on EPA recommendations (Glaser and Forest 1981; USEPA 1996a; and USEPA 1996b). These results were compared to other analytical methods for measuring THM4 concentrations in drinking water.

Linearity Study

The calibration curves appeared linear over a limited range. The r^2 values for three of the four THM4 species were all greater than 0.990, (the exception being $CHBr_3$ with $r^2>0.987$). However, these studies were conducted over a limited concentration range (~0.6 to 20.0 μg/L). To investigate the linearity of the analytes further, higher concentrations of THM4 standards were analyzed up to ~ 37.0 μg/L of each THM (Figure 5.7). The results indicate that the linear range is rather limited for three of the THM4 with 20 to 25.0 μg/L being the highest concentration within an acceptable linear range ($r^2>0.980$). $CHBr_3$ has an acceptable linear range ($r^2>0.980$) up to ~35.0 μg/L.

Method Detection Limit, Accuracy, and Precision Studies

The CMS-GC-ECD and optimized method were used to conduct MDL, accuracy and precision studies of the THM4 species (Table 5.1). Each calibration curve was based on five calibration standards (Figure 5.8). One sample injection was conducted every 60 minutes to minimize carry-over issues. The complete MDL, accuracy and precision study with seven check standards required approximately 12 hours. Ideally, one should analyze check standards with concentrations near the MDL, at least within a factor of 2.5 to 5 times the MDL. Some of the results of MDL studies fall within this range. The check standards here lie within a range from 2 to 19 times the predicted MDL, depending on the particular species. The CMS-GC-ECD does not provide MDL values better than USEPA Method 502.2 (1995b) or USEPA Method 524.2 (1995a). The MDL values ranged from 0.26 to 0.54 μg/L. All of the predicted MDL values are within the acceptable range as stated in the Phase One Report (Emmert et al. 2004b) (target range of 0.5 to 5 μg/L). The GC-ECD used in these studies was an older unit. It is likely that a new instrument (with newer ECD detector and electronic flow control) might provide even better results. The mean % recoveries, which give an estimate of accuracy for the CMS-GC ranged from 100 to 125%. To estimate precision, % RSD was calculated giving values ranging from 3.5 to 8.6%.

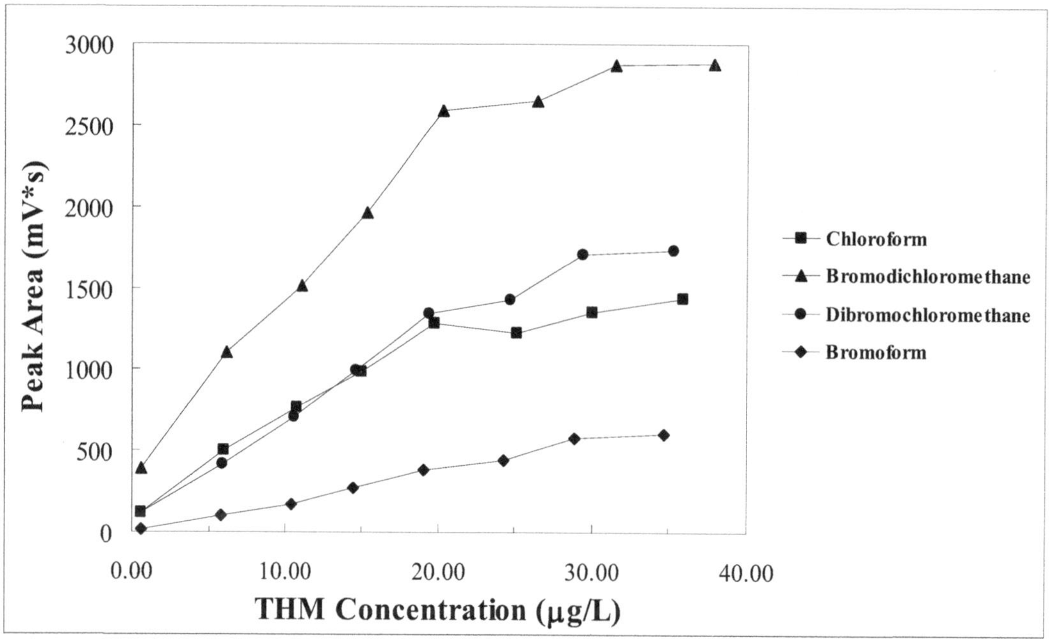

Figure 5.7 Results of linearity studies of THM4 species for CMS-GC-ECD (Source: Brown and Emmert 2006)

Table 5.1
Results from MDL, accuracy and precision studies for the CMS-GC-ECD

THM Species	MDL (μg/L)†	Mean % Recovery	% RSD	r^2	Conc. Range (μg/L)*
$CHCl_3$	0.3	100	4.5	0.997	1-20
$CHBrCl_2$	0.5	104	8.6	0.995	1-20
$CHBr_2Cl$	0.3	125	3.5	0.997	1-20
$CHBr_3$	0.3	124	4.3	0.987	1-20

* Each standard was analyzed once and five standards were analyzed to construct the calibration curves used to make MDL, accuracy and precision estimates.

† Check standard analyzed seven consecutive times at approximately 1.8 μg/L for each THM4 species.

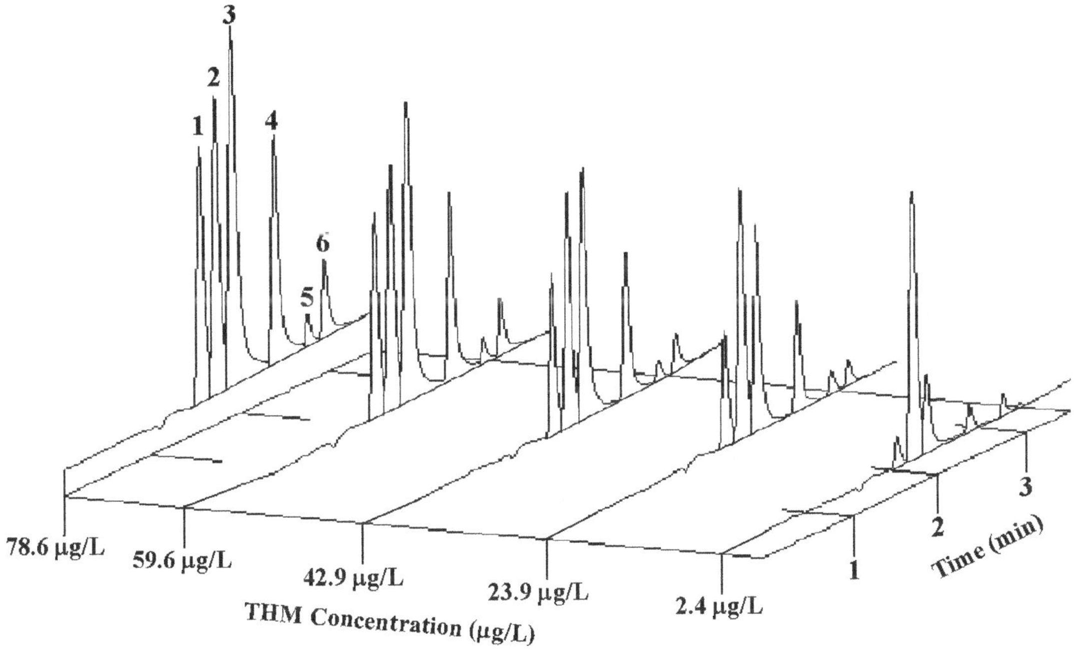

Figure 5.8 Typical chromatograms for THM4 calibration standards (and two possible internal standards) for the CMS-GC-ECD. All components are resolved in less than 4 minutes in Memphis drinking water. Components: (1) $CHCl_3$ 19.8 µg/L, (2) CCl_4 3.8 µg/L (3) $CHBrCl_2$ 20.8 µg/L, (4) $CHBr_2Cl$ 19.4 µg/L, (5) $C_3H_6Br_2$ 11.6 µg/L, (6) $CHBr_3$ 19.1 µg/L. The THM Concentration (x-axis) given for each chromatogram in the diagram is the total concentration of THM4 (Source: Brown and Emmert 2006)

Memphis 1 On-Line Monitoring Study using CMS-GC-ECD

The CMS-GC-ECD monitoring study was conducted from Dec. 16 to 23, 2004 at the University of Memphis sampling site (Memphis 1). The CMS-GC-ECD was interfaced to the drinking water distribution system as shown in the Appendix A. The concentration of each THM4 species was monitored at a sample rate of 1 sample/ hour (though the analyzer was typically taken off-line every 12 hours to run a check standard). The results for each THM4 species were summed and are reported as TTHM4. The results of these studies along with details used in the analysis of the data are presented in Table 5.2.

A plot of the concentration of $CHCl_3$ versus the time is shown in Figure 5.9. For 129 direct comparisons, the concentration ranges from 0.1 - 0.4 µg/L for the EPA 502.2 and from 0.1 - 0.4 µg/L for the CMS-GC method. The average concentration for EPA Method 502.2 is 0.2 ± 0.1 µg/L and for the CMS-GC are 0.2 ± 0.1 µg/L. Figure 5.10 shows a plot of the bias as a function of time for the CMS-GC analyzer compared directly to EPA 502.2. The bias (absolute error) ranged from –0.1 - 0.3 µg/L (averaging 0.1 ± 0.1 µg/L) with % REC_{Bias} values ranged from 68 to 385 % (averaging 177 ± 74 %). The results for $CHBrCl_2$ (Figure 5.11) show 129 direct comparisons and give concentration ranged from 0.1 to 0.5 µg/L (EPA 502.2) and 0.1 to 0.4 µg/L (CMS-GC). The average concentration was 0.3 ± 0.1 µg/L for both EPA 502.2 and CMS-

GC. The bias (Figure 5.12) ranged from –0.2 - 0.3 µg/L (averaging 0.0 ± 0.1 µg/L) and % REC_{Bias} values ranged from 30 to 365 % (averaging 122 ± 53 %). For $CHBr_2Cl$ (Figure 5.13), the concentrations ranged from 0.9 - 1.7 µg/L (EPA 502.2) and 0.1 to 2.0 µg/L (CMS-GC). The average concentration was 1.3 ± 0.2 µg/L (EPA 502.2) and 1.0 ± 0.4 µg/L (CMS-GC). The bias (Figure 5.14) ranged from –1.4 - 0.8 µg/L (averaging –0.2 ± 0.5 µg/L) and % REC_{Bias} values ranged from 5 - 169 % (averaging 86 ± 36 %). For $CHBr_3$ (Figure 5.15), the concentrations ranged from 1.2 - 2.0 µg/L (EPA 502.2) and 0.2 to 1.4 µg/L (CMS-GC). The average concentration was 1.5 ± 0.2 µg/L (EPA 502.2) and 0.7 ± 0.3 µg/L (CMS-GC). The bias (Figure 5.16) ranged from –1.6 - 0.1 µg/L (averaging –0.9 ± 0.4 µg/L) and % REC_{Bias} ranged from 13 - 108 % (averaging 45 ± 21 %). When the data is tested statistically using a t-test (as described earlier with the CMS-FIA methods, the individual concentrations were shown to be different. When the results of the individual species were summed to obtain TTHM4, then the result of the two methods did agree. Overall, the results for TTHM4 (Figure 5.17) are as follows: The concentrations ranged from 1.1 - 4.0 µg/L (EPA 502.2) and 0.2 - 4.5 µg/L (CMS-GC). The average concentration was 2.3 ± 0.9 µg/L (EPA 502.2) and 2.4 ± 0.8 µg/L (CMS-GC). The bias (Figure 5.18) ranged from –3.1 to 2.8 µg/L (averaging 0.1 ± 1.1 µg/L) and % REC_{Bias} ranged from 4.9 - 278 % (averaging 118% ± 55 %). These results are summarized in Table 5.2.

Table 5.2
Memphis 1 monitoring study: monitoring THM4 concentrations using CMS-GC-ECD

	$CHCl_3$		$CHBrCl_2$		$CHBr_2Cl$		$CHBr_3$		TTHM4	
Statistic*Method*	*EPA*	*CMSGC*	*EPA*	*CMSGC*	*EPA*	*CMSGC*	*EPA*	*CMSGC*	*EPA*	*CMSGC*
N	129	129	129	129	129	129	129	129	129	129
Avg. Conc.	0.2	0.2	0.3	0.3	1.3	1.0	1.5	0.7	2.3	2.4
Std. Dev Conc.	0.1	0.1	0.1	0.1	0.2	0.4	0.2	0.3	0.9	0.8
Conc. Range	0.1 to 0.4	0.1 to 0.4	0.1 to 0.5	0.1 to 0.4	0.9 to 1.7	0.1 to 2.0	1.2 to 2.0	0.2 to 1.4	1.1 to 4.0	0.2 to 4.5
Avg. Bias	---	0.1	---	0.0	---	-0.2	---	-0.9	---	0.1
Std. Dev Bias	---	0.1	---	0.1	---	0.5	---	0.4	---	1.1
Bias Range	---	-0.1 to 0.3	---	-0.2 to 0.3	---	-1.4 to 0.8	---	-1.6 to 0.1	---	-3.1 to 2.8
% REC_{Bias}	---	177	---	122	---	86	---	45	---	118
s.d. % REC_{Bias}	---	74	---	53	---	36	---	20	---	55
% REC_{Bias} Range	---	70 to 385	---	30 to 365	---	5 to 169	---	13 to 108	---	5 to 278
t-test* $t_{critical}$ $t_{calculated}$	---	Yes 2.356 12.700	---	Yes 2.356 3.270	---	Yes 2.356 5.496	---	Yes 2.356 16.331	---	No 2.356 1.025

* The $t_{critical}$ and $t_{calculated}$ values listed were calculated for the N listed using a two-tailed t-test with 98% confidence. Typically a $t_{calculated}$ value should be less than the $t_{critical}$. Yes or No as indicated means if statistically different.

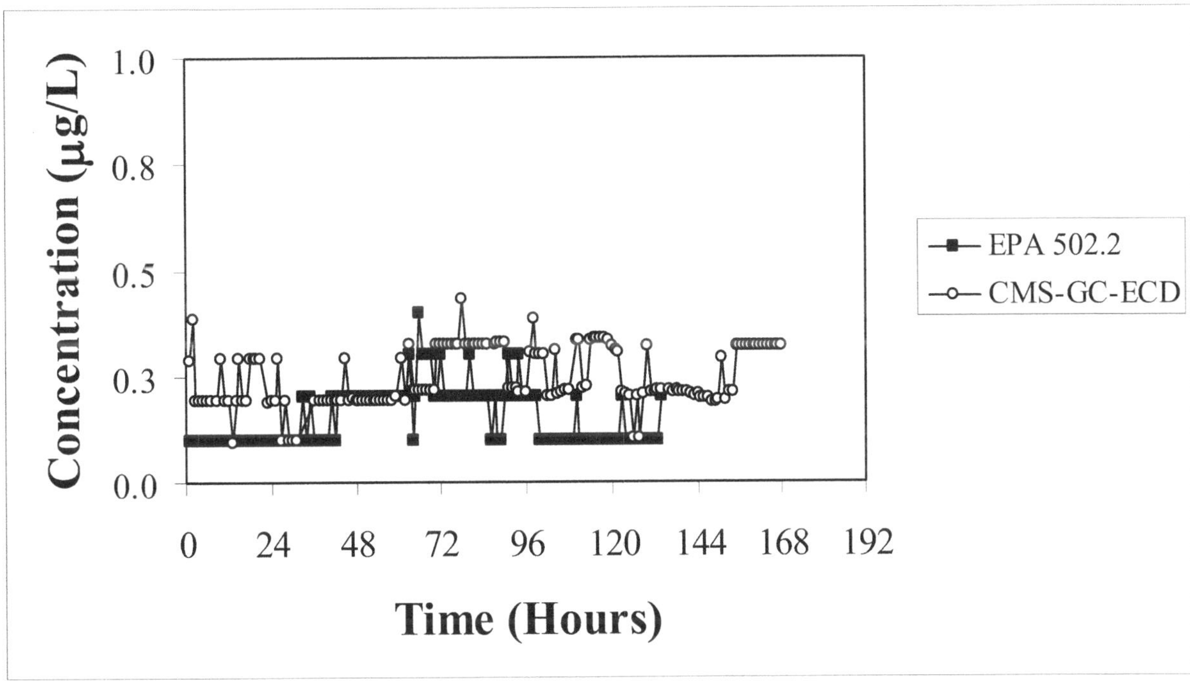

Figure 5.9 Memphis 1 concentration plot of $CHCl_3$ for EPA 502.2 and CMS-GC-ECD

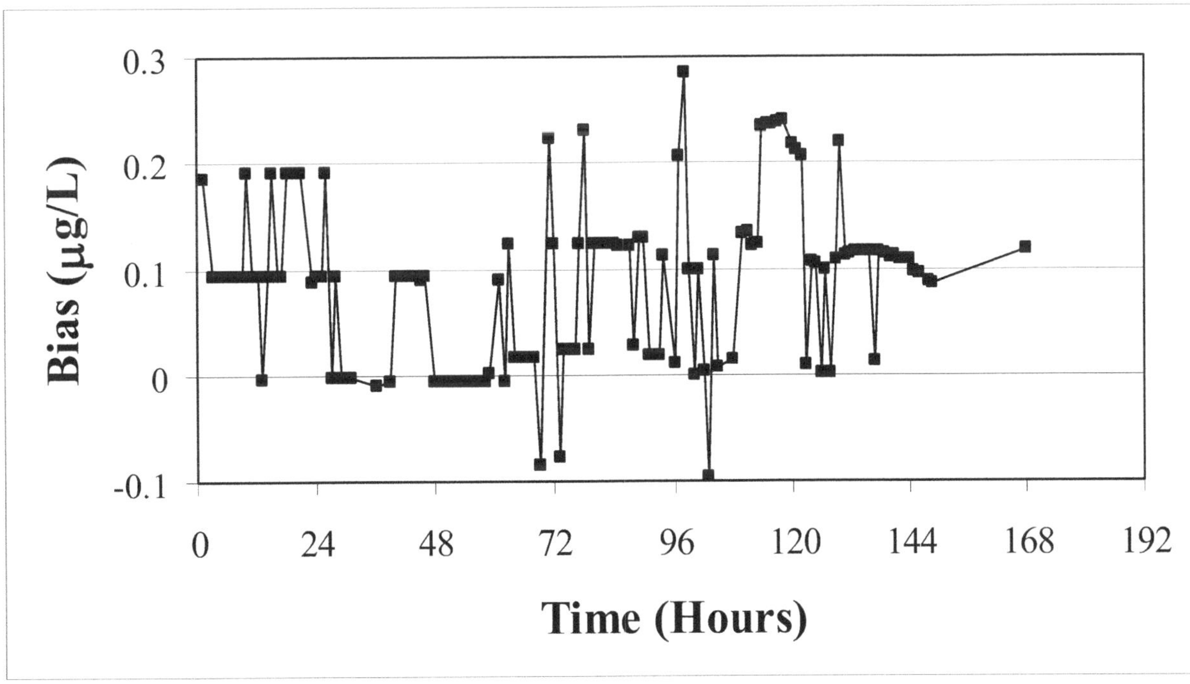

Figure 5.10 Memphis 1 bias plot of $CHCl_3$ comparing CMS-GC-ECD to EPA 502.2

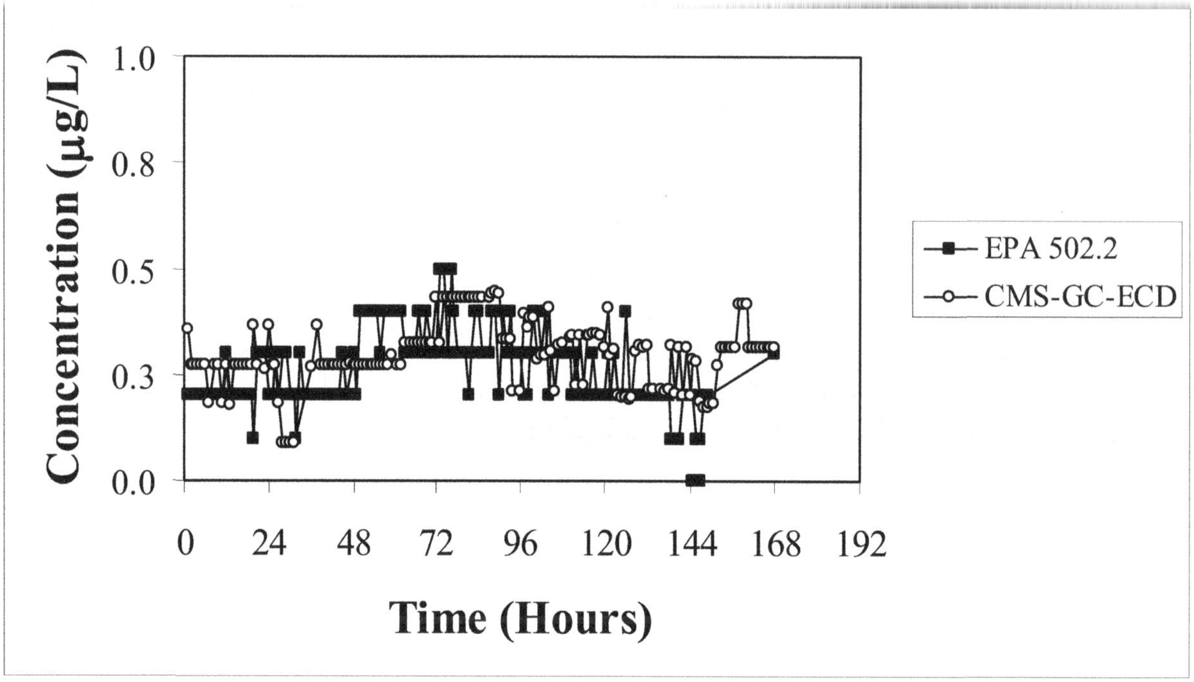

Figure 5.11 Memphis 1 concentration plot of $CHBrCl_2$ for EPA 502.2 and CMS-GC-ECD

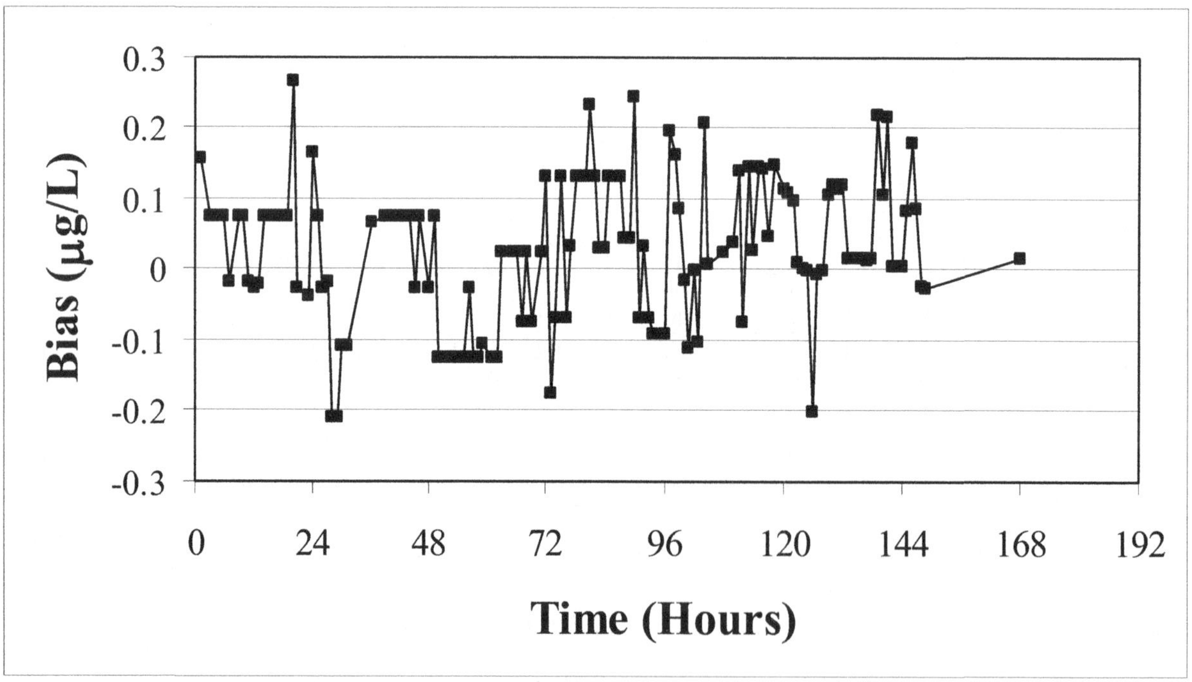

Figure 5.12 Memphis 1 bias plot of $CHBrCl_2$ comparing CMS-GC-ECD to EPA 502.2

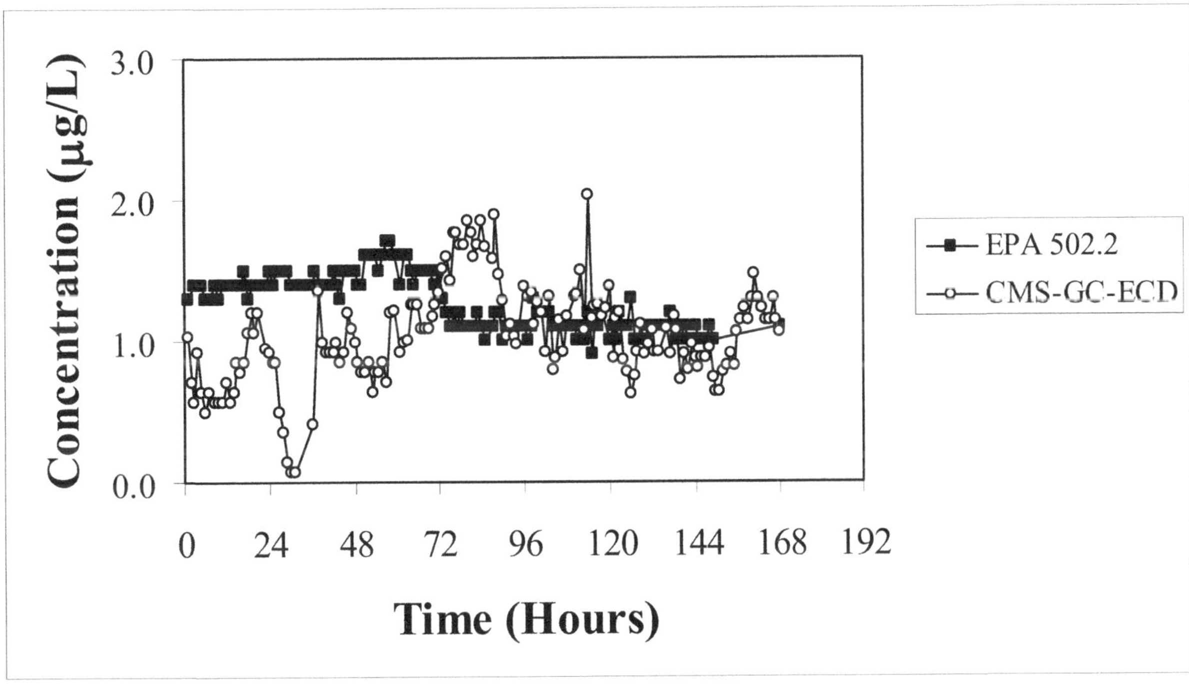

Figure 5.13 Memphis 1 concentration plot of $CHBr_2Cl$ for EPA 502.2 and CMS-GC-ECD

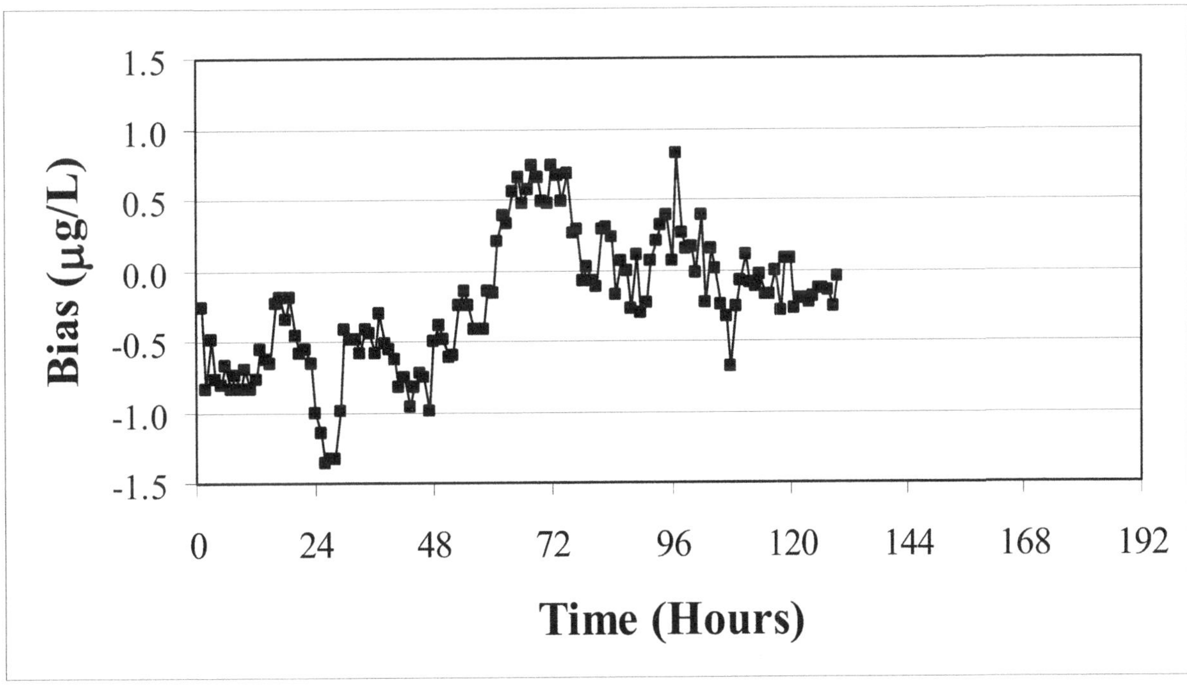

Figure 5.14 Memphis 1 bias plot of $CHBr_2Cl$ comparing CMS-GC-ECD to EPA 502.2

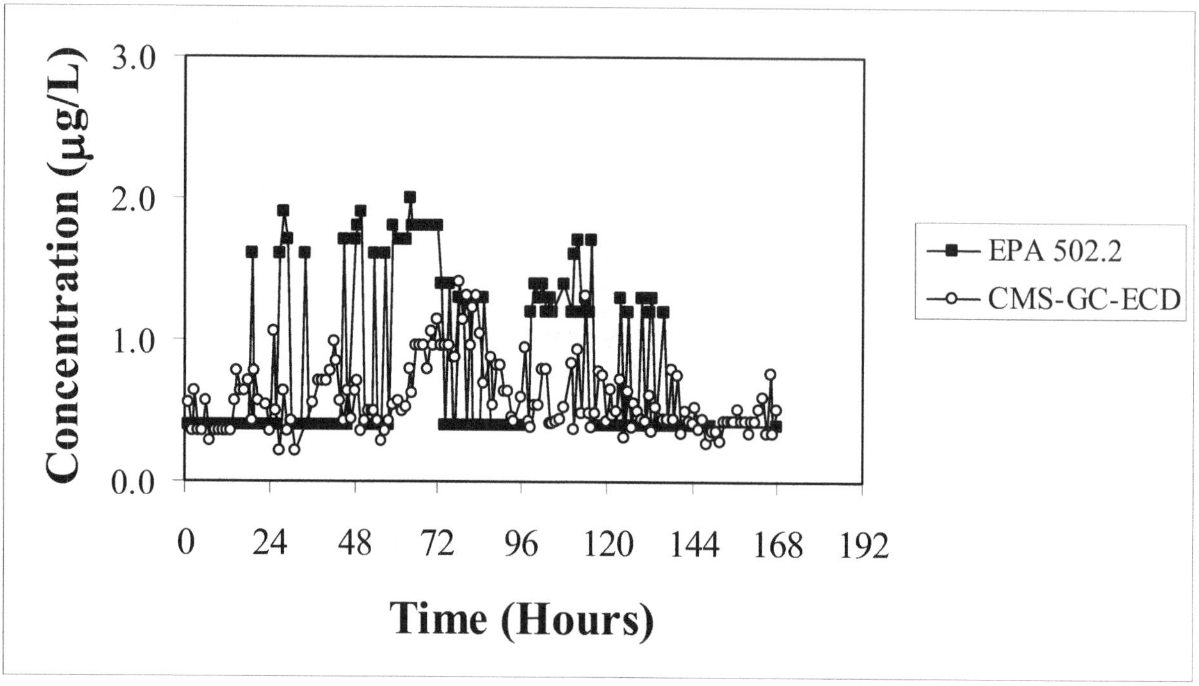

Figure 5.15 Memphis 1 concentration plot of $CHBr_3$ for EPA 502.2 and CMS-GC-ECD

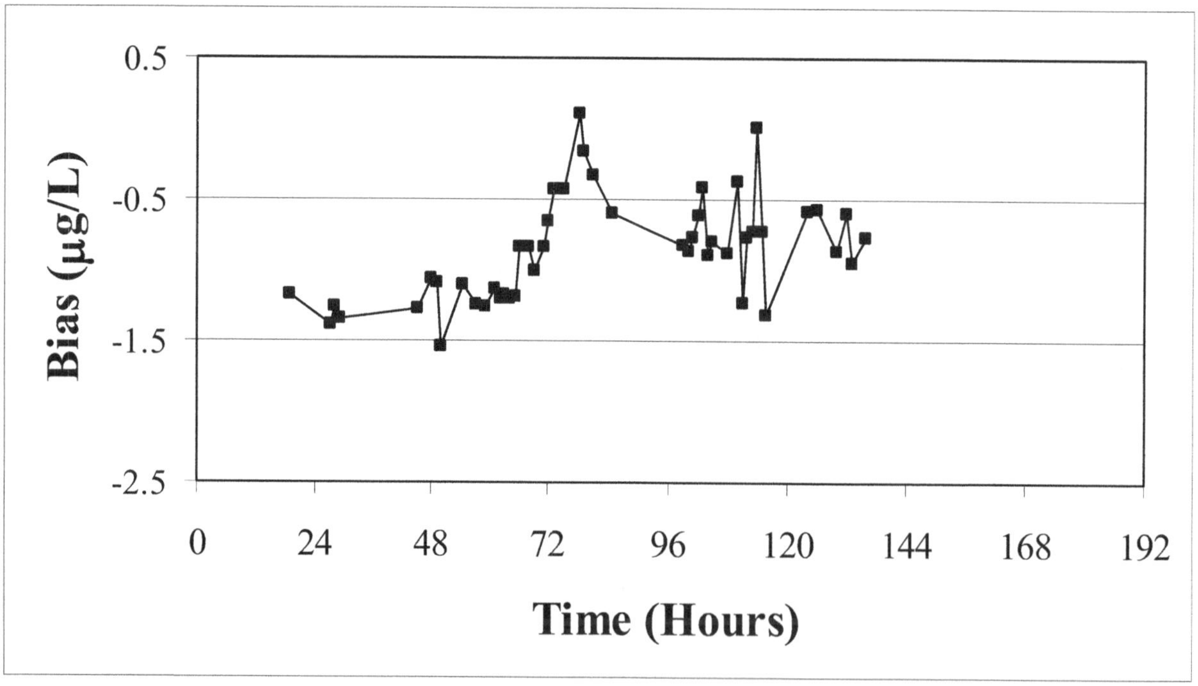

Figure 5.16 Memphis 1 bias plot of $CHBr_3$ comparing CMS-GC-ECD to EPA 502.2

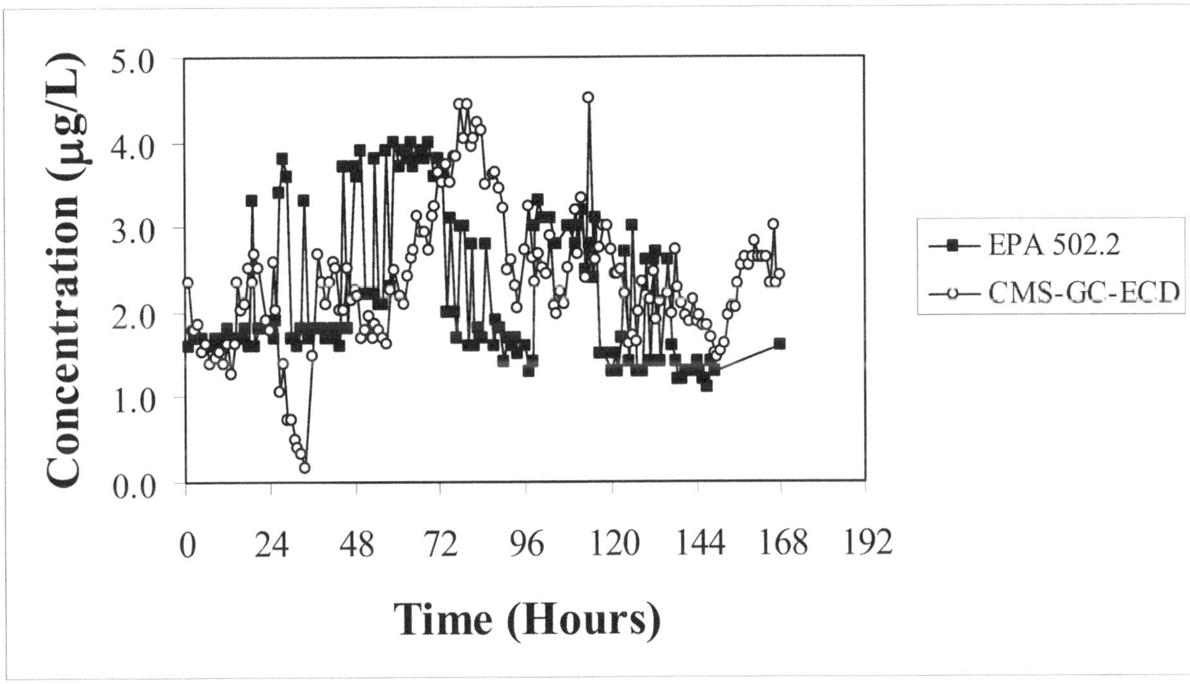

Figure 5.17 Memphis 1 concentration plot of TTHM4 for EPA 502.2 and CMS-GC-ECD

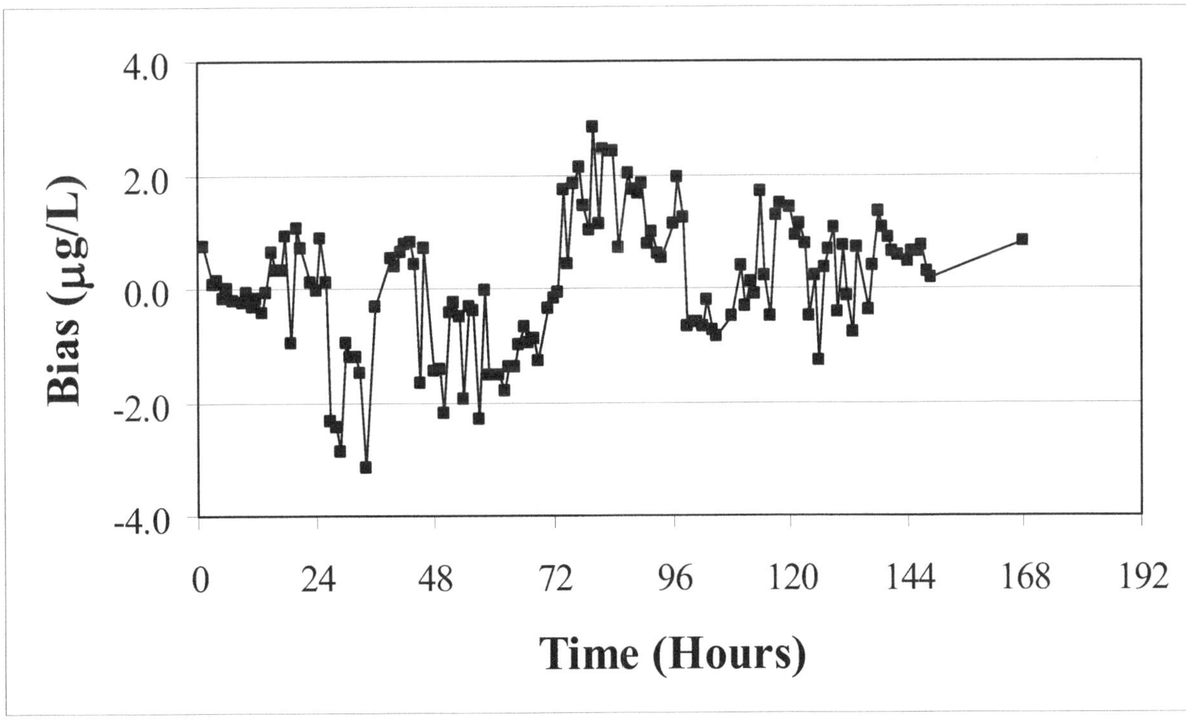

Figure 5.18 Memphis 1 bias plot of TTHM4 comparing CMS-GC-ECD to EPA 502.2

Memphis 2 On-Line Monitoring Study using CMS-GC-ECD

The Memphis 2 monitoring study was conducted from May 19 to 27, 2005. The CMS-GC-ECD was interfaced to the drinking water distribution system as shown in the Appendix A. The concentrations of each THM4 species was monitored at a sample rate of ~1 sample/ hour (though the analyzer was typically taken off-line every 12 hours to run a check standard). The results for each THM4 species were summed and are reported as TTHM4. The results of these studies along with details used in the analysis of the data are presented in Table 5.3.

Table 5.3
Memphis 2 monitoring study: monitoring THM4 concentrations using CMS-GC-ECD

	$CHCl_3$		$CHBrCl_2$		$CHBr_2Cl$		$CHBr_3$		TTHM4	
Statistic/*Method*	*EPA*	*CMSGC*	*EPA*	*CMSGC*	*EPA*	*CMSGC*	*EPA*	*CMSGC*	*EPA*	*CMSGC*
N	91	91	91	91	91	91	91	91	91	91
Avg. Conc.	2.0	1.5	3.4	3.3	5.1	4.5	2.9	1.3	13.3	10.5
Std. Dev Conc.	0.5	0.4	0.7	0.7	1.5	0.7	1.2	0.4	3.6	2.0
Conc. Range	1.1 to 3.3	0.6 to 2.5	2.2 to 5.6	2.0 to 4.6	3.3 to 11.7	3.1 to 6.8	1.4 to 7.1	0.5 to 2.2	8.5 to 26.2	6.4 to 15.2
Avg. Bias	---	–0.4	---	–0.1	---	–0.6	---	–1.7	---	–2.8
Std. Dev Bias	---	0.5	---	0.8	---	1.6	---	1.2	---	3.7
Bias Range	---	–2.1 to 0.7	---	–2.1 to 1.5	---	–7.7 to 1.5	---	–6.1 to 0.1	---	–16.3 to 2.4
% REC_{Bias}	---	80	---	100	---	94	---	49	---	83
s.d. % REC_{Bias}	---	23	---	24	---	24	---	20	---	21
% REC_{Bias} Range		29 to 140	---	54 to 147	---	34 to 144	---	12 to 107	---	37 to 127
t-test* $t_{critical}$ $t_{calculated}$	---	Yes 2.368 8.600	---	No 2.368 1.098	---	Yes 2.368 3.638	---	Yes 2.368 13.025	---	Yes 2.368 7.305

* The $t_{critical}$ and $t_{calculated}$ values listed were calculated for the N listed using a two-tailed t-test with 98% confidence. Typically a $t_{calculated}$ value should be less than the $t_{critical}$. Yes or No as indicated means if statistically different.

A plot of the concentration of $CHCl_3$ versus the time is shown in Figure 5.19. For 91 direct comparisons, the concentration ranged from 1.1 - 3.3 µg/L for EPA 502.2 from 0.6 - 2.5 µg/L for the CMS-GC method. The average concentration for EPA 502.2 was 2.0 ± 0.5 µg/L and for the CMS-GC was 1.5 ± 0.4 µg/L. Figure 5.20 shows a plot of the bias as a function of time for the CMS-GC analyzer compared to EPA 502.2. The bias (absolute error) ranged from –2.1 - 0.7 µg/L (averaging –0.4 ± 0.5 µg/L) with % REC_{Bias} values ranged from 29 - 140 % (averaging 80 ± 23 %). The results for $CHBrCl_2$ (Figure 5.21) show 91 direct comparisons gives concentration ranged from 2.2 - 5.6 μg/L (EPA 502.2) and 2.0 - 4.6 μg/L (CMS-GC). The average concentration was 3.4 ± 0.7 μg/L (EPA 502.2) and 3.3 ± 0.7 μg/L (CMS-GC). The bias (Figure 5.22) ranged from –2.1 - 1.5 μg/L (averaging –0.1 ± 0.8 μg/L) and % REC_{Bias} ranged from 54 - 147 % (averaging 100 ± 24 %). For $CHBr_2Cl$ (Figure 5.23), the concentrations ranged from 3.3 - 11.7 μg/L (EPA 502.2) and 3.1 - 6.8 μg/L (CMS-GC). The average concentration was 5.1 ± 1.5 μg/L (EPA 502.2) and 4.5 ± 0.7 μg/L (CMS-GC). The bias (Figure 5.24) ranged from –7.7 – 1.5 μg/L (averaging –0.6 ± 1.6 μg/L) with % REC_{Bias} values ranged from 34 - 144 % (averaging 94 ± 24 %). For $CHBr_3$ (Figure 5.25), the concentrations ranged from 1.4 - 7.1 μg/L (EPA 502.2) and 0.5 - 2.2 μg/L (CMS-GC). The average concentration was 2.9 ± 1.2 μg/L (EPA 502.2) and 1.3 ± 0.4 μg/L (CMS-GC). The bias (Figure 5.26) ranged from –6.1 - 0.1 μg/L (averaging –1.7 ± 1.2 μg/L) and % REC_{Bias} values ranged from 12 – 107 % (averaging 49 ± 20 %). Overall, the results for TTHM4 (Figure 5.27) concentrations ranged from 8.5 to 26.2 μg/L (EPA 502.2) and 6.4 to 15.2 μg/L (CMS-GC). The average concentration was 13.3 ± 3.6 μg/L (EPA 502.2) and 10.5 ± 2.0 μg/L (CMS-GC). The bias (Figure 5.28) ranged from –16.3 to 2.4 μg/L (averaging –2.8 ± 3.7 μg/L) and % REC_{Bias} values ranged from 37 - 127 % (averaging 83 ± 21 %).

When the two methods were compared statistically, the $t_{critical}$ for the 91 comparisons was 2.368. For $CHCl_3$ the $t_{calculated}$ is 8.600 and implies that there is a statistical difference between the methods for $CHCl_3$. The $t_{calculated}$ for $CHBrCl_2$ is 1.098 compared to 2.368 for $t_{critical}$,, indicating that there is not a significant difference between the two methods. For $CHBr_2Cl$ and $CHBr_3$, the $t_{calculated}$ values were 3.638 and 13.025 respectively, indicating that both methods are statistically different for these species. For TTHM4, the $t_{calculated}$ was 7.305 compared to $t_{critical}$ 2.368. This indicates that the results of the two methods are not statistically different.

Qualitatively, the results of the CMS-GC-ECD monitoring studies are very similar to those of the Memphis 1 study. For each THM4 species, the bias is always slightly negative, but relatively small (ranging from –2 - –0.1 g/L), depending on the THM4 species. Again, the trend is for the magnitude of the bias be smaller for $CHCl_3$ and $CHBrCl_2$ (–0.4 ± 0.5 μg/L and –0.1 ± 0.8 μg/L, respectively) and larger for $CHBr_2Cl$ and $CHBr_3$ (–0.6 ± 1.6 and –1.7 ± 1.2 μg/L, respectively). The bias for TTHM4 was relatively low at –2.8 ± 3.7 μg/L. The bias between the CMS-GC and EPA 502.2 methods ranged from –16.3 to 2.4, μg/L for TTHM4. The absolute error is between the two methods is on the order of 1-3 μg/L—or approximately 26 to 80 times lower than the MCL.

This study was carried out in a very slightly more complex sample matrix. The concentrations of DBP species was expected to be at slightly higher than those of the Memphis 1 studies (based on routine compliance monitoring conducted by MLGW), but still relatively low (~10 to 25 μg/L levels for TTHM4) to most utilities. Qualitatively, it appears the methods "track" well in that changes in the concentrations of THM4 species and TTHM4 concentrations appear to agree—meaning when a change (increase or decrease) is observed in one method, it is usually observed in the other method.

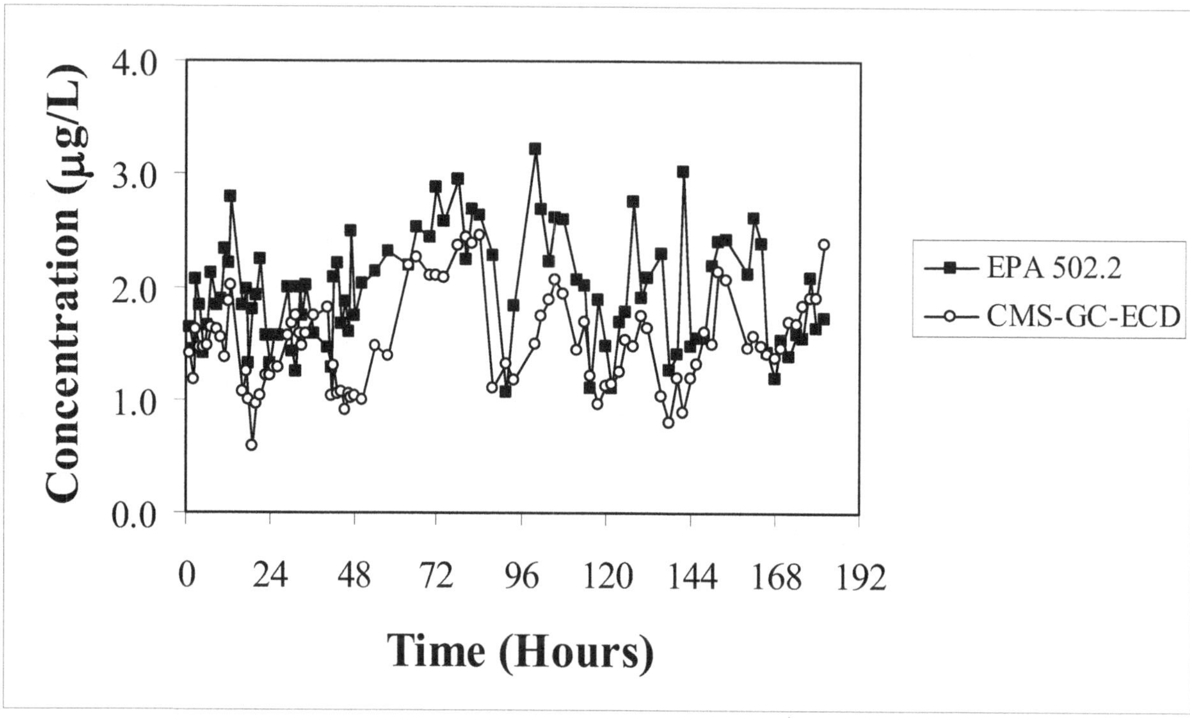

Figure 5.19 Memphis 2 concentration plot of $CHCl_3$ for EPA 502.2 and CMS-GC-ECD

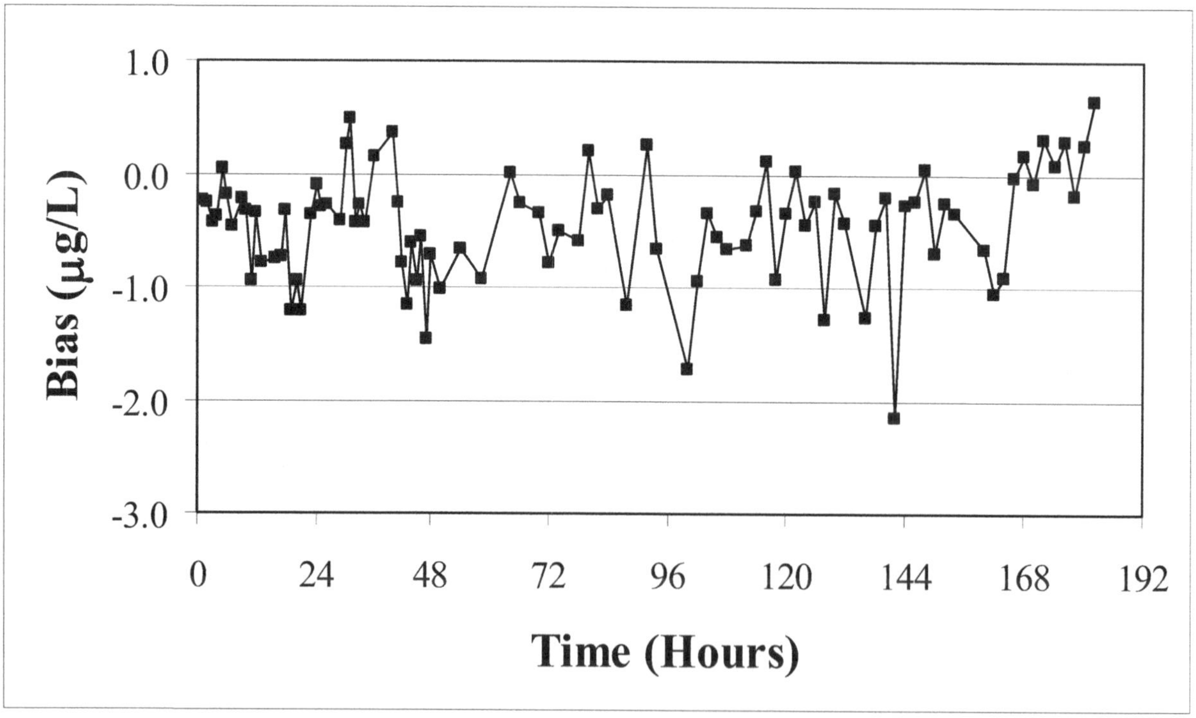

Figure 5.20 Memphis 2 bias plot of $CHCl_3$ comparing CMS-GC-ECD to EPA 502.2

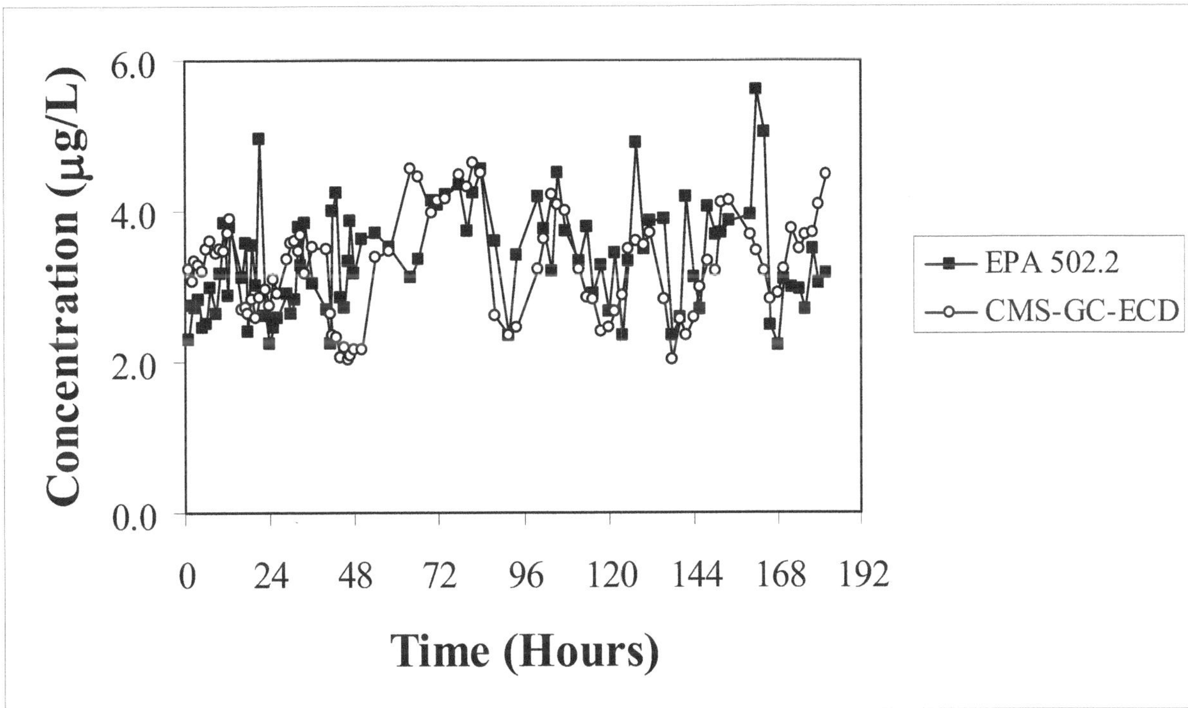

Figure 5.21 Memphis 2 concentration plot of $CHBrCl_2$ for EPA 502.2 and CMS-GC-ECD

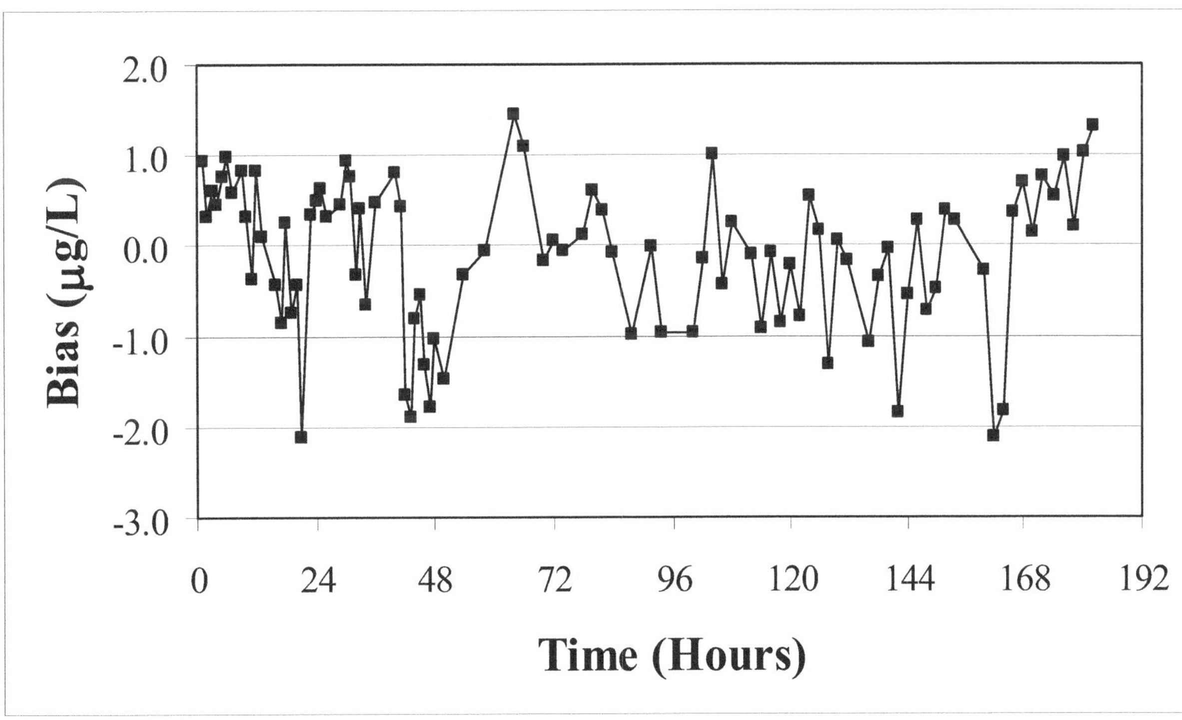

Figure 5.22 Memphis 2 bias plot of $CHBrCl_2$ comparing CMS-GC-ECD to EPA 502.2

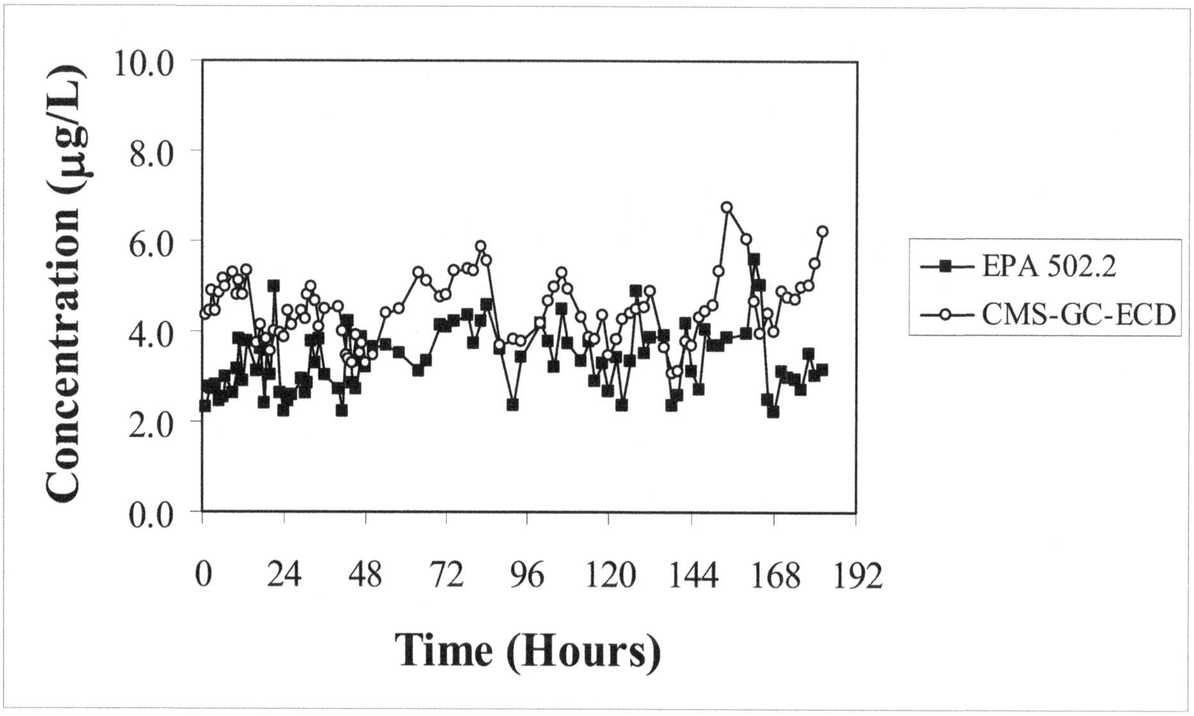

Figure 5.23 Memphis 2 concentration plot of $CHBr_2Cl$ for EPA 502.2 and CMS-GC-ECD

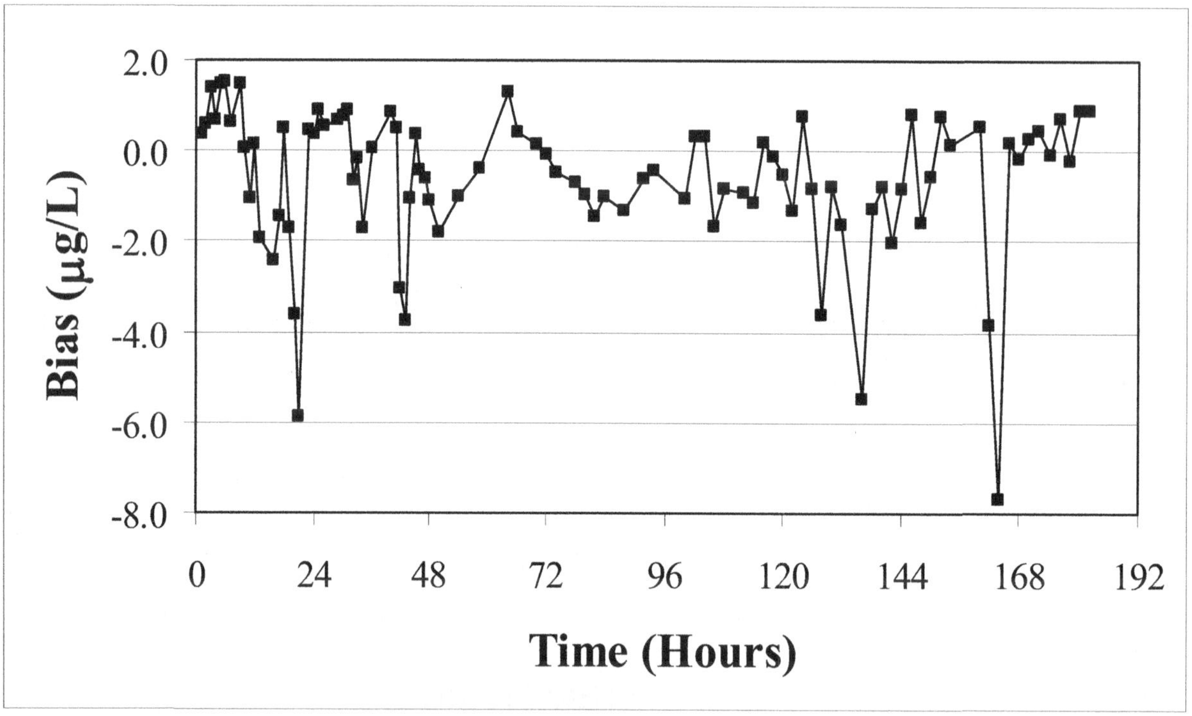

Figure 5.24 Memphis 2 bias plot of $CHBr_2Cl$ comparing CMS-GC-ECD to EPA 502.2

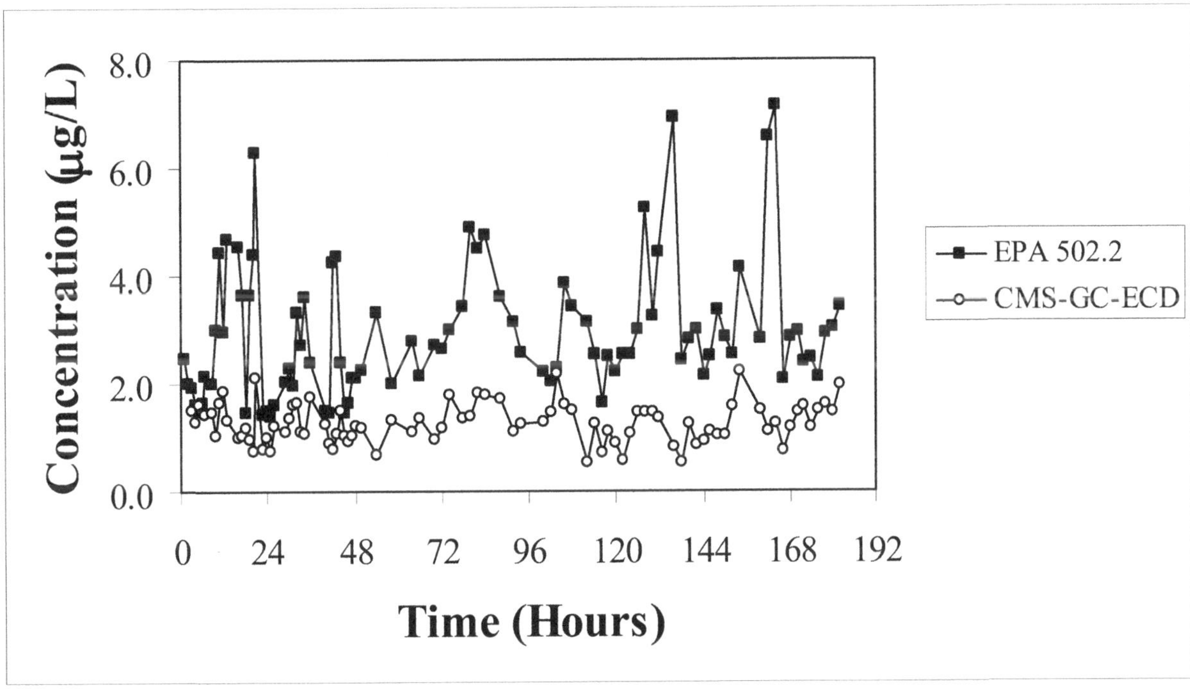

Figure 5.25 Memphis 2 concentration plot of $CHBr_3$ for EPA 502.2 and CMS-GC-ECD

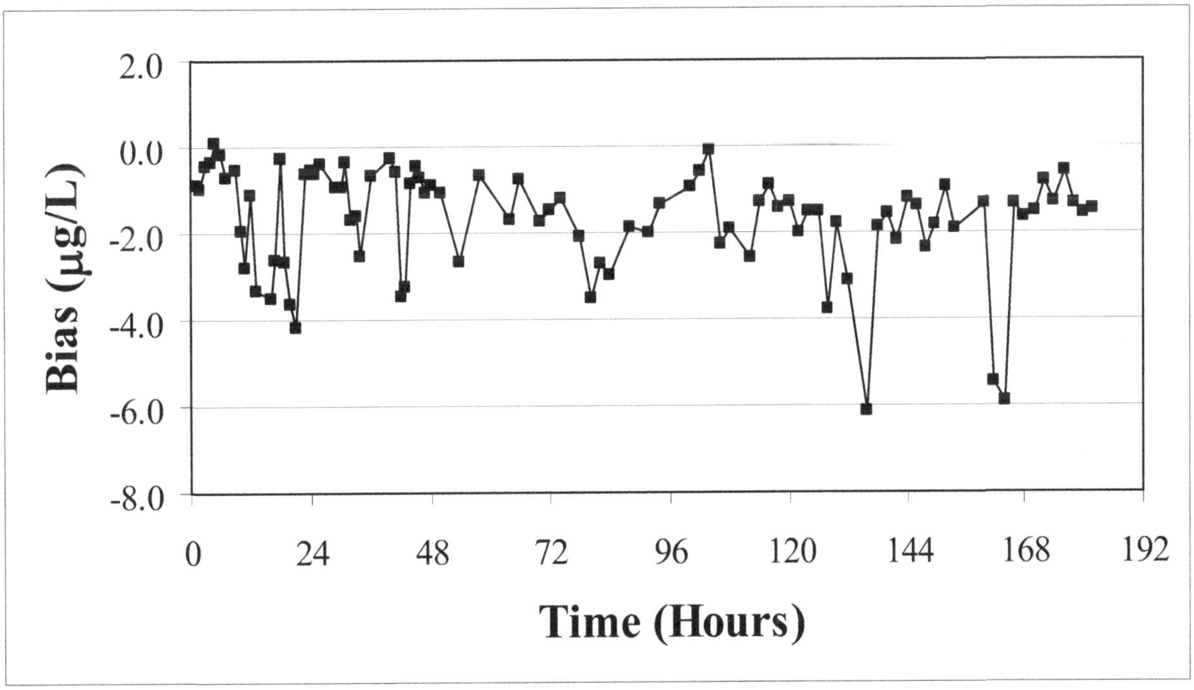

Figure 5.26 Memphis 2 bias plot of $CHBr_3$ comparing CMS-GC-ECD to EPA 502.2

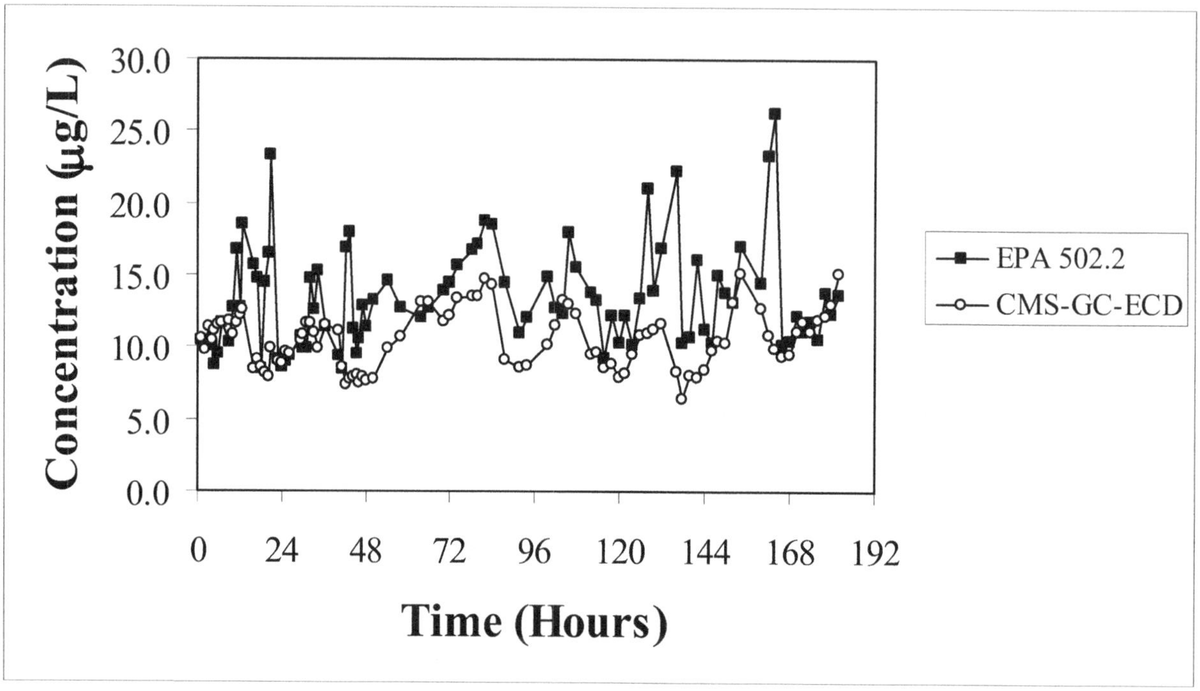

Figure 5.27 Memphis 2 concentration plot of TTHM4 for EPA 502.2 and CMS-GC-ECD

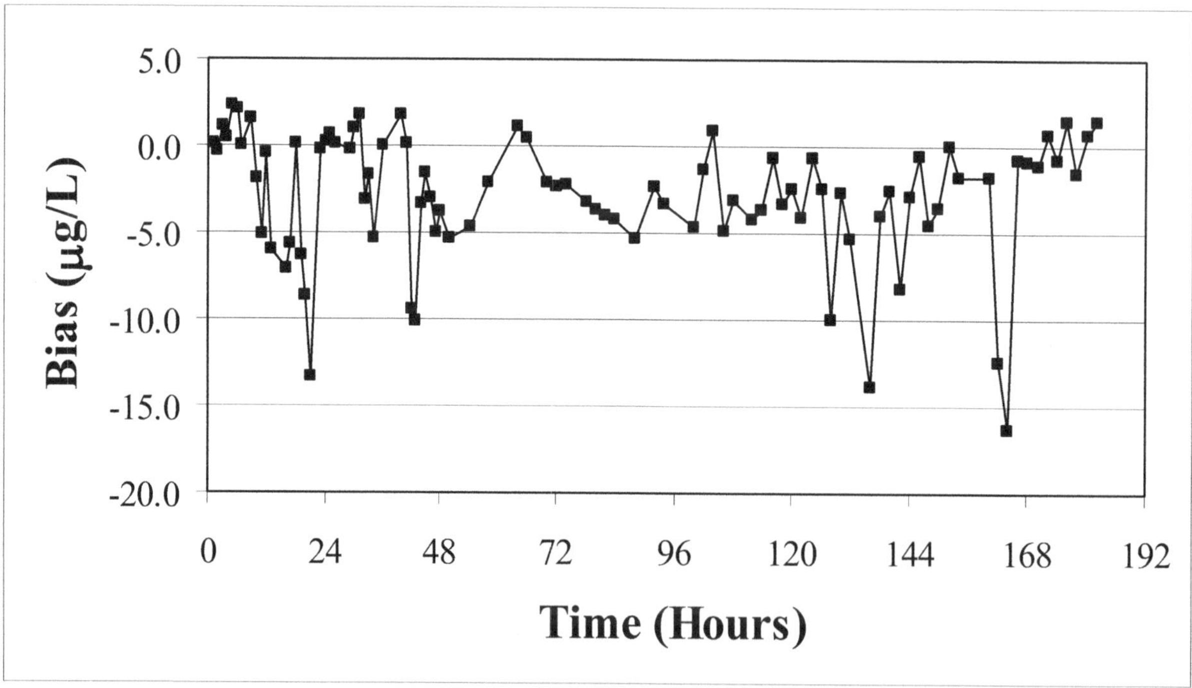

Figure 5.28 Memphis 2 bias plot of TTHM4 comparing CMS-GC-ECD to EPA 502.2

Houston On-Line Monitoring Study using CMS-GC-ECD

During the Houston monitoring studies, side-by-side comparisons were made to both EPA Method 502.2 and EPA Method 524.2. This CMS-GC-ECD monitoring study was conducted from June 21 to 28, 2005 at the Houston, TX sampling site (Houston). The results of these studies will be referred to with respect to the EPA Method used, i.e., Houston (502.2) or Houston (524.2). The CMS-GC-ECD was interfaced to the drinking water distribution system as shown in the Appendix A. The concentrations of each THM4 species were monitored with sampling rate of ~1 sample/ hour (though the analyzer was typically taken off-line every 12 hours to run a check standard). The results for each THM4 species were summed and are reported as TTHM4. The results along with details used in the analysis of the data are presented in Table 5.4 for the Houston (502.2) study and Table 5.5 for the Houston (524.2) study. The raw concentration data for each THM was first corrected for changes in temperature during the study by using the appropriate correction factor as mentioned in earlier in the Chapter and then for mean % recovery, which was calculated at the beginning of the study. This CMS-GC original corrected data is denoted as CMS Org. in both Tables 5.4 and 5.5. The corresponding EPA method is denoted as EPA in the tables. After careful consideration of these two data sets, the analyst evaluated additional "correction factors" in an effort to identify and isolate other possible factors that affect analytical signal. This was done primarily because of the large increase in bias that occurred during hours 30 through 88 for each THM4 species. The check standard data collected during the monitoring study was used to calculate a single estimate of percent recovery—differentiated here as % $REC_{Monitor}$. This factor was used as a correction factor to take into account any variability that might has occurred during monitoring that resulted in the increased bias.

For clarity, it is important to remind the reader that the "percent recovery" calculation is used throughout this project in different ways. First, the "percent recovery" calculation on a check standard is calculated in MDL, accuracy and precision studies. In this case, the results of seven consecutive check standard analyses are averaged to calculate the mean % recovery. The mean % recovery is reported in the MDL, accuracy and precision studies and used as an estimate of detector accuracy. The researchers also use the "percent recovery" calculation in on-line monitoring studies. In these studies, the candidate method (CMS-GC-ECD, in this Chapter) is compared to an appropriate EPA method (EPAA 502.2 and EPA 524.2, in this case). In making this comparison, the implicit assumption is that the EPA method provides the "true value" of the analyte concentration. Using this convention, it is possible at each comparison between two methods to calculate the "percent recovery" here denoted as % REC_{Bias} because it is the % REC calculated during the "bias" studies. The third "percent recovery" calculated was just described above (% $REC_{Monitor}$) as a possible correction that accounts for current detector response, or to put another way possible detector drift due perhaps to electronic drift in the detector or also changes in the sample matrix that might affect analytical signal. % $REC_{Monitor}$ is being explored as a possible correction factor that might be useful in certain situations if one starts to observe large deviations during a study from EPA standard methods. This % $REC_{Monitor}$ corrected data is shown as the CMS-GC Corrected data, which is denoted as CMS Corr. in Tables 5.4 and 5.5.

With each check standard analyzed during the study, a new % $REC_{Monitor}$ correction factor was applied to the specific samples analyzed immediately following the check standard analysis. For example, during hours 1 to 12 of the Houston study, $CHCl_3$ data was corrected by -12.2%, hours 14 to 31 by -5.1%, hours 33 to 54 by -70.5%, hours 56 to 78 by -72.7%, hours 80 to 86 by -86.0%, hours 88 to 112 by -73.5%, hours 113 to 136 by -44.3%, hours 138 to 150 by - 0.7%, and hours 152 to 167 by -13.1%. The appropriate correction factors for the other THM4 were also applied in a similar manner. The purpose of these tests was to evaluate the data with respect to a "snapshot" of detector response measured each shift.

Houston (502.2)

The results of the Houston (502.2) studies are summarized in Table 5.4. Plots of the concentrations of $CHCl_3$ for the EPA 502.2 data, the CMS-GC Original data, and the CMS-GC Corrected data versus time are shown in Figure 5.29. For 97 direct comparisons, the concentration ranged from 18.4 - 34.7 μg/L for the EPA Method, from 9.9 - 52.8 μg/L for (CMS-GC Org. data) and 11.1 - 42.2 μg/L for (CMS-GC Corr. data). The average concentration for EPA 502.2 is 29.0 ± 2.9 μg/L, for (CMS-GC Org. data) is 34.0 ± 11.8 μg/L and for (CMS-GC Corr. data) is 26.8 ± 7.5 μg/L. Figure 5.30 shows a plot of the bias as a function of time for (CMS-GC Org. data) and (CMS-GC Corr. data) compared directly to EPA 502.2. The bias values ranged from –18.6 - 25.2 μg/L (averaging 5.0 ± 11.8 μg/L) with % REC_{Bias} values ranged from 43 - 236 % (averaging 118 ± 43 %) for the (CMS-GC Org. data). The bias ranged from – 17.6 to 18.7 μg/L (averaging –2.1 ± 7.6 μg/L) with % REC_{Bias} values ranged from 46 to 183 % (averaging 93 ± 27 %) for the (CMS-GC Corr. data).

The results for $CHBrCl_2$ (Figure 5.31) show 97 direct comparisons and give concentration ranged from 9.6 - 18.4 μg/L (EPA 502.2), 7.3 - 27.1 μg/L (CMS-GC Org.) and 7.9 - 25.6 μg/L (CMS-GC Corr.). The average concentration was 15.2 ± 1.6 μg/L (EPA 502.2), 19.7 ± 4.5 μg/L (CMS-GC Org.) and 16.7 ± 3.6 μg/L (CMS-GC Corr). The bias (Figure 5.32) ranged from –7.8 - 14.1 μg/L (averaging 4.5 ± 5.1 μg/L) with % REC_{Bias} ranged from 49.4 - 246 % (averaging 132 ± 37 %) for CMS-GC Org. data. The bias (Figure 5.32) ranged from –7.2 - 10.3 μg/L (averaging 1.5 ± 3.8 μg/L) with % REC_{Bias} ranging from 53 - 174 % (averaging 111 ± 26 %) for CMS-GC Corr. data.

For $CHBr_2Cl$ (Figure 5.33), the concentrations ranged from 4.1 - 8.4 μg/L (EPA 502.2), 1.6 - 8.6 μg/L (CMS-GC Org.) and 1.7 - 8.2 μg/L (CMS-GC Corr.). The average concentration was 5.9 ± 0.8 μg/L (EPA 502.2), 6.4 ± 1.5 μg/L (CMS-GC Org.) and 5.4 ± 1.2 μg/L (CMS-GC Corr.). The bias (Figure 5.34) ranged from –4.9 - 3.6 μg/L (averaging 0.5 ± 2.0 μg/L) and % REC_{Bias} ranging from 26.7 - 184% (averaging 111 ± 35 %) for (CMS-GC Org.). The bias (Figure 5.34) for (CMS-GC Corr.) ranged from –4.5 - 2.3 μg/L (averaging – 0.5 ± 1.5 μg/L) with % REC_{Bias} ranging from 29 - 141% (averaging 94 ± 24 %).

The concentrations of $CHBr_3$ were less than the MDL in each case for these studies. Bias calculations were not done using this data. As discussed earlier, the concentrations of the individual THM4 species were summed to calculate TTHM4. The TTHM4 (Figure 5.35) concentrations ranged from 32.1 - 58.6 μg/L (EPA 502.2), 18.9 - 88.2 μg/L (CMS-GC Org.) and 19.7 - 75.9 μg/L. The average concentration was 50.1 ± 4.3 μg/L (EPA 502.2), 60.0 ± 17.5 μg/L (CMS-GC Org.) and 49.4 ± 11.2 μg/L (CMS-GC Corr.). The bias (CMS-GC Org.), shown in Figure 5.36, ranged from –26.2 - 42.5 μg/L (averaging 10.0 ± 18.5 μg/L) and % REC_{Bias} values ranged from 43.0 - 232% (averaging 121% ± 39 %). The bias (CMS-GC Corr.), shown in Figure 5.36, ranged from – 24.9 - 28.6 μg/L (averaging – 0.7 ± 11.7 μg/L) with and % REC_{Bias} ranging from 44.6 - 165% (averaging 99 ± 24 %).

Table 5.4
Houston monitoring study (502.2): comparison of % RECMonitor corrected and original corrected data for CMS-GC to USEPA Method 502.2

	$CHCl_3$			$CHBrCl_2$			$CHBr_2Cl$			$CHBr_3$			TTHM4		
Statistic/ *Method*	*EPA*	*CMS Org.*	*CMS Corr.*	*EPA*	*CMS Org.*	*CMS Corr.*	*EPA*	*CMS Org.*	*CMS Corr.*	*EPA*	*CMS Org.*	*CMS Corr.*	*EPA*	*CMS Org.*	*CMS Corr.*
N	97	97	97	97	97	97	97	97	97	97	97	97	97	97	97
Avg. Conc.	29.0	34.0	26.8	15.2	19.7	16.7	5.9	6.4	5.4	n.d.	n.d.	n.d.	50.1	60.0	49.4
Std. Dev Conc.	2.9	11.8	7.5	1.6	4.5	3.6	0.8	1.5	1.2	n.d.	n.d.	n.d.	4.3	17.5	11.2
Conc. Range	18.4 to 34.7	9.9 to 52.8	11.1 to 42.2	9.6 to 18.4	7.3 to 27.1	7.9 to 25.6	4.1 to 8.4	1.6 to 8.6	1.7 to 8.2	---	---	---	32.1 to 58.6	18.9 to 88.2	19.7 to 75.9
Avg. Bias	---	5.0	-2.1	---	4.5	1.5	---	0.5	-0.5	---	n.d.	n.d.	---	10.0	-0.7
Std. Dev Bias	---	11.8	7.6	---	5.1	3.8	---	2.0	1.5	---	n.d.	n.d.	---	18.5	11.7
Bias Range	---	-18.6 to 25.2	-17.6 to 18.7	---	-7.8 to 14.1	-7.2 to 10.3	---	-4.9 to 3.6	-4.5 to 2.3	---	---	---	---	-26.2 to 42.5	-24.9 to 28.6
% REC_{Bias}	---	118	93	---	132	111	---	111	94	---	n.d.	n.d.	---	121	99
s.d. % REC_{Bias}	---	43	27	---	37	26	---	35	24	---	n.d.	n.d.	--	39	24
% REC_{Bias} Range		43 to 236	46 to 183	---	49 to 246	53 to 174	---	27 to 184	29 to 141	---	---	---	---	43 to 232	45 to 165
t-test* $t_{critical}$ $t_{calculated}$	---	Yes 2.366 4.172	Yes 2.366 2.760	---	Yes 2.366 8.579	Yes 2.366 3.769	---	No 2.366 2.352	Yes 2.366 3.119	---	---	---	---	Yes 2.366 5.306	No 2.366 0.607

* The tcritical and tcalculated values listed were calculated for the N listed using a two-tailed t-test with 98% confidence. Typically a tcalculated value should be less than the tcritical. Yes or No as means if statistically different.

The results of statistical analysis of the CMS-GC Org study was as follows: The $t_{critical}$ for 97 comparisons was 2.366. For $CHCl_3$, $CHBrCl_2$, and TTHM4, $t_{calculated}$ is 4.172, 8.579, 5.306, respectively, indicating a statistical difference between the two methods. For $CHBr_2Cl$, the results are not statistically different with a $t_{calculated}$ value is 2.352. The results for the CMS-GC Corr studies were similar, except that when the values are corrected for % REC_{Monit}, a statistical difference for $CHBr_2$ arises, but the TTHM4 results now agree.

Qualitatively, there are several items to note in the Houston (502.2) study. First, between approximately 36 – 96 hours, there is a marked increase in the bias comparing $CHCl_3$ and TTHM4 measurements. This increase is slight for $CHBrCl_2$ and $CHBr_2Cl$, but apparent. If the assumption is made that this increase is not due to an actual increase in $CHCl_3$ concentration, since such a corresponding increase is not observed in EPA 502.2, the question arises as to what would cause such behavior.

One possibility is that the calibration for the CMS-GC-ECD has changed. To account for this, check standards are run two times daily and were used to "correct" the data for any such drift. When this was done, despite statistical differences, one can discern some improvement in the results. For example, the average bias decreases (improves) from 5.0 to –2.1 for $CHCl_3$, 4.5 to 1.5 for $CHBrCl_2$, and 10 - –0.7 for TTHM4. The standard deviation on the bias goes down in each case. The % REC_{Bias} data shows slight improvement on average as well. It might be that some other chemical species, similar to the two smaller molecular weight species appears in the sample. This species might co-elute and complicate separation. The result is that the analytical signal for these species increases—the species act to produce a positive bias in the response. It is also possible that this increased response is not due to a different species, but could be the result of some sort of "membrane loading" phenomena that is poorly understood, despite the detailed carry-over experiments that were carried out earlier. A third possibility might be that the membrane has active sites to which THM4 species adsorb, and that these sites must be "doped" or perhaps equilibrated before yields improve. A final possibility might be that a compound that will coelute with the THM4 species, lingers in the membrane, and takes several hours to "purge".The true effect likely could be a combination of these factors.

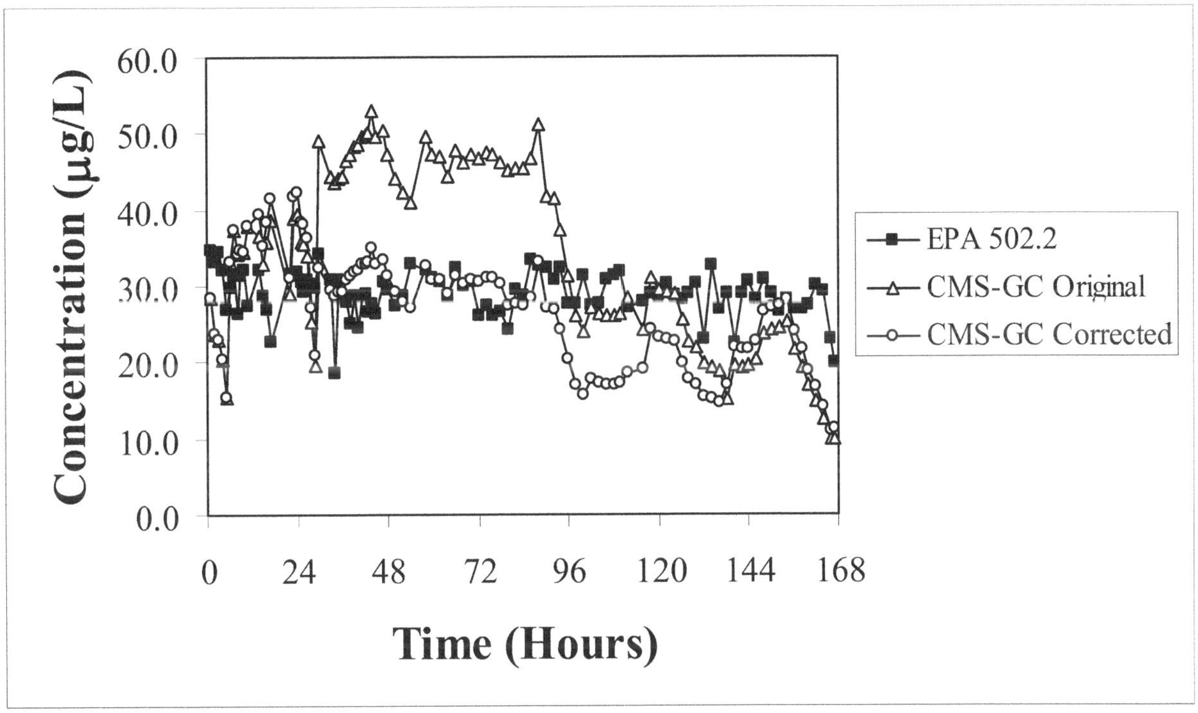

Figure 5.29 Houston concentration plot of $CHCl_3$ for EPA 502.2, CMS-GC-ECD original data and CMS-GC-ECD corrected data

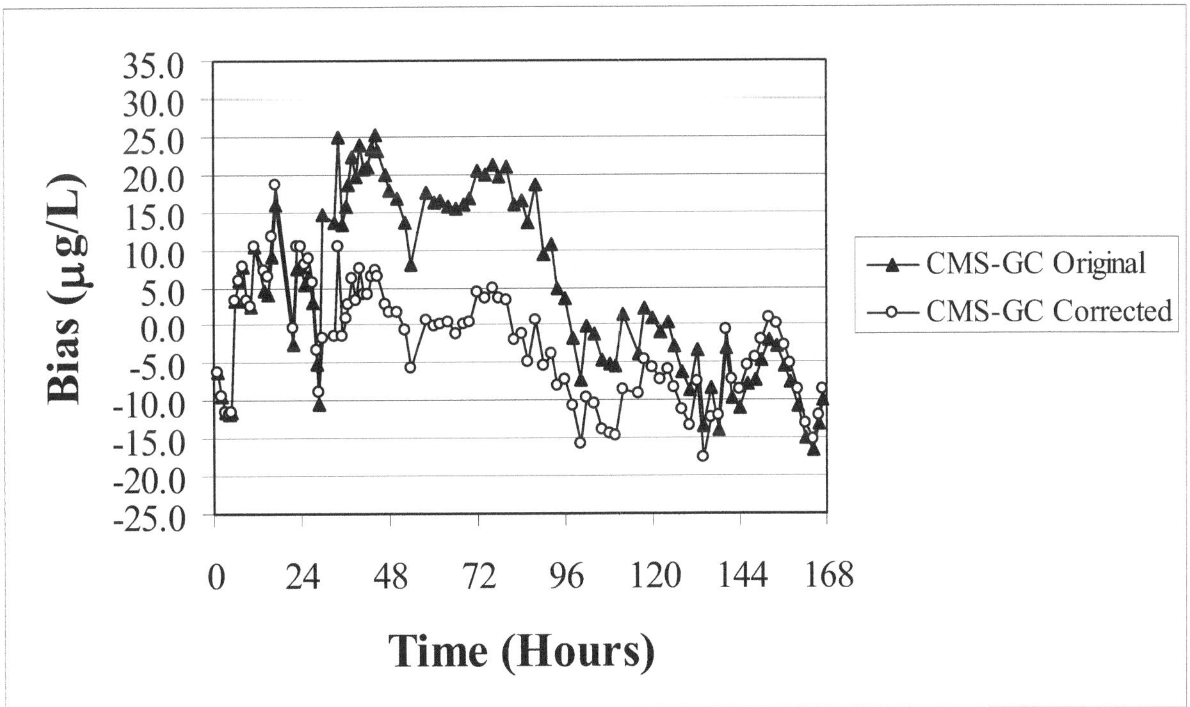

Figure 5.30 Houston bias plot of $CHCl_3$ comparing the CMS-GC-ECD original data and CMS-GC-ECD corrected data to EPA 502.2

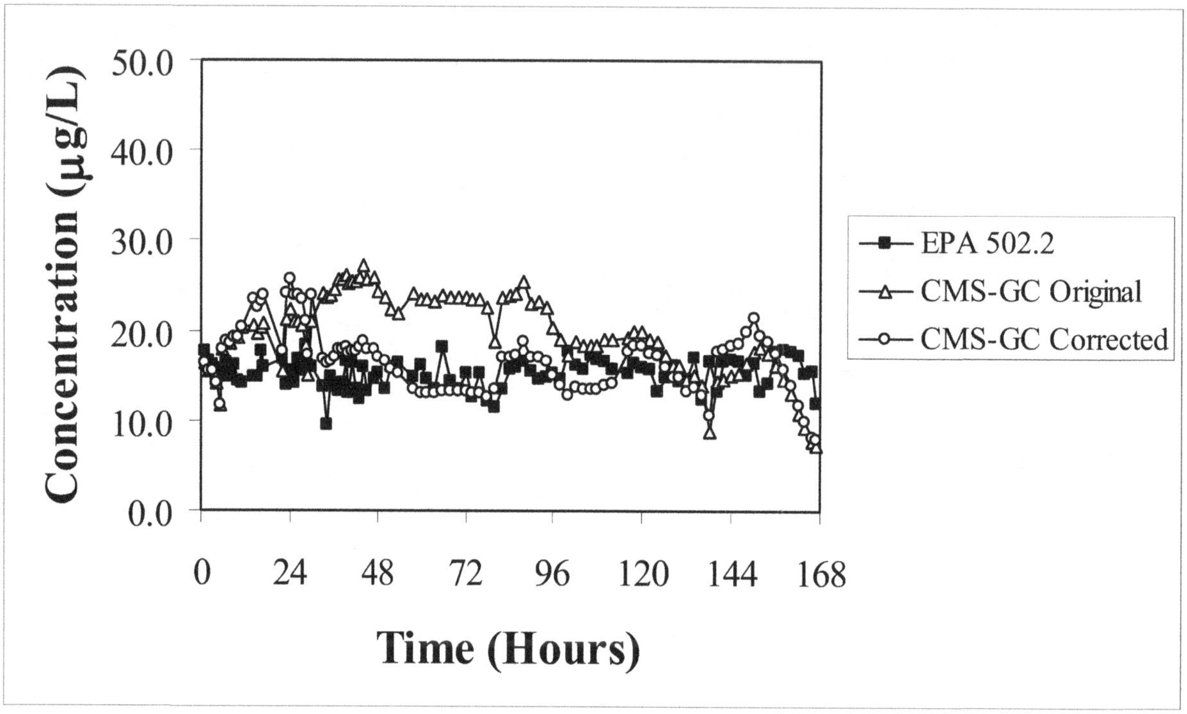

Figure 5.31 Houston concentration plot of $CHBrCl_2$ for EPA 502.2, CMS-GC-ECD original data and CMS-GC-ECD corrected data

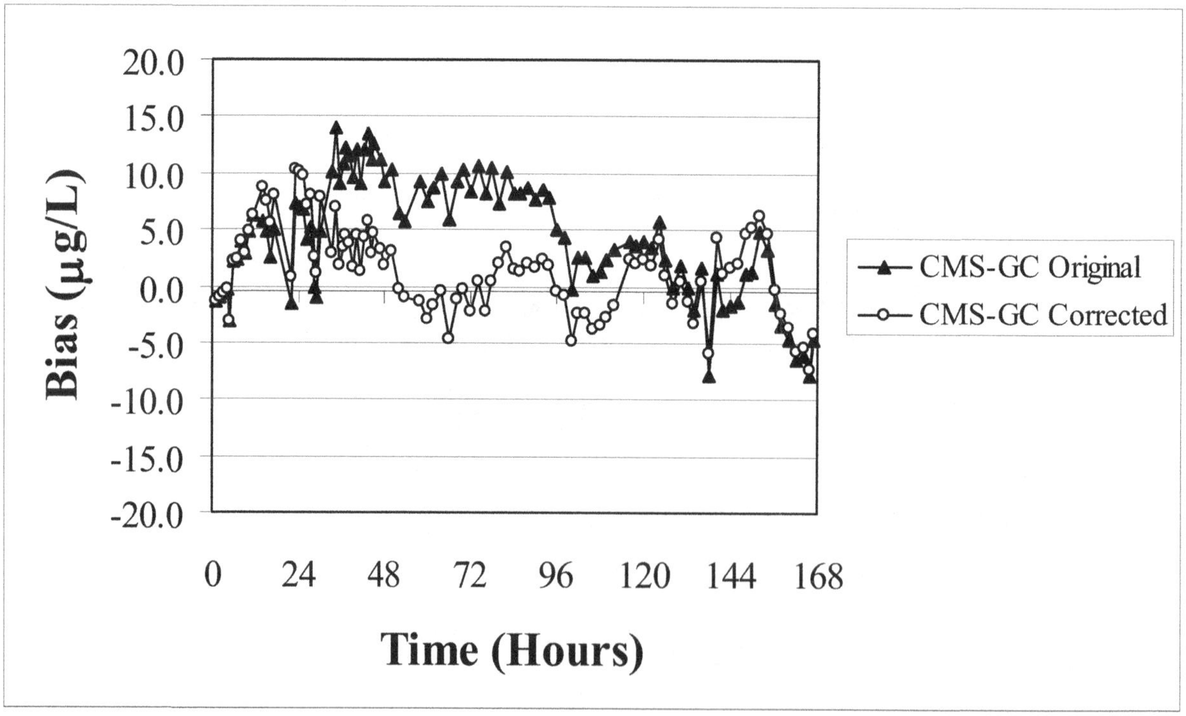

Figure 5.32 Houston bias plot of $CHBrCl_2$ comparing the CMS-GC-ECD original data and CMS-GC-ECD corrected data to EPA 502.2

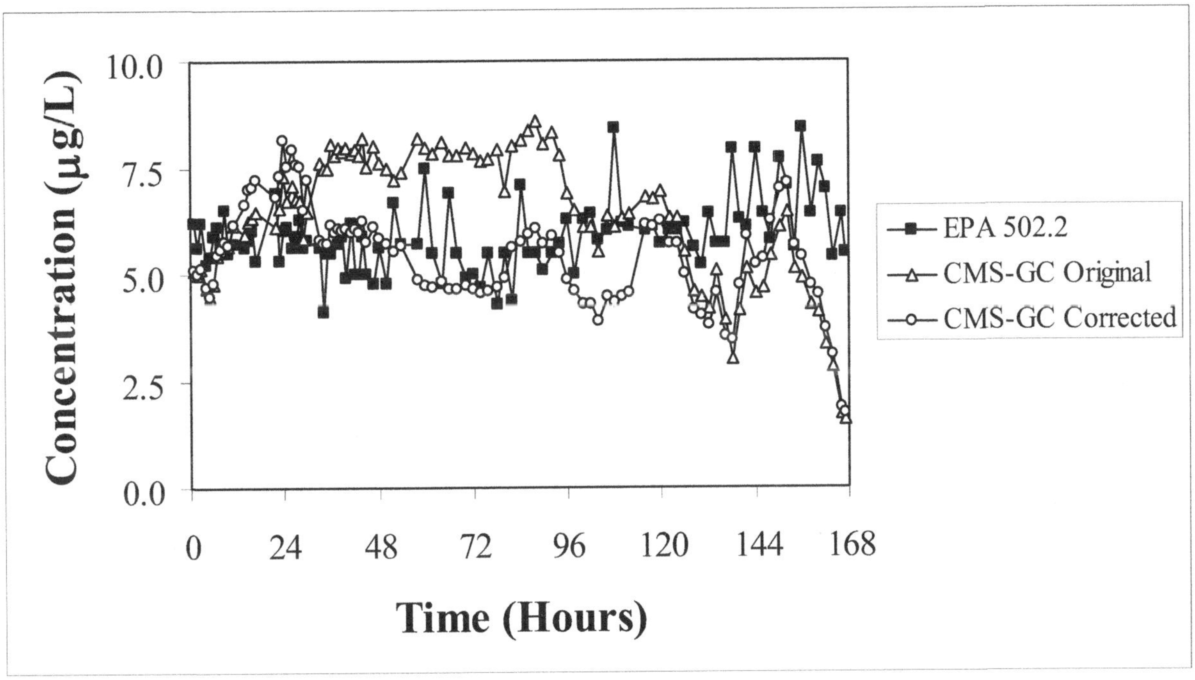

Figure 5.33 Houston concentration plot of $CHBr_2Cl$ for EPA 502.2, CMS-GC-ECD original data and CMS-GC-ECD corrected data

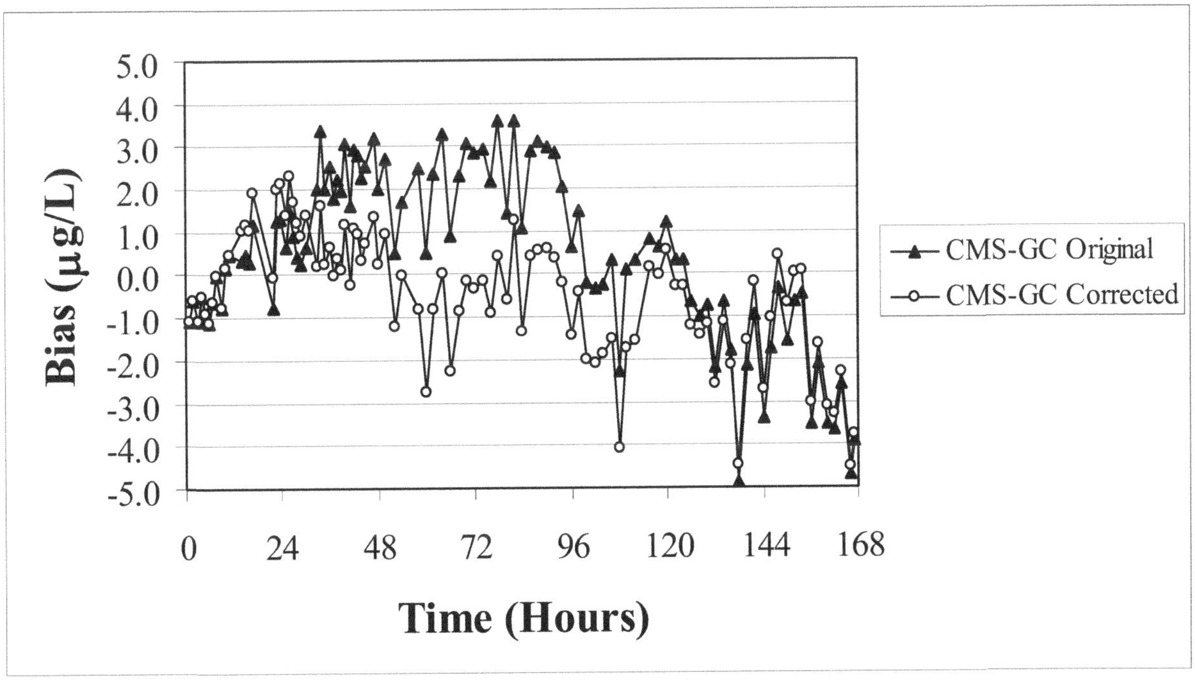

Figure 5.34 Houston bias plot of $CHBr_2Cl$ comparing the CMS-GC-ECD original data and CMS-GC-ECD corrected data to EPA 502.2

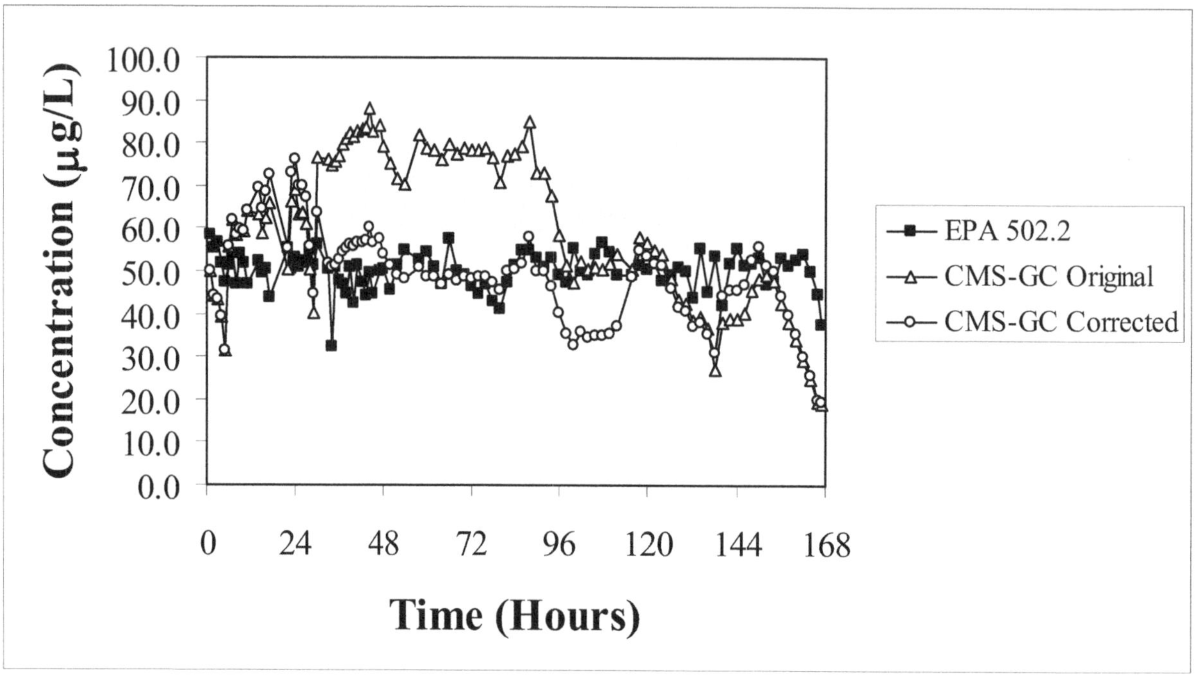

Figure 5.35 Houston concentration plot of TTHM4 for EPA 502.2, CMS-GC-ECD original data and CMS-GC-ECD corrected data

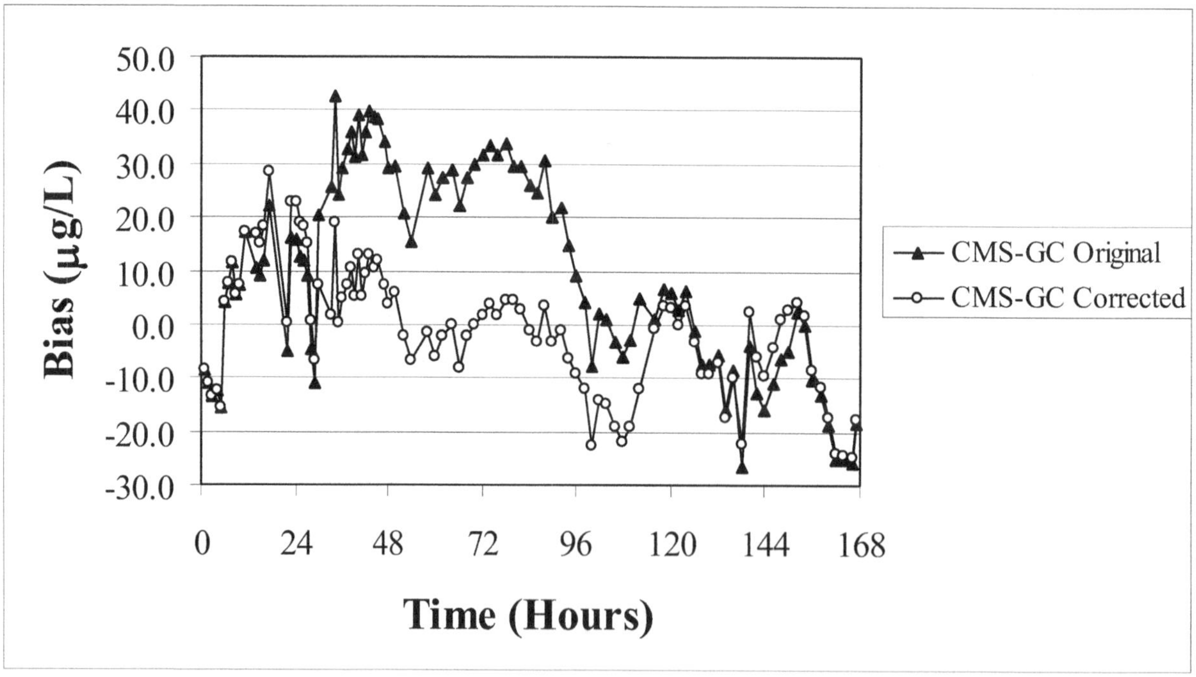

Figure 5.36 Houston bias plot of TTHM4 comparing the CMS-GC-ECD original data and CMS-GC-ECD corrected data to EPA 502.2

Houston (524.2)

Figure 5.37 shows plots of the concentrations of $CHCl_3$ for the EPA 524.2, the CMS-GC Org., and the CMS-GC Corr. studies versus time. For 66 direct comparisons, the concentration ranges from 7.6 - 30.9 µg/L for the EPA 524.4, from 19.5 - 52.8 µg/L for the CMS-GC Org, and 15.4 - 52.6 µg/L for CMS-GC Corr. The average concentration for EPA 524.2 was 16.0 ± 4.5 µg/L, for CMS-GC Org. was 38.6 ± 9.8 µg/L, and for CMS-GC Corr., was 28.5 ± 7.6 µg/L. Figure 5.38 shows plots of the bias as a function of time for CMS-GC Org. and CMS-GC Corr. compared directly to EPA 524.2. The bias (CMS-GC Org.) ranged from 5.2 - 36.3 µg/L (averaging 22.7 ± 9.0 µg/L) with % REC_{Bias} values ranging from 136 - 467 % (averaging 252 ± 74 %). The bias (CMS-GC Corr.) ranged from – 1.5 - 38.8 µg/L (averaging 12.5 ± 8.8 µg/L) with % REC_{Bias} values ranging from 95 - 501 % (averaging 191 ± 79 %).

The results for $CHBrCl_2$ (Figure 5.39) show 66 direct comparisons and give concentration ranges from 3.1 - 18.4 µg/L (EPA 524.2), 14.6 - 27.1 µg/L (CMS-GC Org.) and for 12.8 to 25.6 µg/L (CMS-GC Corr.). The average concentration was 8.1 ± 5.0 µg/L (EPA 524.2), 21.7 ± 3.1 (CMS-GC Org.) and 18.5 ± 3.7 µg/L (CMS-GC Corr.). The CMS-GC Org. bias (Figure 5.40) ranged from –1.4 - 23.1 µg/L (averaging 13.6 ± 6.0 µg/L) with % REC_{Bias} ranging from 92 - 846 % (averaging 378 ± 199 %). The CMS-GC Corr. bias ranged from 0.7 - 19.4 µg/L (averaging 10.4 ± 4.8 µg/L) with % REC_{Bias} values ranging from 104 - 590 % (averaging 310 ± 151 %).

For $CHBr_2Cl$ (Figure 5.41), the concentrations ranged from 0.4 - 5.6 µg/L (EPA 524.2), 4.2 - 8.4 µg/L (CMS-GC Org.) and from 3.8 - 8.2 µg/L (CMS-GC Corr.). The average concentration was 1.6 ± 1.6 µg/L (EPA 524.2), 7.0 ± 1.0 µg/L (CMS-GC) and 5.7 ± 1.1 µg/L (CMS-GC Corr.). The CMS-GC Org bias (Figure 5.42) ranged from 2.0 - 8.0 µg/L (averaging 5.4 ± 1.5 µg/L) with % REC_{Bias} ranging from 136 - 2045 % (averaging 815 ± 505 %). The CMS-GC Corr. bias ranges from 0.2 - 7.6 µg/L (averaging 4.1 ± 1.9 µg/L) with % REC_{Bias} values ranging from 104 - 1715 % (averaging 687 ± 445 %). For $CHBr_3$ (Figure 5.43), the concentrations ranged from 0.4 - 7.6 µg/L (EPA 524.2) and were less than the MDL of 0.3 µg/L (CMS-GC Org.), so no further correction was applied. The average concentration was 2.4 ± 1.5 µg/L (EPA 524.2) and not detected for CMS-GC. The bias, calculated using the MDL values for $CHBr_3$ (Figure 5.44) ranged from –7.3 to –0.1 µg/L (averaging –2.1 ± 1.5 µg/L) with % REC_{Bias} ranging from 3.9 - 85.7 % (averaging 17 ± 12 %).

For TTHM4 (Figure 5.45), the concentrations ranged from 12.0 – 50.3 µg/L (EPA 524.2), 38.5 - 88.2 µg/L (CMS-GC) and 32.9 - 84.0 µg/L (CMS-GC Corr.). The average concentration was 24.2 ± 8.8 µg/L (EPA 524.2), 67.4 ± 13.6 µg/L (CMS-GC Org) and 52.9 ± 11.5 µg/L (CMS-GC Corr.). The CMS-GC Org. bias (Figure 5.46) ranged from –16.7 - 66.9 µg/L (averaging 43.2 ± 12.8 µg/L) with % REC_{Bias} ranging from 151 - 535 % (averaging 300% ± 92 %). The CMS-GC Corr. bias ranged from 2.0 to 64.7 µg/L (averaging 28.7 ± 14.6 µg/L) with % REC_{Bias} values ranging from 104 - 586 % (averaging 242 ± 97 %).

With respect to statistical comparison, the CMS-GC Org. data set has 66 comparisons for $CHCl_3$ and $CHBrCl_2$. The $t_{critical}$ is 2.385. With $CHCl_3$ and $CHBrCl_2$, the $t_{calculated}$ was 20.377 and 18.28, indicating that there is a statistical difference between the two methods. For $CHBr_2Cl$ and $CHBr_3$, the $t_{critical}$ for the 46 and 45 comparisons was 2.412. The $t_{calculated}$ values were 25.007 and 9.595 for $CHBr_2Cl$ and $CHBr_3$, respectively. Again, this indicates there is a statistical difference between the two methods. For TTHM4, with 66 comparisons ($t_{critical}$ = 2.385), the $t_{calculated}$ is 27.372, indicating a statistical difference between the two methods. Despite the application of the

% REC_{Monit} correction factor, similar results to the CMS-GC Org. data were observed. In each case, the two methods were shown to give statistically different results.

The results of the Houston (524.2) were similar in several ways to the Houston (502.2) study as both were conducted at the same time. The marked increase at ~ 36 hours still appears, but seems less pronounced. The duration is much shorter compared to CMS-GC-ECD and EPA 502.2. This seems to support the thought that some similar compound might be co-eluting with $CHCl_3$. Moreover, this positive bias is evident for all THM4 species. The bias improves significantly for $CHCl_3$ and TTHM4, (22.7 to 12.5μg/L and 43.2 to 28.7μg/L, respectively), but hardly changes for the remaining THM4 species. This further indicates that the problem species with these measurements, for whatever reason is $CHCl_3$. Honestly, this is suprising as the experience in the U of M laboratory is that $CHBr_3$ is typically more problematic. The reverse is true in this study, where the $CHBr_3$ results appear to be in the most agreement. As discussed in the Houston (502.2) studies, there are several possibilities for this increase in bias such as a change (degradation) in the system calibration, the introduction of an interfering/co-eluting speices, perhaps the "poisoning" of the membrane or an increase in pressure causing an increase in flow rate, which would increase apparent THM4 concentrations. The Houston (524.2) studies again illustrate a need to revise the strategy for carrying out monitoring studies. In retrospect, pre-monitoring surveys should be done to confirm chromatography prior to extended monitoring studies. In summary, use of CMS-GC corrected values did give very slight improvement for each THM4 species, especially $CHCl_3$ which is reflected in the TTHM4 measurement. Using the % $REC_{Monitor}$ correction did not seem to be as effective as it was when applied to the EPA 524.2.

Table 5.5
Houston monitoring study (524.2): comparison of % $REC_{Monitor}$ corrected and original corrected data for CMS-GC to USEPA Method 524.2

	$CHCl_3$			$CHBrCl_2$			$CHBr_2Cl$			$CHBr_3$			TTHM4		
Statistic/ *Method*	*EPA*	*CMS Org.*	*CMS Corr.*	*EPA*	*CMS Org.*	*CMS Corr.*	*EPA*	*CMS Org.*	*CMS Corr.*	*EPA*	*CMS Org.*	*CMS Corr.*	*EPA*	*CMS Org.*	*CMS Corr.*
N	66	66	66	66	66	66	46	46	46	45	45	45	66	66	66
Avg. Conc	16.0	38.6	28.5	8.1	21.7	18.5	1.6	7.0	5.7	2.4	n.d.	n.d.	24.2	67.4	52.9
Std. Dev Conc.	4.5	9.8	7.6	5.0	3.1	3.7	1.6	1.0	1.1	1.5	n.d.	n.d.	8.8	13.6	11.5
Conc. Range	7.6 to 30.9	19.5 to 52.8	15.4 to 52.6	3.1 to 18.4	14.6 to 27.1	12.8 to 25.6	0.4 to 5.6	4.2 to 8.4	3.8 to 8.2	0.4 to 7.6	0.3 to 0.3	0.3 to 0.3	12.0 to 50.3	38.5 to 88.2	32.9 to 84.0
Avg. Bias	---	22.7	12.5	---	13.6	10.4	---	5.4	4.1	---	-2.1	-2.1	---	43.2	28.7
Std. Dev Bias	---	9.0	8.8	---	6.0	4.8	---	1.5	1.9	---	1.5	1.5	---	12.8	14.6
Bias Range	---	5.2 to 36.3	-1.5 to 38.8	---	-1.4 to 23.1	0.7 to 19.4	---	2.0 to 8.0	0.2 to 7.6	---	-7.3 to -0.1	-7.3 to -0.1	---	16.7 to 66.9	2.0 to 64.7
% REC_{Bias}	---	252	191	---	378	310	---	815	687	---	17.0	17.0	---	300	242
s.d. % REC_{Bias}	---	74	79	---	199	151	---	505	445	---	12.0	12.0	---	92	97
% REC_{Bias} Range	---	136 to 467	95 to 501	---	92 to 846	104 to 590	---	136 to 2045	104 to 1715	---	3.9 to 85.7	3.9 to 85.7	---	151 to 535	104 to 586
t-test* $t_{critical}$ $t_{calculated}$	---	Yes 2.385 20.37	Yes 2.385 11.52	---	Yes 2.385 18.28	Yes 2.385 17.66	---	Yes 2.412 25.01	Yes 2.412 14.48	---	Yes 2.412 9.595	Yes 2.412 9.595	---	Yes 2.385 27.37	Yes 2.385 15.99

* The $t_{critical}$ and $t_{calculated}$ values listed were calculated for the N listed using a two-tailed t-test with 98% confidence. Typically a $t_{calculated}$ value should be less than the $t_{critical}$. Yes or No as means if statistically different.

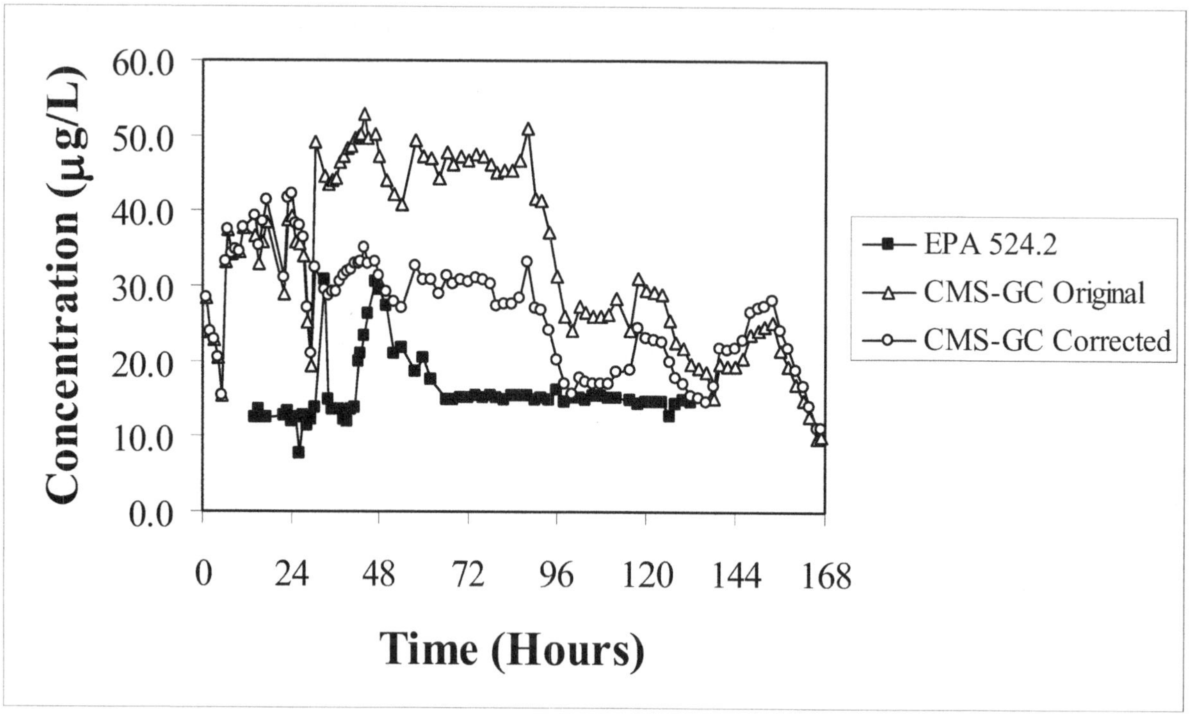

Figure 5.37 Houston concentration plot of $CHCl_3$ for EPA 524.2, CMS-GC-ECD original data and CMS-GC-ECD corrected data

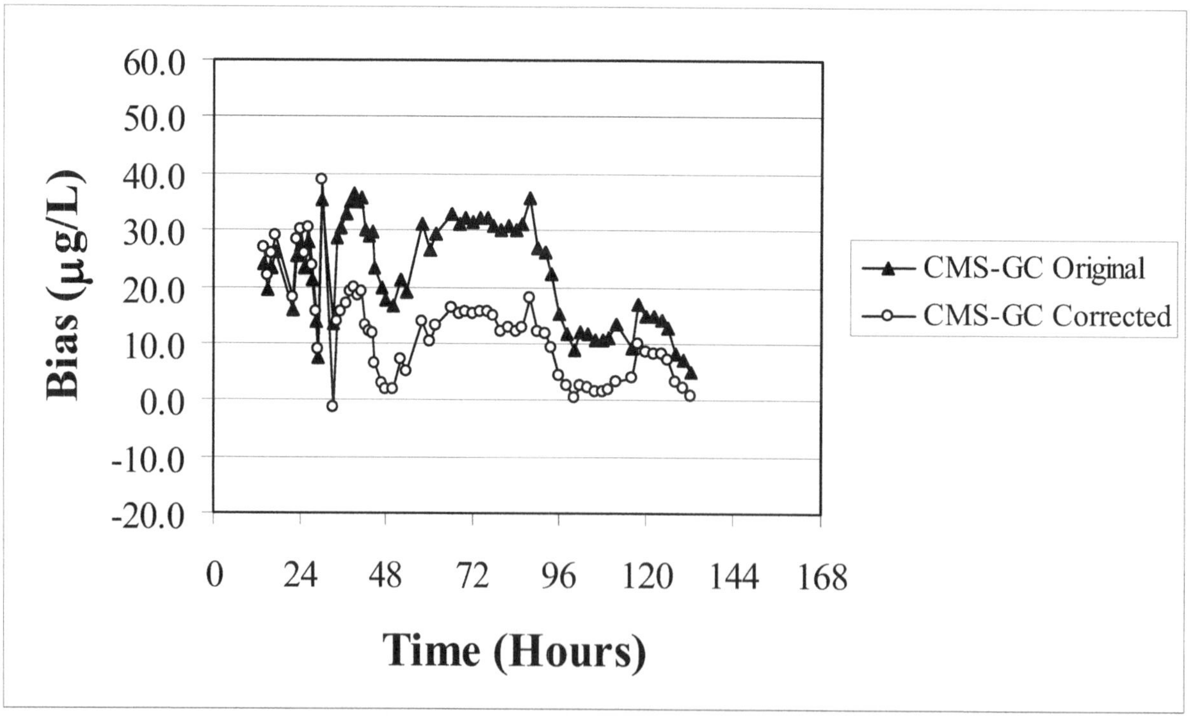

Figure 5.38 Houston bias plot of $CHCl_3$ comparing the CMS-GC-ECD original data and CMS-GC-ECD corrected data to EPA 524.2

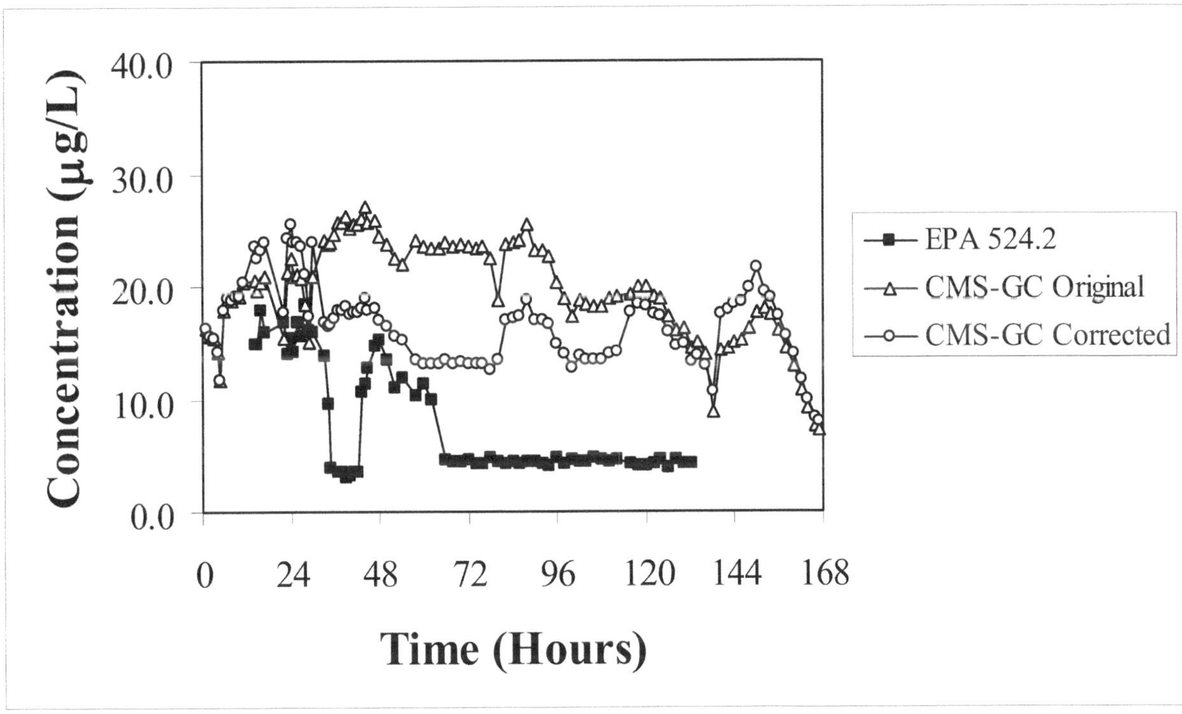

Figure 5.39 Houston concentration plot of $CHBrCl_2$ for EPA 524.2, CMS-GC-ECD original data and CMS-GC-ECD corrected data

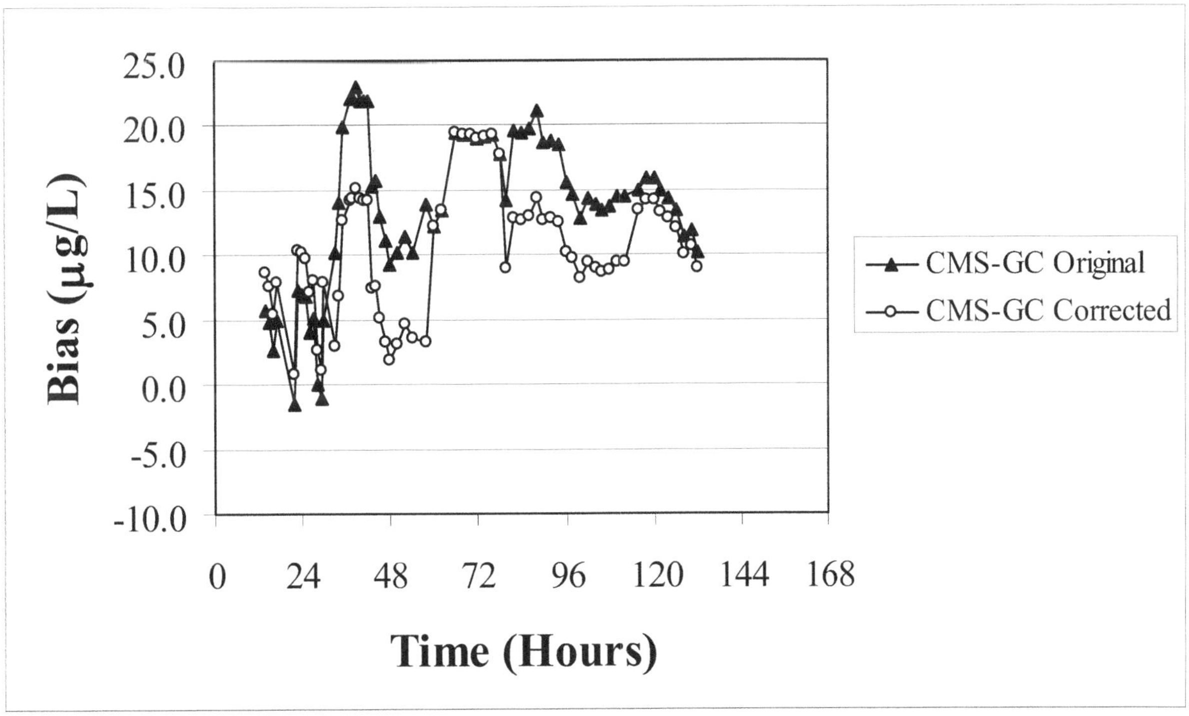

Figure 5.40 Houston bias plot of $CHBrCl_2$ comparing the CMS-GC-ECD original data and CMS-GC-ECD corrected data to EPA 524.2

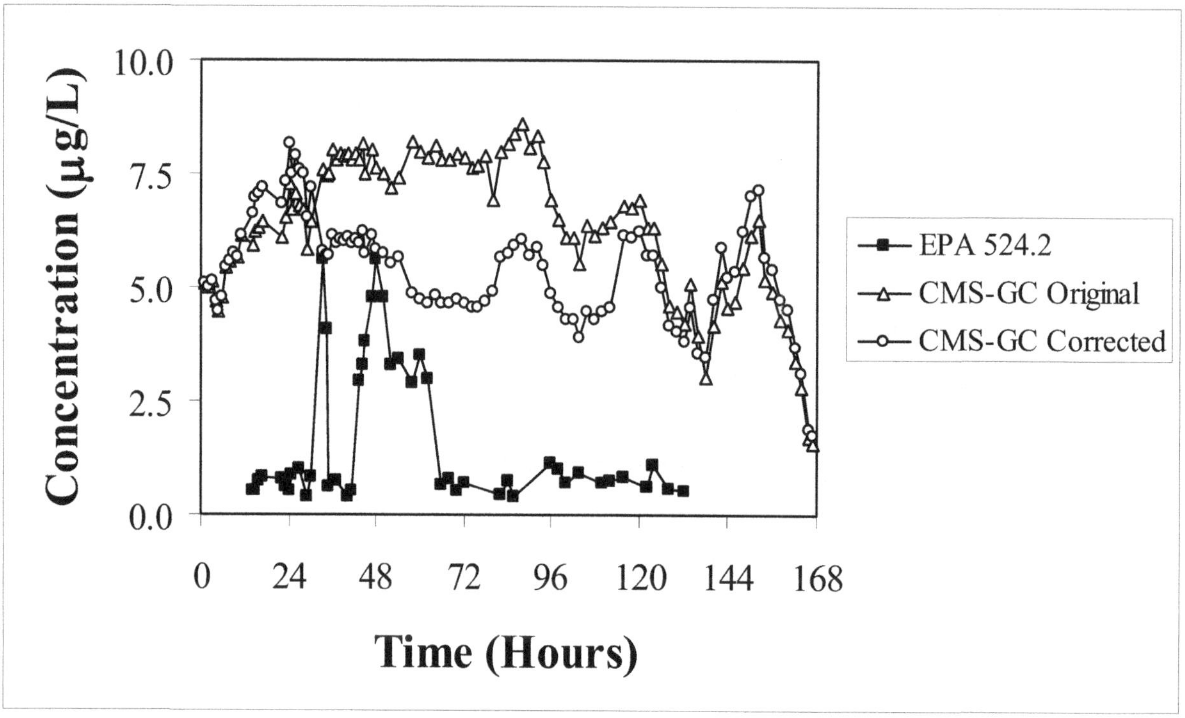

Figure 5.41 Houston concentration plot of $CHBr_2Cl$ for EPA 524.2, CMS-GC-ECD original data and CMS-GC-ECD corrected data

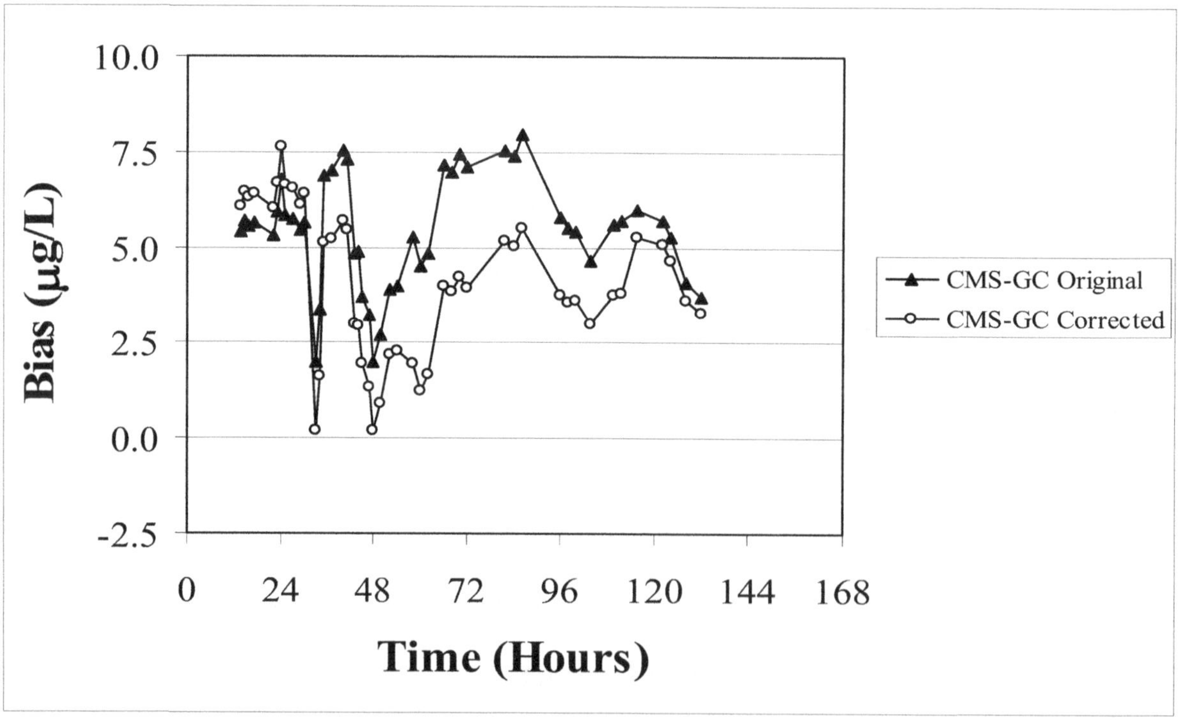

Figure 5.42 Houston bias plot of $CHBr_2Cl$ comparing the CMS-GC-ECD original data and CMS-GC-ECD corrected data to EPA 524.2

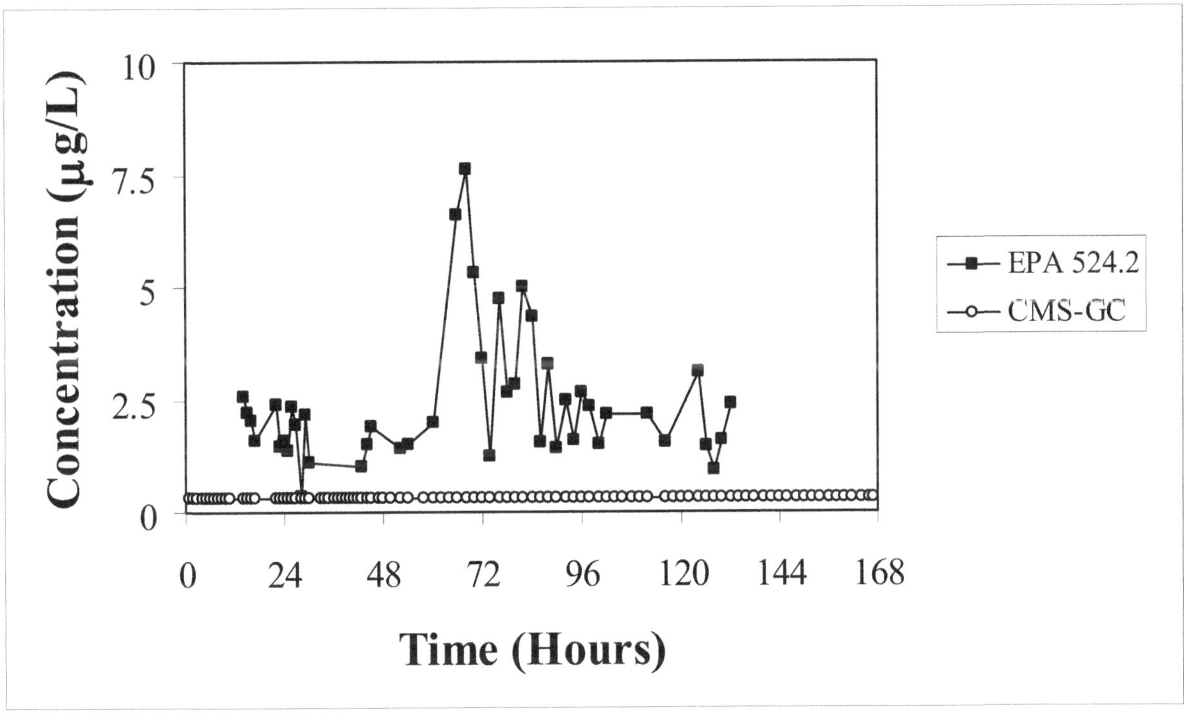

Figure 5.43 Houston concentration plot of $CHBr_3$ for the EPA 524.2 and CMS-GC-ECD original data

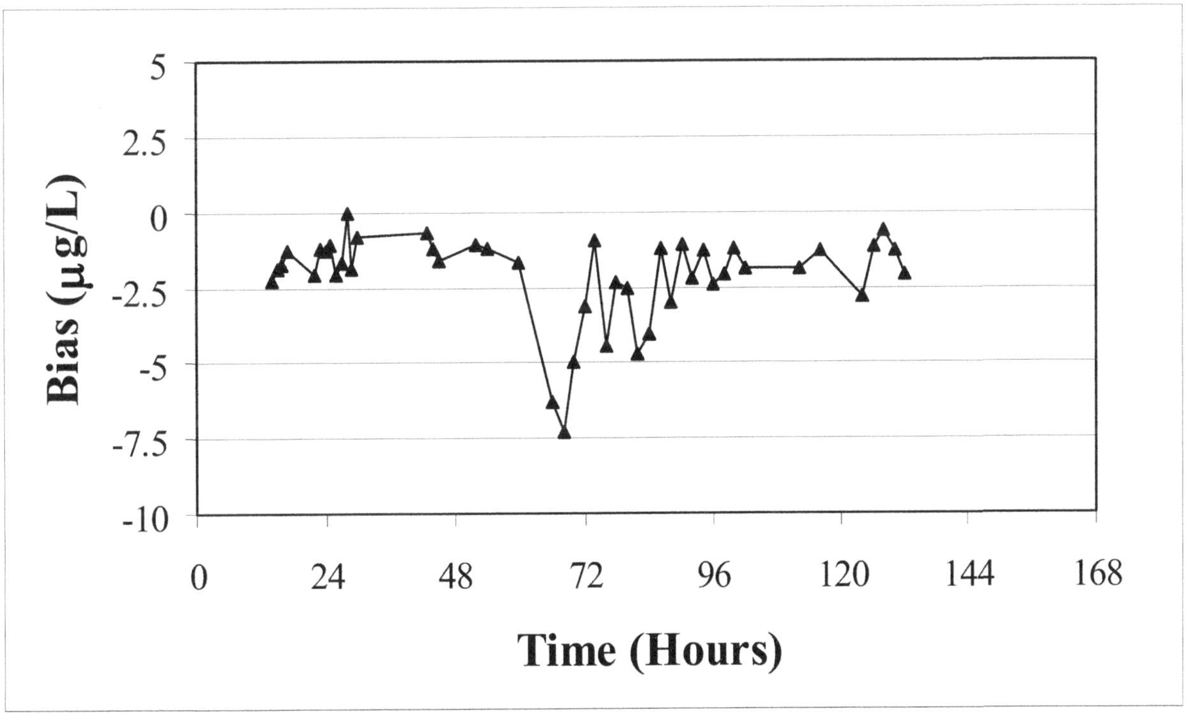

Figure 5.44 Houston bias plot of $CHBr_3$ comparing CMS-GC-ECD original data to EPA 524.2

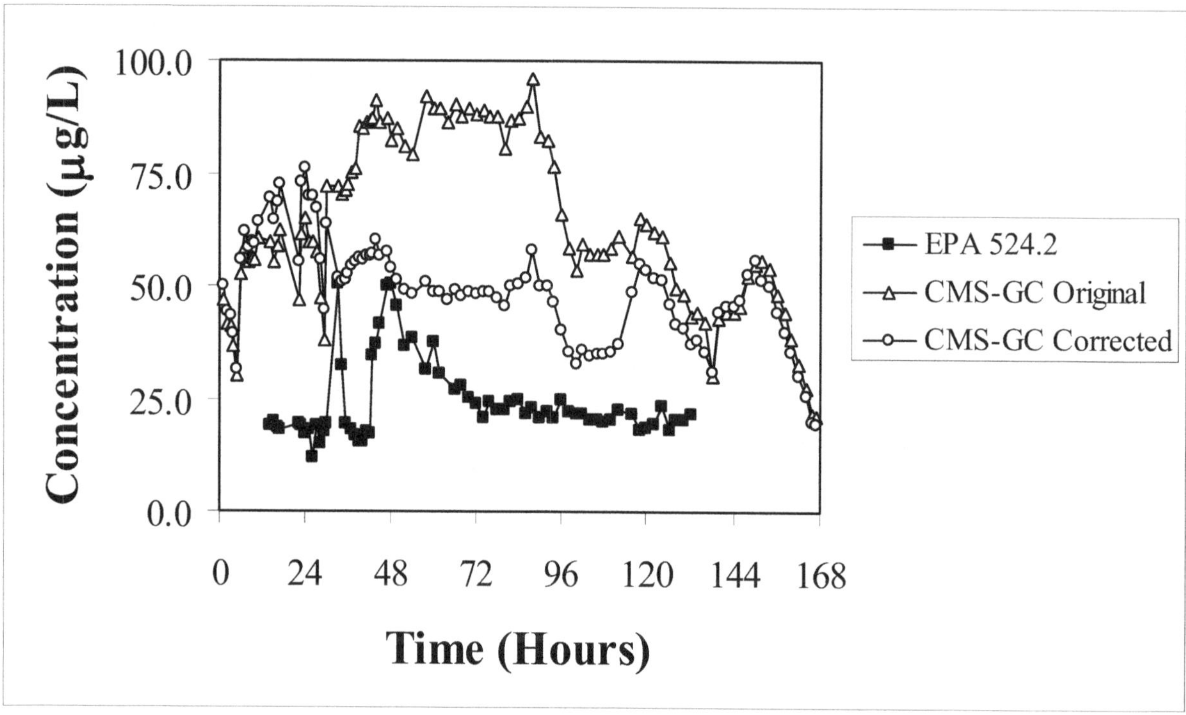

Figure 5.45 Houston concentration plot of TTHM4 for EPA 524.2, CMS-GC-ECD original data and CMS-GC-ECD corrected data

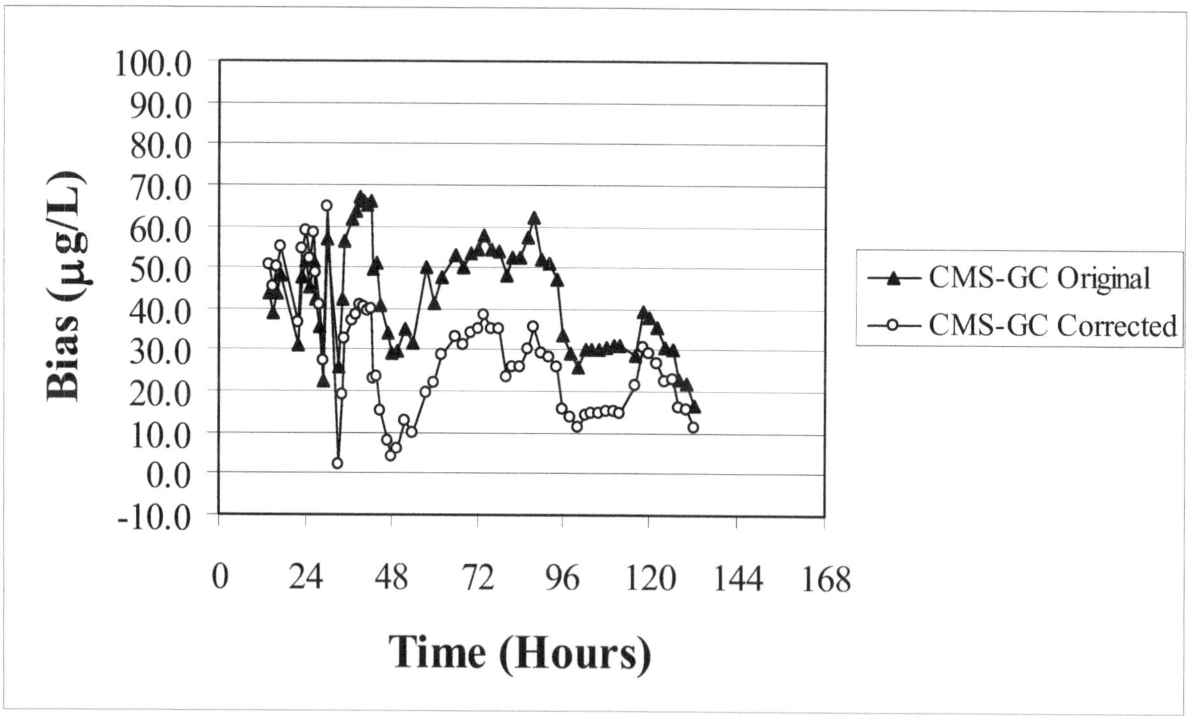

Figure 5.46 Houston bias plot of TTHM4 comparing the CMS-GC-ECD original data and CMS-GC-ECD corrected data to EPA 524.2

CONCLUSIONS AND RECOMMENDATIONS FOR THE CMS-GC ANALYZER

The CMS-GC method is the most promising technique for on-line monitoring of THM4 concentrations in drinking water distribution systems. The CMS-GC-ECD improved upon the SCMS-GC-ECD method through the use of the CMS as a sampling device. While the SCMS probe needs a separate flow-cell to conduct on-line monitoring studies; however, the CMS is the flow-cell, so adapting to an on-line sampling site is simpler. When considering cost, the CMS is a less expensive sampling device than the SCMS; however, the CMS requires in lab construction whereas the SCMS is commercially available. The CMS is a device that can be redesigned and constructed to meet many applications as shown in Chapter 4. As mentioned previously, this is a GC approach that requires two gas cylinders, and thus use of the word portable to describe the technique is with reserve.

Recommendation 5.1: A more robust peristaltic pump (continuous duty-medical) would increase the sample flow rate through the CMS and likely increase analytical response for the THM4. This would also minimize any variations that could be caused by changes in the water distribution pressure.

Recommendation 5.2: Automated internal standard addition need to be further explored. Preliminary studies have indicated that this is a more complex problem than one might realize. Ideally, analyst could use both calibration curve and internal standardization during on-line studies. Furthermore it would also be desirable to use automated standard addition to assess the impact of different drinking water matrices.

Recommendation 5.3: CMS devices need to be constructed with longer silicone rubber membranes as this should increase analytical signal. Longer membranes will likely change sample carry-over behavior as well. Therefore, further optimization will likely be necessary if longer membrane devices are constructed.

Recommendation 5.4: Silicone rubber membranes of different dimensions, such as outer/inner diameters and wall thicknesses should be evaluated to possibly improve performance. Different membrane materials should also be explored.

Recommendation 5.5: Larger sample injection volumes (larger sample loops) should be explored as larger sample injection volumes might improve MDL values for THM4 species as well as other species of interest in drinking water analysis.

Recommendation 5.6: The CMS-GC-ECD need to be applied to other species of interest in drinking water analysis such as volatile organic compounds.

Recommendation 5.7: The CMS device has been used in this Report in conjunction with FIA and GC analysis. It is likely that the CMS device might be useful with spectroscopic techniques as well as other GC detectors. The CMS device needs to be interfaced to a MS to allow for confirmatory analysis.

Recommendation 5.8: Carry-over reduction is needed to quicken analysis time, so it is recommended that air purging of the CMS device (air pumped into the Tefzel shell to purge volatiles) be explored. It might also be possible to improve sensitivity using a reagent to "salt out" THM4 species inside the CMS device.

CHAPTER 6
ION CHROMATOGRAPHY WITH MEMBRANE SUPPRESSED CONDUCTIVITY DETECTION FOR HAA5 SPECIES

INTRODUCTION

A previous AwwaRF Report (Lopez-Avilla 2000) explored using a microextraction ion chromatography-membrane suppressed conductivity detection (IC-MSCD) method for HAA9 analysis. The microextraction step was adapted from USEPA Method 552 (1990) and involved back extracting the HAA9 species into reagent water after the initial extraction into MTBE. Prior to developing the microextraction procedure, the IC-MSCD method was optimized and the MDL values for each HAA9 species ranged from 0.75 to 3.0 μg/L. Using the microextraction procedure (with an 8.4 times concentration factor), the reported MDLs ranged from 0.045 μg/L for MCAA to 1.10 μg/L for TBAA. The mean percent recoveries ranged from 71 to 107%, percent relative standard deviations of 1.9 to 9.3% for check standard concentrations in the range of 1 to 4 μg/L. The method was evaluated in four real drinking water samples from California by comparing the results obtained to those using Standard Method 6233B (1992) and USEPA 552.1 (1992). The microextration-IC method agreed within ±5 μg/L for all four water samples.

Lopez-Avilla noted that the microextraction-IC method requires an experienced analyst to carry out the microextraction procedure and therefore, does not really qualify as a "simplified method". Bromochloroacetic acid (BCAA) co-eluted with nitrate ion, making quantification of BCAA difficult. All drinking water samples analyzed showed nitrate ion interference with BCAA. Small organic acids, such as glycolic, acrylic, oxalic, pyruvic, glyuoxalic, glutaric, and succinic acids were found not to interfere. Sulfate ion was problematic if extreme care was not taken in separating the aqueous and organic layers.

From the standpoint of on-line monitoring, the implementation of automated microextraction procedures is complicated. However the separation and quantification of the HAA9 species is promising. The use of microextraction/pre-concentration of sample was avoided in this research by increasing the sample injection volume. This also meant that the IC-MSCD method could be adapted to on-line monitoring. By simply fitting a sample injection valve with a large injection loop and connecting it to the drinking water distribution system, online monitoring of HAAs would be possible.

Liu et al. (2003) proposed a novel method for the simultaneous determination of nine HAAs in drinking water by IC with large volume direct injection (LVI; 500 μL). The MDLs for the HAA9 species ranged from 1.91 μg/L for MCAA to 48.98 μg/L for TCAA. Acceptable linearity was observed over the 10 to 6000 μg/L concentration levels. Mean % recoveries ranged from 91.3 to 107% and % RSD ranged from 2 to 5%. In another study, Liu and Mou (2003) proposed a revised LVI-IC-MSCD method that simultaneously quantified HAA9 species, perchlorate ion and some common anions in one analysis. They evaluated two high capacity anion exchange columns: a carbonate selective column and a hydroxide-selective hydrophilic column. The MDLs for HAA5 species were as follows: MCAA (1.46 μg/L), MBAA (2.49 μg/L), DCAA (2.16 μg/L), DBAA (2.16 μg/L) and TCAA (1.85 μg/L). The %RSD values were: MCAA (3.4%), MBAA (8.1%), DCAA (5.2%), DBAA (3.16%) and TCAA (2.9%). The mean % recoveries were: MCAA (98.8%), MBAA (99.1%), DCAA (98.5%), DBAA (98.0%) and TCAA (103.2%).

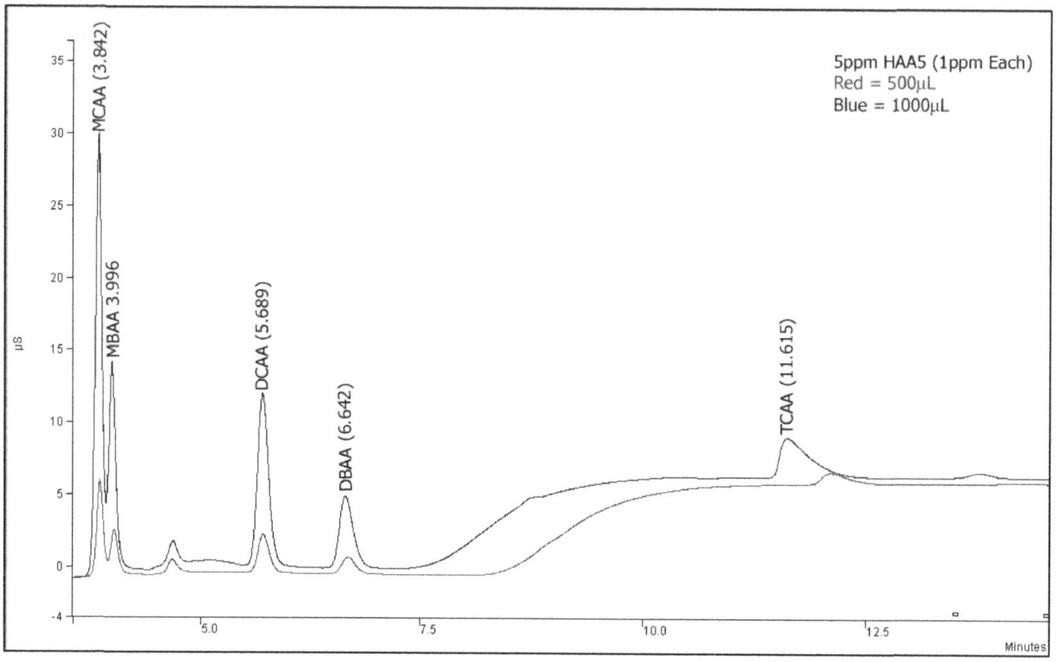

Figure 6.1 Qualitative comparison of IC-MSCD chromatograms using LVI and XLVI

Table 6.1
Results from MDL, accuracy and precision studies for the IC-UV method using XLVI for HAA5 using BOR and CAR eluents

HAA5	MDL (µg/L)†		Mean % Recovery		% RSD		r^2		Conc. Range (µg/L)*
	BOR	CAR	BOR	CAR	BOR	CAR	BOR	CAR	
MCAA	61	84	71	58	36	62	0.974	0.977	50 – 350
DCAA	19	54	68	96	12	24	0.972	0.989	50 – 350
TCAA	79	489	96	288	35	72	0.969	0.790	50 – 350
MBAA	189	74	75	104	107	30	0.997	0.990	50 – 350
DBAA	25	47	100	83	11	24	0.986	0.978	50 – 350

* Each calibration standard was analyzed one time. Typically 4 to 5 calibration standards were analyzed for each calibration curve.

† The check standard (75 µg/L of each HAA) was analyzed seven consecutive times.

More promising results were obtained using the IC-MSCD method. Initial studies were carried out using XLVI in conjunction with IC-MSCD and a CAR eluent. The MDL values ranged from 1.5 to 10 µg/L. Mean % Recoveries ranged from 96.7 to 104% and % RSD values ranged from 3.1 to 8.1%. Typical calibration curves for the IC-MSCD method are shown in Figure 6.2

Additional studies used the XLVI IC-MSCD method with a BOR eluent. The advantage of the BOR eluent is that it exhibits a lower background conductivity compared to CAR eluents—thus lower MDL values. The results of preliminary MDL, accuracy and precision studies are presented in Table 6.2.

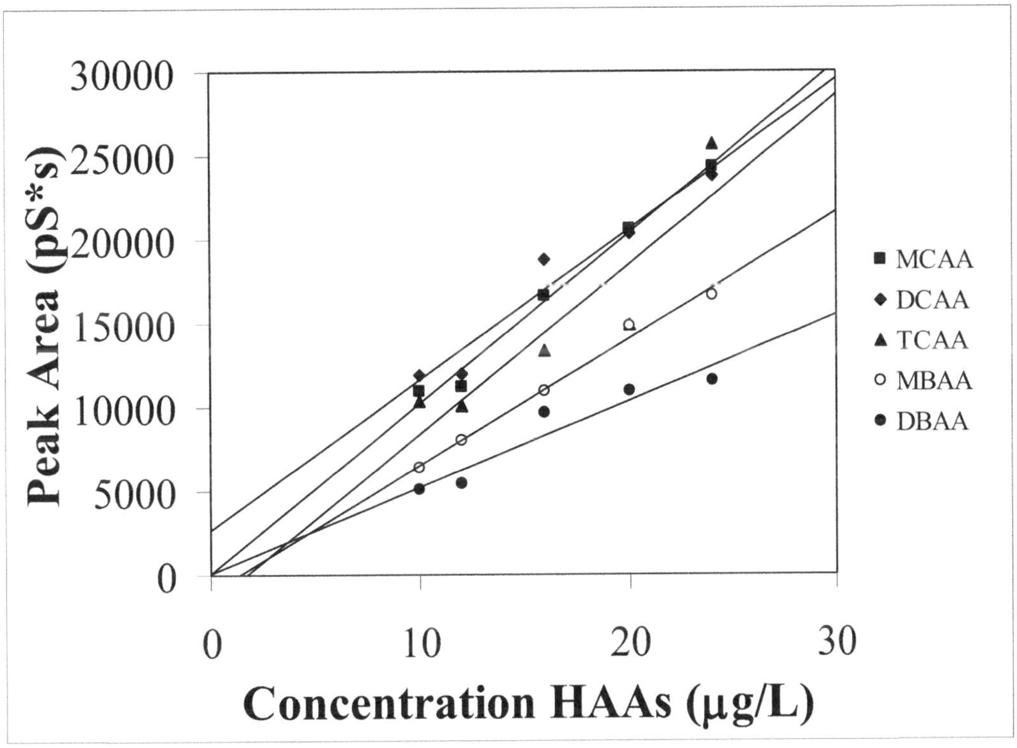

Figure 6.2 Calibration curves for the XLVI IC-MSCD method for HAA5 using a CAR eluent

Table 6.2
Results from MDL, accuracy and precision studies of the IC-MSCD method using XLVI for HAA5 using BOR and CAR eluents

HAA5	MDL (µg/L)†		Mean % Recovery		% RSD		r^2		Conc. Range (µg/L)*
	BOR	CAR	BOR	CAR	BOR	CAR	BOR	CAR	
MCAA	1.4	0.6	103	96.8	4.0	2.0	0.991	0.986	10 – 24
DCAA	2.2	8.8	99.4	123	6.5	2.07	0.977	0.951	10 – 24
TCAA	1.9	4.5	121	103	4.6	12.7	0.953	0.827	10 – 24
MBAA	6.1	1.7	82.5	97.6	21.4	4.9	0.977	0.990	10 – 24
DBAA	4.4	5.5	99.8	110	12.6	14.4	0.856	0.906	10 – 24

* CAR – Each calibration standard was analyzed one time. BOR – The highest calibration standard was analyzed twice. Typically five calibration standards were analyzed for each calibration curve.

† The check standard (11 µg/L each HAA) was analyzed seven consecutive times.

The results using the BOR eluent were slightly improved over those obtained with the CAR eluent. In each case, the results obtained by U of M researchers are similar to those of Liu and Mou (2003). Considering that Liu and Mou (2003) analyzed their HAA check standards at a concentration level of 300 µg/L, the results from the U of M laboratory might be considered more realistic estimates of what to expect from the method. The check standards in the U of M study were 11 µg/L for each HAA concentration level. Table 6.3 presents the results of MDL, accuracy and precision studies for the optimized IC-MSCD method for HAA5. These studies were carried out in the presence of the common ions and inorganic DBPs and compare favorably to other reported methods. The MDL values range from 1 to 3 μg/L. The mean percent recoveries are acceptable and the % RSD are less than 5% for three of the five HAA species. Calibration curves for the XLVI-IC-MSCD method using the BOR eluent are shown in Figure 6.3. A typical chromatogram is shown in Figure 6.5, and illustrates the major issue with this method – small peaks from HAA5 in the presence of large peaks from common ions.

Table 6.3
Results from MDL, accuracy and precision studies in the presence of common anions using the BOR eluent and the optimized XVLI-IC-MSCD method

HAA Species‡	MDL (μg/L)†	Mean % Recovery	% RSD	r^2	Conc. Range (μg/L)*
MCAA	2.7	97	7.9	0.986	10 – 24
DCAA	1.2	91	3.8	0.959	10 – 24
TCAA	1.8	108	4.8	0.897	10 – 24
MBAA	1.1	102	3.1	0.980	10 – 24
DBAA	2.1	102	5.8	0.892	10 – 24

* Each calibration standard was injected once. There were six standards for each calibration curve.
† The check standard was analyzed seven times.
‡ The study was done using the Onguard®-Ag in the presence of the following common ions found in drinking water: fluoride ion (1.0 mg/L), chloride ion (4.5 mg/L), bromide ion (1.0 mg/L), bromate ion (0.15 mg/L), nitrite ion (0.25 mg/L), nitrate ion (1.0 mg/L), sulfate ion (3.8 mg/L), orthophosphate (1.4 mg/L).

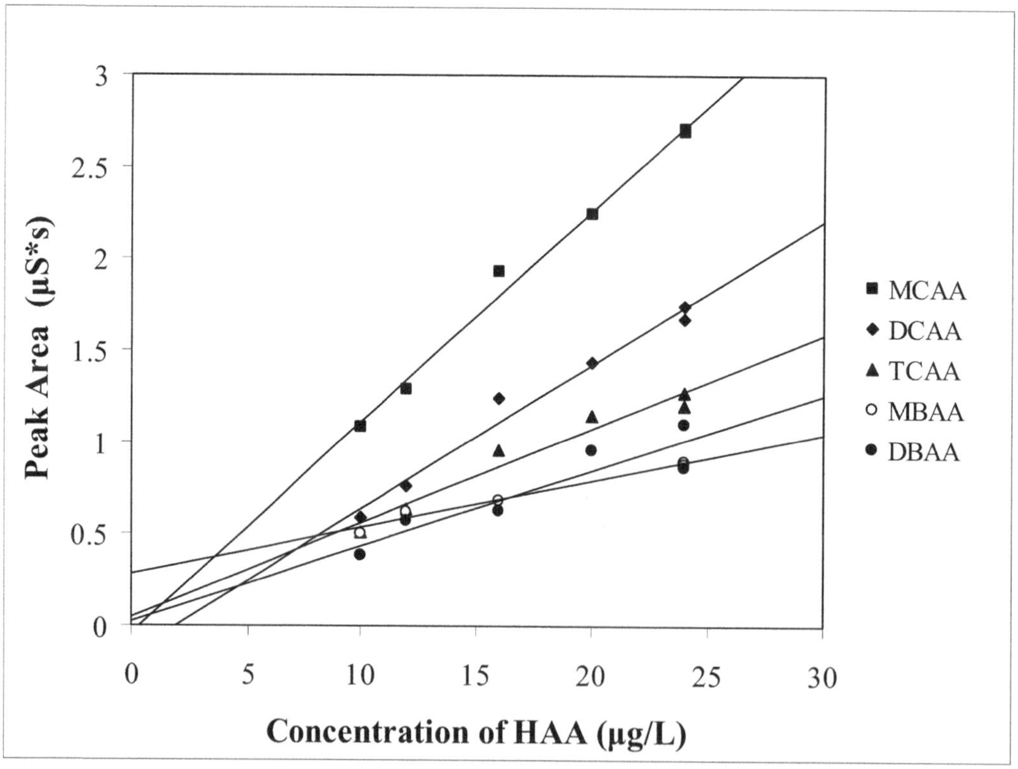

Figure 6.3 XLVI IC-MSCD calibration plots for HAA5 using the BOR eluent

IC-MSCD / Borate Eluent Interference Studies

An important concern with IC-MSCD is the effect of interfering species. In particular, the common anions including fluoride, chloride, nitrite, nitrate, sulfate ions and orthophosphate, which are expected to be present in typical drinking water. In addition, in some regions, bromide ion and bromate ion might be problematic. However, a synthetic sample that was prepared to approximate the Memphis drinking water anion matrix did not exhibit resolution problems. The concentrations of common ions added were based on the results reported by MLGW Water Quality Laboratory (bromide ion and bromate ion were added for completeness, but are typically not at such a large concentrations in Memphis drinking water).

Figure 6.4 shows a typical chromatogram from this analysis, which illustrates one of the drawbacks of IC-MSCD when applied to low level HAA5 analysis. Note that while it was possible to resolve the HAA5 species from the common ions and other inorganic DBPs, the chromatogram exhibits high responses for the common ions (typically at mg/L concentration levels) and low response for the HAA5 species (typically at the μg/L concentration levels). This means that use of IC-MSCD requires very good separation and will likely need to be optimized to take into account local matrices with respect to common ion concentrations and other inorganic DBPs. The most serious problems seem to arise from nitrite ion interfering with DCAA and sulfate ion interfering with TCAA. Additional efforts were devoted to finding avenues around these interferences.

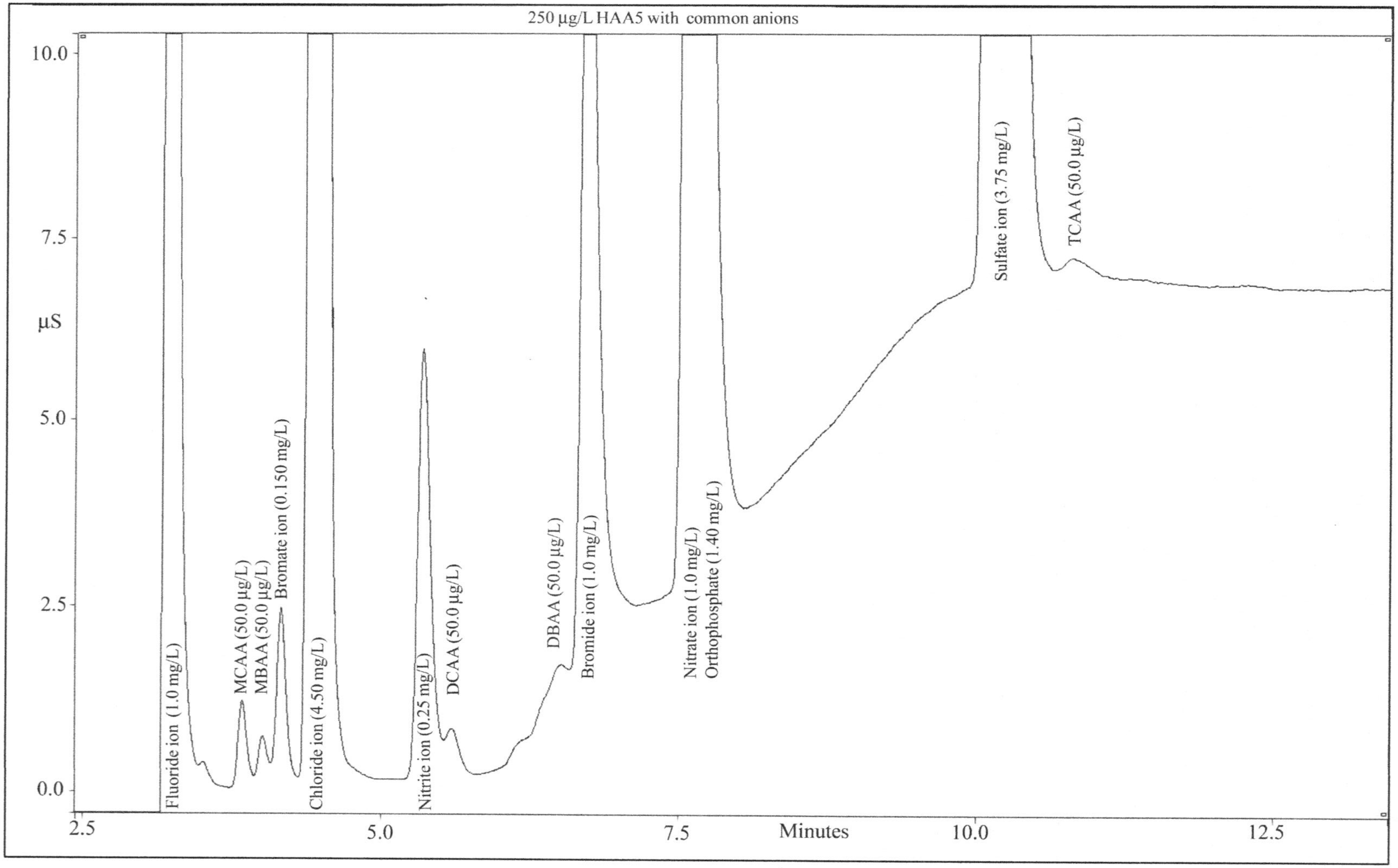

Figure 6.4 XLVI IC-MSCD chromatogram using the BOR eluent showing the resolution of HAA5 species in the presence of the common anions, bromide ion and bromate ion

Linearity Study

Studies to evaluate the linear range of the XLVI-IC-MSCD method for measuring HAA5 concentrations in drinking water in the presence of possible interferences were performed. The linearity of the method is acceptable over the range of 10 to 24 μg/L for each HAA5 species. Additional linearity studies were also conducted over a broader concentration range (Table 6.4). Figure 6.5 shows a plot of the peak area versus concentration of HAA species over the concentration range of 10 to 1000 μg/L in the presence of possible interferences. These results seem to indicate a relatively large linear range for the IC-MSCD method.

Table 6.4
Results from the linearity study using the optimized IC-MSCD method for HAA5 in the presence of common anions, bromate ion and bromide ion

HAA Species*	Conc. Range (μg/L)	Replications of a Standard	Total Number of Standards	r^2
MCAA	10 – 1000	1	20	0.998
DCAA	10 – 1000	1	20	0.996
TCAA	10 – 175	1	9	0.700
MBAA	10 – 1000	1	20	0.990
DBAA	10 – 1000	1	20	0.999

* The study was done using an Onguard®-Ag in the presence of the following common ions found in drinking water: fluoride ion (1.0 mg/L), chloride ion (4.5 mg/L), bromide ion (1.0 mg/L), bromate ion (0.15 mg/L), nitrite ion (0.25 mg/L), nitrate ion (1.0 mg/L), sulfate ion (3.8 mg/L), orthophosphate (1.4 mg/L).

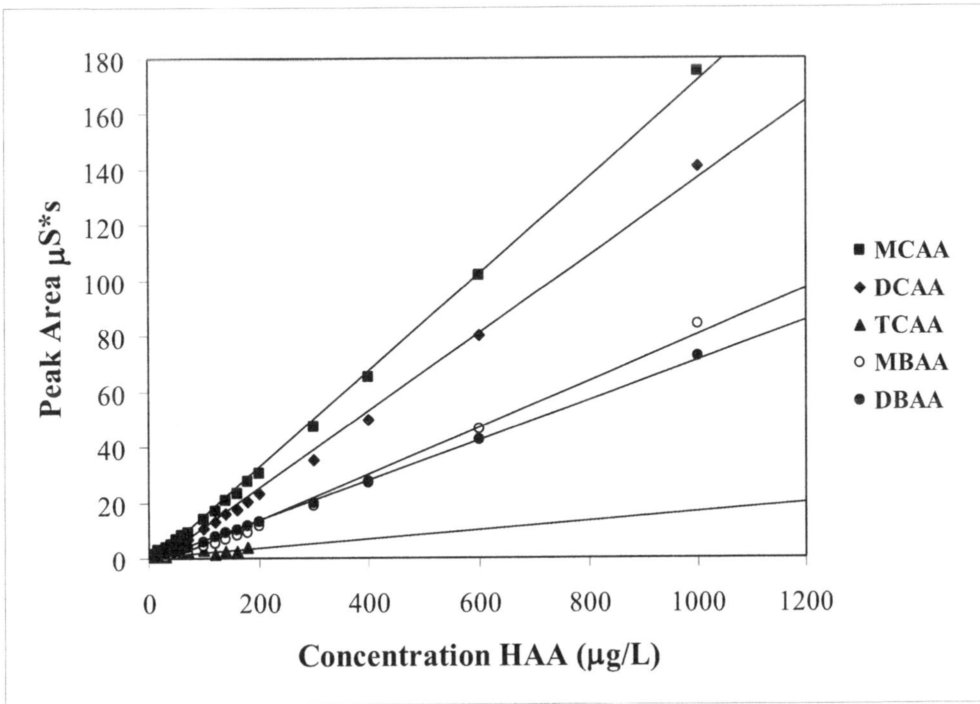

Figure 6.5 Results from the XLVI IC-MSCD linearity study in the presence of possible interferences

XLVI-IC-MSCD Analysis of Real World Water Samples

Drinking water samples were collected from various locations in the region. These samples were analyzed for HAA5 using IC-MSCD and EPA Method 552.3. The results of these studies are mixed, and the purpose was to get an idea of the types of problems that might be faced when the methods are applied to on-line monitoring. Samples were collected from middle and west Tennessee, and southern Louisiana. The goal of the sampling was to obtain drinking water that originated in rivers and ground water to provide some variation in the drinking water matrix. There appears to be significant problems with nitrite and sulfate ion interferences as discussed earlier. The results of these studies are collected in Table 6.5. Additional analysis and discussion of the results follow. Taken as a whole, it is difficult to see any trends in the data on these preliminary studies aimed at comparing the IC-MSCD method for HAA5 to EPA Method 552.3. A cursory glance of the numbers leaves the impression that the two analytical methods do not provide comparable results. However this initial impression may be misleading. Consider evaluating each sample site independently—one site at a time—and see if any trends are apparent.

Table 6.5
Comparison of IC-MSCD to EPA 552.3 for HAA5 species and THAA5

Sample	Method	MCAA	DCAA	TCAA	MBAA	DBAA	Total HAA5
TN1	EPA	n.d.	14.8	17.8	n.d.	0.97	33.6
	IC	1.9	5.2	n.d	n.d.	n.d.	7.1
TN2	EPA	6.8	20.2	23.0	n.d.	0.94	50.9
	IC	n.d	4.0	n.d	n.d.	n.d.	4.0
TN3	EPA	n.d	12.3	18.2	n.d.	0.96	31.5
	IC	n.d	n.d.	n.d.	n.d.	n.d.	n.d
LA1	EPA	n.d	19.3	20.2	n.d.	0.93	40.3
	IC	n.d	11.8	n.d.	n.d.	n.d.	11.8
TN4	EPA	n.d	1.4	n.d.	n.d.	1.19	2.6
	IC	n.d	2.1	n.d.	n.d.	n.d.	2.1
PEO-098	EPA	12.3	12.9	n.d.	13.1	25.1	53.4
	IC	16.6	5.7	n.d.	11.2	12.0	45.5

For example, the TN1 sample site is characterized by a sample with a fairly high concentration of sulfate ion. The raw water is pumped from the Caney Fork River to a reservoir where it is chlorinated and distributed. Sulfate ion, at high concentrations, interferes with the IC-MSCD method for TCAA. This is especially apparent in the TN1 sample. The TN1 sample, as with all the samples in this study, (Table 6.5 and Figure 6.6) was analyzed in duplicate using EPA 552.3 and the IC-MSCD methods. THAA5 is calculated in each method by summing the concentration of each HAA5 species. EPA 552.3 gives 33.6 μg/L for THAA5 while the IC-MSCD gives 7.1 μg/L---apparently not very good agreement. Consider that in these results, it is impossible to quantitate TCAA because of sulfate ion interference using the IC-MSCD method.

In this case, approximately 46% (almost one-half) of THAA5 are not accounted for by the IC-MSCD method. Certainly, there will be inconsistent results when these two methods are compared. This is exactly the observation, so if sulfate ion concentration is high and the sample contains a relatively high concentration of TCAA, the results of the two methods will not agree.

The observations noted in the result from the TN1 sample are more apparent in the TN2 water sample (Table 6.5 and Figure 6.7). In the TN2 sample, sulfate ion interference is even more pronounced and thus the current IC-MSCD method cannot quantitate TCAA. The result is a large difference between the EPA 552.3 results and the IC-MSCD results (50.9 µg/L vs 4 µg/L, respectively). With high concentrations of sulfate ion and a high percentage of HAA5 being TCAA (~45%), the results from the two methods do not agree.

Similar results are observed with the TN3 sample (Table 6.5 and Figure 6.8) and LA1 (Table 6.5 and Figure 6.9). In each case, high sulfate ion concentration interferes with the quantitation of TCAA. The analysis of the TN3 samples was further complicated by the presence of a high concentration of some unidentified species that complicated analysis of DCAA. Regardless, the trend holds with this sample in that relatively high concentrations of sulfate ion coupled with relatively high concentration of TCAA lead to a large difference in the results of the two methods.

Analysis of the TN4 water sample (Table 6.5 and Figure 6.10) shows a different situation. With the TN4 water, TCAA does not appear to be present appreciably. In this case, the two methods compare favorably.

A performance evaluation standard was analyzed using the IC-MSCD method. The results are presented in Table 6.6. Again, there is interference between TCAA and sulfate ion, as well as interference between nitrite ion and DCAA. The IC-MSCD method passed the performance evaluation standard for three of the five HAA species, but failed to quantify DCAA and TCAA effectively.

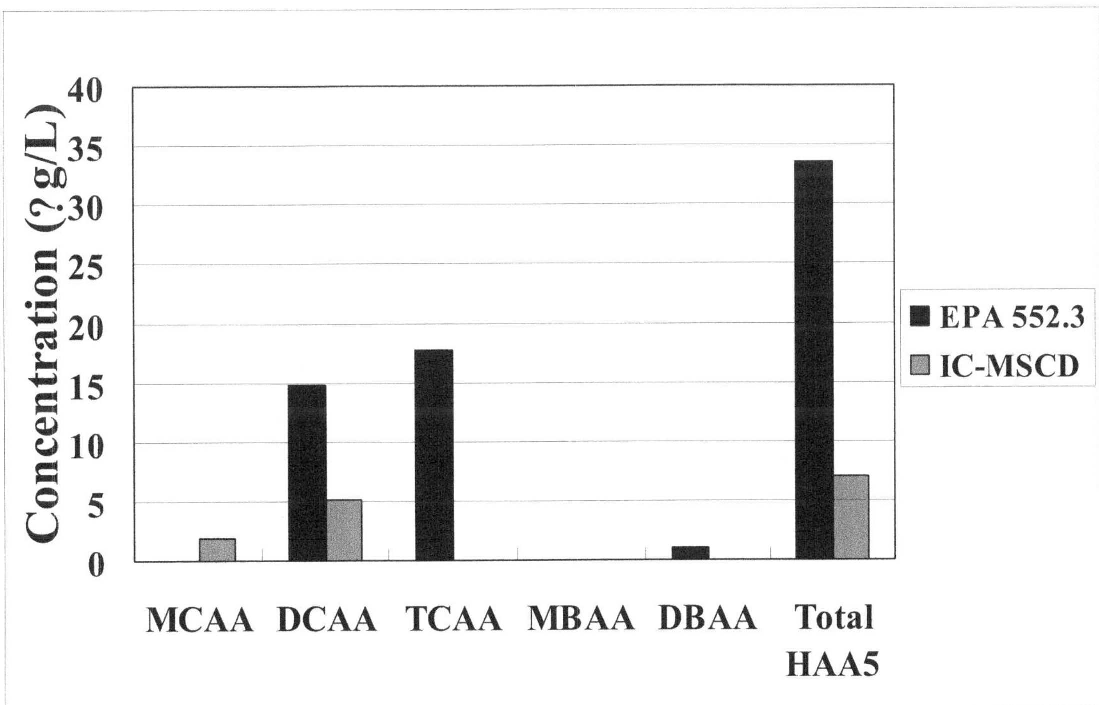

Figure 6.6 Comparison of EPA 552.3 and XLVI IC-MSCD for the TN1 water sample

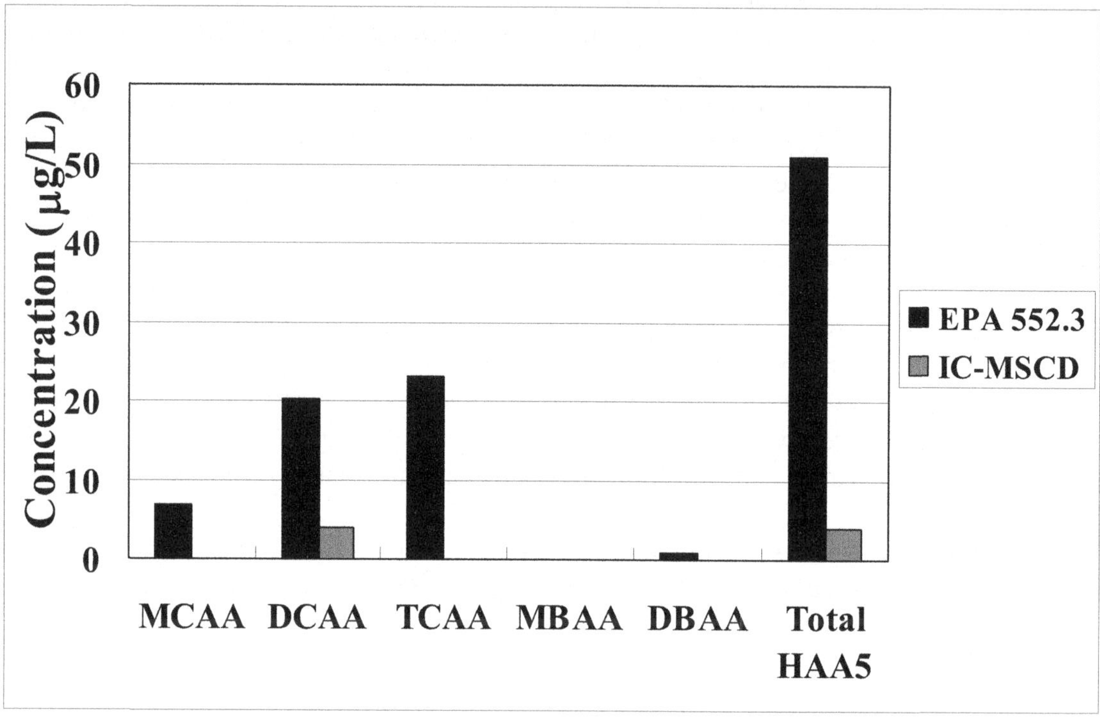

Figure 6.7 Comparison of EPA 552.3 and XLVI IC-MSCD for the TN2 water sample

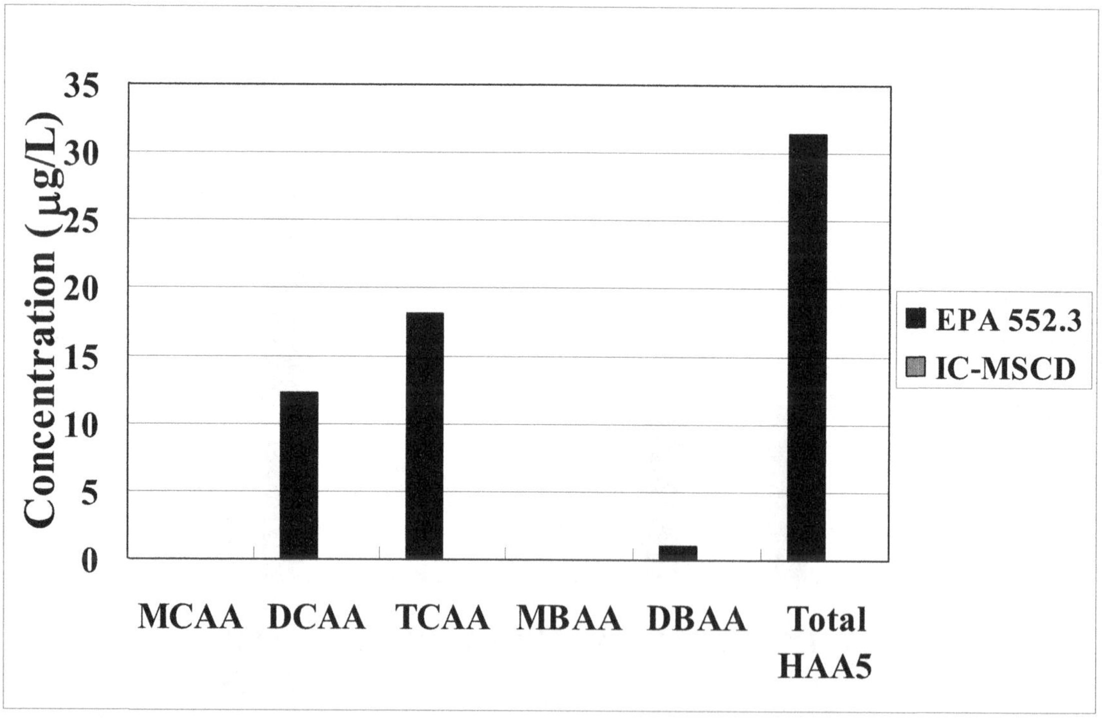

Figure 6.8 Comparison of EPA 552.3 and XLVI IC-MSCD for the TN3 water sample

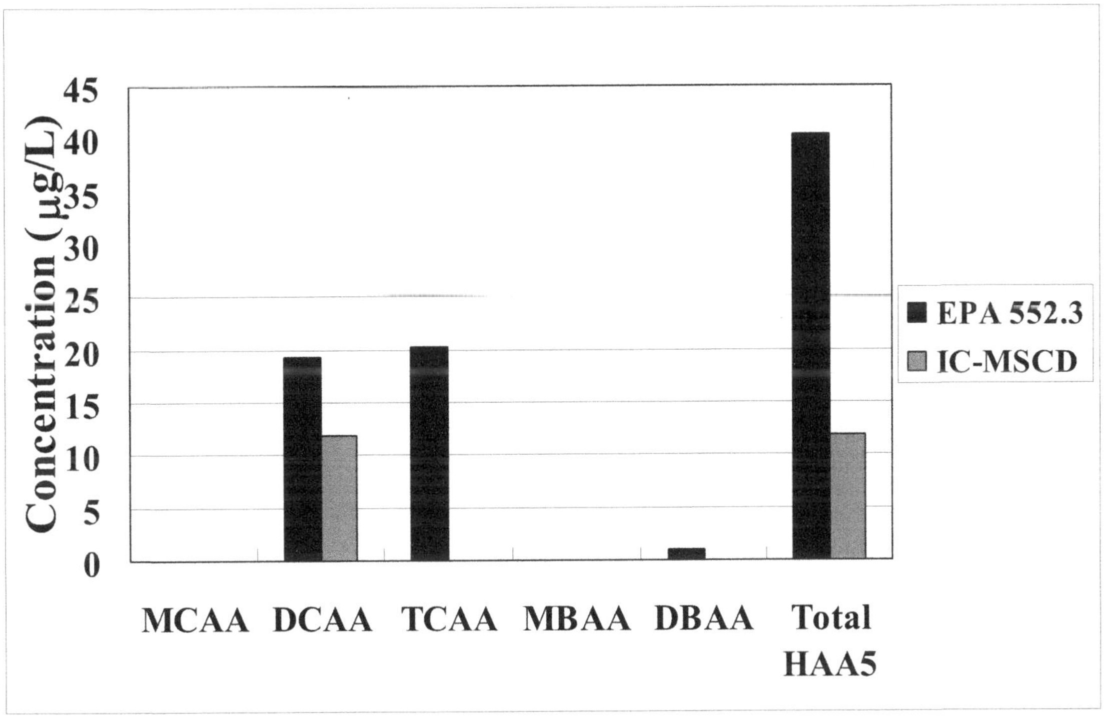

Figure 6.9 Comparison of EPA 552.3 and XLVI IC-MSCD for the LA1 water sample

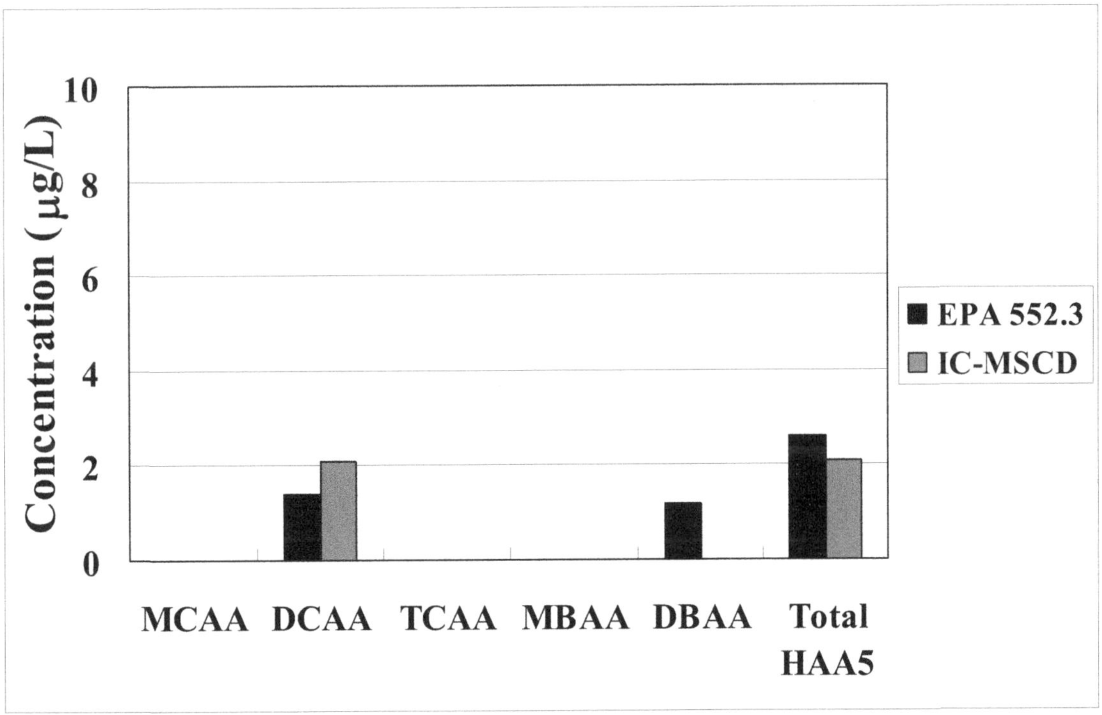

Figure 6.10 Comparison of EPA 552.3 and XLVI IC-MSCD for the TN4 water sample

Table 6.6
Results of a performance evaluation sample for IC-MSCD

HAA species	PE accepted result	Accepted Conc. Range	U of M IC-MSCD Results (µg/L)*	PE Result: PASS/FAIL
MCAA	11.3	3.37 – 17.6	17	PASS
DCAA	18.8	11.7 – 24.6	5.7	FAIL
TCAA*	26.9	10.6 – 36.5	n.d.	FAIL
MBAA	12.0	4.93 – 17.30	11	PASS
DBAA	13.8	6.75 – 19.0	12	PASS

* The current IC-MSCD method suffers from a SO_4^{2-} interference that may be alleviated using commercially available cartridges

Summary of Preliminary XLVI-IC-MSCD Studies

In summary, the results of these analyses point out a major limitation of the IC-MSCD method. When working with real world samples that originate from different sources (surface and ground waters), the composition of the sample matrix may play a larger role in the success or failure of the method. These studies illustrate that the IC-MSCD method worked well for three of the five HAA5 species. There were significant problems with nitrite and sulfate ion interferences. It was thought commercially available pre-treatment cartridges would take advantage of certain chemistries to remove unwanted species that complicate analysis. For example, Ag (I) cartridges can be used to remove chloride ion interference and a Ba (II) cartridge may be used to minimize sulfate ion interference. So while the IC-MSCD method showed promise for monitoring HAA5 (as illustrated by the MDL, accuracy, precision and linearity results), when it is taken from the laboratory to real world samples, analysis is complicated by the matrix containing common ions and inorganic DBPs. In particular, the presence of sulfate and nitrite ion complicates the chromatography; therefore, limiting the IC-MSCD method for two of the most important HAA species—TCAA and DCAA.

Alternative Methods for Minimizing Interferences

One study was aimed at using an alternative high capacity anion exchange column (Dionex IonPac AG9-HC guard column (50 x 4 mm) and IonPac AS9-HC high capacity analytical column (250 x 4 mm)) to improve separation between HAA species and interferences. The second approach involved preliminary studies evaluating the utility of using commercially available removal cartridges (Ba (II) cartridges from Dionex and Alltech) to minimize the effects (through removal) of interfering anions like sulfate ion.

The first approach to get around this sulfate ion interference was to use the high capacity IC column (AS9-HC) to solve the problem of co-elution issues that were occurring between nitrite ion and DCAA, and sulfate ion and TCAA. A method was developed that resolved all components except MCAA and bromate ion. In most situations, bromate ion is not a large problem. The DCAA-nitrite ion interference was effectively minimized as was the TCAA-sulfate ion interference. At this point, the results appeared very promising for solving the IC-MSCD issues. However, after the IC-MSCD method with the high capacity column was developed further, it was not possible to work at HAA5 concentrations in the range 1 to 25 μg/L levels with acceptable MDL, accuracy and precision, even though at higher concentrations, it was clear the column offered better resolution. At concentrations similar to previously presented IC-MSCD studies, the high capacity column could not detect the HAA species reliably (Figure 6.11).

The second approach was to use commercially available pre-treatment cartridges to remove sulfate ion interference. Two commercial sources provided cartridges for evalutation, Dionex and Alltech. Dionex offers cartridges in the 2.0 to 2.2 meq (1 mL) and 5.0 to 5.5 meq (2.5 mL) sizes, and Alltech offers comparable sizes. These Ba (II) cartridges require activation with addition of Ca (II) ions in the form of a solution of calcium chloride. The calcium ions exchange with the Ba (II) ions on the resin, then the Ba (II) ions precipitate as barium sulfate (removing sulfate ion in the process). An Ag (I) pre-treatment cartridge used for chloride ion removal is placed after the Ba (II) cartridge to remove the excess chloride ions from the $CaCl_2$ activation. A sample was analyzed with and without pretreatment using the tandem sulfate ion/chloride ion pretreatment system. An approximate five-fold reduction of peak area for the sulfate ion peak was observed (Figure 6.12). The idea behind these experiments was that the removal of such a large amount of sulfate ion makes it feasible to resolve the sulfate ion and TCAA peaks. An additional attraction would be that with the removal of sulfate ion interference would give the analyst a bit of "wiggle room" to alter the chromatography gradient, thus the DCAA and nitrite ion components could be more fully resolved.

Unfortunately, the results of these studies were not satisfactory using either of these two approaches. While it was possible to optimize the analysis of TCAA in the presence of high concentrations of sulfate ion through a combination of pretreatment cartridges and optimized separation, the resolution of DCAA from normal concentrations of nitrite ion was not possible. The researchers carried out some very preliminary pretreatment studies aimed at removing nitrite ion in some manner, but these were not successful. For example, several oxidizing agents were used hoping to oxidize nitrite ion to nitrate ion including ozone, ozone and hydrogen peroxide and ozone in basic solution (to produce hydroxyl radical). None of the pretreatment techniques showed promise. To the best of our knowledge, there is no commercially available pretreatment for nitrite ion removal.

CONCLUSIONS AND RECOMMENDATIONS FOR XLVI-IC-MSCD METHODS

Despite the best efforts of these researchers, it was decided that the IC-MSCD method was going to require too many modifications to allow for routine on-line monitoring. The chromatography is too complex even in the relatively clean matrix of Memphis drinking water. Each time the method is applied to different matrices, there would likely need to be extensive method development and optimization done for satisfactory results. It is possible that the method would fail altogether in systems with high concentrations of common ions and inorganic DBPs.

These results seem inconsistent with some of the findings discussed in the literature of the Phase One Final Report (Emmert et. al 2004b). Liu and Mou (2003) report using a similar method with satisfactory results. Closer examination shows that the authors reported single digit MDL values, accuracy and precision estimates derived from a check standard approximately 300 times higher that the MDL values. This is inconsistent with good laboratory procedures. As analytical chemists, the researchers in our laboratory strive to work within a factor of two of the reported MDL values, when possible, so the results are reported for check standards that are similar in concentration to those of the estimated MDL value. This "factor of two" limit is perhaps a bit demanding. The understanding in the laboratory is that the EPA recommends the analyst work within a factor of 2.5 to 5 of the reported MDL value. Whichever factor one chooses, the results reported by Liu and Mou (2003) are not acceptable in that they are a factor of ~300 times higher in their check standard concentrations than their reported MDL values.

However, the results reported here are consistent with those of Lopez-Avila (2000) who suggests that IC-MSCD methods do not provide suitable "simplified" methods for HAA5 analysis in real water samples. It should be noted that there are possibilities of automating sample pretreatment using flow injection and sequential injection methods. In the future, it might be possible to use IC-MSCD for on-line monitoring studies, but at the present time, this method is not recommended. However, the IC-MSCD research does provide a foundation upon which to base the post column reaction of NCA followed by fluorescence detection of the HAA5-NCA product species. This work is discussed in detail in Chapter 7.

Recommendation 6.1: IC-MSCD methods are not recommended for on-line monitoring. Sensitivity and selectivity issues are problematic. For example, the situation in most cases would be to measure µg/L levels of HAA species in the presence of mg/L levels of common ions. This would mean that a high degree of resolution would be necessary.

Recommendation 6.2: The relative simplicity of the IC-MSCD method is attractive, but complex matrix effects will likely complicate analysis. The analyst would need to optimize the method to take into account common ions present in local distribution systems. A more selective detector is needed instead of the fundamentally non-selective conductivity detector.

Recommendation 6.3: Sample pre-treatment using FIA and sequential injection analysis may be possible in the future, but this sort of pretreatment approach further complicates analysis and increases capital expenses.

Recommendation 6.4: IC separation needs to be combined with selective detection. For example, as shown in Chapter 7, it is possible to use IC separation followed by post-column reaction with NCA with detection of the fluorescent NCA-HAA reaction products. This PC-IC method should provide the necessary sensitivity and selectivity to allow for on-line HAA monitoring.

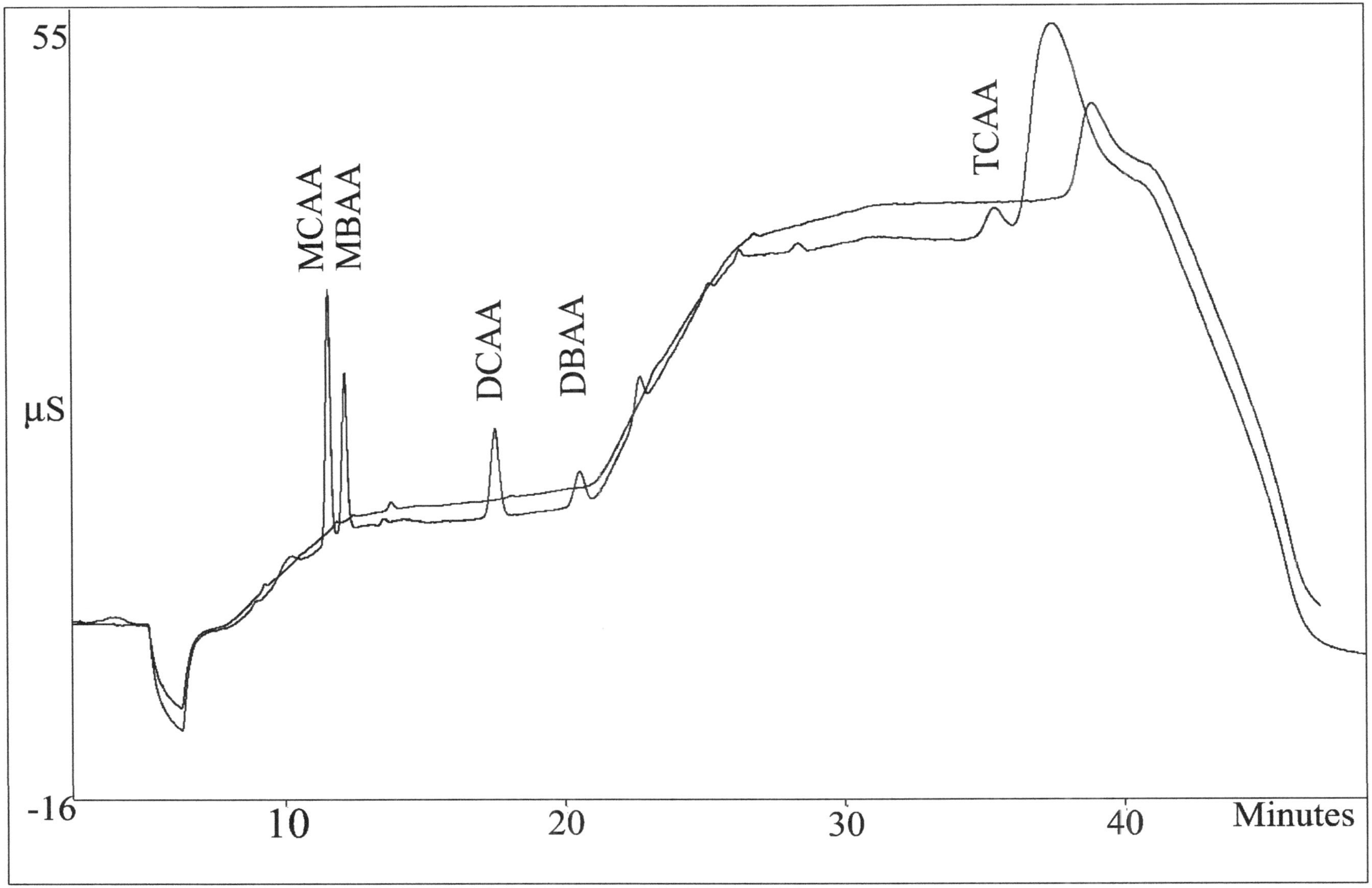

Figure 6.11 XLVI IC-MSCD chromatogram using the AS9-HC IC column

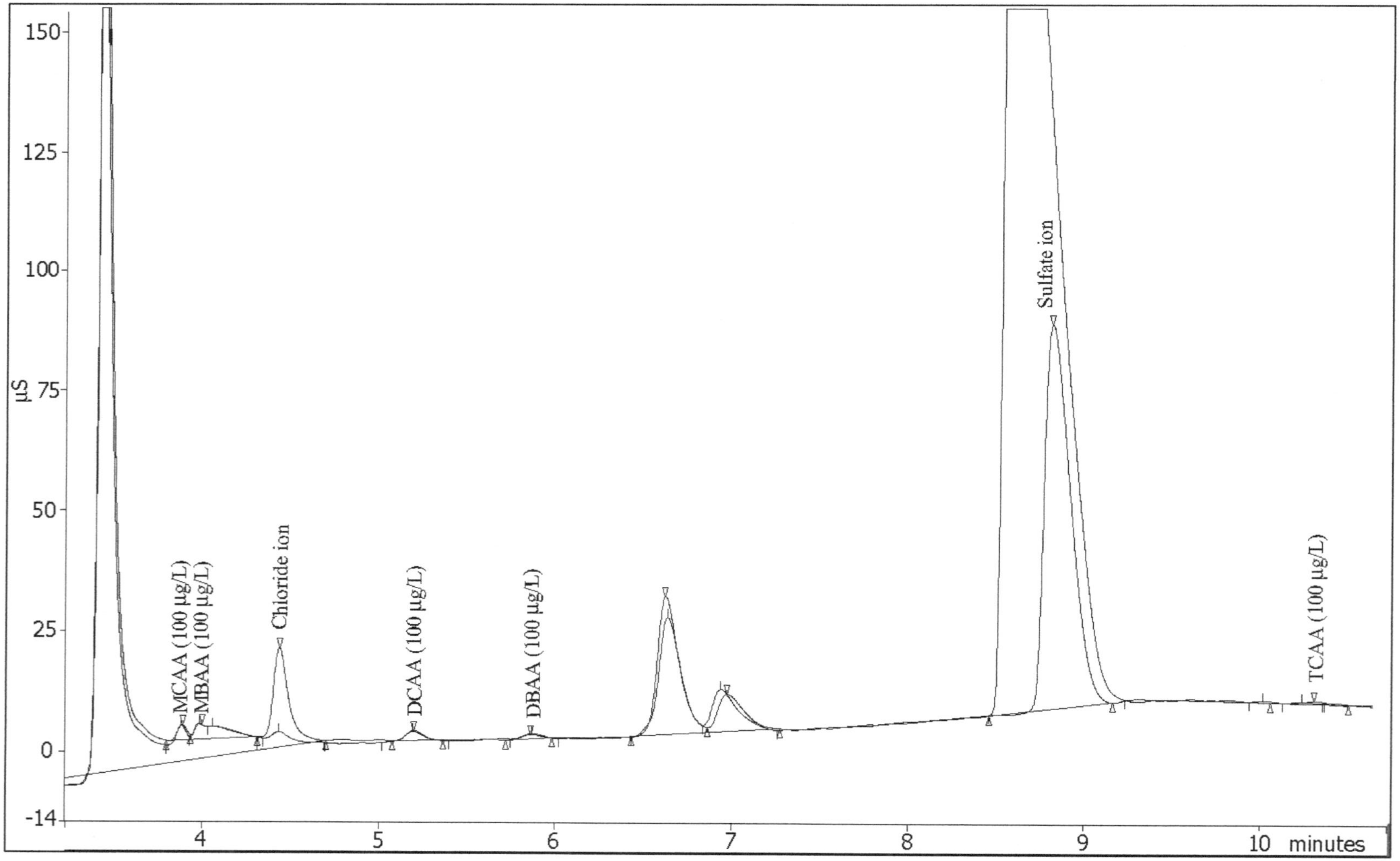

Figure 6.12 Memphis water spiked with 500 µg/L HAA5 before and after calcium (II) activation

CHAPTER 7
POST COLUMN-ION CHROMATOGRAPHY USING THE NCA REACTION WITH HAA5 SPECIES FOLLOWED BY FLUORESCENCE DETECTION

INTRODUCTION

The objective of this research was to combine IC separation of HAA5 with a post column reaction using the NCA chemistry (PC-IC) to develop a sensitive and selective method for HAA5 (Simone et al. 2006). It was desirable that this method be readily adaptable to on-line monitoring directly from the drinking water distribution system. It is not the goal to replace the USEPA Methods, but to compliment them by providing researchers with on-line screening methods that can be used in near real-time monitoring. The results of optimization, MDL, accuracy, precision, linearity and interference studies are presented. In addition, the fully automated PC-IC analyzer (Figure 7.1) is compared to USEPA Method 552.3 (2003a) in side-by-side on-line monitoring studies carried out in the Memphis and Houston drinking water distribution systems.

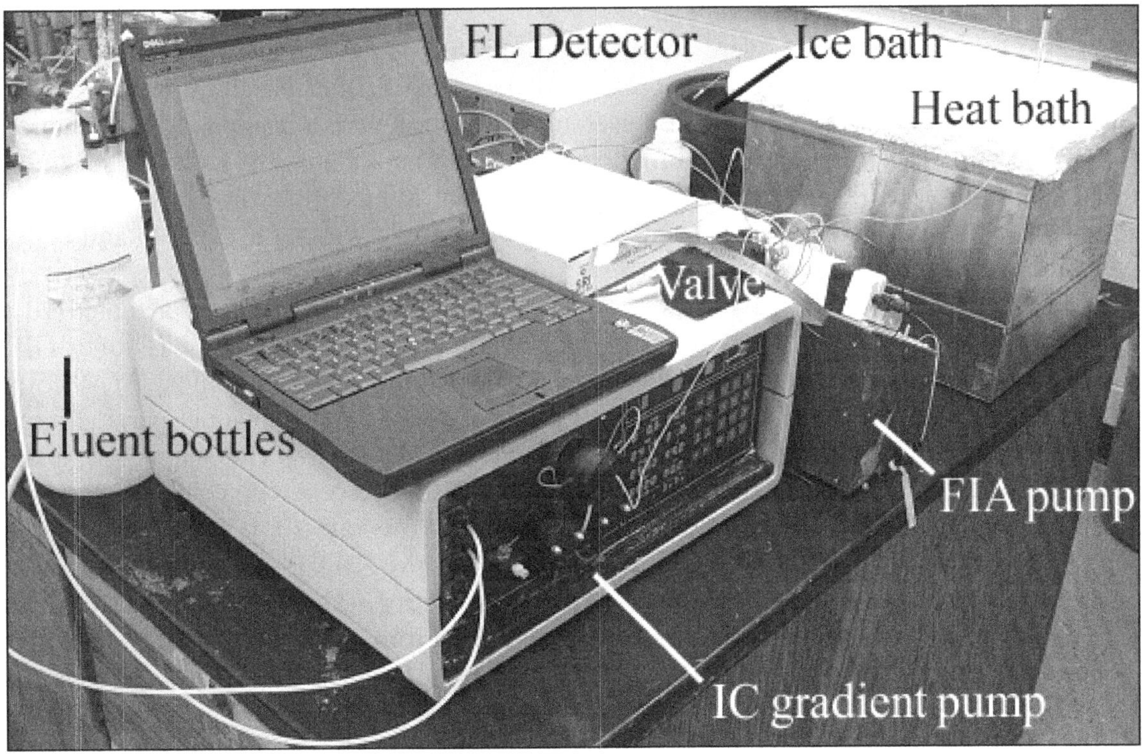

Figure 7.1 Picture of the optimized PC-IC instrument

EXPERIMENTAL

All chemicals used for standards and reagents were of 97% purity or greater. The HAA standards were purchased from Sigma-Aldrich. Sodium hydroxide was purchased from Fisher

Scientific and boric acid was purchased from Matheson Coleman and Bell. All reagents, standards and eluents were prepared in reagent grade water with a resistivity of 18.2 MΩcm and total organic carbon (TOC) concentrations of ≤10 μg/L produced by a Millipore Mill-Q Plus/Mill-RO® Water Purification System. Glassware was meticulously cleaned using concentrated detergent and then rinsed with distilled water three times followed by reagent water several times. The IC eluents were prepared by dissolving boric acid (H_3BO_3) and sodium hydroxide (NaOH) into reagent grade water followed by degassing with nitrogen for 20 minutes. A stock standard solution (SSS) containing 100 mg/L of each HAA5 species was prepared by weighing 10.0 mg of each HAA and 10 mg of ammonium chloride into a single 100.00 mL volumetric flask and diluting to volume with reagent water. The SSS was then diluted by a factor of 100 with reagent water to produce a primary dilution standard (PDS) containing 1.00 mg/L of each HAA5 species. The PDS was then diluted by a factor of 10 with reagent water to make the secondary dilution standard (SDS) containing 100 μg/L of each HAA5 species. The PDS and SDS were prepared daily from the SSS and diluted accordingly to make calibration standards. The NCA reagent solution (37.5% NCA) was prepared by adding NCA to reagent water. Slight heating was used to facilitate dissolution of the solid NCA. The 2M NaOH solution was prepared by dissolving NaOH pellets into reagent water.

Instrumentation

Two different analyzers were constructed, a prototype system and an optimized system (Figure 7.2). The prototype PC-IC used a relatively inexpensive filter-based fluorescent detector. The excitation source was a mercury lamp used in conjunction with a bandpass filter centered on 360 nm ± 45 nm. The emission was a broadband emission with the use of a glass filter (420 nm cutoff). The optimized PC-IC used a state-of-the-art dual monochromator-based detector with scanning capabilities (Waters 2475). The excitation wavelength was set to 365 nm and emission wavelength was set to 455 nm.

The prototype PC-IC analyzer was comprised of an microelectrically actuated 10 port sample injection valve (VICI, Inc) with a 2000 μL sample loop, a gradient pump (Dionex GPM-1 gradient pump), guard and analytical anion exchange columns (Dionex IonPac AG-9 (50 x 4 mm) and Dionex IonPac AS9-SC (250 x 4 mm), respectively), a two-channel peristaltic pump (Buchler Polystaltic), a knitted open-tubular (KOT) mixing device/trap (SeQuant AB), a heated water bath (Thelco Precision Scientific Model 83), and a fluorescence detector (Waters Model 420). The eluent gradient was produced at a constant flow rate of 0.7 mL/min from two components: a borate buffer (BOR) (400 mM H_3BO_3: 80 mM NaOH) and reagent grade water. The gradient was programmed as follows: The initial conditions were 40 mM H_3BO_3 : 8.0 mM NaOH for 2.0 min. The sample injection time was 4 min. From 2.0 to 4.0 min the concentration of the eluent increased to 120 mM H_3BO_3 and 24 mM NaOH and was held until 5.0 min. From 5.0 to 8.5 min, the eluent concentration increased to 400 mM H_3BO_3 and 80 mM NaOH and held until 10.0 min. From 10.0 to 15.0 min, the eluent concentration returned to initial conditions. At 17.5 min, flow was stopped for 25 min to allow for post column reaction to occur, and then resumed at 42.5 min.

The post-column reaction system consisted of the peristaltic pump and a mixing manifold. The flow rate of the NCA reagent was 0.25 mL/min and the 3M NaOH was 0.25 mL/min, for a total flow rate (including the BOR IC eluent flow rate) of 1.2 mL/min. The NCA and NaOH were premixed prior to merging with the BOR IC eluent. The post column trap was a

20 m KOT combined with 11 meters of a lab-constructed serpentine coil for a total length of the post column trap of 31 meters. This large trap was needed to contain the entire chromatogram from the IC column so that it could be heated during the stopped-flow step. The entire reaction trap was immersed in a heated water bath at 94°C. The fluorescent NCA-HAA5 products were detected with a fluorescence detector. MCAA and MBAA coeluted at 46.8 min. One of the primary goals of this research was to be able to measure total HAA5 concentrations at a sampling rate of 1 sample per hour. It was not possible to resolve the monohalogenated species with the current PC-IC method and obtain this goal. As a compromise, the monohalogenated species were allowed to co-elute. DCAA eluted at 48.8 min, followed by DBAA at 49.4 min and TCAA was at 53.1 min. The total run time was 55 minutes.

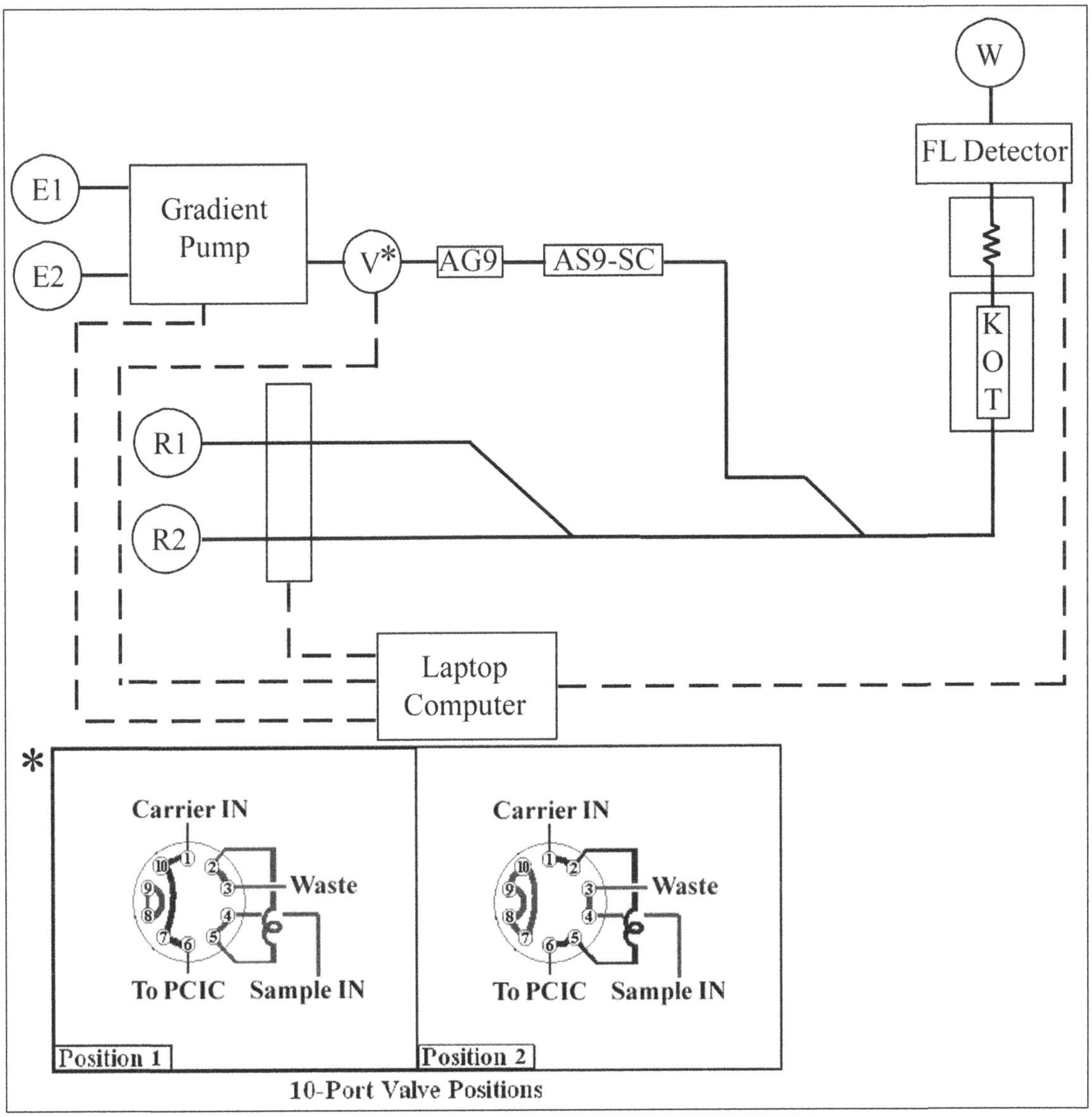

Figure 7.2 Schematic diagram of the optimized PC-IC analyzer. The solid lines are liquid tubing and dashed lines are electronic connections. E1, borate buffer; E2, reagent water; V, 10-port injection valve; R1, 2M NaOH; R2, 37.5% NCA; KOT, knitted open tubular reaction coils; FL, fluorescence detector; and W is waste

The optimized PC-IC analyzer was similar in design to the prototype PC-IC, except that a more advanced fluorescence detector (Waters Model 2475 dual wavelength with scanning capabilities) was used and all PC-IC functions (sample injection, gradient and peristaltic pump operation, and data collection) were fully automated. The gradient eluent was the same as used with the prototype instrument except that the flow rate was held constant at 0.5 mL/min. The gradient program was as follows: The initial conditions were 4 mM H_3BO_3: 0.8 mM NaOH for 4.0 min. The sample injection time was 7.0 minutes. From 4.0 to 10.0 min the concentration of the eluent increased to 240 mM H_3BO_3: 48 mM NaOH and held constant 26.0 min. From 26.0 to 30 min, the eluent concentration increased to 292 mM H_3BO_3: 58.4 mM NaOH. The flow was stopped for 10.0 min. At 40.0 min, the flow was restarted and the concentration increased to 360 mM H_3BO_3: 72.0 mM NaOH at 50.0 min. From 50.0 to 55.0 min, the concentration of the eluent returned to the initial conditions (4 mM H_3BO_3: 0.8 mM NaOH).

The post-column reaction system was a FIALab 2500 four channel peristaltic pump equipped with a mixing manifold. The NCA reagent (37.5%) and NaOH reagent (2M) flow rates were 0.15 mL/min. The NCA and NaOH reagents were premixed using a serpentine coil before merging with the BOR IC eluent. The post column trap was a KOT geometry mixing device/trap comprised of a 20 m, a 10 m and a 5 m units connected in series for a total length of 35 meters (SeQuant AB). This device was used to trap and mix the NCA/ NaOH stream with the HAA5 species eluting from the column. The KOT was immersed in the heated water bath (94°C) and flow was stopped for 30 min to allow formation of the fluorescent NCA-HAA5 reaction products. MCAA and MBAA coeluted at 43 min, DCAA at 45 min, DBAA at 46 min, and TCAA at 51 min. The total run time was 60 minutes.

For all PC-IC studies, the retention time for a component is defined as the time from when sample is injected until it is detected by the fluorescence detector. This time includes the separation time as well as any post column reaction time.

RESULTS AND DISCUSSION

Preliminary MDL, Accuracy and Precision Studies for Prototype PC-IC

In order to obtain MDL values in the desired concentration range (~single μg/L), preliminary studies showed that it was necessary to use stopped flow techniques. In stopped flow techniques, the idea is to stop all pumps so that the "plugs" of eluent containing the separated HAA-NCA-FL reaction products are trapped in a reaction coil and heated to allow formation of higher concentrations of the particular HAA-NCA-FL product. Using the stopped-flow technique seems simple enough; just stop the pumps, either manually or through electronic control. However, in practice a reaction coil must be long enough to trap the entire chromatogram (or at least the section containing the HAA-NCA-FL product species). It is also desirable that the reaction coil be designed in such a way to prevent dispersion of the FIA gradient-like chromatogram flowing to the detector. For example, at a total flow rate of 1.2 mL/min, which includes the IC flow and post column-FIA reagent flow, a reaction coil of ~30 m is needed to trap the entire chromatogram. In the past, the researchers have used both the "super-serpentine" approach and/or the KOT approach. Reactions coils made using these methods provide for a way to place a large length of tubing in a small compartment, like a temperature controlled heat bath. More importantly, they act in a way to minimize dispersion compared to a geometry made simply by coiling the tubing in a circle. Unfortunately, making these larger reaction coils by hand can be quite difficult when lengths of greater than ~10 m are needed.

Fortunately some of these devices are available from vendors like Global FIA and SeQuant, though purchasing these larger devices could be quite expensive ($1-2K in some cases). SeQuant offers the commercially available KOT design that is used in this work. The KOTs can be made to order, requesting longer than 20 m, but in practice it is more desirable to use multiple 10 m KOTs to reduce the chance of failure. The KOTs can be assembled in series to produce traps with larger volume traps. The PC-IC constructed for this study used a KOT trap/reaction coil assembled from a 20, 10 and 5 m KOT in series to produce a total length of 35 meters. A 20 m KOT sells for ~$1000. Figure 7.3 shows a 20 m KOT and close up view showing the knitted design. In summary, the KOT provides two advantages. The KOT allows one to place a very large trap in a compact space such as a heating bath, and the unique design of the KOT acts to minimize dispersion of the FIA-gradient like PC-IC chromatogram.

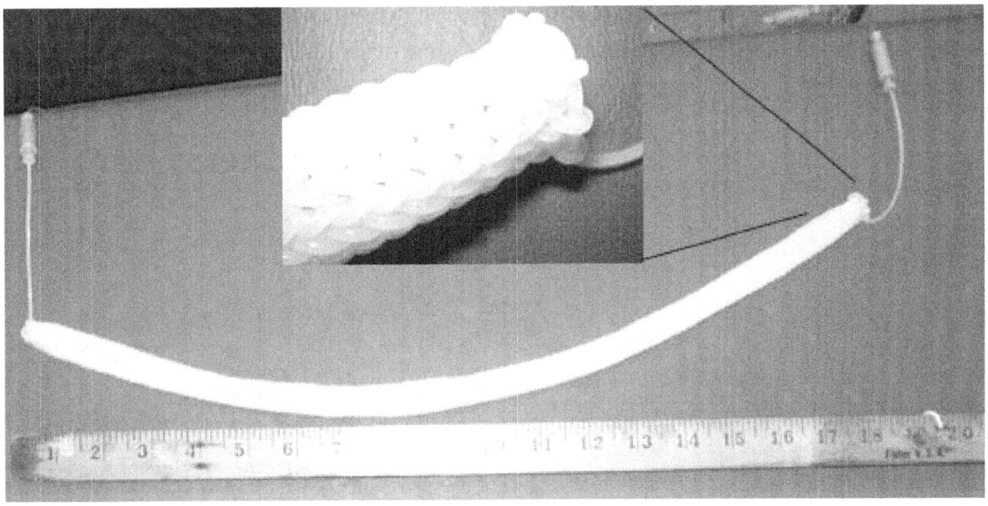

Figure 7.3 The 20 m SeQuant Knitted Open Tubular (KOT) reaction coil

Preliminary MDL, accuracy and precision studies were conducted using the prototype PC-IC analyzer for the HAA5 species. A preliminary calibration curve is show in figure 7.4. The MDL for each HAA5 component was calculated using the USEPA recommended procedure (Glaser and Forest 1981; USEPA 1996a). The mean percent recoveries and percent relative standard deviations (% RSD) were calculated based on USEPA recommendations (USEPA 1996b). The results of these MDL, accuracy and precision studies for the prototype PC-IC are presented in Table 7.1. With this relatively inexpensive, low-end, instrument, the MDL estimates ranged from 1.7 - 7.6 μg/L. These predicted MDL values were within a factor of 1 - 2.6 times the check standard concentrations. The mean percent recoveries ranged from 83 - 127 %. The % RSD estimates ranged from 14 - 25 %. Further studies were conducted with the prototype using two other fluorescence detectors (an additional Waters 420 and a Varian Fluorichrom (Model 430020-02)) with similar results. Additional experiments were designed to automate both the gradient and peristaltic pumps in an effort to reduce the relatively high % RSD estimates. This was done using PeakSimple TTL outputs to turn off and on the gradient pump and to control a solid state relay, thus either supplying or cutting off power to the peristaltic pump. The results of the study with the partially automated PC-IC gave similar results. It was concluded that the best way to improve the performance of the PC-IC analyzer would be to use a state-of-the-art fluorescence detector.

The Waters 2475 fluorescence detector has a smaller flow cell compared to the older fluorescence detectors used with the prototype PC-IC analyzer. This resulted in backpressure problems when the prototype chromatographic conditions were used. The chromatographic conditions were modified to alleviate these problems. The total flow rate passing through the detector was reduced from 1.2 mL/min to 0.8 mL/min. Specifically, the flow rate of the IC gradient pump was reduced from 0.7 mL/min to 0.5 mL/min and the flow rate of peristaltic pump was reduced from 0.25 mL/min for both NCA and NaOH to 0.15 mL/min for each reagent.

The results of the prototype PC-IC method demonstrates the quality of data that researcher's could expect if they are using an older fluorescence detector (perhaps already available in the laboratory) as opposed to a state-of-the-art fluorescence detector. For example, if a utility is experiencing problems meeting the MCL levels for THAA5 (60 μg/L), they might need to conduct optimization studies of their processes over a concentration range of ~10 - 100 μg/L. In this case, a less expensive fluorescence detector could be used to carry out the method with reasonable results.

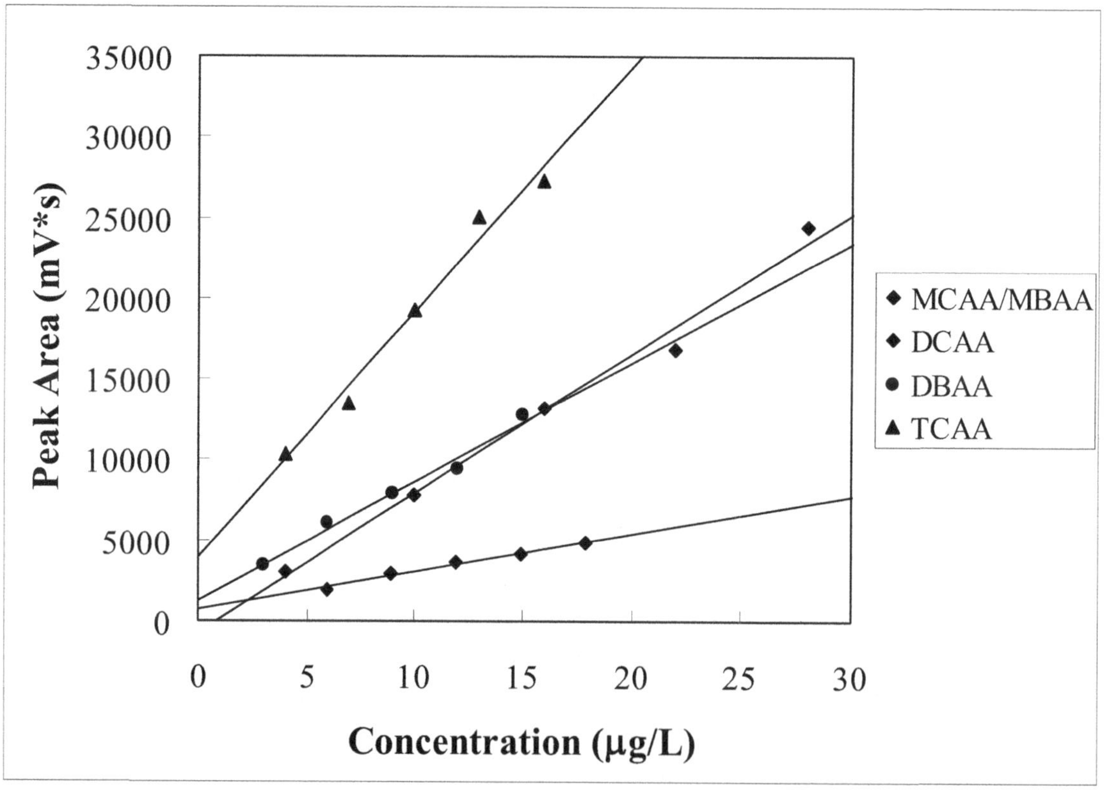

Figure 7.4 Prototype PC-IC calibration curves for each HAA5 species

Table 7.1
Results from MDL, accuracy and precision studies for the prototype PC-IC

HAA5 Species	MDL (μg/L)†	Mean % Recovery	% RSD	r^2	Conc. Range (μg/L)*
MCAA/MBAA	3.4	106	14	0.988	4.0 – 28
DCAA	7.6	127	25	0.982	6.0 – 18
DBAA	1.7	83.3	15	0.983	3.0 – 15
TCAA	3.9	116	19	0.980	4.0 – 16

* Each standard was analyzed once. There are five calibration standards for each curve.
† The check standard (MCAA/MBAA (7.0 μg/L); DCAA (7.5 μg/L); DBAA (4.5 μg/L); TCAA (5.6 μg/L)) was analyzed seven consecutive times.

Optimization of PC-IC Method for HAA5

Before conducting detailed MDL, accuracy, precision and linearity studies with the revised instrument, a series of optimization studies were carried out. Specifically, the effects of sample injection volume, NaOH reagent concentration, heating time/KOT length, and fluorescence excitation and emission wavelengths were studied.

Optimization of Sample Injection Volume

The sample injection volume was varied with a constant HAA5 concentration (100 μg/L) and the analytical signal was recorded (Figure 7.5). Four different sample loop volumes were used (1000 μL, 1500 μL, 2000 μL and 2500 μL). There were no significant changes in the peak shape as the sample injection volume was changed with volumes 1000 μL through 2000 μL as long as a "loading gradient" was used. The loading gradient refers to programming the eluent strength to be very weak (low concentration) for a particular time period. This allows the relatively large sample injection volume to be concentrated into a narrow band before higher eluent concentrations begin the separation process on the column. With the 2500 μL sample injection volume, the chromatogram exhibited significantly skewed peaks, especially for TCAA. With the exception of DBAA, all other components exhibited a decrease in analytical signal. The 2000 μL sample injection volume provided the best overall analytical signal for each HAA species while also showing no qualitative change in peak shape.

Optimization of the Sodium Hydroxide Reagent Concentration

Early studies have shown that NCA is not stable as a reagent at the high pH of the hydroxide solution. This being the case, these two reagents need to be prepared separately. Very preliminary studies have been done to evaluate the rate of the formation of the fluorescence signal as a function of pH. More detailed studies are discussed in the next chapter. Because of the identified effect of pH, the effect of NaOH reagent concentration on analytical signal was evaluated over the concentration range of 1.0 M to 4.0 M using a 100 μg/L HAA5 standard. Figure 7.6 shows a plot of the analytical signal as a function of hydroxide ion concentration. The study indicates that the maximum analytical signal was observed for all HAA species when a hydroxide ion concentration of 2.0 M was used.

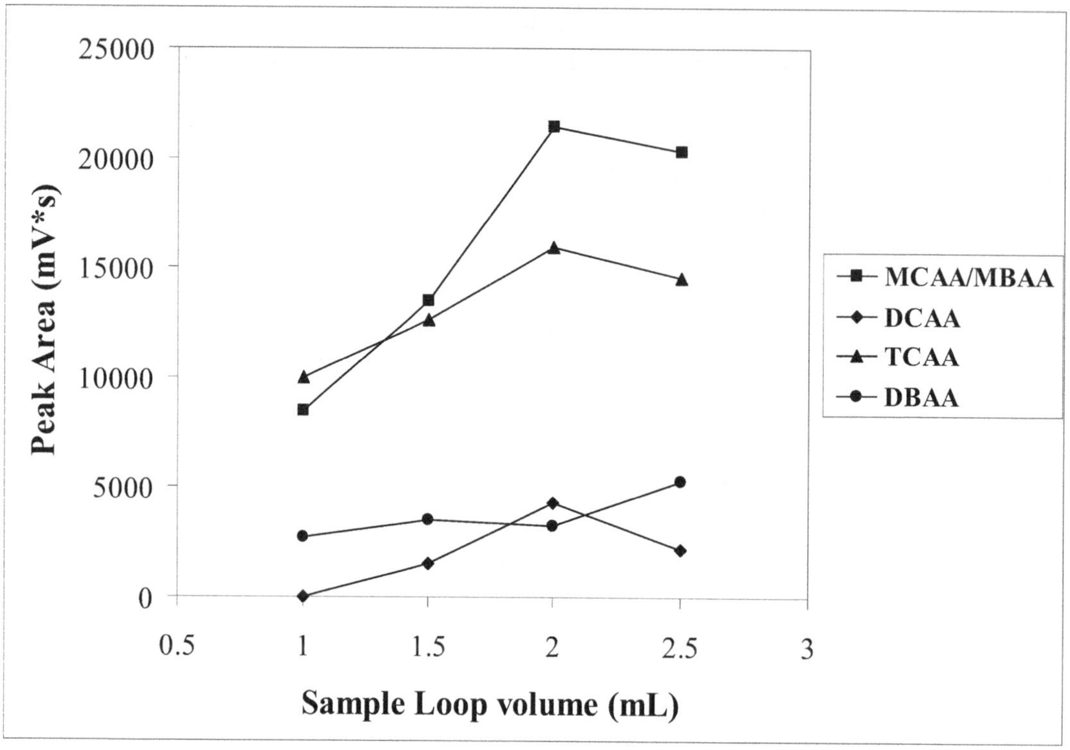

Figure 7.5 Optimization plot of the sample injection volume for PC-IC

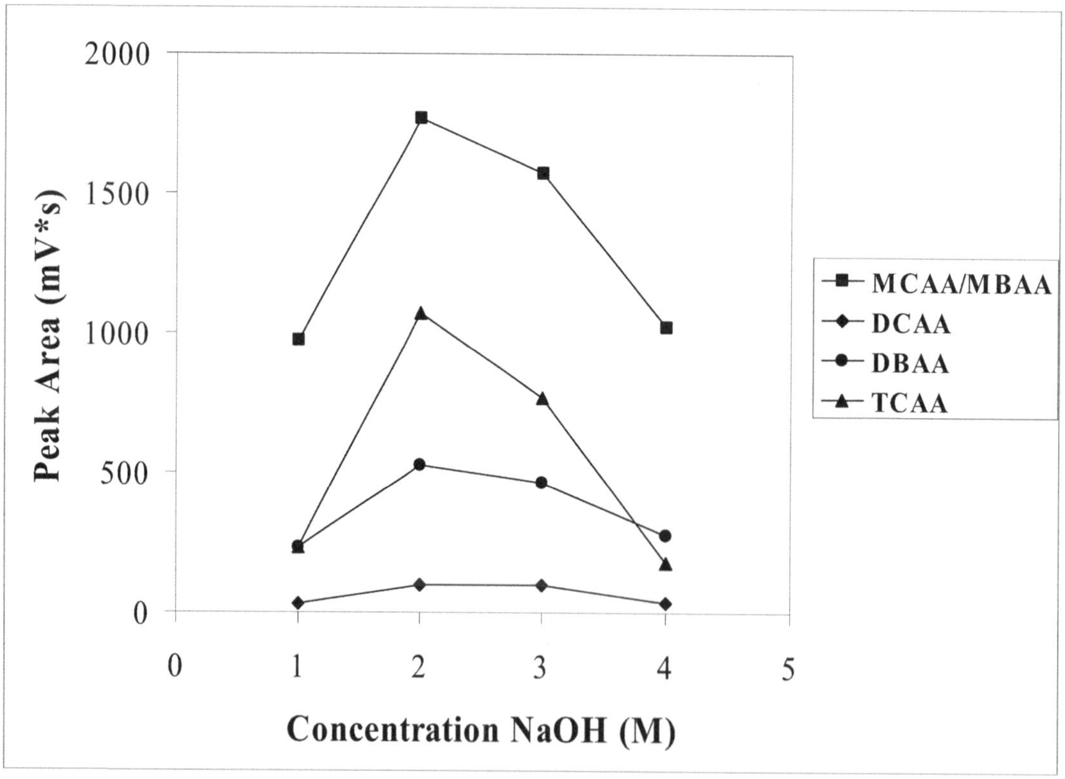

Figure 7.6 Optimization plot of NaOH concentration for PC-IC

Related to the NaOH reagent, studies were designed to evaluate the effect of carbonate coating of NaOH pellets (an impurity) on reproducibility since the AS9-SC IC column is carbonate ion selective. Method performance was compared using NaOH prepared from NaOH pellets and 50% NaOH solution. The certificate of analysis on the NaOH pellets indicates a 0.4% sodium carbonate (Na_2CO_3) impurity. For 2 L of eluent prepared using pellets, 6.4 g of NaOH pellet is used. An impurity of 0.4 % sodium carbonate results in 0.026 g (0.25 mM) of Na_2CO_3 present. The 0.25 mM Na_2CO_3 is 14% of the typical IC concentration (1.8 mM Na_2CO_3) used for anions by IC on the AS9-SC at a full strength borate buffer eluent. For initial conditions and during the chromatographic run, the concentration is much lower, and thus likely to be an insignificant effect.

To test if the carbonate was a significant factor for retention time variations, a standard containing 50µg/L of each HAA5 species was analyzed in triplicate with two different eluents. One eluent prepared as 400 mM boric acid (H_3BO_3) and 80 mM NaOH with pellets and the other eluent, 400 mM H_3BO_3 and 80 mM NaOH with 50% NaOH solution. The results from the study show that the retention time of the individual HAA5 species did not differ more than 0.5% using relative error with this change in reagent source. From a quantitative standpoint, the differences in the analytical signal (peak area) using the two different eluents was observed. There was an overall loss in analytical signal compared to the eluent prepared using NaOH pellets. DCAA was affected the most since it has the smallest response to begin with. For the MCAA/MBAA, DCAA, DBAA and TCAA the relative difference was -5.2, -27, 0.2 and -8.5 %, respectively.

Optimization of the Heating Time/KOT Length

It takes approximately 20 minutes for the sample to flow through the PC-IC, emerge from the 35 m KOT, and be detected by the detector. This time is sufficient to allow for detection of single µg/L concentration levels of MCAA, MBAA, DBAA and TCAA, but it is not sufficient for detection of single µg/L concentration levels of DCAA. Stopped-flow methods were used to improve MDL values. Both the gradient pump and the peristaltic pump were programmed to stop for specified time periods and the chromatogram was trapped in the KOT contained in a heated water bath. The optimum stopped-flow time was determined by injecting a HAA5 standard (50 µg/L) while varying the stopped-flow time. A plot of analytical signal versus stopped-flow time is shown in figure 7.7. The analytical signal of DCAA increased slightly from a stopped-flow time of 0 min (continuous flow) to a stopped-flow time of five minutes. Over the stopped-flow times of five to 15 minutes, the analytical signal for DCAA did not increase appreciably. The remaining HAA5 species exhibited increased analytical signals with increased stopped-flow times. In an effort to keep an on-line monitoring rate of one sample per hour, the decision was made to adopt a stopped-flow time of 10 minutes.

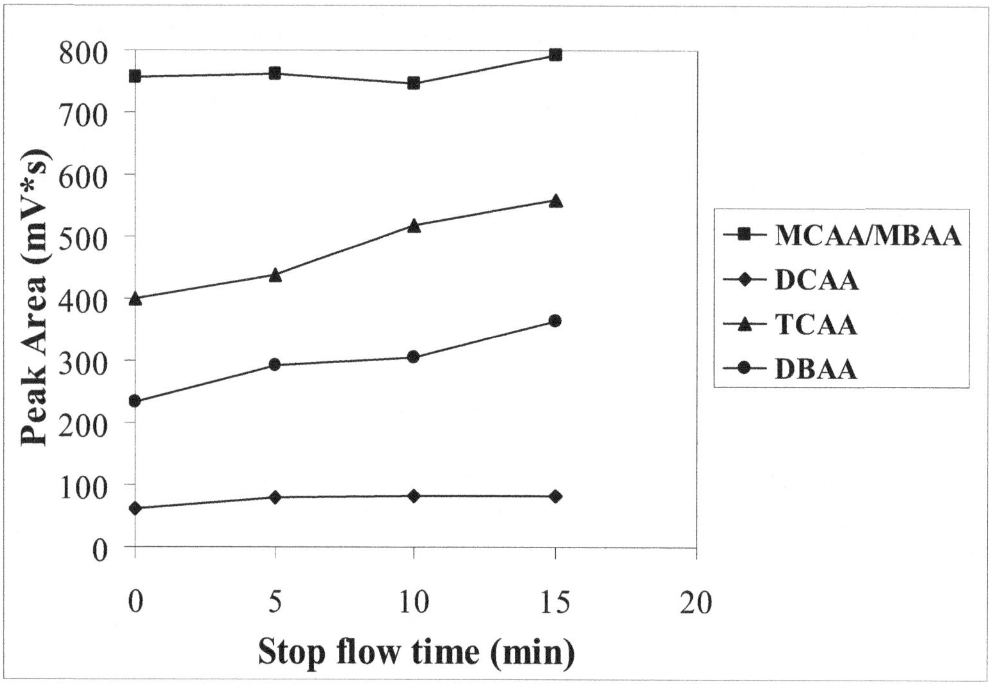

Figure 7.7 Optimization plot of stop flow heat time for PC-IC

Optimization of the Fluorescence Excitation and Emission Wavelengths

The prototype PC-IC used a filter-based fluorescence detector that gave little control over excitation and emission wavelengths. The optimized PC-IC analyzer used a dual wavelength scanning fluorescence detector that allowed for more detailed examination of the excitation and emission characteristics of the NCA-HAA5 products. NCA alone will produce some background fluorescence. With the optimized PC-IC, it was possible to more closely examine the background fluorescence of NCA. The monochromator-based fluorescence detector was used to excite at 365 nm and 370 nm. The background fluorescence of NCA increased in going from 365 to 370 nm. Thus an excitation wavelength of 365 was chosen as the optimum wavelength of excitation as it gave the lowest background fluorescence of NCA.

The dual channel wavelength emission feature of the Waters 2475 fluorescence detector was used to find the optimum emission wavelength for the NCA-HAA5 products (Figure 7.8). Emission at 450 nm was used as the standard wavelength on one channel for all these optimization studies. Using the second channel, the emission was varied from 430 nm to 470 nm in 10 nm increments. The peak area for each HAA5 species measured using the variable emission wavelength was divided by the peak area at 450 nm for each HAA5 species. The peak area at 460 nm gave the largest analytical signal over the wavelength region 430 - 470 nm. Additional experiments were carried out over the range 450 - 460 nm and the results showed that emission was greatest at 455 nm. This wavelength was chosen as the optimum wavelength of emission to be used for the remaining experiments.

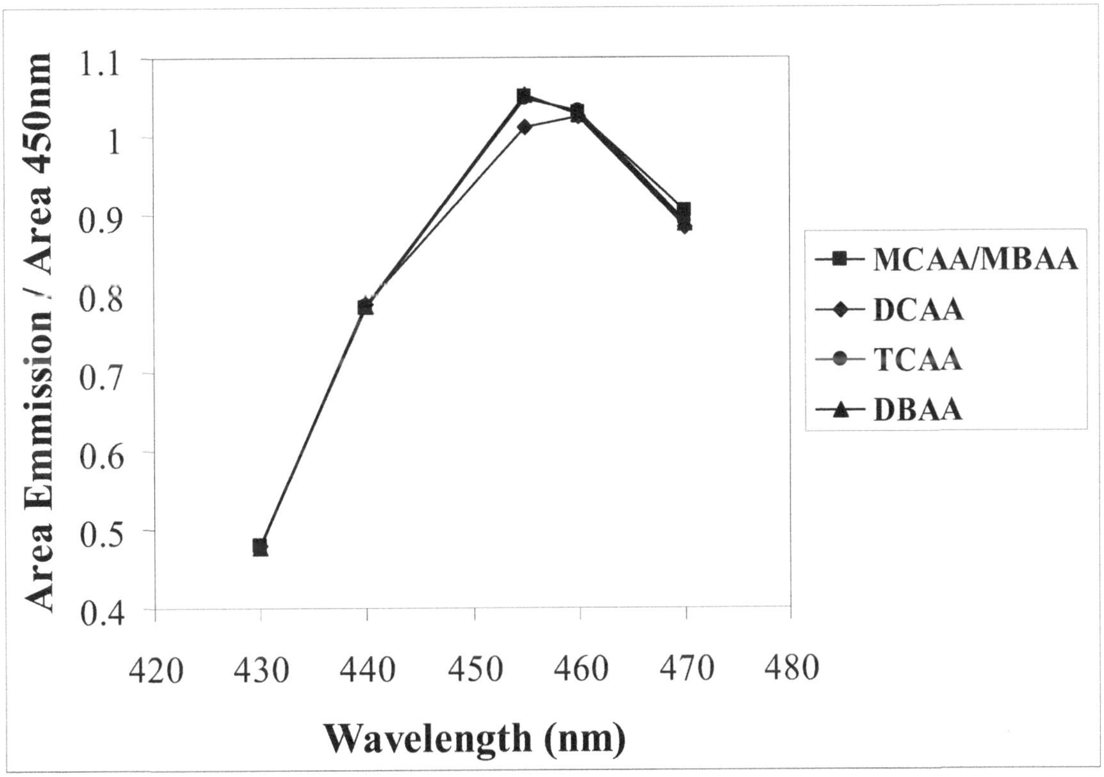

Figure 7.8 Optimization plot of fluorescence emission wavelength for PC-IC

MDL, Accuracy, Precision and Linearity Studies with the Optimized PC-IC Analyzer

The optimized PC-IC analyzer and method were used to conduct detailed MDL, accuracy and precision studies of the HAA5 species. Each calibration curve was based on five calibration standards. The MDL for each HAA5 component was calculated using USEPA recommended procedures (Glaser and Forest 1981; USEPA 1996a). The mean percent recovery and percent relative standard deviation (% RSD) were calculated based on USEPA recommendations (USEPA 1996b). With these studies, a check standard was analyzed using seven consecutive analyses as indicated in Table 7.2. The mean % recovery was calculated by averaging the % recoveries for each of these analyses. The percent recovery was calculated by dividing the experimental concentration result by the known concentration of the check standard and multiplying by 100. The % RSD was calculated by dividing the standard deviation of the concentration by the average concentration and multiplying by 100.

MDL, Accuracy and Precision Studies

The results of detailed MDL, accuracy, precision and linearity studies for the optimized PC-IC analyzer and method are shown in Table 7.2. Figure 7.9 shows typical chromatograms from a calibration curve (Figure 7.10). Ideally, the check standard concentrations should be within a factor of 2.5 to 5 times the MDL. In this study, the predicted MDL values are within a factor of 3.5 to 7 times the MDL using the USEPA method. The MDL ranges from 1.2 - 2.1 μg/L. None of the MDL values for each HAA5 species are superior to those of EPA 552.3

(reported MDL values ranging from 0.012 - 0.17 µg/L). However, all of the predicted MDL values are within the desired range proposed by the Phase One Report (Emmert 2004b) for on-line monitoring methods (target range of 0.5 - 5 µg/L). The mean % recoveries can be used as an estimate of accuracy, and ranged from 79.4 - 105 %. The precision, estimated using the % RSD, ranged from 5.5 - 11 %.

Linearity Studies

The calibration curves appear linear over the limited range used for the MDL, accuracy and precision studies (1 - 25 µg/L). To investigate the linearity of the analytes further, higher concentrations of HAA5 standards were analyzed (up to the range 100 - 200 µg/L depending on the HAA5 species. All showed linear responses up to these higher concentration levels. It should be noted that the linear range of up to 230 µg/L for MCAA/MBAA is the combined concentration and response from these two coeluting species (Figure 7.11).

Table 7.2
Results from MDL, accuracy and precision studies for PC-IC

HAA5 Species	MDL (µg/L)†	Mean % Recovery	% RSD	r^2	Conc. Range (µg/L)*
MCAA/MBAA	2.1	84.3	5.9	0.998	2.3 – 45
DCAA	1.4	79.4	7.1	0.963	1.3 – 26
DBAA	2.1	80.2	11	0.992	1.2 – 25
TCAA	1.2	105	5.5	0.996	1.1 - 21

* Each standard was analyzed once. There are five calibration standards for each curve.

† The check standard (MCAA/MBAA (13.6 µg/L); DCAA (7.7 µg/L); DBAA (7.4 µg/L); TCAA (6.3 µg/L)) was analyzed seven consecutive times.

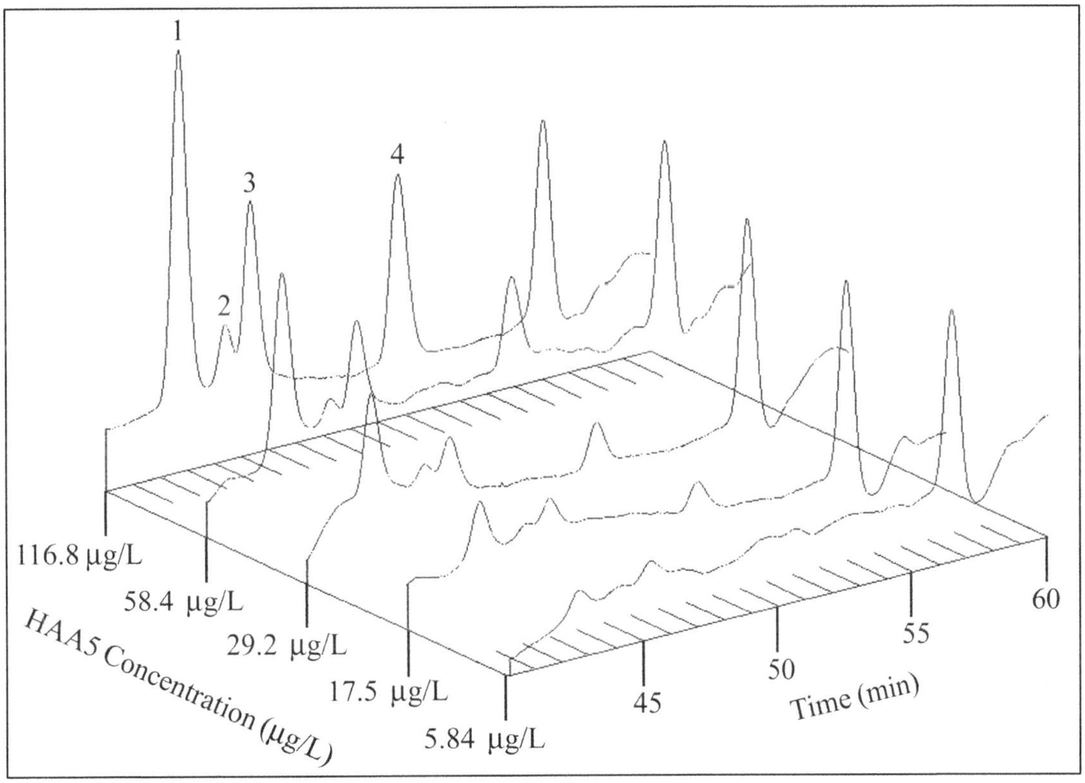

Figure 7.9 Typical HAA5 calibration chromatograms for PC-IC. The total HAA5 concentration is labeled on the x-axis. (1) MCAA/MBAA 45.4 μg/L, (2) DCAA 25.6 μg/L, (3) DBAA 24.8 μg/L, (4) TCAA 21.0 μg/L (Source: Simone et al. 2006)

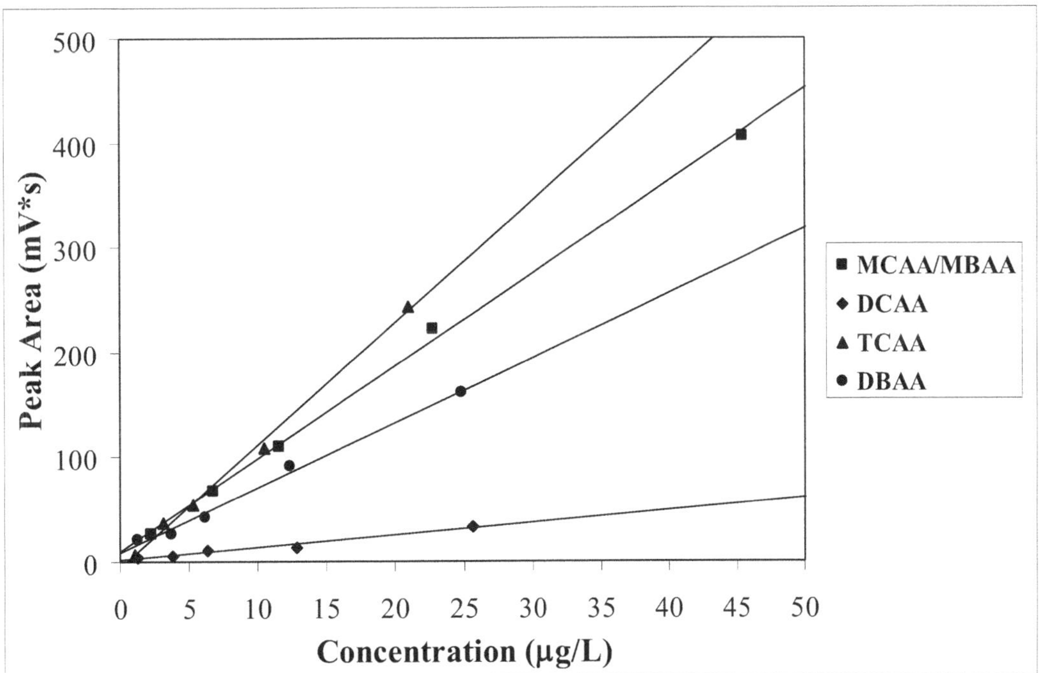

Figure 7.10 Calibration plots of HAA5 for the optimized PC-IC instrument

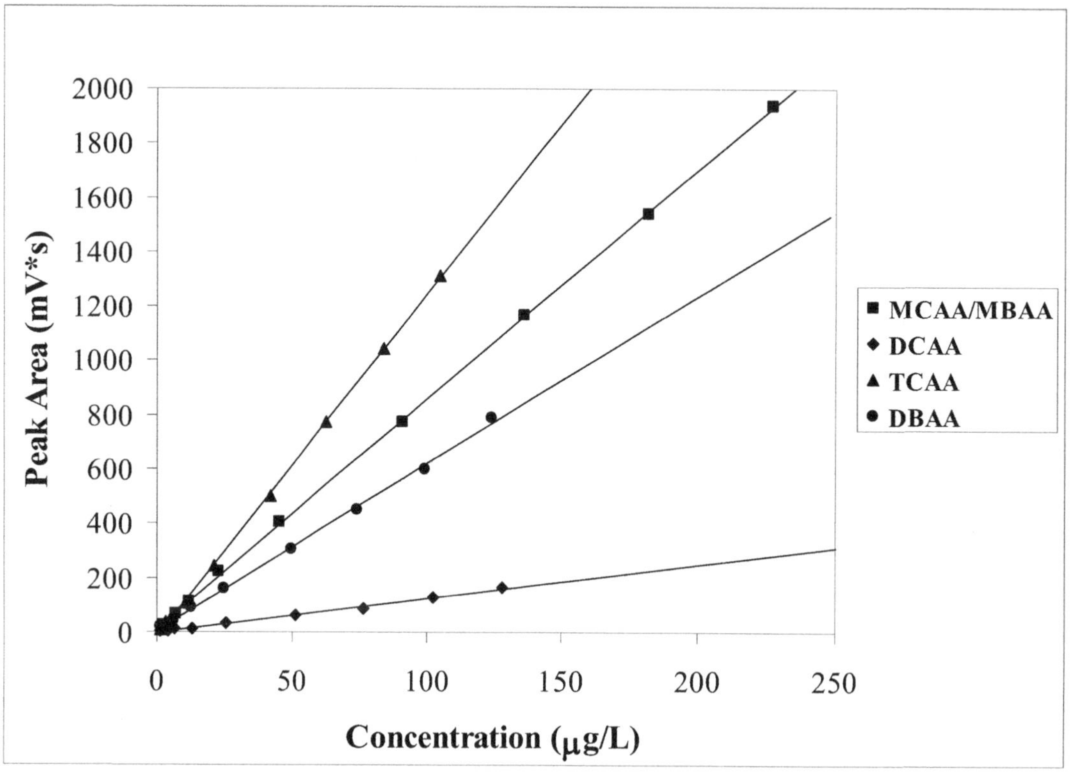

Figure 7.11 Plot of linearity study for HAA5

Interference Studies

Interference studies were carried out to evaluate the effect of common ions and other species that might be present in drinking water samples. Specifically, studies were designed to evaluate the interference of common ions including fluoride ion (F^-), chloride ion (Cl^-), bromide ion (Br^-), nitrite ion (NO_2^-), nitrate ion (NO_3^-), phosphate ion (PO_4^-), and sulfate ion (SO_4^{2-}). In addition other possible DBPs were studied as possible interferences, including bromate ion (BrO_3^-) and four additional HAA species: bromochloroacetic acid (BCAA), bromodichloroacetic acid (BDCAA), dibromochloroacetic acid (DBCAA) and tribromoacetic acid (TBAA). The effect of TOC concentration was also evaluated.

Interference from Common Ions in Drinking Water

One disadvantage of conductivity detection is that it is fundamentally non-selective. Any charged species will register some response. Since real world drinking water samples may contain many common anions, it is important that they should not interfere in the measurement of HAA5 species. A simulated drinking water sample was prepared to mimic the typical concentrations of common anions and other possible DBPs. The simulated sample was spiked with 50 μg/L of each HAA5 species. The concentrations of the common anions were based upon those reported in the yearly Memphis water quality report. The resulting solution was analyzed with an IC-MSCD method, and the optimized PC-IC method (Figure 7.12). The fundamental disadvantage exhibited by IC-MSCD when faced with trace levels of HAAs (~50 μg/L each) in drinking water is the comparatively small response for HAA5 species and large response for the common anions. However, the PC-IC method produces a comparatively large response for HAA5 species and minimal response for the common anions. This selectivity for HAA5 species over common anions and other possible ionic disinfection by-products is one of the primary advantages of the PC-IC approach.

The effects of common ions were investigated further. On-line monitoring studies were carried out in different drinking water systems and in simulated matrices. In two real world systems, the concentration of HAA5 species were measured in the presence of relatively low chloride ion and sulfate ion concentrations (7.4 and 10.7 mg/L, respectively) and in slightly higher chloride ion and sulfate ion concentrations (30 and 80 mg/L, respectively). In this monitoring study, PC-IC compared very well EPA 552.3, with an average bias of less than 5 μg/L for the individual HAA5 species and 7.5 μg/L for THAA5.

Additional MDL studies were done in simulated drinking water samples with chloride ion and sulfate ion concentrations of 100 mg/L each. In the presence of high anion concentrations, the MDLs for all but DCAA were similar (ranging from ~6 - 11 μg/L, compared to 1 - 2 μg/L). DCAA suffered a slight loss in MDL (~50 μg/L compared to 1.4 μg/L), but was still below the MCL of 60 μg/L. The mean % recoveries ranged from 97 - 213 % compared to 79.4 - 105 %. The % RSD values ranged from 17 - 46 % at the individual HAA5 concentrations ranging from 8.6 - 17 μg/L, compared to 5.5 - 7.1 % at 6.3 - 13.6 μg/L. Clearly, there are concerns that still need to be further explored concerning the impact of common ion interferences and effect on analytical signal. It may be possible to use pretreatment cartridges to minimize interferences from common ions should they prove problematic (Barron and Paull 2004a, 2004b; Barron et al. 2005).

Interference from other chlorinated and HAA species

Additional studies were carried out to evaluate the response of other possible HAA species. Geme et. al (2005) have shown that other halogenated species can react with NCA to form fluorescent products. For example, THMs, haloacetonitriles and haloacetone would be expected to interfere in batch or FIA based chemistries where separation of the mixture of halogenated organics has not occurred. Interference from these neutral species is not expected to be a problem in IC measurements as these species would not be retained by the IC column. Preliminary studies confirmed these observations. However, there are four other HAA species (typically referred to, along with the HAA5 species, as HAA9) that could be present and might possibly act as interfering species. Four standards were prepared, each having 50 μg/L of THAA5 plus one of the remaining four HAAs comprising HAA9. Figure 7.13 shows that BDCAA, DBCAA, and TBAA do not interfere with TCAA, and are baseline resolved from each other. BCAA however, co-elutes with DCAA and DBAA, this could potentially be a problem should that HAA9 species be present. This research demonstrates that the PC-IC method could be used to quantify all HAA9 species and further effort was devoted to this end.

TOC Interference Studies

The effect of TOC concentrations on the PC-IC method was investigated. HAA5 standards (50 μg/L) were prepared in the presence ~0 (reagent water), 1, 5, 10, and 20 mg/L

TOC standards (APHA, AWWA, and WEF 1992) and analyzed using the PC-IC method (Figure 7.14). As the TOC concentration increased from ~0 to 20 mg/L, there was an approximately 20 % decrease in the analytical signal. In real drinking water samples, the TOC values are expected to remain relatively constant over the course of a typical monitoring study (estimated at ~ one week). The effects of TOC could be minimized by preparing samples with TOC concentrations mimicking that of the water to be monitored.

Evaluating Sample Carry-over Effects

Studies were designed to evaluate the effect of sample carry-over or possible memory effects in the PC-IC method. A 200 μg/L THAA5 standard was injected, followed by successive injection of three reagent water blanks. There were no significant carry-over effects observed and there was no measurable effect on HAA5 quantification when a water sample was injected each hour during monitoring studies (Figure 7.15).

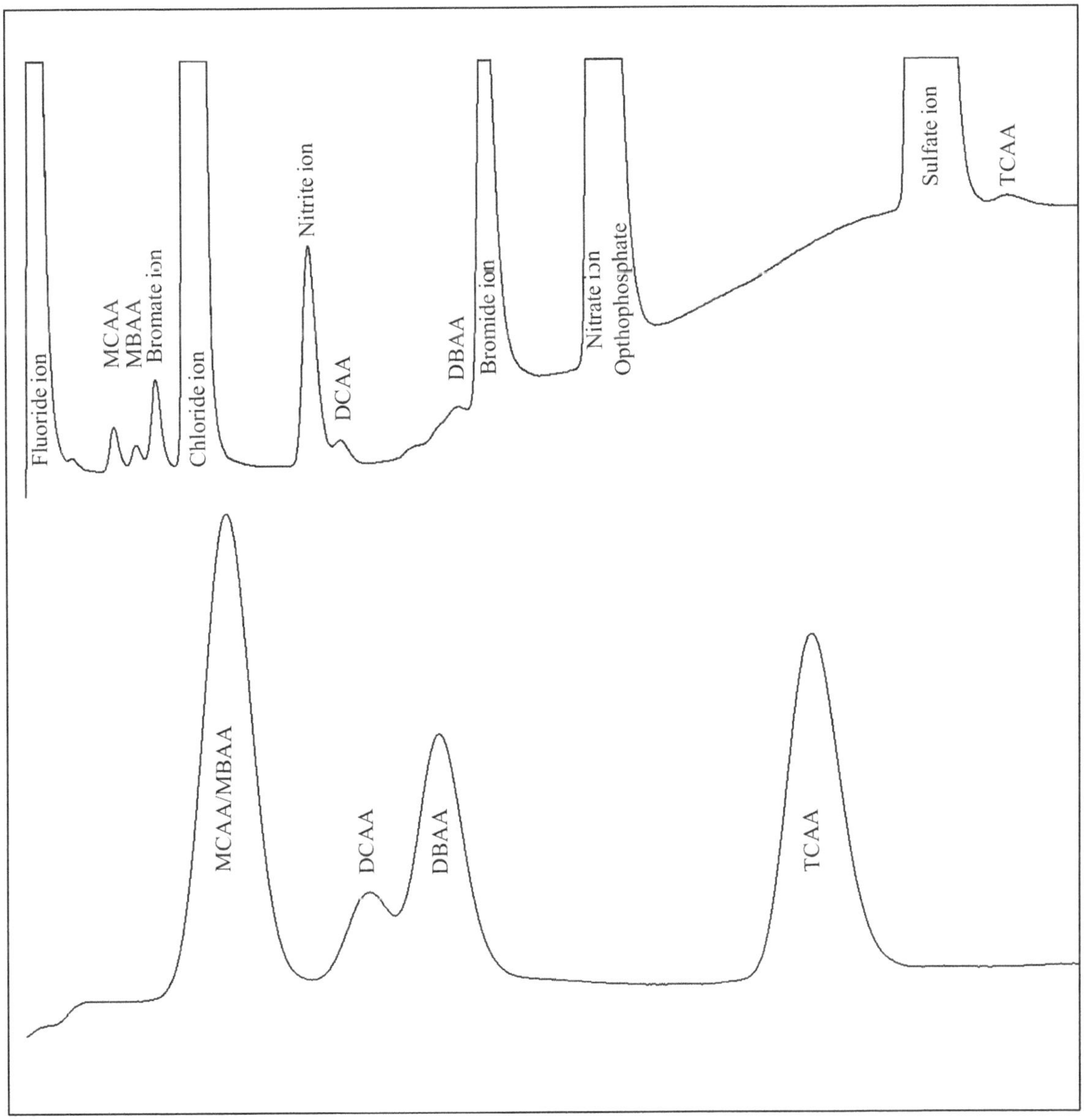

Figure 7.12 Comparison of a PC-IC chromatogram (bottom) with an XLVI-IC-MSCD chromatogram (top). HAA5 concentrations: 50 μg/L each of MCAA, DCAA, TCAA, MBAA, DBAA. Common anions and their concentrations (mg/L): fluoride ion (1.0), chloride ion (4.5), bromide ion (1.0), nitrite ion (0.25), nitrate ion (1.15), phosphate ion (1.40), sulfate ion (3.75) bromate ion (0.150 mg/L)

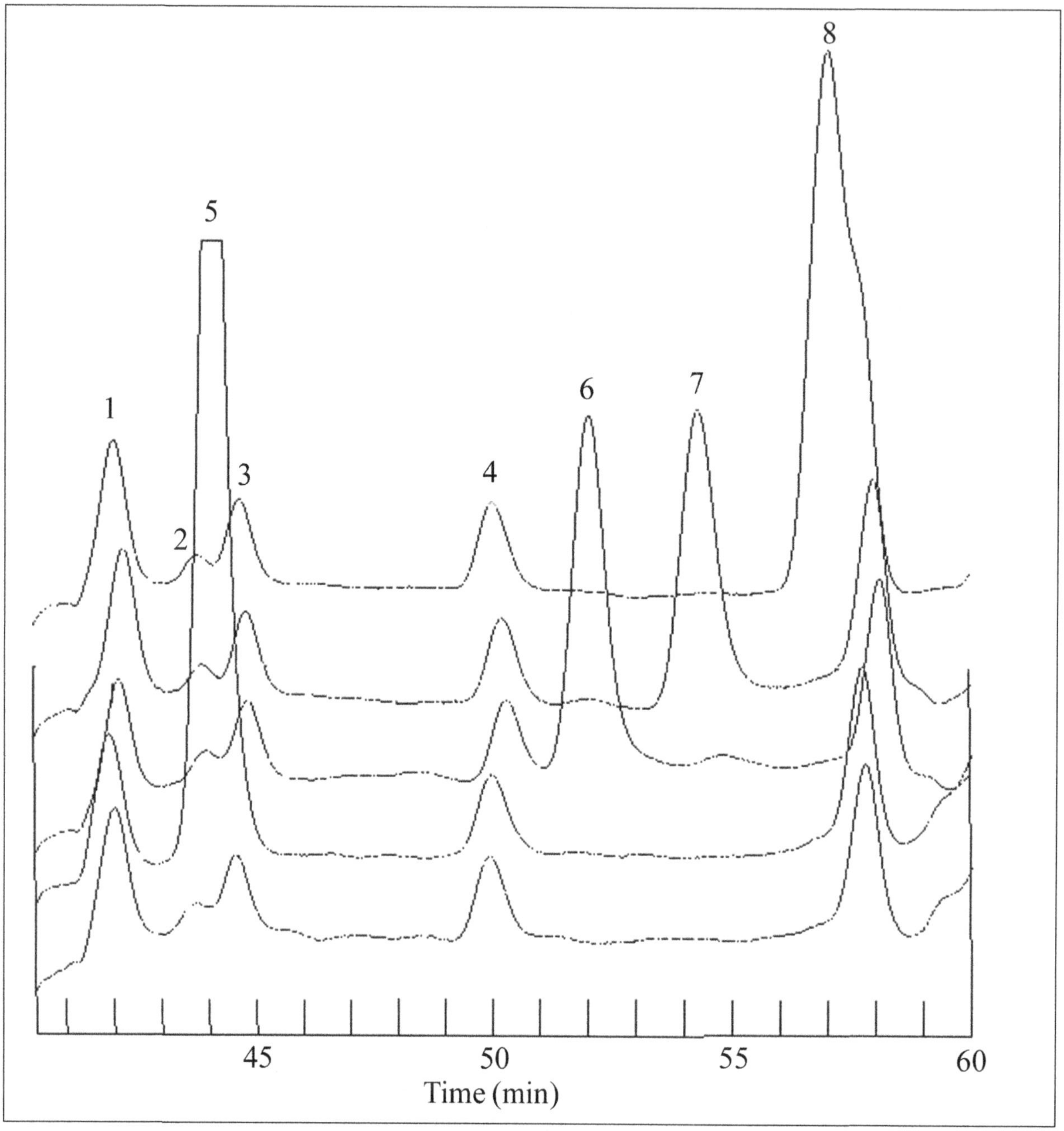

Figure 7.13 PC-IC interference studies for HAA9. (1) MCAA/MBAA (22.7 µg/L), (2) DCAA (12.8 µg/L), (3) DBAA (12.4 µg/L), (4) TCAA (10.5 µg/L), (5) BCAA (80 µg/L), (6) BDCAA (25 µg/L), (7) DBCAA (25 µg/L), (8) TBAA (44 µg/L) (Source: Simone et al. 2006)

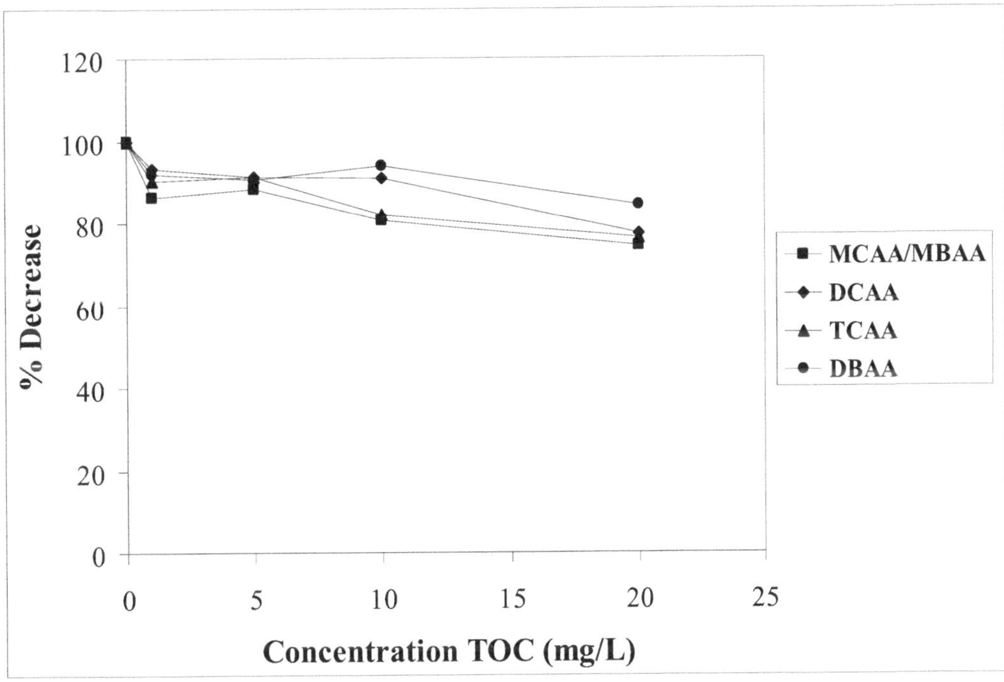

Figure 7.14 Effect of TOC concentration on the analytical signal of each HAA5 species

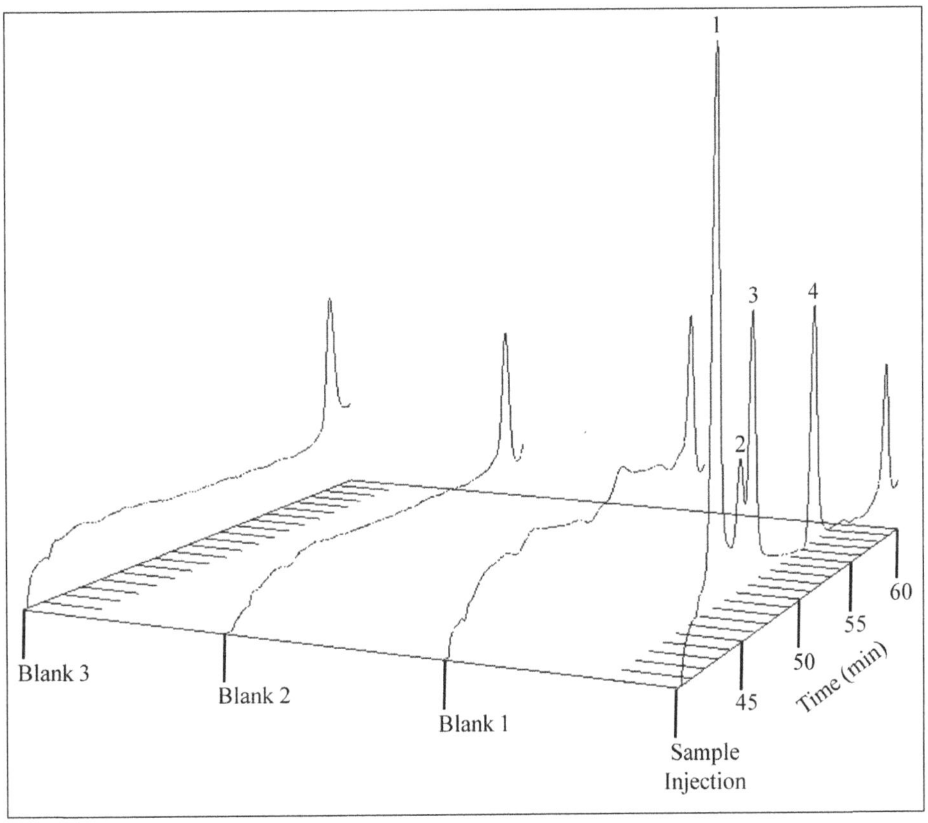

Figure 7.15 Carry-over study of PC-IC (1) MCAA/MBAA (90.8 μg/L), (2) DCAA (51.2 μg/L), (3) DBAA (49.6 μg/L), (4) TCAA (42.0 μg/L)

Memphis 2 On-Line Monitoring Study using PC-IC

The PC-IC portion of the Memphis 2 monitoring study was conducted from May 19 to 27, 2005. The PC-IC was interfaced to the drinking water distribution system using the adapter described in the Appendix A As noted earlier, the monohalogenated species (MCAA and MBAA) were not resolved and are reported as a sum of concentrations. Samples were monitored at a rate of ~1 sample/ hour (though as with the other monitoring methods, the analyzer was typically taken off-line each 12 hours to analyze a check standard). The results for each HAA species were summed to determine the total concentration of HAA5 species (THAA5). The results of these studies along with details used in the analysis of the data are presented in Table 7.3.

Table 7.3
Memphis 2 monitoring study: monitoring HAA5 concentrations using PC-IC

	MCAA/MBAA		DCAA		TCAA		DBAA		THAA5	
Statistic/*Method*	*EPA*	*PCIC*	*EPA*	*PCIC*	*EPA*	*PCIC*	*EPA*	*PCIC*	*EPA*	*PCIC*
N	57	57	68	68	68	68	68	68	64	64
Avg. Conc.	<1.9	3.6	0.5	2.2	0.3	2.3	1.6	3.3	1.8	9.3
Std. Dev Conc.	---	2.3	0.1	1.8	0.1	2.2	0.8	2.5	0.9	6.7
Conc. Range	---	2.1 to 13.4	0.5 to 0.7	1.4 to 12.2	0.3 to 0.5	1.2 to 13.6	0.6 to 3.5	2.1 to 12.4	0.3 to 3.9	1.0 to 32.5
Avg. Bias	---	1.7	---	1.6	---	2.0	---	1.7	---	7.5
Std. Dev Bias	---	2.3	---	1.8	---	2.2	---	2.4	---	6.4
Bias Range	---	0.2 to 11.5	---	0.9 to 11.7	---	0.7 to 13.3	---	-1.4 to 10.1	---	0.3 to 28.8
% REC_{Bias}	---	187	---	427	---	747	---	241	---	663
s.d. % REC_{Bias}	---	121	---	365	---	741	---	168	---	546
% REC_{Bias} Range	---	109 to 704	---	230 to 2448	---	236 to 4534	---	59 to 920	---	112 to 3100
t-test*		Yes		Yes		Yes		Yes		Yes
$t_{critical}$	---	2.395	---	2.383	---	2.383	---	2.383	---	2.383
$t_{calculated}$		5.448		7.415		7.257		6.007		9.408

* The $t_{critical}$ and $t_{calculated}$ values listed were calculated for the N listed using a two-tailed t-test with 98% confidence. Typically a $t_{calculated}$ value should be less than the $t_{critical}$. Yes or No means if statistically different.

A plot of the concentration of monohalogenated HAA species (MCAA + MBAA) as a function of time is shown in Figure 7.16. For 57 direct comparisons, the concentration of MCAA/MBAA was less than the MDL of 1.9 µg/L for EPA 552.3 for all samples analyzed. The PC-IC reported a concentration range from 2.1 to 13.4 for MCAA+MBAA. The average concentration was less than the MDL (1.9 μg/L) for EPA 552.3 and 3.6 ± 2.3 μg/L for PC-IC. A plot of the bias as a function of time (Figure 7.17) ranged from 0.2 to 11.5 μg/L for the PC-IC (using the MDL value for EPA 552.3 as the "true value". The average bias was 1.7 ± 2.3 μg/L. The % REC_{Bias} ranged from 109 to 704% with an average of 187 ± 121%. For 68 direct comparisons, the concentration of DCAA (Figure 7.18) ranged 0.5 - 0.7 μg/L (EPA 552.3) and 1.4 - 12.2 μg/L (PC-IC). The average concentration was 0.5 ± 0.1 μg/L (EPA 552.3) and 2.2 ± 1.8 μg/L (PC-IC). The bias (Figure 7.19) ranged from 0.9 - 11.7 μg/L (averaging 1.6 ± 1.8 μg/L) with % REC_{Bias} ranged from 230 - 2448 % (averaging 427 ± 365 %). With TCAA, (Figure 7.20), the concentration ranged from 0.3 - 0.5 μg/L (EPA 552.3) and 1.2 - 13.6 μg/L (PC-IC). The average concentration was 0.3 ± 0.1 μg/L (EPA 552.3) and 2.3 ± 2.2 μg/L (PC-IC). The bias (Figure 7.21) ranged from 0.7 to 13.3 μg/L (averaging 2.0 ± 2.2 μg/L) with % REC_{Bias} ranged from 236 - 4534 % (averaging 747 ± 741 %). With DBAA (Figure 7.22), the concentrations ranged from 0.6 - 3.5 μg/L (EPA 552.3) and 2.1 - 12.4 μg/L (PC-IC). The average concentration was 1.6 ± 0.8 μg/L (EPA 552.3) and 3.3 ± 2.5 μg/L (PC-IC). The bias (Figure 7.23) ranged from –1.4 - 10.1 μg/L (averaging 1.7 ± 2.4 μg/L) and % REC_{Bias} ranged from 59 - 920 % (averaging 241 ± 168 %). THAA5 concentrations (Figure 7.24) ranged from 0.3 to 3.9 μg/L (EPA 552.3) and 1.0 to 32.5 μg/L (PC-IC). The average TTHM4 concentration was 1.8 ± 0.9 μg/L (EPA 552.3) and 9.3 ± 6.7 μg/L (PC-IC). The bias (Figure 7.25) ranged from 0.3 to 28.8 μg/L (averaging 7.5 ± 6.4 μg/L) and % REC_{Bias} ranged from 112 to 3100% (averaging 663 ± 546 %).

Statistically, it is not suprising that the two methods do not give statistically the same results. The value of $t_{critical}$ for the 57 comparisons is 2.395 and for MCAA/MBAA the $t_{calculated}$ was 5.448. The $t_{critical}$ values for the 68 comparisons (DCAA, TCAA, and DBAA) is 2.383. The $t_{calculated}$ value for DCAA is 7.414, for TCAA it is 7.257, and for DBAA it is 6.007. In each case, the result of the methods are not statistically the same. The same is true for THAA5 where $t_{critical}$ for the 64 comparisons is 2.383 and the $t_{calculated}$ is 9.408.

This was the first study that the PC-IC method was used. Many aspects of the operation of the analyzer were still being evaluated. One of the first things noticed was that after ~15-18 hours, it was necessary to replace the peristaltic pump tubing. The effect of worn tubing is evident in the plots over the first 72 hours of the study. Once a regular schedule of tube replacement was established, the results of the study show better agreement. For example, when the first 72 hours of the study are neglected, the average bias for the individual species and THAA5 all get smaller. Specifically, MCAA/MBAA average bias drops from 1.7 ± 2.3 μg/L to 0.9 ± 1.0 μg/L; DCAA average bias drops from 1.6 ± 1.8 μg/L to 1.2 ± 1.0 μg/L; TCAA average bias drops from 2.0 ± 2.2 μg/L to 1.4 ± ± 1.2 μg/L; DBAA average bias drops from 1.7 ± 2.4 μg/L to 1.1 ± 1.1 μg/L; and THAA5 average bias drops from 9.3 ± 6.7 μg/L to 7.3 ± 2.7 μg/L. Slight improvements are also seen in the % REC_{Bias} estimates, though they are not as pronounced as the bias improvements. Statistically, the results still are not in agreement.

On the surface, these results may not appear promising. Closer examination shows that this is not the case. This method provides a way to measure the concentrations of HAA species directly from the water distribution systems. A sample can be measured each hour. Compared to the EPA method, the PC-IC method is much less labor intensive. Perhaps the largest advantage in the PC-IC method is the lack of interference from common anions. Improvements are needed, but these result of the Memphis 2 study encouraged the researchers to continue in this area.

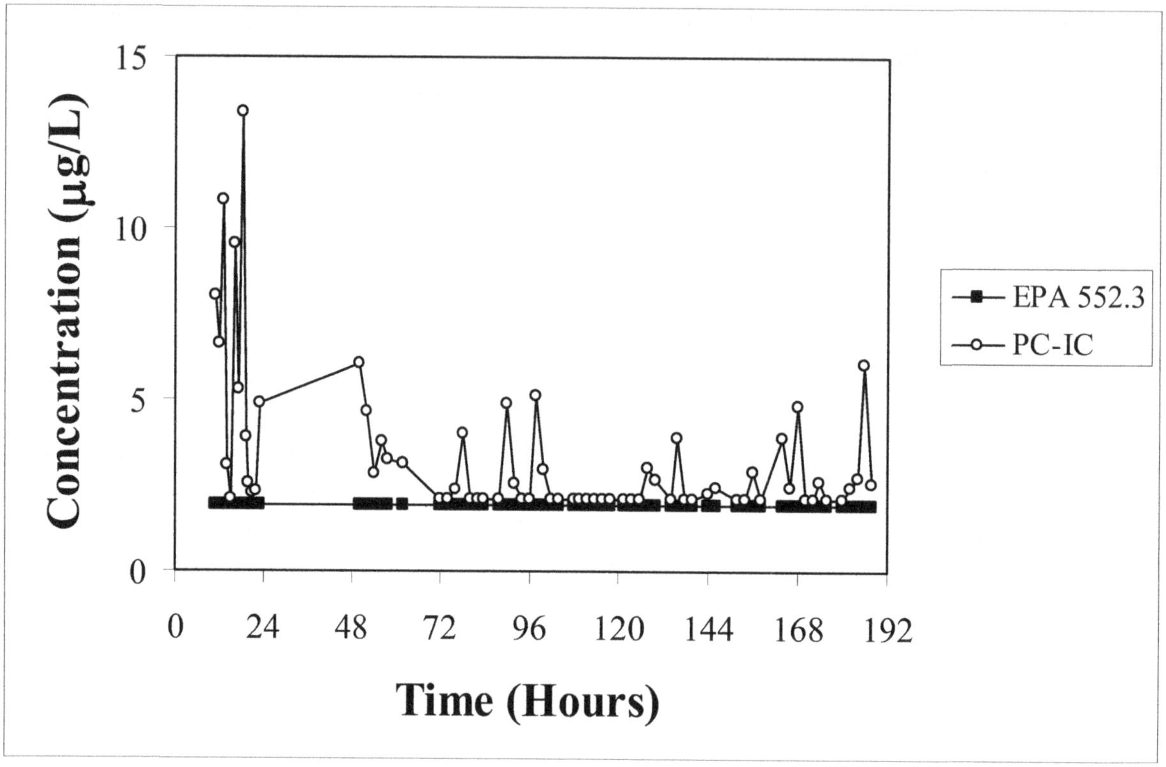

Figure 7.16 Memphis 2 concentration plot of MCAA/MBAA for EPA 552.3 and PC-IC

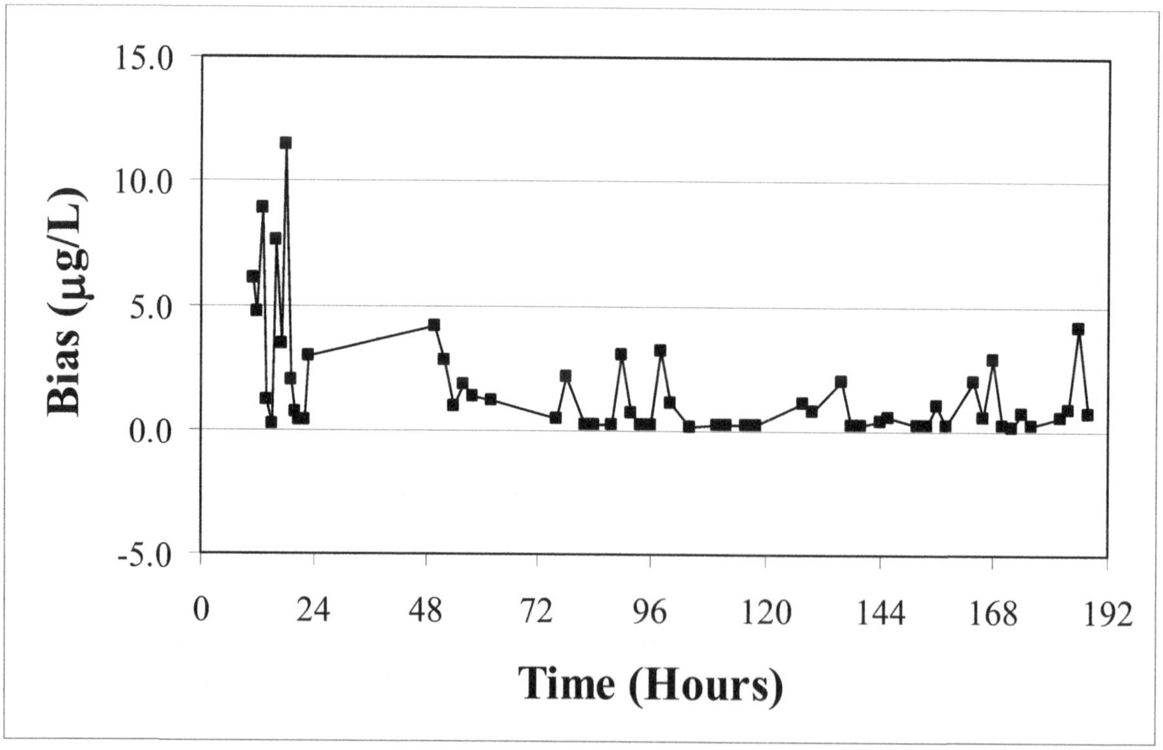

Figure 7.17 Memphis 2 bias plot of MCAA/MBAA comparing PC-IC to EPA 552.3

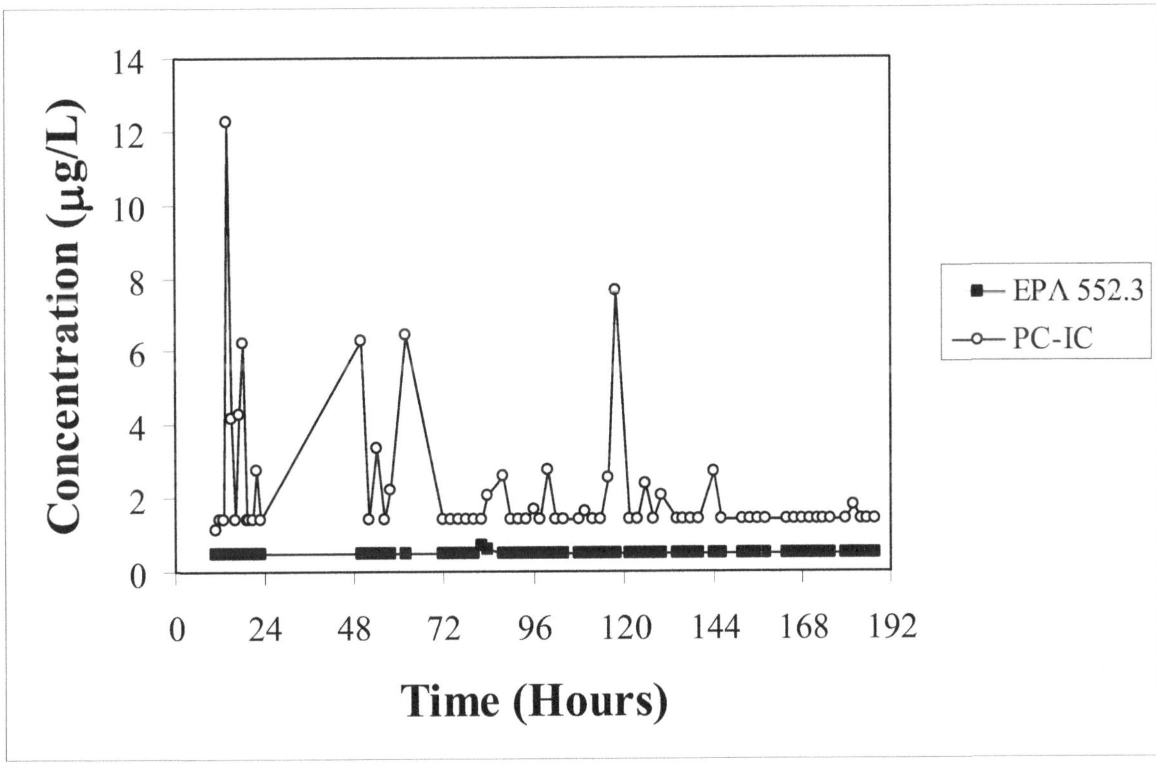

Figure 7.18 Memphis 2 concentration plot of DCAA for EPA 552.3 and PC-IC

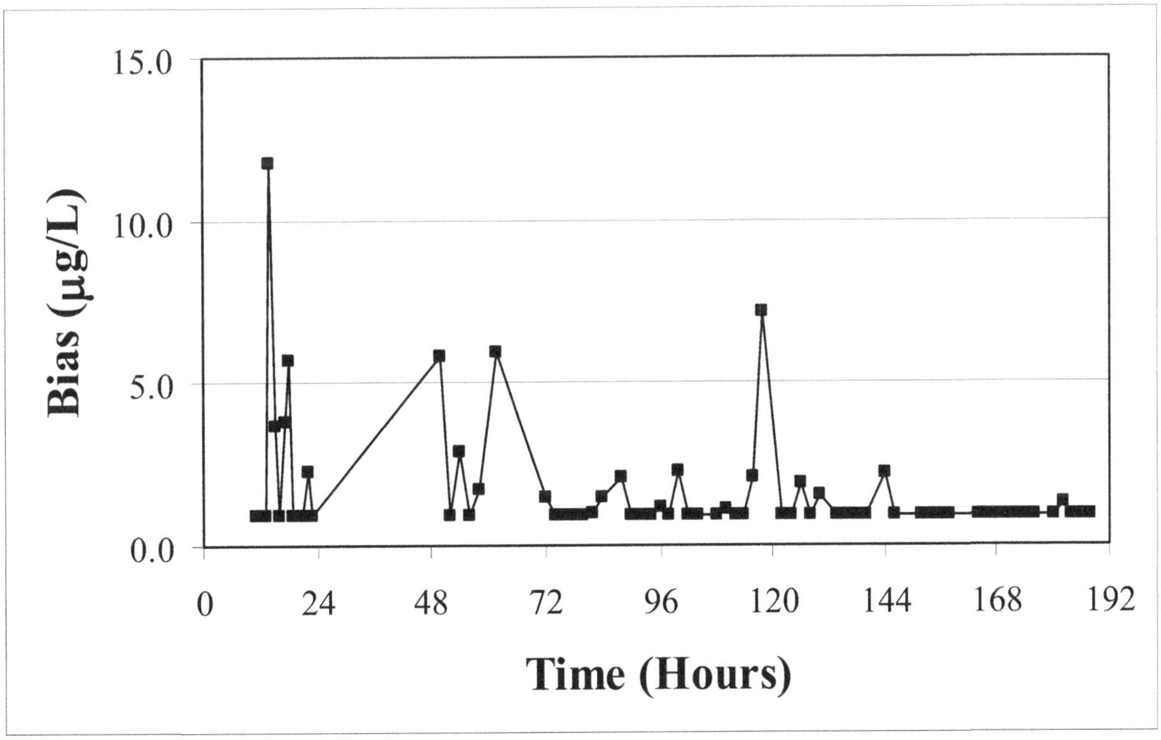

Figure 7.19 Memphis 2 bias plot of DCAA comparing PC-IC to EPA 552.3

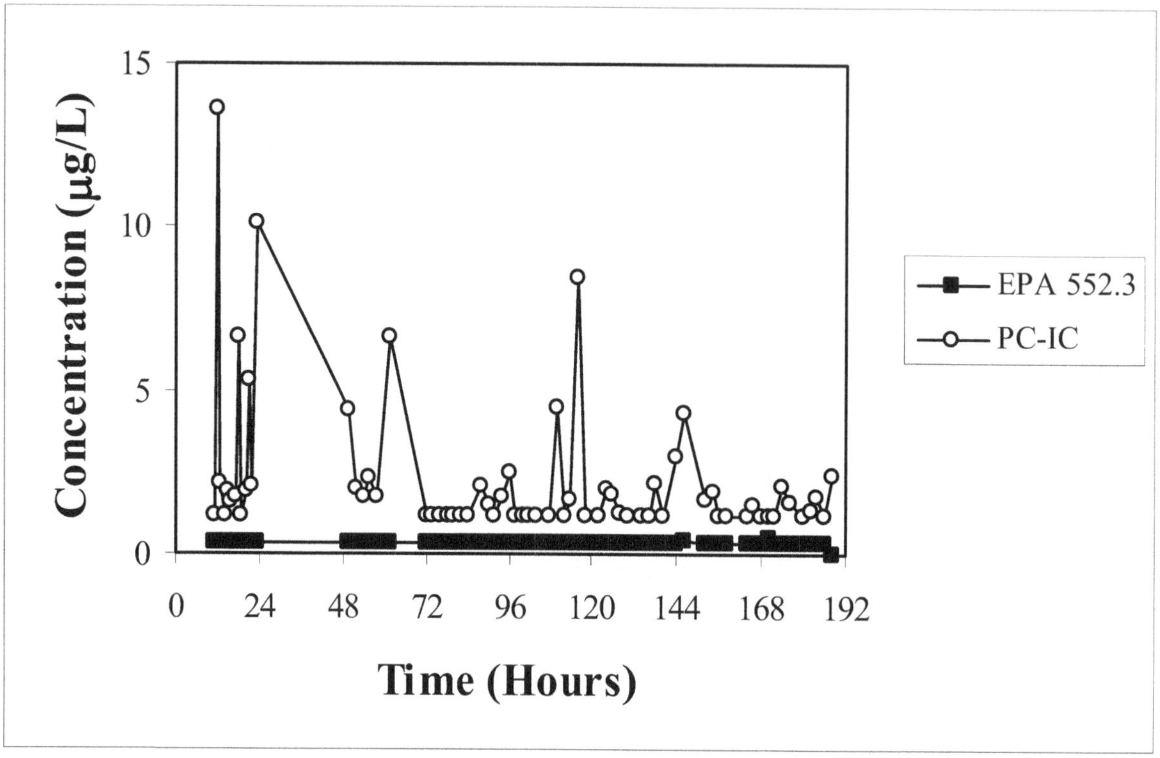

Figure 7.20 Memphis 2 concentration plot of TCAA for EPA 552.3 and PC-IC

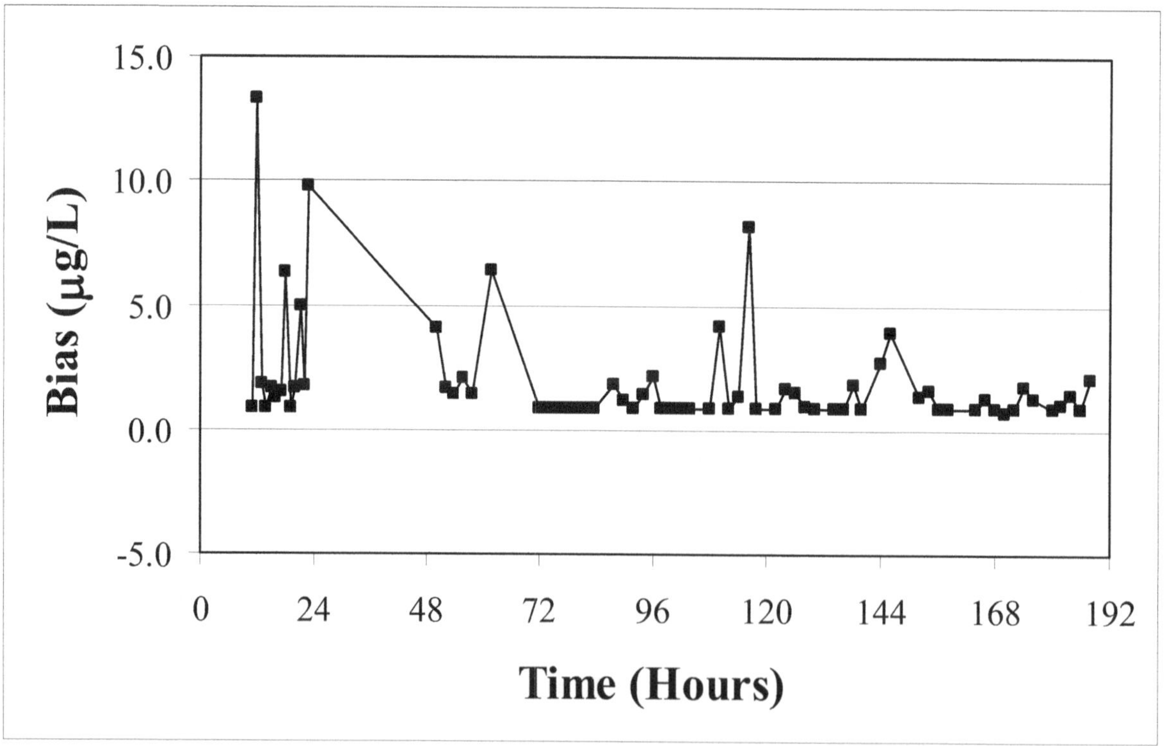

Figure 7.21 Memphis 2 bias plot of TCAA comparing PC-IC to EPA 552.3

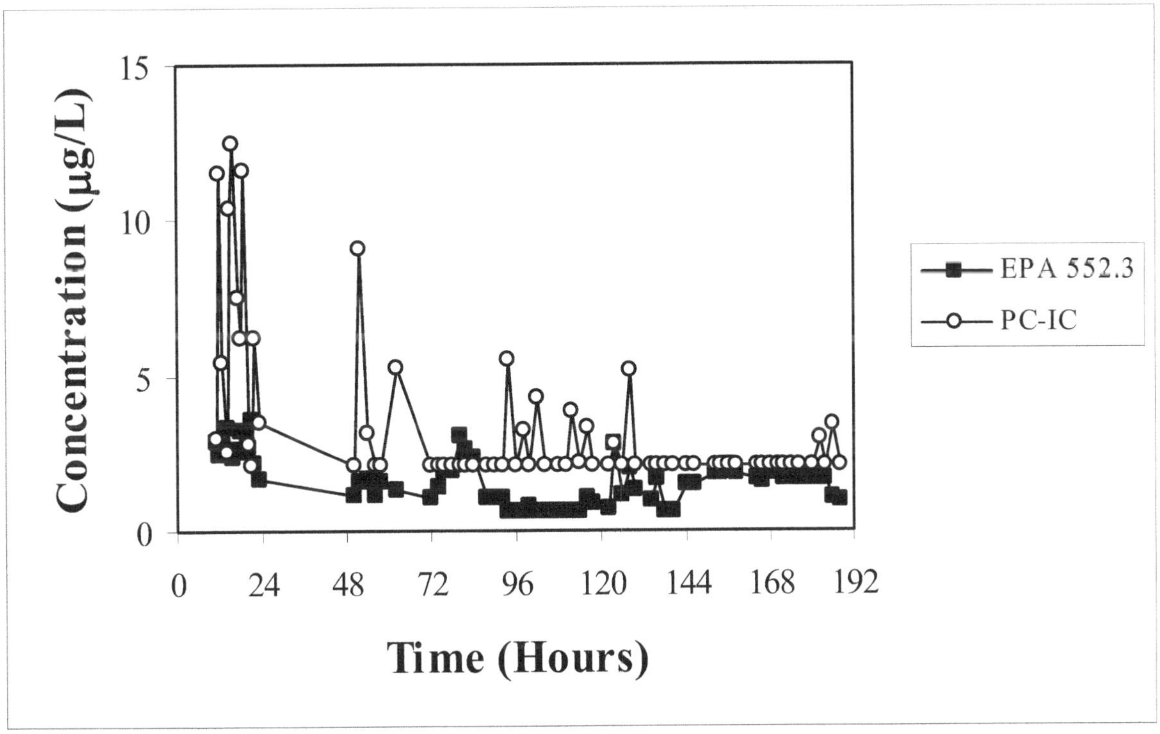

Figure 7.22 Memphis 2 concentration plot of DBAA for EPA 552.3 and PC-IC

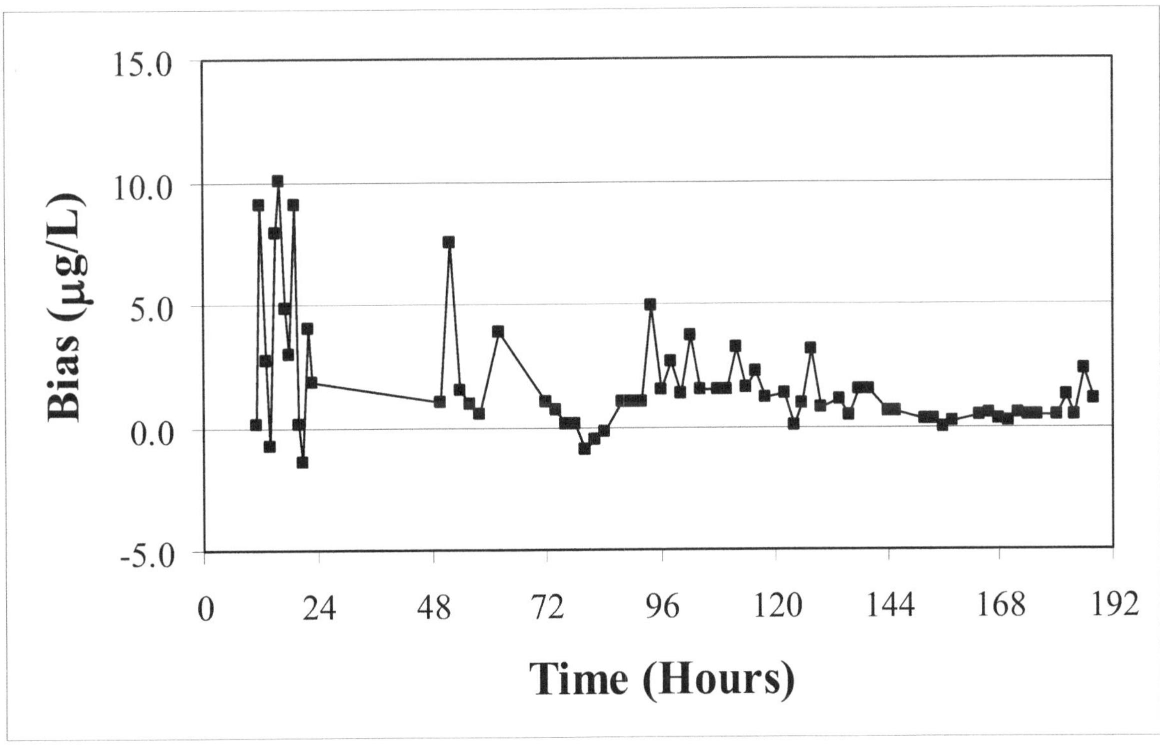

Figure 7.23 Memphis 2 bias plot of DBAA comparing PC-IC to EPA 552.3

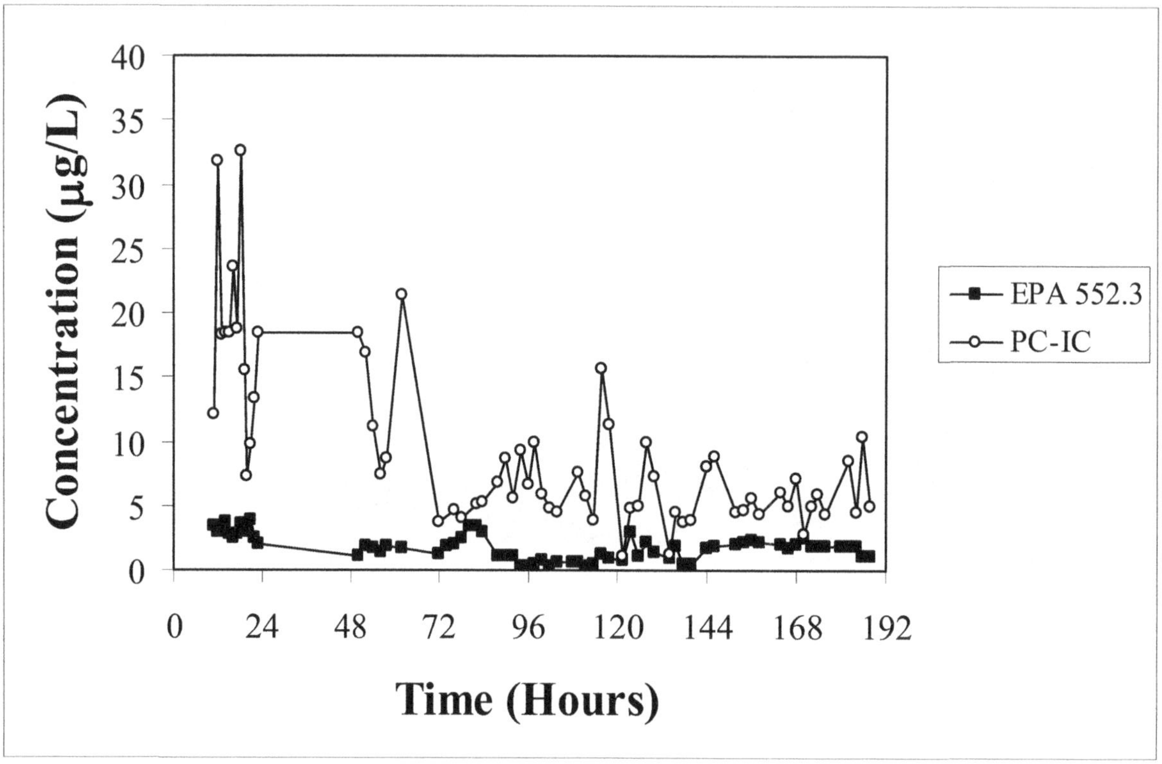

Figure 7.24 Memphis 2 concentration plot of THAA5 for EPA 552.3 and PC-IC

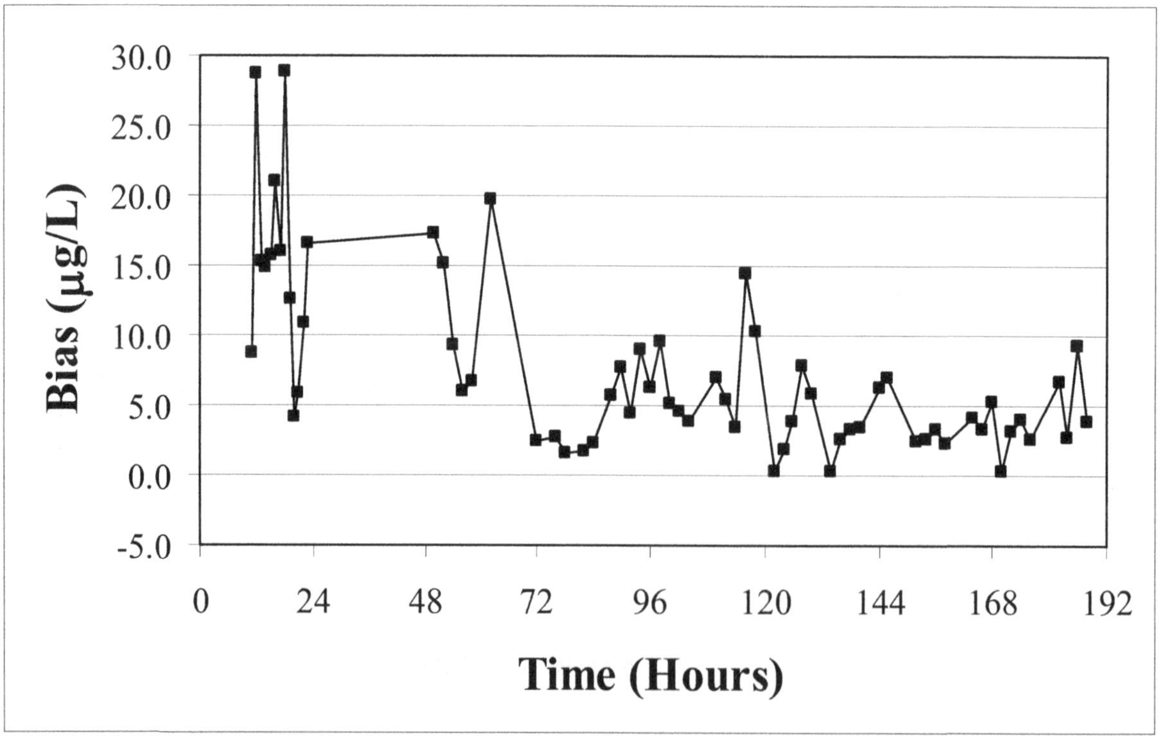

Figure 7.25 Memphis 2 bias plot of THAA5 comparing PC-IC to EPA 552.3

Houston On-Line Monitoring Study using PC-IC

The Houston on-line monitoring studies were conducted June 21-28, 2005 at the the City of Houston Department of Public Works sampling site (Houston). The PC-IC was interfaced to the drinking water distribution system using the adaptor described in the Appendix C. The studies were carried out in the same manner as described in the Memphis studies with other instruments presented in Chapters 4 and 5. The results of the Houston study along with detailed analysis of the data are presented in Table 7.4

A plot of the concentration of monohalogenated HAA species (MCAA + MBAA) as a function of time is shown in Figure 7.26. For 56 direct comparisons, the concentrations ranged from 1.9 - 8.0 (EPA 552.3) and 7.2 - 51.1 (PC-IC). The average concentration of MCAA + MBAA was 3.2 ± 1.8 μg/L (EPA 552.3) 38.9 ± 7.6 μg/L (PC-IC). A plot of the bias as a function of time (Figure 7.27) ranged from 5.3 - 49.2 μg/L for the PC-IC compared directly to EPA 552.3. The average bias was 35.7 ± 7.3 μg/L. The % REC_{Bias} ranged from 378 to 2689% with an average of 1503 ± 620%. The results for DCAA (Figure 7.28) from 56 direct comparisons and showed concentration ranges from 18.4 - 40.0 μg/L (EPA 552.3) and 9.1 - 87.8 μg/L (PC-IC). The average concentration was 29.2 ± 5.3 μg/L (EPA 552.3) and 39.6 ± 16.1 μg/L (PC-IC). The bias (Figure 7.29) ranged from –26.7 - 50.7 μg/L (averaging 10.4 ± 14.8 μg/L) with % REC_{Bias} ranging from 26 - 275 % (averaging 136 ± 50 %). For TCAA, (Figure 7.30), the concentrations ranged from 6.6 - 11.2 μg/L (EPA 552.3) and 1.2 - 2.1 μg/L (PC-IC). The average concentration was 9.0 ± 1.2 μg/L (EPA 552.3) and 1.2 ± 0.1 μg/L (PC-IC). The bias (Figure 7.31) ranged from –10 to –5.4 μg/L (averaging –7.8 ± 1.2 μg/L) with % REC_{Bias} ranging from 10.7 - 20.9% (averaging 13.8 ± 2.1 %). For DBAA (Figure 7.32), the concentrations ranged from 0.6 - 7.6 μg/L (EPA 552.3) and were less than the MDL (2.1μg/L) for all PC-IC measurements. The average concentration was 4.7 ± 2.7 μg/L (EPA 552.3) and less than the MDL (2.1 μg/L) for PCIC. The bias (Figure 7.33), calculated using the PC-IC MDL, ranged from –5.5 - 1.5 μg/L (averaging -2.6 ± 2.7 μg/L) with % REC_{Bias} ranged from 28 - 367% (averaging 100 ± 108 %). The THAA5 (Figure 7.34) concentration ranged from 34.0 - 61.6 μg/L (EPA 552.3) and 42.3 - 136 μg/L (PC-IC). The average concentration was 45.3 ± 5.7 μg/L (EPA 552.3) and 78.7 ± 18.6 μg/L (PC-IC). The bias (Figure 7.35) ranged from –10.4 - 80.3 μg/L (averaging 33.4 ± 18.0 μg/L) and % REC_{Bias} ranged from 81 - 279% (averaging 175 ± 41 %).

The two methods were compared statistically as described earlier. The $t_{critical}$ for the 56 comparisons 2.396. For MCAA/MBAA, DCAA, TCAA and DBAA the $t_{calculated}$ values are 36.811, 5.256, 49.551 and 7.444, indicating a statistical difference between the two methods for each of these species. THAA5 gave a $t_{calculated}$ of 13.870, indicating that the two methods are statistically different here as well.

The data exhibits a very large bias values (approximately 36, 10, –8, and –3) for the HAA5 species. These results are mirrored in the THAA5 results. It is possible that the errors come from at least two different sources. For example, there could be other species in the water sample that complicate the chromatography and interfere with the HAA5 determinations. This could be the case, in particular for the monohalogenated species and DCAA since both exhibit a large positive bias. A more integrative model takes into account that large positive bias values are exhibited by monohalogenated species and DCAA, while a large negative bias is seen in the TCAA results. It might be possible that the monohalogenated species and DCAA are reacting (while the EPA sample is in storage, awaiting extraction) further to produce TCAA. The PC-IC analyzer is able to detect this as it is sampling on-line in near-real time. It should be noted that it is possible for both of these factors to act in unison to increase the bias between the two methods.

Correction for BCAA Interference

As discussed, it is likely that there are other species that might act as interferents in the PC-IC method in Houston water. An immediate possibility is that other HAA species might be present. In fact, studies have been done that suggest that the chromatography needs to be optimized to account for all HAA9 species. It was presented earlier in this chapter that BCAA will co-elute with DCAA and possibly DBAA. (Unfortunately, this was learned after the Houston studies were finished). Earlier (Chapter 4) it was discussed how duplicate Houston samples remaining from the study were analyzed to ascertain the presence of other HAA species using EPA 552.3. In short, using spiking studies, BCAA was detected and quantified in Houston water. The retention time was matched with standards and the internal standard of EPA 552.3 was used to approximate the concentration of BCAA.

The estimated concentration of BCAA was added to the concentration of DCAA to obtain the total concentraton DCAA and BCAA. The concentration of DCAA was then divided by the total concentration and multiplied by 100 to obtain the % DCAA of what is reported by PC-IC, in a "back of an envelope" type calculation. This percentage was then applied to the PC-IC data for DCAA concentration to account for the BCAA interference. This new concentration was compared to EPA 552.3 using bias, % REC_{Bias} and student's t-test to evaluate the effect of this correction (Table 7.5).

The original (Org.) data has only been corrected for mean % recovery obtained from the check standards. The corrected (Corr) data is the Org. data corrected for the % DCAA in the PC-IC peak. The EPA 552.3 DCAA results ranged from 18.4 to 40.0 μg/L (averaging 29.2 ± 5.3 μg/L), and the results for EPA 552.3 THAA5 results ranged 34.0 to 61.6 μg/L (averaging 45.3 ± 5.7 μg/L). The "BCAA corrected" DCAA concentrations (Figure 7.36) show concentration ranges from 9.1 - 87.8 μg/L (Org.) and 6.3 - 69.4 μg/L (Corr.). The average concentration was 39.6 ± 30.7 μg/L (Org.) and 30.7 ± 15.3 μg/L (Corr.). The bias for the Org. data (Figure 7.37) ranged from –26.7 - 50.7 μg/L (averaging 10.4 ± 14.8 μg/L) and the Corr. ranged from –29.3 - 36.6 μg/L (averaging 1.4 ± 13.5 μg/L). The % REC_{Bias} ranged from 26 - 275 % (averaging 136 ± 50 %) for the Org. data and ranged from 19 to 224 % (averaging 104 ± 45%) for the Corr. data. The THAA5 correction (Figure 7.38) that results form this were as follows: For 56 direct comparison to EPA 552.3, the THAA5 concentrations ranged from 42.3 - 136 μg/L (Org.) and 31.6 - 118 μg/L (Corr.). The average concentration was 78.7 ± 18.6 μg/L (Org.) and 69.7 ± 17.7 μg/L (Corr.). The bias (Figure 7.39) for the Org. data ranged from –10.4 - 80.3 μg/L (averaging 33.4 ± 18.0 μg/L) and the bias for the Corr. data ranged from –13.1 - 64.9 μg/L (averaging 24.4 ± 17.5 μg/L).The % REC_{Bias} for the Org. data ranged from 81 - 279% (averaging 175 ± 41 %) compared to % REC for Corr. data ranging from 77 - 266% (averaging 155 ± 40 %).

When the results were compared statistically, the $t_{critical}$ (n=56) is 2.396. For DCAA Org. the $t_{calculated}$ is 5.256, indicating a statistical difference (as discussed earlier). But when the BCAA correction factor is applied, the $t_{calculated}$ value for DCAA Corr. is 0.799 and the methods are not statistically different with repect to DCAA concentrations. However, when the results for THAA5 are evaluated, the $t_{calculated}$ value is 10.4226, indicating there is still an overall statistical difference between the result reported for THAA5 for the two methods.

After correction for BCAA interference, DCAA is not statistically different from EPA 552.3. This makes sense as BCAA co-elutes with the current column and gradient programming, and also reacts with NCA to produce the fluorescent product. Correcting for the BCAA interference should, and does, make DCAA agree statistically with EPA 552.3. On a qualitative level, the bias agrees within 5 μg/L and the % REC_{Bias} is also 100 ± 5%. Unfortunately, after correcting for BCAA interference, THAA5 is still statistically different. But, this is also not unexpected. The large positive bias between the monohalogentaed species and TCAA is still comparatively large. It is possible that there are still unidentified species that complicate the TCAA chromatography. It is also suggested that the monohalogenated species might be undergoing further reaction to form TCAA. This model accounts for both the large positive bias with monohalogentaed species and the large negative bias exhibited by TCAA. It also makes chemical sense in that it implies that the monohalogenated species are being further chlorinated over time since the PC-IC samples are analyzed immediately and the EPA sample are analyzed after a holding time. It is still possible that other factors (in addition to the possible monohalogenated to TCAA conversion) play a role in the differences exhibited. For example, the Houston chromatogram drinking water chromatogram shown in Figure 7.40 clearly indicates the presence of other, possibly chemically similar, species in the PC-IC chromatogram that are not present in the check standards or Memphis water. Note the large, apparently unresolved peaks, just prior to TCAA elution. TCAA is now a shoulder on this peak. The component is likely chemically similar to TCAA: it reacts with NCA, thus appearing in the PC-IC chromatogram; It is negatively charged, as it is retained by an anion exchange column. More detailed experiments need to be carried out to further investigate these observations.

Table 7.4
Houston monitoring study: monitoring HAA5 concentrations using PC-IC

	MCAA/MBAA		DCAA		TCAA		DBAA		THAA5	
Statistic/*Method*	*EPA*	*PCIC*	*EPA*	*PCIC*	*EPA*	*PCIC*	*EPA*	*PCIC*	*EPA*	*PCIC*
N	56	56	56	56	56	56	56	56	56	56
Avg. Conc.	3.2	38.9	29.2	39.6	9.0	1.2	4.7	<2.1	45.3	78.7
Std. Dev Conc.	1.8	7.6	5.3	16.1	1.2	0.1	2.7	---	5.7	18.6
Conc. Range	1.9 to 8.0	7.2 to 51.1	18.4 to 40.0	9.1 to 87.8	6.6 to 11.2	1.2 to 2.1	0.6 to 7.6	---	34.0 to 61.6	42.3 to 136
Avg. Bias	---	35.7	---	10.4	---	-7.8	---	-2.6	---	33.4
Std. Dev Bias	---	7.3	---	14.8	---	1.2	---	2.7	---	18.0
Bias Range	---	5.3 to 49.2	---	-26.7 to 50.7	---	-10.0 to -5.4	---	-5.5 to 1.5	---	-10.4 to 80.3
% REC_{Bias}	---	1503	---	136	---	13.8	---	100	---	175
s.d. % REC_{Bias}	---	620	---	50	---	2.1	---	108	---	41
% REC_{Bias} Range	---	378 to 2689	---	26 to 275	---	10.7 to 20.9	---	28 to 367	---	81 to 279
t-test* $t_{critical}$ $t_{calculated}$	---	Yes 2.396 36.811	---	Yes 2.396 5.256	---	Yes 2.396 49.551	---	Yes 2.396 7.444	---	Yes 2.396 13.870

* The $t_{critical}$ and $t_{calculated}$ values listed were calculated for the N listed using a two-tailed t-test with 98% confidence. Typically a $t_{calculated}$ value should be less than the $t_{critical}$. Yes or No means if statistically different.

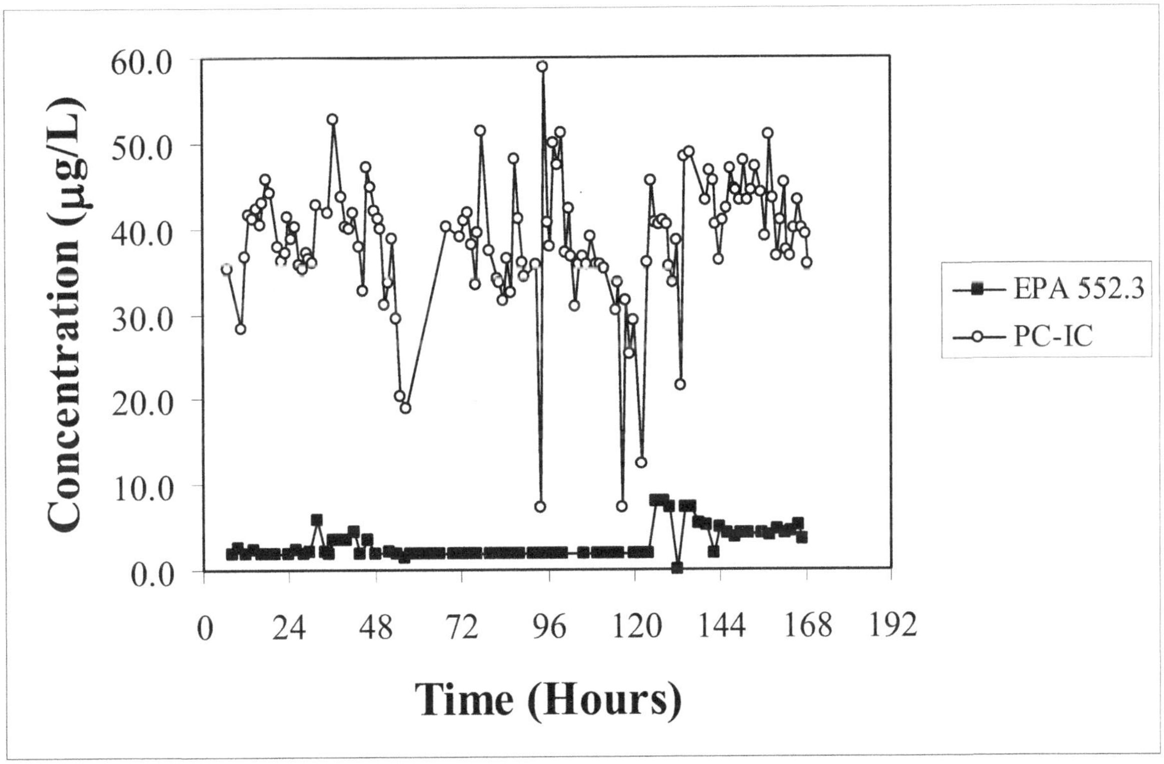

Figure 7.26 Houston concentration plot of MCAA/MBAA for EPA 552.3 and PC-IC

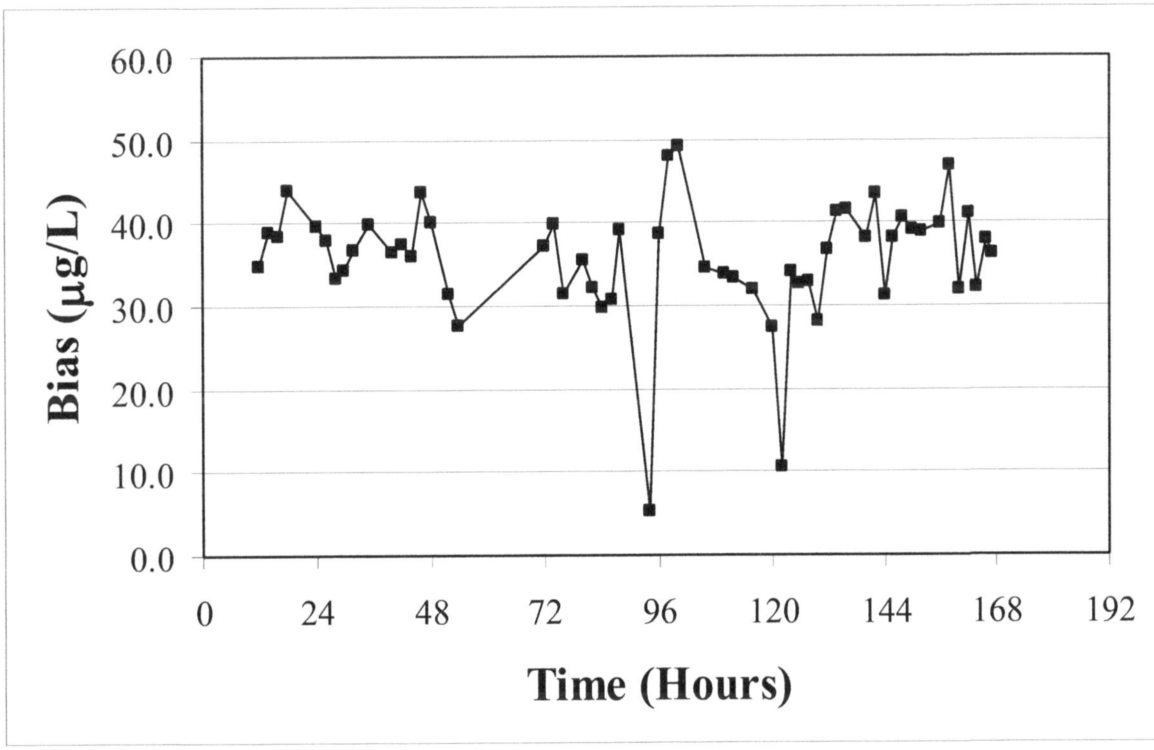

Figure 7.27 Houston bias plot of MCAA/MBAA comparing PC-IC to EPA 552.3

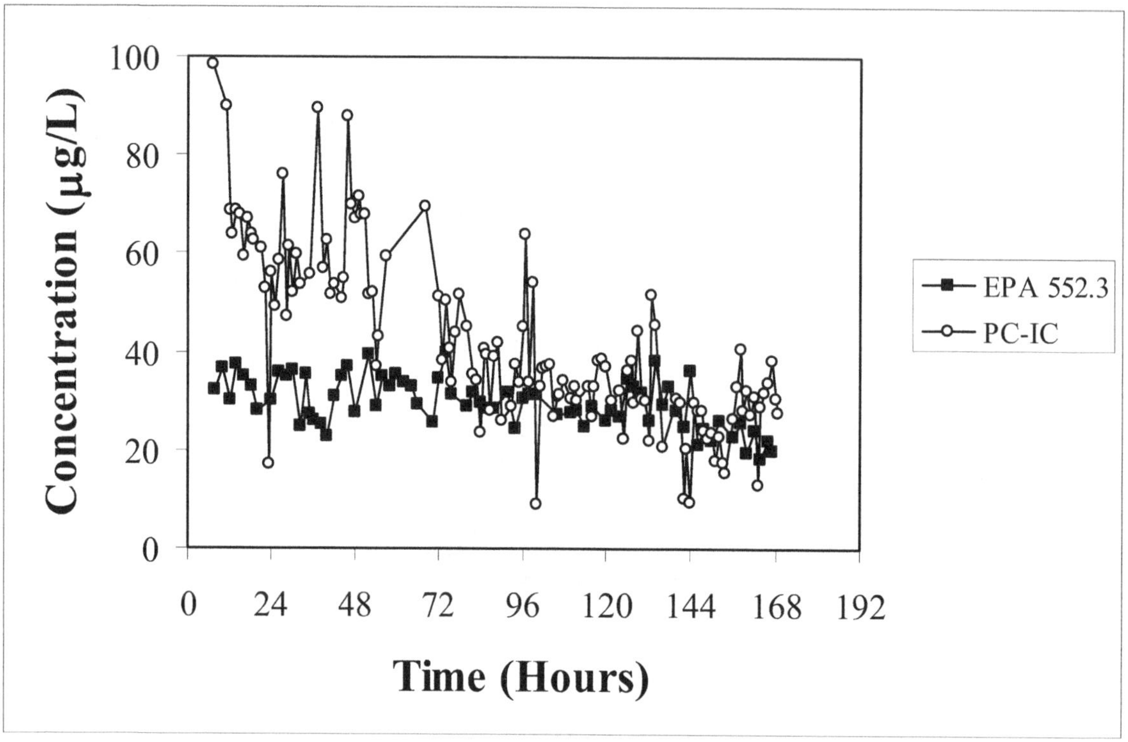

Figure 7.28 Houston concentration plot of DCAA for EPA 552.3 and PC-IC

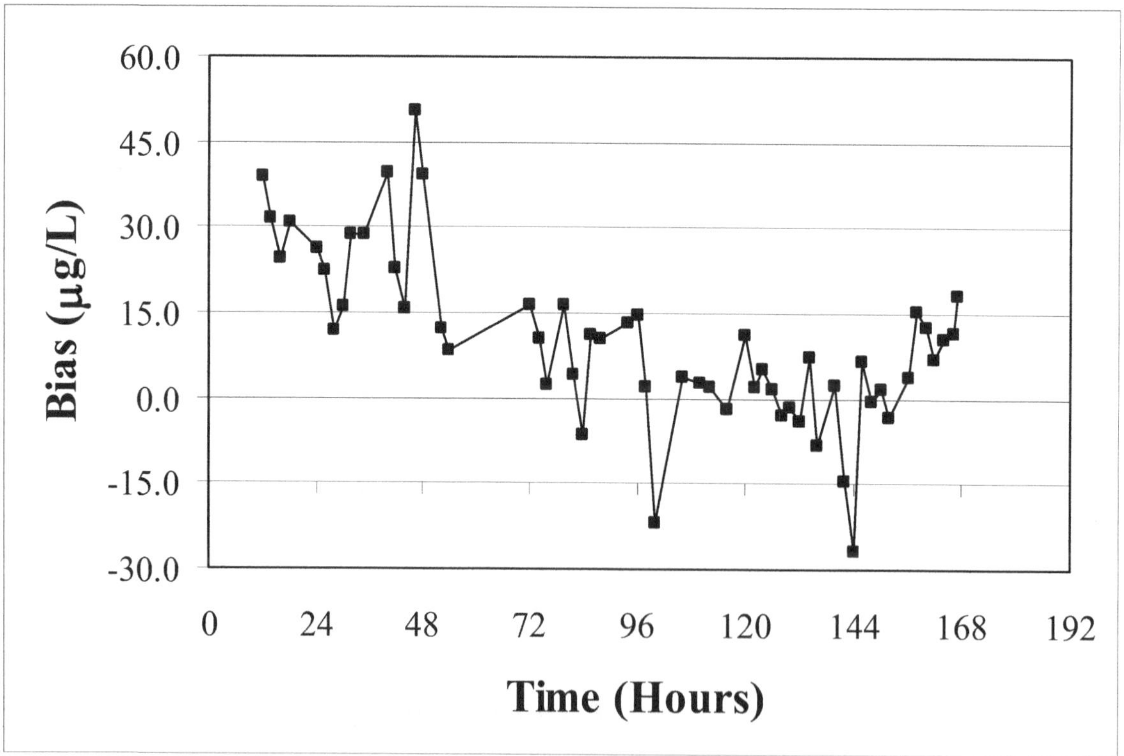

Figure 7.29 Houston bias plot of DCAA comparing PC-IC to EPA 552.3

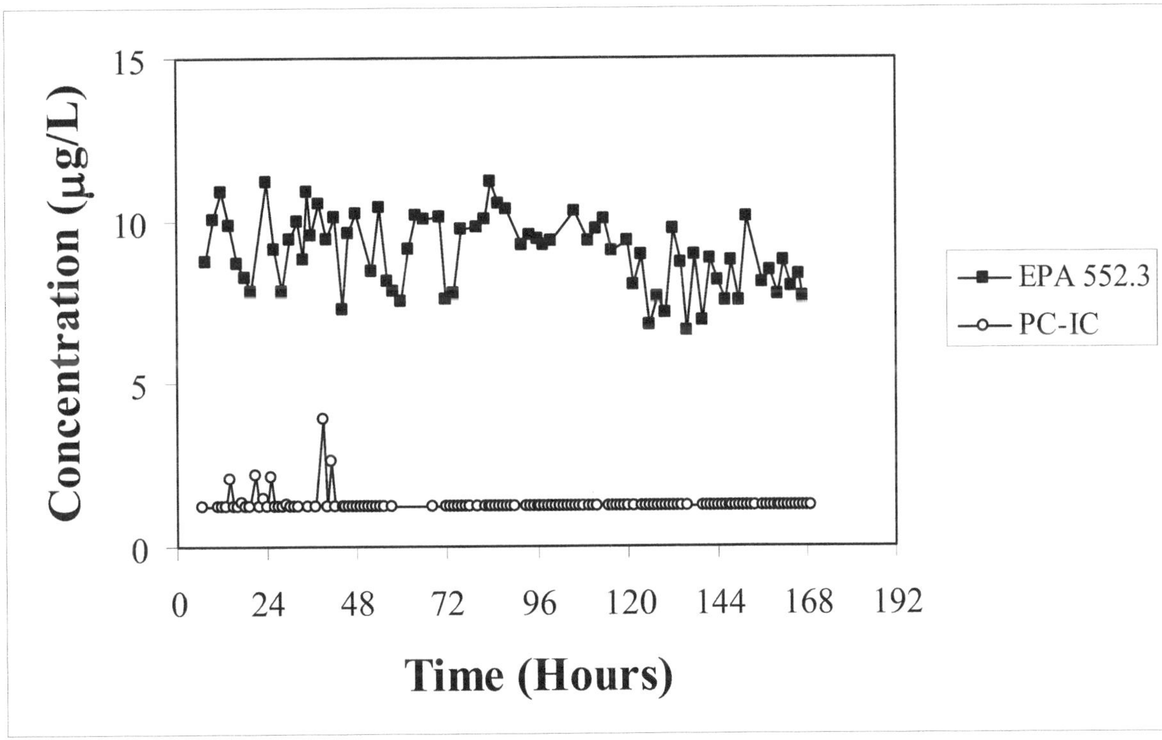

Figure 7.30 Houston concentration plot of TCAA for EPA 552.3 and PC-IC

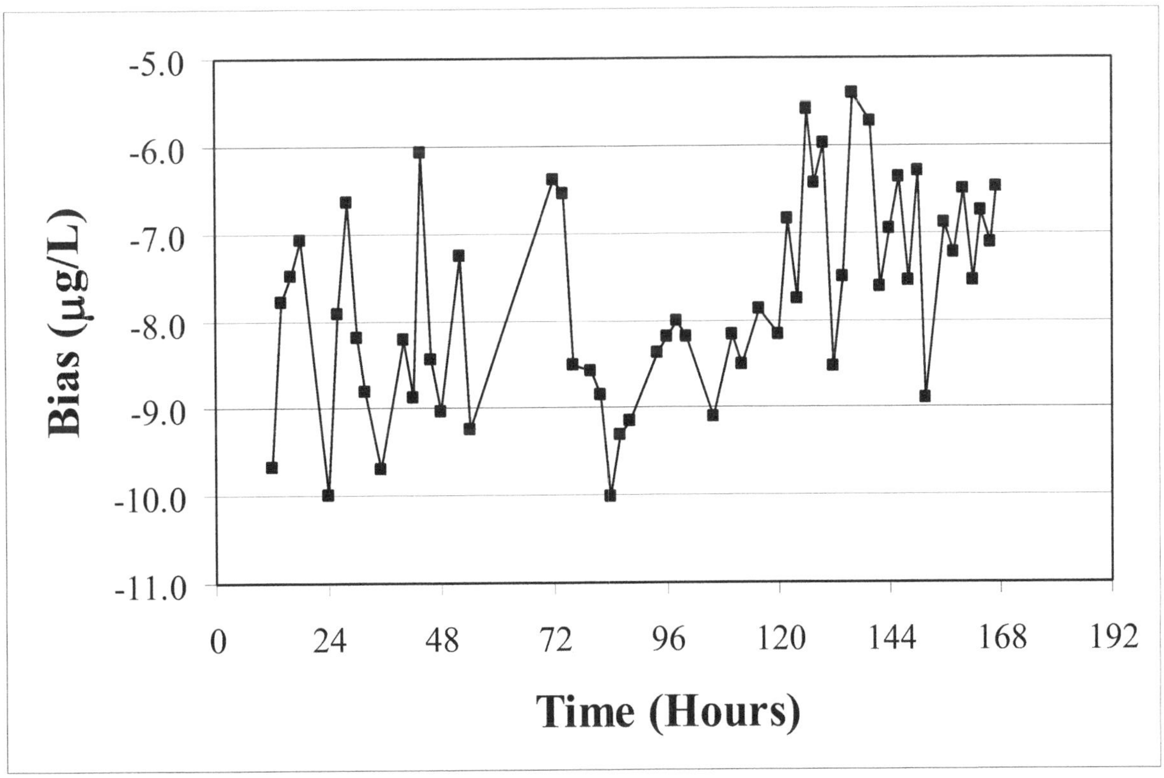

Figure 7.31 Houston bias plot of TCAA comparing PC-IC to EPA 552.3

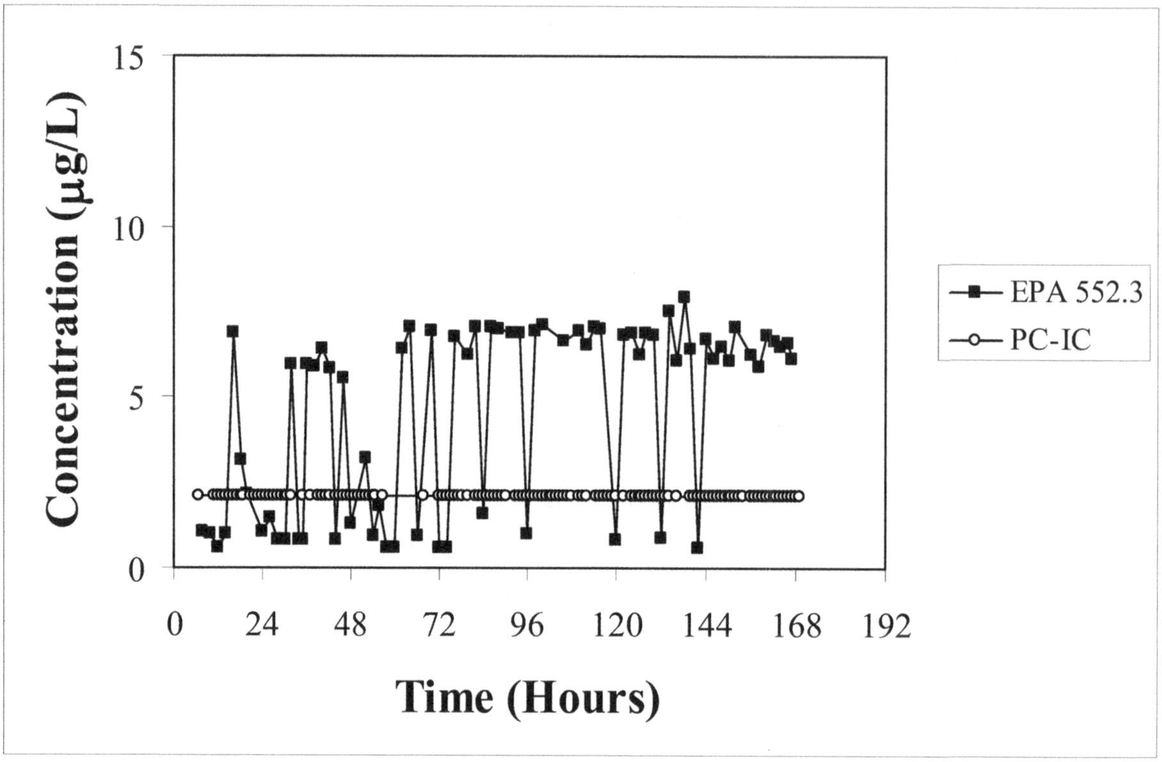

Figure 7.32 Houston concentration plot of DBAA for EPA 552.3 and PC-IC

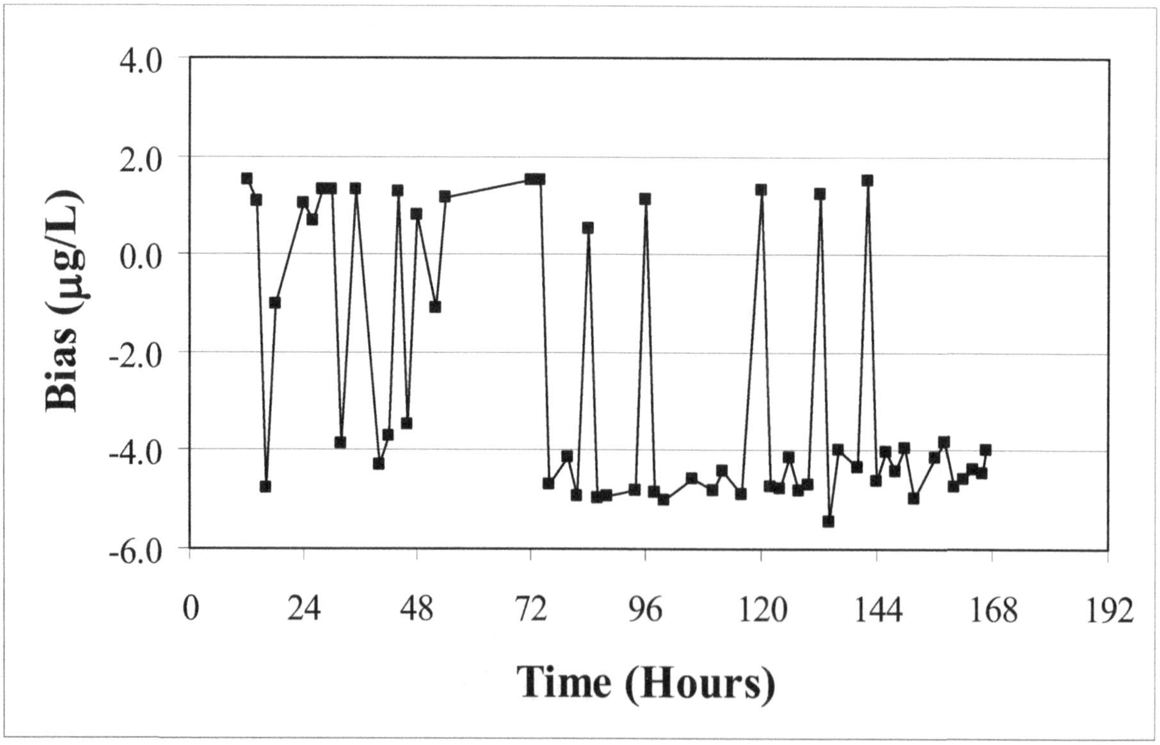

Figure 7.33 Houston bias plot of DBAA comparing PC-IC to EPA 552.3

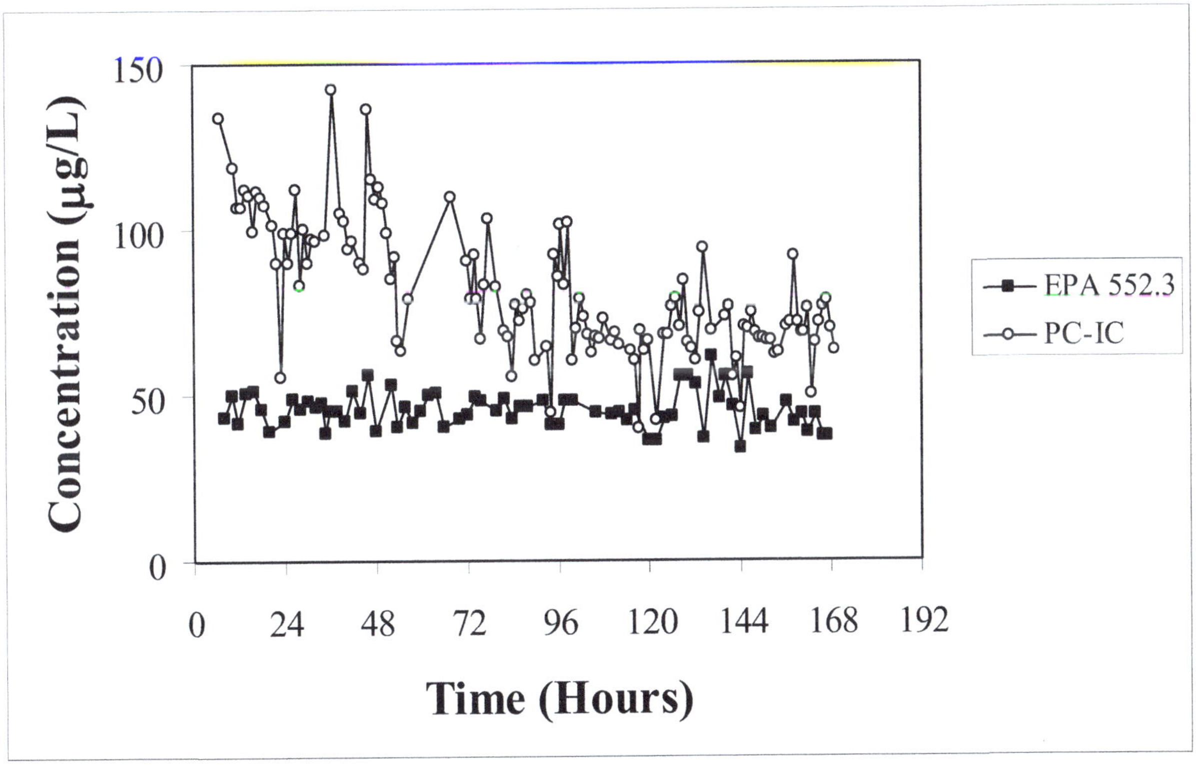

Figure 7.34 Houston concentration plot of THAA5 for EPA 552.3 and PC-IC

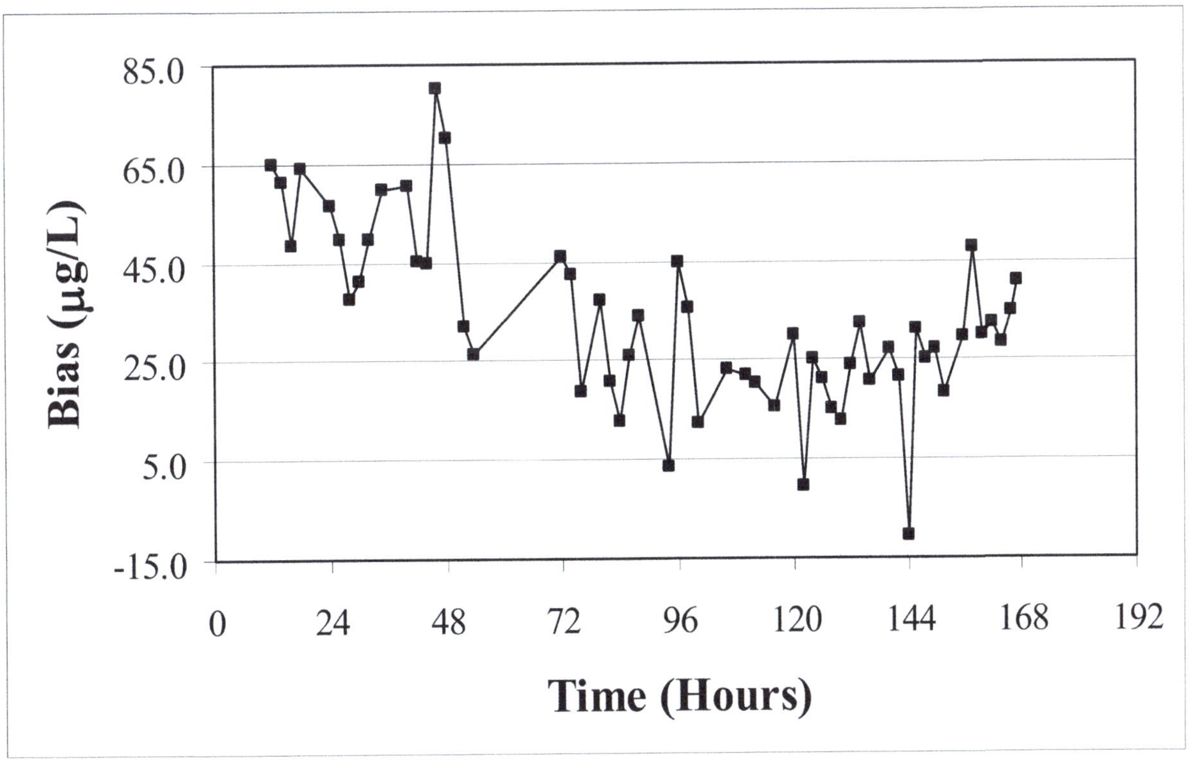

Figure 7.35 Houston bias plot of THAA5 comparing PC-IC to EPA 552.3

Table 7.5
Houston monitoring study: correction of DCAA and THAA5 concentrations for BCAA interference for PC-IC

	EPA 552.3		PCIC DCAA		PCIC THAA5	
Statistic/*Method*	*DCAA*	*THAA5*	*Org.*	*Corr.*	*Org.*	*Corr.*
N	56	56	56	56	56	56
Avg. Conc.	29.2	45.3	39.6	30.7	78.7	69.7
Std. Dev Conc.	5.3	5.7	16.1	15.3	18.6	17.7
Conc. Range	18.4 to 40.0	34.0 to 61.6	9.1 to 87.8	6.3 to 69.4	42.3 to 136	31.6 to 118
Avg. Bias	---	---	10.4	1.4	33.4	24.4
Std. Dev Bias	---	---	14.8	13.5	18.0	17.5
Bias Range	---	---	-26.7 to 50.7	-29.3 to 36.6	-10.4 to 80.3	-13.1 to 64.9
% REC_{Bias}	---	---	136	104	175	155
s.d. % REC_{Bias}	---	---	50	45	41	40
% REC_{Bias} Range	---	---	26 to 275	19 to 224	81 to 279	77 to 266
t-test*	---	---	Yes	No	Yes	Yes
$t_{critical}$			2.396	2.396	2.396	2.396
$t_{calculated}$			5.256	0.799	13.870	10.4226

* The $t_{critical}$ and $t_{calculated}$ values listed were calculated for the N listed using a two-tailed t-test with 98% confidence. Typically a $t_{calculated}$ value should be less than the $t_{critical}$. Yes or No means if statistically different.

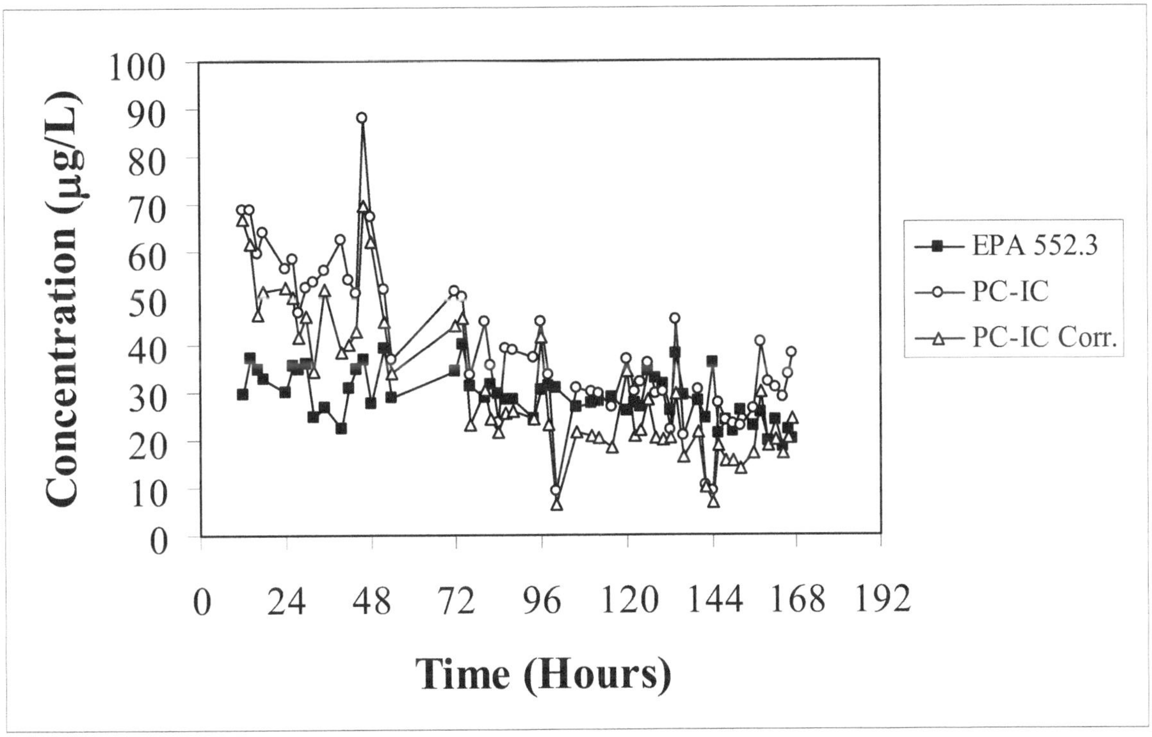

Figure 7.36 Houston concentration plot of DCAA for EPA 552.3, PC-IC and PC-IC corrected for BCAA interference

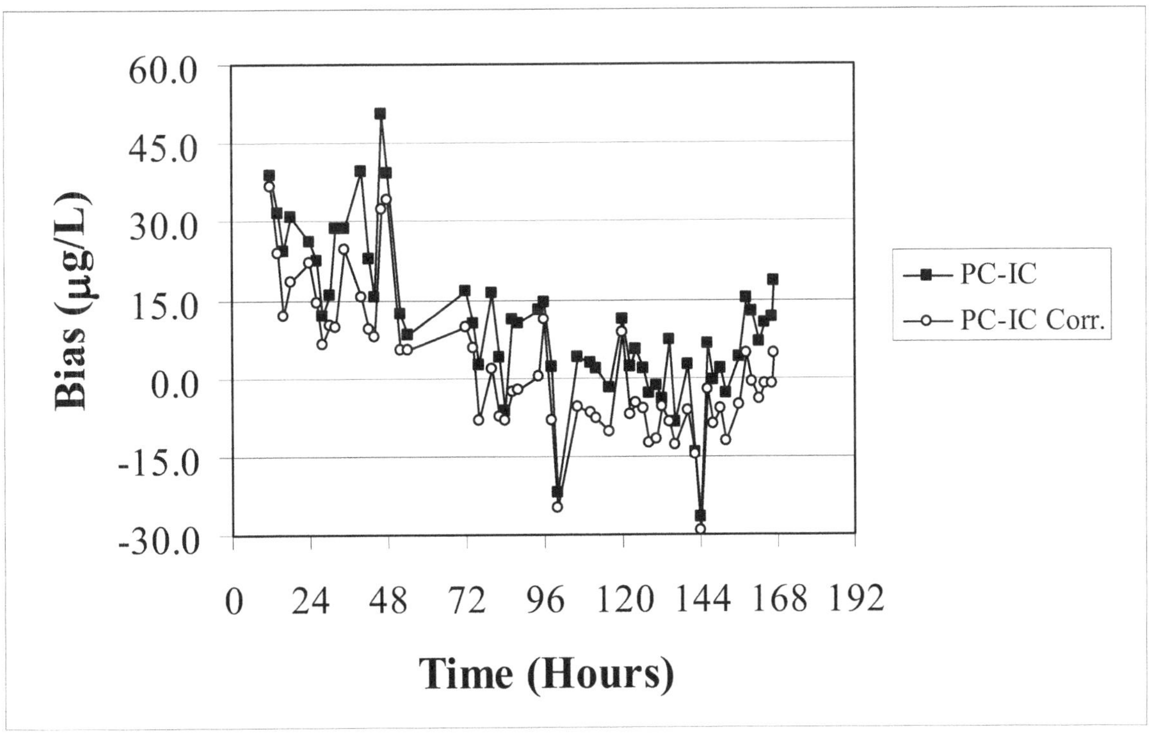

Figure 7.37 Houston bias plot of DCAA comparing PC-IC and PC-IC corrected for BCAA interference to EPA 552.3

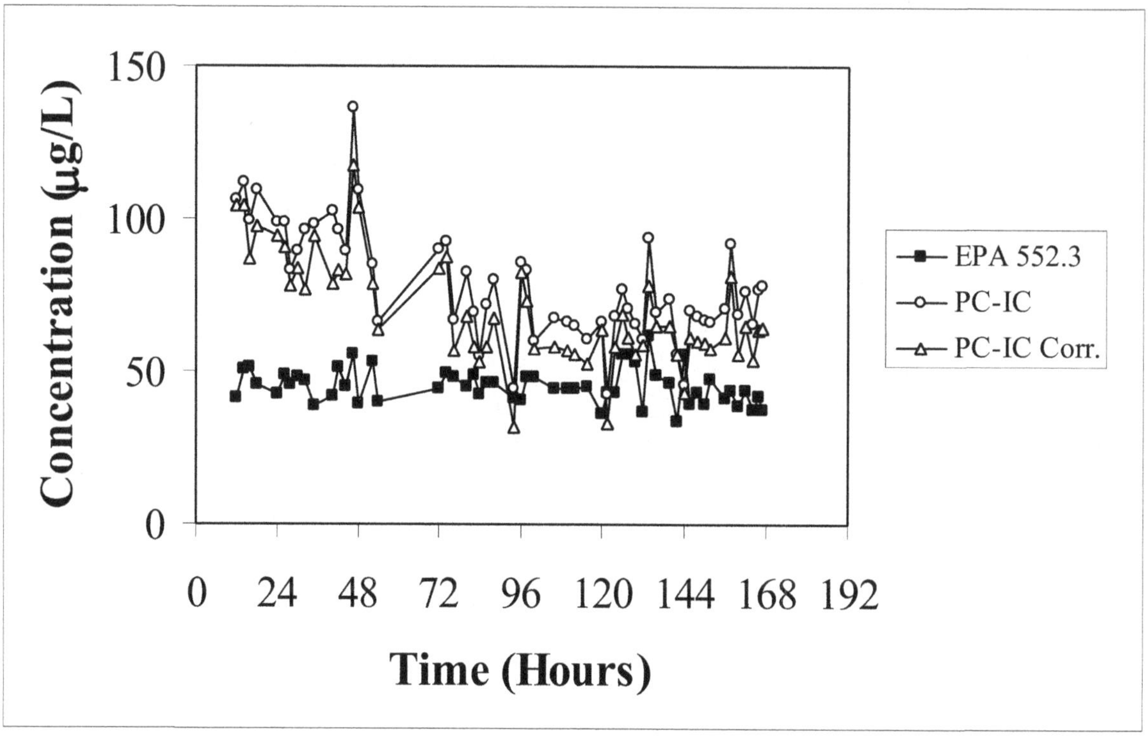

Figure 7.38 Houston concentration plot of THAA5 for EPA 552.3, PC-IC and PC-IC corrected for BCAA interference

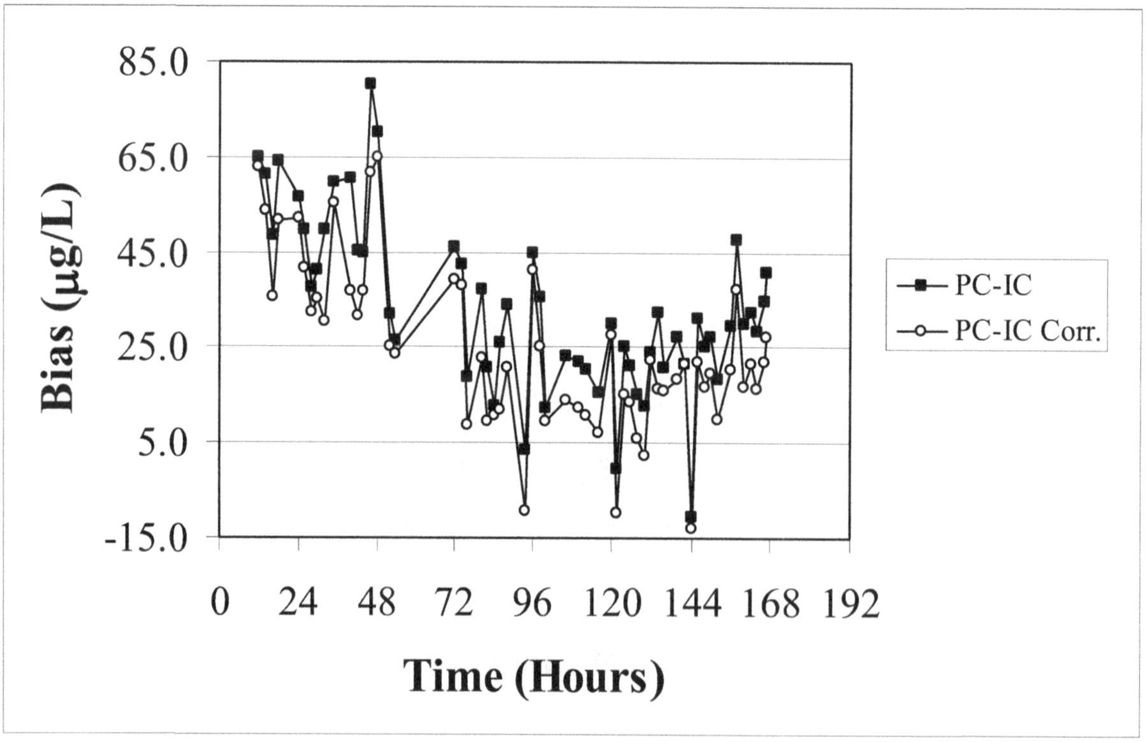

Figure 7.39 Houston bias plot of THAA5 comparing PC-IC and PC-IC corrected for BCAA interference to EPA 552.3

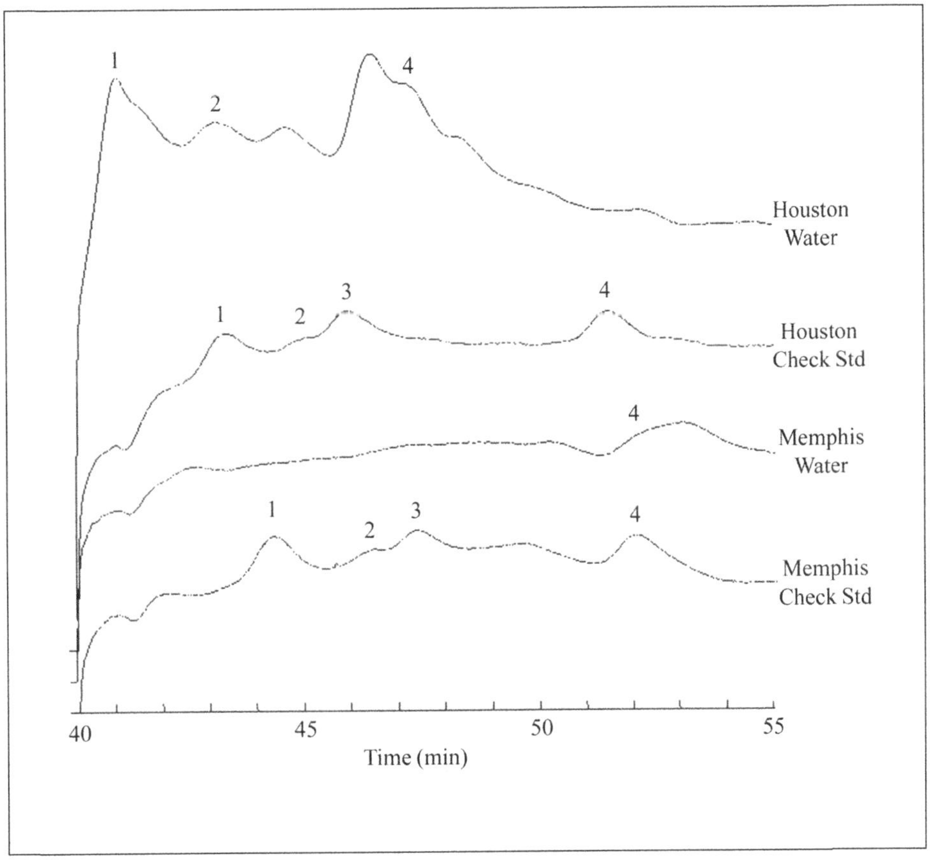

Figure 7.40 PC-IC chromatograms from the Houston and Memphis 2 monitoring studies. (1) MCAA/MBAA, (2) DCAA, (3) DBAA, (4) TCAA

Summary of PC-IC

Figure 7.40 illustrates other issues with the PC-IC method that need to be addressed. Specifically, these problems can be classified as resolution problems and also column "overloading" problems. Houston water has very high concentrations of chloride ion (~30 mg/L), sulfate ion (~50 mg/L), and bicarbonate (~100 mg/L; ~1.7mM) compared to Memphis (4.6 mg/L chloride ion, 6.2 mg/L sulfate ion, and bicarbonate is not routinely detected). In Figure 7.40, the retention times of the HAAs are shifted significantly forward ~3 min for Houston samples because as these anions are loaded onto the column, they take up active sites on the IC column needed for achieving separation. There are a finite number of these exchange sites and as they are taken up, the species move through the column faster because they interact less with the stationary phase. The monohalogenated species are the most affected by this overloading problem, as they elute before the chloride and sulfate anions, and are less strongly bound to the column. DCAA is less affected by chloride ion, but still affected by sulfate ion as it elutes between the two anions. TCAA is the least affected by these anions as it elutes after they would normally elute. Thus, while the common anions do not react with NCA, and thus do not appear in the chromatogram, they can still affect the chromatography of the PC-IC analyzer.

It is also interesting to note in Figure 7.40 (Houston water) the shoulder present on the monohalogenated species peak. PC-IC chromatograms in reagent water, which are considered ideal situations, have shown that MCAA and MBAA are unresolved (Figure 7.40, Houston and Memphis Check Std). This shoulder might also be another anionic halogenated DPB. There were similar resolution problems with DCAA, as previously mentioned, and these were corrected with some success. As discussed earlier, the same is true for TCAA. TCAA also shows resolution problems, but, the other HAA9 species (BDCAA, DBCAA, and TBAA) elute after TCAA. The peak eluting before TCAA might be some other anionic halogenated DBP. All of these co-elution problems illustrate that more work needs to be done to evaluate the PC-IC method for HAA9 species. It is possible that the unidentified species could be iodinated forms of the HAA species. Additional, more complete, interference studies need to be done to evaluate other possible species. These co-elution problems also show that for any on-line monitoring studies, detailed pre-monitoring screening studies need to be done for the specific local matrix. Pre-monitoring studies conducted with batch samples would allow researchers to taken into account and possibly identify any interfering species.

Despite its complications, the PC-IC shows much improved selectively over IC-MSCD and opens possibilities for using IC separation with HAA5 analysis. Another advantage is the elimination of the sample preparation step. The IC-MSCD method required both a chloride ion and sulfate ion removal pretreatment cartridges, where the PC-IC does not require these steps, and can be directly connected to the distribution system, in a straightforward manner. The PC-IC analyzer shows promise as an on-line monitoring tool.

CONCLUSIONS AND RECOMMENDATIONS FOR PC-IC NCA-FL

Two relatively simple and inexpensive analyzers are presented demonstrating the two levels of analyzers that could be easily constructed. With HAA5 levels at the 1-10 μg/L range, a state-of-the-art detector is likely needed. However if one is experiencing problems with meeting the MCL levels for total HAA5 (60 μg/L), and might need to monitor concentrations over the range ~10-100 μg/L, then, a less expensive filter based fluorescence detector could be used. With respect to expense, the two instruments differ a great deal, but both provide the ability to collect data for optimization studies.

The PC-IC is a method that meets many of the characteristics of an ideal analyzer (Emmert et al. 2004b) for online monitoring of HAA5 concentrations directly from the drinking water distribution system. As presented, the PC-IC is a not really portable, but was moved several times in these studies suggesting that it is at least "luggable". The PC-IC is a fully automated analyzer, but does require manual integration of the chromatograms as well as minor daily maintenance for routine operation to obtain reliable results. The sampling rate is ~1 sample per hour with MDL values in the desired 0.5 to 5 μg/L range, with %RSD values of <10% for 4 of the 5 HAA5 species. It has qualitatively shown promise when compared to USEPA Method 552.3. The two side-by-side comparisons were tested in a relatively clean matrix which stressed the low level detection abilities of the instrument and a more complicated and analytically demanding matrix. But the PC-IC method was not designed to replace the USEPA Methods for compliance monitoring. The PC-IC method presented here does not meet the requirements for compliance monitoring, but may be an attractive alternative to researchers and water professionals who could use an automated analyzer to collect data in near real time in the laboratory or at a remote location in the distribution system.

Recommendation 7.1: The PC-IC was recommended for further study. It appears to show promise as an on-line monitoring tool. The authors recommend more study on the full resolution of HAA9, iodonated HAAs, and possibly other species be performed.

Recommendation 7.2: Three 10 m KOTs and one 5 m KOT should be used to assemble the reaction trap in further studies as this spreads the chance of failure over three smaller (and less expensive) KOTs through clogging.

Recommendation 7.3: For post-column addition of reagents, only a peristaltic pump was available. Further study using commercially available post-column HPLC pumps or the Valco M6 pump is recommended. This is suggested because the peristaltic pump tubing is prone to leaking and has a limited lifetime (~12 to 15 hours).

Recommendation 7.4: Further study should be devoted towards the effects of higher TOC, alkalinity and pH on the NCA-FL chemistry.

Recommendation 7.5: Further study to find a suitable internal standard that could be automatically added to the sample while on-line needs to be done. This would simplify calibration when done alone and reinforce calibration when done in conjunction with the calibration curve approach. This approach would allow for correction of any possible drift in the FL detector over time.

CHAPTER 8
DEVELOPING STRAGETIES FOR ON-LINE MONITORING STUDIES

COMPARING THE METHODS IN THIS STUDY TO THE IDEAL METHOD

The characteristics of the Ideal Method were discussed in detail in Chapter One. To summarize, any method will be considered acceptable if it satisfies the following criteria: (1) easily adaptable for on-line monitoring; (2) portable; (3) fully automated; (4) capable of real-time or near real-time monitoring directly from drinking water distribution systems; (5) inexpensive (<$20K); (6) capable of transmitting the data over computer networks to a central server; (7) sampling rates of at least one sample per hour; (8) MDL in range of 0.5 - 5 μg/L, or better with mean % recoveries of 100 ± 5% and % RSD less than ± 10%; (9) minimal interference from TOC concentrations over the TOC concentration range of 1-20 mg/L. The different analytical methods that were explored in this study were compared to the characteristics of the Ideal Method. These results are summarized in Table 8.1.

Comparison of OPTGC for THM4 to the Ideal Method

The on-line monitoring purge and trap gas chromatograph (OPTGC) was an interesting analyzer for several reasons. The instrument was commercially available via special order from SRI Instruments Inc, and it could be readily automated with the PeakSimple software. MDL, accuracy and precision estimates were acceptable. However, when long-term monitoring studies were attempted, the dry electrolytic conductivity detector (DELCD) did not exhibit the required detector stability. The instability of the DELCD arose from the bending of an electrode in the DELCD. This bending or drooping of the electrode over time led to changes in the analytical response after approximately 13 to 18 hours of continuous use. Two attempts were made to modify the DELCD to alleviate these stability problems. One revision was the installation of the DELCD, so that the electrode was vertical, but this modification of the original design did not provide better results. The other involved constructing a support for the electrode. This configuration gave improved results after an extended initial operating period of approximately 24 hours. Unfortunately, each time the OPTGC is powered down (with heated zones allowed to cool to ambient temperature), it has to be re-stabilized for ~24 hours before being calibrated and placed on-line.

The problems with the OPTGC lie solely with the stability of the DELCD. For more routine analytical work, the DELCD is an attractive option. For a small utility, it is versatile in that it can be used in on-line mode or easily switched to a more traditional manually operated purge and trap device. The 24 hour lag time is a disadvantage, but one that can be planned for in advance. An additional concern with the OPTGC when using the DELCD is the limited lifetime of the DELCD electrode. The DELCD electrode will likely require replacement after 1-2 months of continuous operation. The OPTGC system does show promise in that it is commercially available, easy to interface to drinking water distribution systems and fully automated. The researchers suggest that if the OPTGC was fitted with an electron capture detector (ECD), it might be an outstanding choice for on-line monitoring of THM4 concentrations. In results presented later in this report, the OPTGC-DELCD was used in more extensive on-line monitoring studies. However, in the preliminary monitoring studies, it was not used for logistical reasons. At that time, other analyzers offered more promise, and there were limits to the number of analyzers we could operate and maintain. Future studies will be aimed at further exploring the OPTGC-DELCD method.

Comparison of SCMS-GC for THM4 to the Ideal Method

The supported capillary membrane sampler (SCMS) probe is a versatile sampling device for GC, having been interfaced to several GC detectors. In particular, the SCMS-GC with an ECD, a pulsed discharge photoionization detector (PDPID) and an ion trap mass spectrometer (MS) were all exhibited uses for THM4 analysis. In particular, the SCMS-GC-ECD analyzer meets many of the requirements of an Ideal Analyzer, and seems particularly attractive for on-line monitoring of THM4 concentrations. The ECD seems to be the most promising with the goals of the study in mind. The SCMS-GC-ECD is relatively portable (though it does require carrier and make-up gas cylinders). The SCMS-GC can be constructed with commercially available components for reasonable expense. The SCMS-GC-ECD was subjected to one extended monitoring study in Memphis with promising results. The sampling rate for the SCMS-GC-ECD method meets the goal of at least one sample per hour, and when not confined by doing side-by-side studies, can achieve a sampling rate of ~2 samples per hour. The MDL estimates fall in the approximate range of 0.5 - 5μg/L. The mean percent recoveries fall slightly outside the desired range of 100 ± 5% at 104 to 120%. The estimated precision (% RSD) is within the desired value of less than ± 10%., with the actual range being 5.5-10%.

Even though, the SCMS-GC-ECD has satisfactory results the necessity of using a calibration vessel complicated analysis. Significant effort was devoted to redesigning the FL/FC sampling device. Yet fundamentally, the SCMS approach would always require calibration in a separate vessel than the system to be sampled. Ultimately, the SCMS-GC-ECD analyzer and associated methods were not chosen for further study. Logistically, it was only possible for the researchers to use either the SCMS-GC or the capillary membrane sampling-gas chromatography (CMS-GC) as both used the same GC-ECD instrument for separation and detection. SCMS-GC also ranks slightly lower than CMS-GC in our opinion. The calibration vessel necessity coupled with problems with FL/FC sampling led it to be eliminated. Ideally, it would have been desirable to carry out side-by-side studies that included SCMS-GC-ECD and CMS-GC-ECD. The SCMS-GC-ECD method does show some promise as a viable method for on-line monitoring of THM4 species in drinking water distribution systems. Further studies should be done to explore this promising method. In particular, the elimination of the need for a calibration vessel and some form of automated internal standard addition should be examined.

Comparison of CMS-FIA NCA-FL for TTHM4 and THAA5 to the Ideal Method

The capillary membrane sampling flow injection analysis with nicotinamide-fluorescence (CMS-FIA NCA-FL) analyzer is unique among the methods that were explored here in that it was the only one capable of monitoring both TTHM4 and THAA5 on-line. Certainly CMS-FIA NCA-FL shows promise as a relatively simple and inexpensive option for water professionals. As a result of this characteristic combined with the result of on-line monitoring studies, the CMS-FIA NCA-FL analyzer is ranked as of the methods of choice for on-line monitoring of drinking water. A disadvantage of this method is that it cannot selectively measure the concentrations of individual THM4 or HAA5 species. However, the MDL, mean % recoveries and % RSD estimates ranged from 1.1 to 1.6 μg/L, 111 to 115%, and 2.1 to 3.2%, respectively.

Sample throughput was relatively high at 2 samples/hr (one TTHM4 and one THAA5). However if this method is not confined by the necessity of conducting side-by-side comparisons, it could easily achieve a sampling rate of approximately 3 - 4 samples per hour for TTHM4 or THAA5 analysis. The analyzer is field portable, fully automated and user friendly. The analyzer was built in laboratory from commercially available components. The operator skill level was relatively low when compared to other methods in this study. CMS-FIA has compared favorably in side-by-side studies done with EPA Method 502.2, EPA 524.2 and EPA 552.3. Three, ~seven day monitoring studies were done in Memphis and Houston drinking water with acceptable, though certainly not ideal results. We have examined closely the analytical issues that arose from the on-line monitoring studies and once reasonable corrections factors were applied, the agreement between CMS-FIA and EPA methods improved markedly. While not meant to replace the EPA methods for compliance monitoring, the proposed methods showed promise when compared favorably with EPA Methods 502.2, 524.2 and 552.3 and considering the goals of the study. When applied in appropriate situations, such as when operators are trying to optimize treatment processes, the CMS-FIA NCA-FL analyzer can provide on-line monitoring data in a simple and inexpensive manner directly from the distribution system. More effort will be devoted to better understanding the chemistry of possible interferences later in this chapter. Steps will be taken to further optimize this method for improved results when used on-line in near real-time.

Comparison of CMS-GC-ECD for THM4 to the Ideal Method

The CMS-GC-ECD is one of the methods of choice to emerge from our first round of on-line monitoring studies. The CMS-GC-ECD method meets many of the desired characteristics of an ideal analyzer for on-line monitoring of THM4 concentrations directly from drinking water distribution systems. The CMS-GC-ECD is a portable, fully automated analyzer capable of reporting data from remote locations as long as connection to wired or wireless communication is available. The CMS sampling device is inexpensive to construct and interface to a GC-ECD. The sampling rate is ~ 1 sample per hour when run in side-by-side studies with EPA methods. However, when unrestrained by other methods, this analyzer can easily achieve sampling rates of 2 - 3 samples per hour. The MDL values are in the ~0.5 μg/L range with % RSD values of <10% for all THM4 species. It has compared favorably (within the context of the study) in side-by-side comparison studies within the above criteria to EPA Method 502.2 in two separate ~seven day monitoring studies done in Memphis, TN drinking water. Less promising results were observed when similar studies were carried out in Houston, TX. However, we will propose in this Chapter a strategy for improving these results and minimizing these problems. When we applied reasonable correction factors to take into account analytical issues that were encountered in Houston, the results showed promise. The CMS-GC-ECD analyzer, with minor modifications in the manner that it is applied, can be successfully applied to on-line monitoring. Attention must be paid to identifying and accounting for possible interferences prior to conducting a detailed monitoring study. For example, detailed, pre-monitoring surveys need to be carried out prior to conducting extended on-line monitoring studies. As with the CMS-FIA methods, the CMS-GC was not designed to replace EPA methods for compliance monitoring. But the CMS-GC approach offers attractive alternatives for monitoring THM4 concentrations in near real-time in the laboratory and in real distribution systems. Another promising aspect of CMS-GC is that is could likely be used for analysis of emerging DBPs such as iodinated or nitrated analogues of THM4 species.

Comparison of IC-MSCD for HAA5 to the Ideal Method

The ion chromatography-membrane suppressed conductivity detection (IC-MSCD) method is not recommended for on-line monitoring of HAA5 species in drinking water distribution systems. Based upon these studies, IC-MSCD requires too many "on the fly" modifications to allow for routine on-line monitoring. The chromatography is too complex, even in the relatively simple matrix of Memphis drinking water. Each time the method is applied to different matrices, there would likely need to be extensive method development and optimization done to achieve satisfactory results. It is entirely possible that the method would simply fail altogether in some systems with high concentrations of common ions and inorganic DBPs.

These results seem inconsistent with some of the findings discussed in the literature of the Phase One Report (Emmert et. al 2004b). Liu and Mou (2003) report using a similar method with satisfactory results, but there are some questions about the reliability of these MDL, accuracy and precision estimates. The results reported here are consistent with those of Lopez-Avila (2000) who suggests that IC-MSCD methods do not provide suitable "simplified" methods for HAA5 analysis in real water samples. It should be noted that there are possibilities of automating sample pretreatment using flow injection and sequential injection methods. Others (Barron and Paull 2004a, 2004b; Barron et al. 2005) are using preconcentration/pretreatment with some success, however, this becomes cumbersome when applied to on-line monitoring. Another promising technique proposed by the laboratory of Mitra (Kou 2004; Wang et al. 2005a, 2005b) seems to eliminate many problems associated with on-line HAA analysis. The separation mode in this work was HPLC with UV detection. However, this technique is still in development and at present requires some degree of sample pretreatment.

In the future, it might be possible to use IC-MSCD for on-line monitoring studies, but at the present time the decision was made to not pursue it further. The IC-MSCD approach is not recommended, but this research did provide the foundation upon which the post column-ion chromatography (PC-IC) method using the NCA-FL chemistry was based.

Comparison of PC-IC NCA-FL for HAA5 to the Ideal Method

The PC-IC method is one of the methods of choice for HAA analysis to be developed in this study. Of all methods examined, it is the only one that offers a selective way to measure HAA5 concentrations without the need for complex extraction/derivatization steps. The PC-IC analyzer and method meets many of the characteristics of an ideal analyzer (Emmert et al. 2004b) for online monitoring of HAA5 concentrations. It is a "luggable" bench top analyzer that can likely be streamlined for more portability; it is largely automated (though it does require manual integration of the chromatograms to obtain useful results). The sampling rate is ~1 sample per hour with MDL values in the desired 0.5 to 5 μg/L range, with % RSD values of <10% for 4 of the 5 HAA5 species. In this mode, it cannot separate the monohalogenated species and thus they are reported as a total concentration. It has qualitatively compared favorably in a side-by-side comparison study within the above criteria to USEPA Method 552.3. The Memphis 2 side-by-side comparison tested PC-IC in a relatively simple matrix which stressed the low level detection abilities of the instrument. In Houston, the PC-IC instrument was tested in a very demanding matrix. In each case, when adjustments were made to the analyzer (regular pump tube changes) or when reasonable correction factors were applied (taking into account known interferences), then the PC-IC method gave promising on-line monitoring results. As with the other methods discussed here, the PC-IC method was not designed to replace the USEPA methods for compliance monitoring and nor does it meet the requirements for compliance monitoring. However, it may be an attractive alternative to researchers and water professionals who could use an automated analyzer to collect data in near real time in the laboratory or at a remote location in the distribution system. Ideally, this method needs to be optimized to allow for selective measurement of the individual HAA5 species, or even possibly HAA9 species. Simplification in the instrument design to increase portability would also be desirable.

Table 8.1
Evaluating the Candidate Analytical Methods in Terms of the Ideal Method Criteria

Analytical Method	Species	MDL (μg/L)	Mean % Recovery	% RSD	On-line monitoring / Real or Near -Real Time	Comm. Aval. / EPA Comparison Basis	Equipt. Required / Est. Skill Level‡	Instru. cost ($)* / Automated and Portable†	Anal. Time / Sample Rate	Link To Netwk. / Type
"IdealAnalyzer"	THM4 or/and HAA5	0.5-5.0	100 ± 5%	<10%	YES / Any	YES / YES	Instrum. Anal. / Low	<20K / YES	Minimal / 1 per hr	YES / Any
OPTGC-DELCD	$CHCl_3$	0.2	128	2.1	Adapted / Near-Real Time	YES / YES	Instrum. Anal. / Medium	15-17K / YES	15 min / 1 per hr	YES / FTP
	$CHBrCl_2$	0.2	121	2.2						
	$CHBr_2Cl$	0.1	116	1.2						
	$CHBr_3$	0.2	110	2.8						
SCMS-GC-ECD	$CHCl_3$	1.0	120	8.4	Adapted / Near-Real Time	YES / YES	Instrum. Anal. / Medium	15-17K / YES**	15 min / 1 per hr	YES / FTP
	$CHBrCl_2$	1.2	115	10						
	$CHBr_2Cl$	0.6	112	5.5						
	$CHBr_3$	0.6	104	6.6						
SCMS-GCOV	$CHCl_3$	1.6	91	16	Adapted / Near-Real Time	YES / YES	Instrum. Anal. / Med-High	10-12K / YES	5-6 min / 1-1.5 per hr	YES / FTP
	$CHBrCl_2$	1.4	94	13						
	$CHBr_2Cl$	1.6	99	14						
	$CHBr_3$	1.1	119	8.6						
SCMS-GC-MS	$CHCl_3$	0.4	104	2.2	Adaptable / Near-Real Time	YES / YES	Instrum. Anal. / High	64-65K / NO	15 min / 1 per hr	YES / FTP
	$CHBrCl_2$	3.2	98	14						
	$CHBr_2Cl$	2.1	102	6.8						
	$CHBr_3$	1.1	109	2.7						

(continued)

Table 8.1 (Continued)

Analytical Method	Species	MDL (μg/L)	Mean % Recovery	% RSD	On-line monitoring / Real or Near -Real Time	Comm. Aval. / EPA Comparison Basis	Equipt. Required / Est. Skill Level‡	Instru. cost (S)* / Automated and Portable†	Anal. Time / Sample Rate	Link To Netwk. / Type
"IdealAnalyzer"	THM4 or/and HAA5	0.5-5.0	100 ± 5%	<10%	YES / Any	YES / YES	Sd / Low	<20K / YES	Minimal / 1 per hr	YES / Any
CMS-GC-ECD	$CHCl_3$	0.3	100	4.5	Adapted / Near-Real Time	YES / YES	Instrum. Anal. / Medium	~8K / YES‡	15 min / 1 per hr	YES / FTP
	$CHBrCl_2$	0.5	104	8.6						
	$CHBr_2Cl$	0.3	125	3.5						
	$CHBr_3$	0.3	124	4.3						
CMS-FIA NCA-FL	TTHM4	1.1	115	2.1	Adapted / Near-Real Time	YES / YES	Instrum. Anal. / Low-Med	~9K / YES	40 min / 1-1.5 per hr	YES / FTP
	THAA5	1.6	111	3.2						
IC-MSCD (borate eluent)	MCAA	1.4	103	4.0	Unlikely / Near-Real Time	YES / YES	Instrum. Anal. / Med-High	18-25K / YES	50-60 min 1-1.2 per hr	YES / LAN
	DCAA	2.2	99.4	6.5						
	TCAA	1.9	121	4.6						
	MBAA	6.1	83	21						
	DBAA	4.4	100	13						
PC-IC	MCAA/ MBAA	2.1	84.3	5.9	Adapted / Near-Real Time	YES / YES	Instrum. Anal. / Med-High	~30K / YES	60 min / 1 per hr	YES / FTP
	DCAA	1.4	79.4	7.1						
	TCAA	1.2	105	5.5						
	DBAA	2.1	80	11						

* Instrument cost will certainly vary from vendor to vendor. The estimated instrument costs are based on "low-end" instrumentation and specialized materials designed specifically to fit the needs of the particular analysis.

† The meaning of portable is used freely. Normally a GC based system is not considered portable due to the need of gas tanks.

‡ The use of an ECD is generally restricted by a federal/state license. It is not generally accepted practice to move them from location to location as would be necessary for a portable instrument. In addition, there are issues related to venting the ECD exhaust in some instances.

COMPARING CANDIDATE ANALYTICAL METHODS TO EACH OTHER

It is important to place the results of the monitoring studies in context of routine drinking water monitoring. The concentration ranges of interest for most users of these methods are from ~0.5 to ~100 μg/L levels. This range is established by the desired MDL values of the Ideal Method and the upper concentrations defined by THM4 and HAA5 MCL values. The maximum contaminant levels (MCLs) for TTHM4 and THAA5 are 0.080 mg/L (80μg/L) and 0.060 mg/L (60 μg/L), respectively.

Three different monitoring studies were done in three slightly different water sources. The Memphis 1 study is representative of a relatively simple drinking water matrix that comes from pristine groundwater. The concentration of disinfectant is minimal and the Memphis 1 site is located very near the pumping station. At this point, the concentrations of DBP species will be expected to be very low (1-10 μg/L range). The Memphis 1 site taxes the methods MDL, but has a lowest likelihood of exhibiting possible interferences due to a relatively simple matrix. The Memphis 2 sample site was chosen because the water source is less pristine, but still comparatively simple compared to most raw water. This site is located further within the Memphis distribution system and THM4 concentrations are expected to be slightly higher but on the low-end of normally observed DBP concentrations (~5-25 μg/L). These concentrations are intermediate in expected DBP concentrations with slightly higher levels, though still relatively low in terms of possible interference effects. The Houston sampling site represents the most demanding sample matrix in terms of possible interferences. It is the least demanding in terms of taxing the MDL as the concentrations of DBP species are expected to be in the region of the MCL values (~25-100 μg/L). Table 8.2 summarizes the typical water quality data for these three sampling sites.

The results of the on-line monitoring studies for TTHM4 and THAA5 species are shown in Figure 8.1 and 8.2. In these plots of either the concentrations of TTHM4 or THAA5 versus time, the results from the three simultaneously conducted monitoring studies with three different methods are presented for the three sample sites. This is the same data that is presented in detail in the individual chapters, but the plots here are meant to give a broader view of the study. Certainly, there is room for improvement, but qualitatively, this data shows promise. The plots illustrate that the methods for TTHM4 were done at three different concentration levels: Low (1-10μg/L), low-mid (10-25μg/L) and high (40-60 μg/L). The same is true for THAA5, though certainly this analyte is more difficult to analyze compared to TTHM4 species. Arguably, the three different methods all appear to show the same trends. With Memphis 1 and Memphis 2 studies, the results are in good agreement. However in the Houston studies, differences between the methods increased. It is logical to assume that most of the difference exhibited here is a result of the more complex sample matrix of the Houston sampling site. Generally speaking, there are certain parameters that change appreciably in comparing Memphis and Houston sites. Houston water is expected to have a higher ionic strength, as exhibited by the conductivity values, higher hardness, TDS, TOC, total alkalinity, and pH. It is possible that these different parameters might affect the analytical signal and further research needs to be done to evaluate the effect of these different parameters.

Table 8.2
Typical water quality information for the three different on-line monitoring sampling sites: Memphis 1, Memphis 2, and Houston

Analyte*	Memphis 1	Memphis 2	Houston
Cations			
Calcium (Ca (II))	8.8	14.8	43.0
Magnesium (Mg (II))	3.6	6.3	2.6
Manganese (Mn(II))	0.003	0.002	0.01
Sodium (Na (I))	8.3	10.4	21.8
Total Iron Fe (III)	0.04	0.03	<0.03
Anions			
Bromide ion	---	---	0.05
Carbonate ion	---	---	1
Bicarbonate ion	---	---	97
Chloride ion	4.6	7.4	30
Fluoride ion	1.1	1.0	0.5
Nitrate ion (N)	ND	ND	0.49
Nitrite ion (N)	ND	ND	0.02
Orthophosphate ion	1.4	1.3	0.02
Sulfate ion	6.2	10.7	53
Misc			
Color (Pt-Co units)	ND	ND	2
UV254 (abs units)	---	---	0.096
Total Diss. Solids (Calc.)	54	67	213
Suspended Solids	---	---	<0.2
Total Organic Carbon	0.4	0.7	3.9
Conductivity (μmhos/cm)	110	136	350
Total Alkalinity	41	68	80
Hardness	36	60	124
Turbidity (NTU)	0.30	0.17	0.08
pH (pH scale)	7.3	7.3	8.0
Temperature (°C)	18	18	22

*All units are in mg/L unless otherwise specified

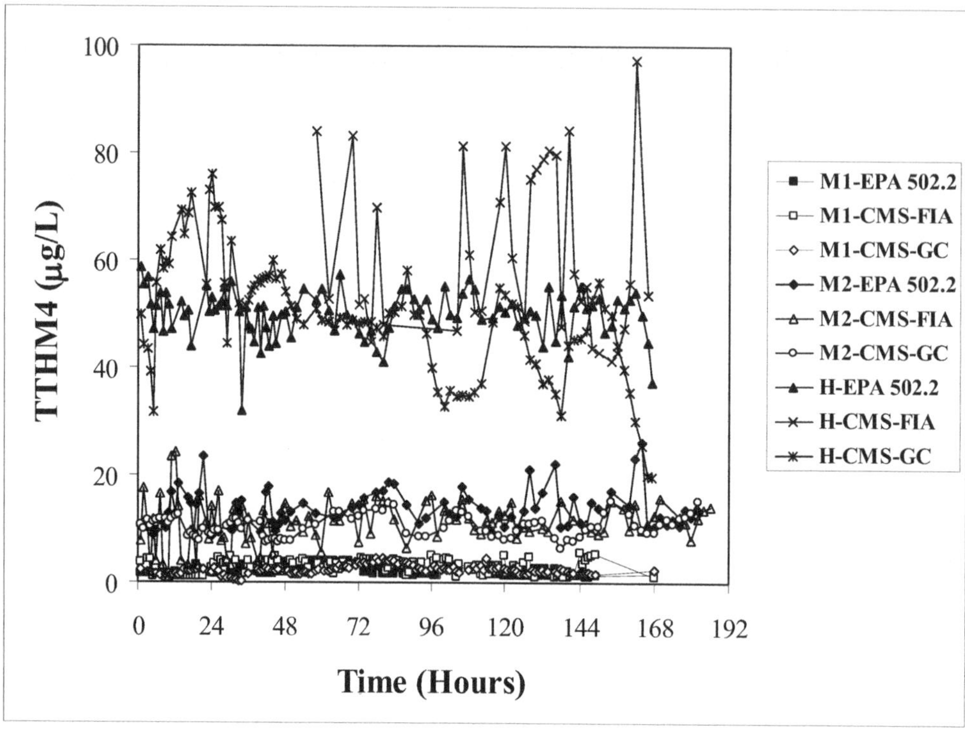

Figure 8.1 Combined plots of TTHM4 concentrations from the Memphis 1 (M1), Memphis 2 (M2) and Houston (H) monitoring sites

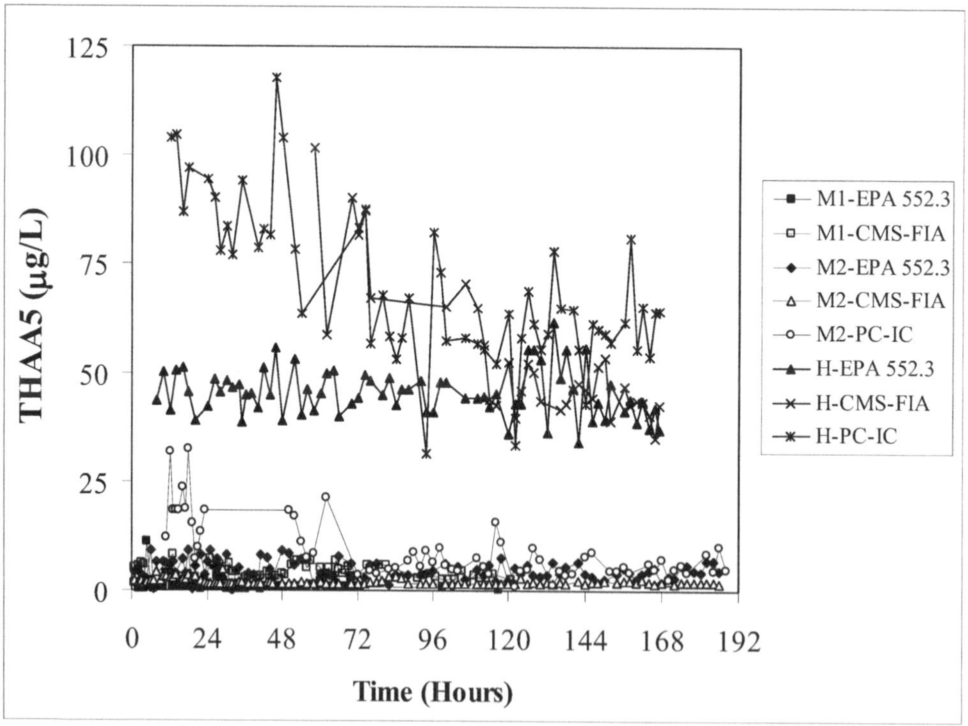

Figure 8.2 Combined plots of THAA5 concentrations from the Memphis 1 (M1), Memphis 2 (M2) and Houston (H) monitoring sites

REVISED STRATEGIES FOR ON-LINE MONITORING STUDIES

After comparing all the methods, it was proposed that the methods of choice for on-line monitoring are CMS-GC-ECD for THM4 species and TTHM4, CMS-FIA NCA-FL for TTHM4 and THAA5, and PC-IC NCA-FL for HAA5 species and THAA5. In addition, the OPTGC method also showed promise, so was used in the second round of monitoring studies to monitor THM4 species and TTHM4. Ideally, these methods should be run simultaneously during a monitoring study and will provide for independent methods for monitoring TTHM4 and THAA5. For high quality work, the on-line monitoring methods need to be run in conjunction with appropriate EPA methods. In addition, check standards need to be analyzed, so that corrections can be made to improve accuracy and take into account slight changes in the analytical signal over extended monitoring periods.

Moreover, it is suggested that prior to beginning an on-line monitoring study, researchers should conduct preliminary surveys of the particular system to be studied. Specific attention needs to be paid to identifying possible interferences and evaluating the demands of local sample matrices. Strategies are described for each method of choice.

For CMS-GC studies, it is suggested that EPA 524.2 be used to identify possible interferents and chromatography issues. Grab samples of the target distribution system could be analyzed to establish any chromatography issues with the actual matrix. The GC-MS can check peak purity as well as identify any compounds that complicate chromatography or otherwise simply appear in the sample. Any chromatography issues should be corrected to minimize any analytical issues prior to carrying out the actual monitoring study. Similar studies need to be done for HAA5 species. For example, LC-MS, if this is available, could provide useful information about possible interferents in preliminary surveys. Alternatively, and perhaps more practically, EPA 552.3 can be used to identify other HAA species (and similar species) that are present in the target water. Analyzing EPA 552.3 extracts by GC-MS could also lead to important information about the matrix to be faced in an on-line study. The information from these studies could alert the analyst to possible interferences and chromatography issues prior to starting the actual study. The compounds detected in the surveys could also be used to identify any possible interferences in the CMS-FIA methods for TTHM4 and THAA5 species.

Pre-monitoring surveys should be carried out using batch (grab) samples prior to conducting full extended monitoring studies. An adage to apply in this regard might be stated as follows: *The simpler the monitoring method, the more complex the pre-monitoring studies need to be.* This means that the simpler (non-chromatographic based methods) such as CMS-FIA, are more susceptible to interferences as no separation of species occurs. Because the analytes are detected as one unresolved unit, it is even more important to identify possible interferences. The chromatography of the more complex methods (CMS-GC and PC-IC) makes them less susceptible to interference issues. In each case, the pre-monitoring survey can aid researchers in accounting for interferences and taking steps to minimize them. The extension of the PC-IC method to the analysis all HAA9 species could provide much more insight about possible interfering species in the CMS-FIA methods, while also minimizing any problems with comparing PC-IC to EPA 552.3.

SUMMARY OF RECOMMENDATIONS

Recommendation 8.1: Conduct instrument specific MDL, accuracy and precision studies. Conduct preliminary survey of possible interferences using available resources. For example, EPA Methods, preferably GC-MS techniques, or other EPA methods should be used to identify possible interferences. Extension of the PC-IC method to account for all HAA9 species would also provide useful information.

Recommendation 8.2: Use the results from preliminary survey to account for possible interferences in CMS-FIA NCA-FL method where the target analytes are not subjected to separation from possible interferences.

Recommendation 8.3: Use the results of preliminary surveys of possible interferences to confirm that there are not co-elution issues in the chromatographic methods. Develop, modify and validate site specific chromatography to account for any issues.

Recommendation 8.4: Run side-by-side comparisons to appropriate standard method (EPA 502.2, 524.2, 552.3 etc) as often as practical, but at least twice every 12 hours. Use the % $REC_{Monitor}$ to correct measurements, where necessary and if deemed appropriate.

Recommendation 8.5: Develop methods for automated internal standard addition for the chromatographic methods, and automated standard addition for the FIA based methods. In each case, these alternative calibration methods might solve some matrix problems. Unfortunately, our experience shows that this approach may be more complicated than it appears and will undoubtedly result in some dilution of the sample. Additional effort needs to be done to explore alternative calibration techniques. This work needs to focus on accurate and precise sample injection prior to the sample entering the appropriate sampling device.

Recommendation 8.6: Additional effort needs to be devoted to identifying why the analytical signal tends to drift over time. It is important to gain a better understanding of what causes the drift. Then the drift needs to be minimized or at least accounted for when calculating the concentration of DBPs by using regular analysis of check standards to compensate for any drift.

REVISITING AND REVISING THE ANALYZERS FOR ON-LINE MONITORING

Following the first round of monitoring studies, additional effort was put into improving the different on-line analyzers based upon what we had learned about their deficiencies. An overview of these modifications and improvements will be presented. Prior to carrying out additional on-line monitoring studies, premonitoring surveys of Memphis and Houston drinking water distribution systems were conducted. Several samples were analyzed using available resources to identify possible interfering compounds. Analytical issues were identified and steps were taken to minimize any interference issues. Three different sets of studies were done using Memphis and Houston samples, focusing attention on each of the proposed on-line monitoring methods, CMS-GC, CMS-FIA, and PC-IC. It was also possible to devote further effort to exploring OPTGC-DELCD methods.

Instrument Modifications for CMS-GC and OPTGC Methods for THM4 Species

In this section, instrument changes and other sampling modifications will be discussed for all GC based methods including CMS-GC and OPTGC methods. These methods, once modified, will be compared in additional monitoring studies presented in Chapter 9.

Changes to CMS-GC-ECD Method and Instrumentation

After completing the on-line monitoring studies in Phase Two, several of the most addressable proposed recommendations were explored depending on instrument capabilities and available funding. It was decided that it would be possible to increase the sample flow rate and improve performance. A Cole-Parmer peristaltic pump (pump head model 7013 with recommended long life Tygon® tubing) capable of achieving flow rates ranging from 0.1 to 1000 mL/min was used to further optimize the CMS-GC method. This pump can support up to two channels. The tubing used with the pump is rated at ~1000 hours, so it is sufficient for conducting extended on-line monitoring studies. Neoprene tubing is also offered, which is reported to last ~10,000 hours. Two optimization studies, the sample flow rate and the sampling time studies, were repeated to evaluate the effect when utilizing this pump.

The revised sample flow rate study showed that a flow rate of ~10.0 mL/min gave the highest analytical signal with minimum increase with higher rates. Though this flow rate did provide higher analytical signals for the THM4 species it also resulted in the formation of air pockets inside the CMS, so a flow rate of 5.0 mL/min was chosen for further studies. At this flow rate the analytical signal was at least 20 % higher than with the original flow rate of ~2.5 mL/min. A sampling time study was also performed, but the highest analytical signal obtained was at the same time (15 minutes) as previous studies. Thus, with the same sampling time, we can get about 20% higher analytical signal by using the new peristaltic pump. Sampling times past this provided minimal gain in analytical signal and would make calibration more difficult and labor intensive.

An attempt was made to make the CMS longer by using different techniques to insert the silicone tubing into the Tefzel® shell. When constructing the CMS previously, the silicone tubing was pulled through the Tefzel® shell using a thin copper wire, yet this limited the length of capillary membrane we were able to thread through the outer tube. A technique involving a vacuum pump pulling an end of the silicone tubing through the shell was attempted. While this seems straightforward the air also being pulled with the tubing caused contamination; therefore, this was not a viable technique. A second technique involved a peristaltic pump whereby the silicone tubing was inserted beneath reagent water to be pulled through. The tubing could only be pulled through several centimeters before stopping; therefore, it was decided to keep the original 120 cm length for the CMS. We are still trying to find novel ways to make the CMS device with a longer capillary membrane possibly connecting two or more in series

Another recommendation for the CMS-GC was to minimize the carry-over effect. Increasing the flow rate did not noticeably change the amount of carry-over, yet did decrease the time it took to purge the CMS (from 60 minutes to 45 minutes). Purging with air was attempted, yet resulted in the accumulation of particulates in the CMS; therefore, making it difficult to pump water back through the device. It was decided to continue calibrating at one sample per hour. It should be noted that the CMS-GC can analyze samples two times an hour if the expected concentration of sample or standard is relatively constant. A carry-over effect would be problematic only when analyzing a high standard then immediately analyzing a low standard. For on-line monitoring, carry-over should have a minimal effect, since the concentration of analytes does not tend to drop suddenly. Because the sample is constantly being flowed through the CMS, the amount of compounds presented to the sample loop from the CMS remains relatively constant. For example, if the CMS is purged with reagent water following an analysis of a standard the amount of THMs present within and on the membrane walls rapidly decreases. Likewise, if the concentration of a sample was to change from a high level to a slightly lower level then the analytical signal obtained during subsequent analyses should decrease quantitatively as long as the sampling time of at least 15 minutes is met.

Studies were not attempted using a high ionic strength rinse solution, but it is assumed that this would be beneficial for calibration studies (decreases could purge time between standards). Adapting this technique to on-line monitoring would be more difficult.

Other changes to the "portable" CMS-GC-ECD analyzer (CMS-GC1) include: (1) installing a longer megabore capillary column (Restek MXT-1 30m x 0.53 mm x 5.0 µm f.t.) to extend chromatography; (2) placing the injection valve inside the GC oven; (3) eliminating all heated transfer lines by interfacing the CMS directly to the valve inside the GC oven; and (4) no desiccator vessel was used. These changes not only helped to make the CMS-GC-ECD a more compact system, but also reduced overall set up time for on-line monitoring studies. With the new column the resolution between THMs was increased from ~0.5 minutes to ~1.0 - 1.5 minutes

An additional CMS-GC-ECD analyzer (CMS-GC2) was constructed using a Varian 3400 GC-ECD (bench top scale), a CMS device (120 cm), a similar capillary column and a similar injection valve (installed inside the GC oven). This analyzer was built in order to compare to the "portable" CMS-GC-ECD during on-line monitoring studies. The chromatography (retention times and resolution) is comparable to the "portable" instrument. The Varian was modified to allow for control using PeakSimple software.

Changes to OPTGC Method and Instrumentation

The only changes made to the OPTGC were to the chromatography, which was extended to allow maximum resolution between THM4. Another detector was not able to be obtained, so the DELCD was used. A short on-line monitoring study was performed to ensure that the DELCD could remain stable enough for extended on-line studies. In practice, the OPTGC must be set up approximately one day before the study begins, so that it will stabilize for the on-line study. For calibration and on-line monitoring, the OPTGC continued to use the Autoclude® peristaltic pump with a ~2.5 mL/min flow rate. This was due to the lack of an extra Cole-Parmer pump for the OPTGC. All in all, this simpler pump worked well enough, but the results could likely be improved with a more advanced pump.

Changes to On-line Sampling Procedure and Configuration

The monitoring studies conducted in 2004 and 2005 used pressure driven flow through the CMS by use of a water regulator, but this has the possibility of being inconsistent due to pressure fluctuations in the distribution system. For the Memphis EX and Houston EX studies, the Cole-Parmer pump was used for both calibration and on-line monitoring for the CMS-GC-ECD analyzers. The OPTGC used the Autoclude® pump for both calibration and monitoring.

The details of these connections to the drinking water distribution system are presented in Appendix A.

Premonitoring Surveys Related to CMS-GC and OPTGC Methods for THM4 Species

The quantification of THM4 in a relatively simple drinking water matrix such as the Memphis matrix is not usually an issue; however, the low levels of THM4 present in Memphis water also present a chance for greater error in determination even if slight interferences are present. There are three main concerns when considering using the CMS-GC-ECD or OPTGC for measuring THM4 concentrations in Memphis drinking water. First, there is the possibility that another volatile, halogenated compound, such as a dichloroethane may pervaporate through the silicone tubing in the CMS or OPTGC and be co-eluted with a THM4 species using existing chromatography. Second, a volatile, non-halogenated compound such as toluene may perturb the pervaporation of a THM(s) to a point were detection and quantification may be altered. Third, the CMS or OPTGC may be affected from extended exposure to unknown interferents that are not volatile. If present this type of compound could effectively adsorb or "foul" the membrane (blockage), thus preventing successful THM4 pervaporation. This factor is not a large concern with Memphis drinking water, yet may be very important when analyzing a more complex matrix such as Houston drinking water.

For the Memphis EX premonitoring survey, samples were collected at approximately the same time and from the same source and analyzed immediately by the methods mentioned below. Twelve samples were analyzed at the University of Memphis, two using PT-GC-MS, four using PT-GC-ELCD/PID, four using two different CMS-GC instruments, and two samples using OPTGC. Three additional samples were analyzed at the MLGW water quality lab using a certified USEPA 502.2 instrument. There are two main advantages for using these various methods and analytical instruments. (1) The different methods and instruments all have different parameters, settings (especially their chromatography), and method detection limits (MDLs), accuracy and precision values, so if these methods can resolve interfering compounds that may be present in the Memphis samples from THM4 then it may be possible to alter the CMS-GC-ECD or OPTGC chromatography to obtain similar results. (2) Any possible interferences can be identified by retention time, and by their mass spectra. Peak purity for each THM4 species can be also determined with the GC-MS to ensure that co-elution is not taking place.

All instruments and methods used to analyze the Memphis drinking water samples were able to detect all four THMs. Quantitatively; the TTHM4 concentration reported by EPA 502.2 for was 2.1 µg/L (this measurement was confirmed at the MLGW laboratory where 502.2 result for TTHM4 was 2.7 µg/L). The two CMS-GC instruments (CMS-GC1 and CMS-GC2) reported 1.6 µg/L and 1.8 µg/L, respectively, and the OPTGC reported 2.2 µg/L. Qualitatively, there were no common VOCs detected by any of the methods. The PT-GC-MS also detected all four THMs with peak purities above 90%. Even though chromatography issues did not arise, it was decided to extend the chromatography for the CMS-GC-ECD methods to allow for better resolution between the THM4 species.

The quantification of THM4 in a more complex water matrix such as Houston is more difficult; however, very few VOC interferences at high concentrations are reported by previous water quality reports published by the City of Houston. There are similar concerns from the Memphis premonitoring surveys when using the CMS-GC-ECD and OPTGC analyzers for measuring THM4 concentrations in Houston drinking water. The most important factor is the possibility that interfering halogenated VOCs and other VOCs may have an effect on pervaporation, thus peak shape and peak retention time. Like the Memphis EX premonitoring survey, several methods were employed to correct for chromatography if needed or address any issues that may arise with the CMS-GC and OPTGC methods.

While Memphis drinking water has relatively low concentrations of $CHCl_3$ and $CHBrCl_2$, Houston water has higher (10-40 ug/L each) concentrations of these THMs. Therefore to correctly determine the concentrations of these analytes, a larger calibration range was chosen. A range of 1.0-40.0 μg/L for each THM was used for the PT-GC-ELCD/PID. The other methods were calibrated up to 20.0 ug/L for each THM for these studies. Again multiple methods and instruments were used to perform this pre-monitoring survey.

The University of Memphis (U of M) instruments used were all recalibrated to bracket for THM concentrations found in Houston drinking water. Only three THMs were detected and quantified using these instruments and methods at the U of M. $CHBr_3$ concentrations were less than the MDL of the methods. EPA 502.2 gave results for a total THM4 concentration of 29.8 μg/L. Similar studies conducted at MLGW were inconclusive due to $CHCl_3$ being out of calibration range for the MLGW instrument for Houston water. CMS-GC1 reported 27.2 μg/L and CMS-GC2 reported 26.8 μg/L. For OPTGC, the TTHM4 concentration was 26.4 μg/L. Chlorobenzene was detected using both the University of Memphis and MLGW 502.2 methods. The concentration was estimated to be between 0.2 and 0.5 μg/L. EPA 502.2 at the University of Memphis detected 1,1,1,2-tetrachloroethane, which could possibly be 2-chlorotoluene instead due to very similar retentions. EPA 502.2 at MLGW detected 2-chlorotoluene at a concentration of 0.1 μg/L. 1,2,4-trichlorobenzene (0.06 μg/L), and 1,2,3-trichlorobenzene (0.08 μg/L) were also detected. For the PT-GC-MS, all four THMs were detected with peak purities all above 92%, additionally chlorobenzene, and a halogenated toluene were detected. 1,2,4-trichlorobenzene and 1,2,3-trichlorobenzene were not detected with PT-GC-MS; regardless these two compounds would not interfere with THM4 species due to the fact that they elute several minutes after $CHBr_3$.

Of the compounds considered based upon the premonitoring survey and the historical data collected in the distribution system, chlorobenzene appears the most likely possible interferent as it elutes between $CHBr_2Cl$ and $CHBr_3$ when using either the PT-GC-MS or EPA 502.2. Slight co-elution between $CHBr_2Cl$ and chlorobenzene was noticed for CMS-GC1, yet not for CMS-GC2 or OPTGC. The chromatographic conditions were altered slightly to improve resolution.

Additional Studies of the Kinetics of NCA with DBPs and Possible Interferents

An important goal of this research period was to increase our understanding of the fundamental chemistry of the fluorescence reaction of THM4 and HAA5 species with nicotinamide (NCA). This chemistry is the basis for two of the analytical methods that we have developed: CMS-FIA and PC-IC. We have explored how the rate of the formation of the FL intensity of the NCA-DBP products changes as a function of hydrogen ion concentration (H^+). All studies were done at 90°C in an effort to mimic the conditions during the CMS-FIA and PC-IC methods. These studies are important in that many reactions exhibit pH dependency. Should such dependency exist (and we will demonstrate that it does shortly), it may give us important clues as to the chemical mechanism of the DBP-NCA reactions. Moreover, a more thorough evaluation of possible interferents has been done, so that they might be minimized. Particular attention has been focused on disinfectant species such as free available chlorine (FAC), chlorine dioxide, monochloramine (NH_2Cl) and dichloramine ($NHCl_2$). For the THM4 and HAA5 species, the preliminary rate law we propose is used to make predictions about the rate of the formation of the product species.

Fluorescence Spectra of THM4

Our previous work (Geme et al. 2005; Simone et al. 2006) for monitoring the DBPs in drinking water used a broadband fluorescence spectrophotometer from which we were able to determine little with respect to excitation and emission wavelengths for the FL reaction. Using a fluorescence spectrometer, we were able to measure the excitation and emission spectra of solutions prepared by reacting THM4 and HAA5 species with NCA in batch samples with concentrations made to mimic those in the flow cell of the CMS-FIA analyzer. The fluorescence excitation and emission spectra of individual THM4 species at concentrations of a 100 μg/L each are shown in Figure 8.3. Excitation wavelengths (λ_{excite}) for individual THM4 species ranged from 364 to 372 nm, with an average λ_{excite} of 368 nm. Emission wavelengths ($\lambda_{emission}$) ranged from 456 to 460 nm, with an average $\lambda_{emission}$ of 458 nm. The THM4 spectra are very similar in appearance and intensity. The average deviation in fluorescence intensity is ~12% considering the four THM4 species. The largest difference in the fluorescence intensity was ~20% between $CHBr_2Cl$ and $CHBrCl_2$. The brominated species exhibited higher fluorescent intensity compared to chlorinated species.

Fluorescence Spectra of HAA5

As with the THM4 studies above, individual solutions of HAA5 species (100 μg/L) were reacted with NCA at conditions designed to mimic the conditions in the CMS-FIA flow cell. The λ_{excite} and $\lambda_{emission}$ spectra of NCA-HAA5 products are shown in Figure 8.4. The λ_{excite} for individual HAA5 species ranged from 360 to 372 nm, with an average λ_{excite} of 366 nm. The $\lambda_{emission}$ values for HAA5 species ranged from 452 to 460 nm, with an average $\lambda_{emission}$ of 456 nm. The average percent difference in the fluorescence intensity of the HAA5 species showed ~7.3% deviation with the largest difference of ~13% between MBAA and TCAA.

Kinetic studies of the reaction

There are different ways to study and propose a rate law for a chemical reaction. The most common way involves varying the concentration of reactants in a systematic manner while temperature is held constant. Additionally, it is possible to measure the concentrations of reactants or products as a function of time. To determine the rate constant, plots of concentration versus time are prepared and the rate constant is determined from the slope. Many chemical reactions exhibit hydrogen ion (H^+) dependencies on rate. If an H^+ dependency exists, details of the chemistry are exposed. The approach for these studies was to observe changes in the rate of the formation of the fluorescence intensity produced by the NCA-THM4 or NCA-HAA5 products as a function of H^+. Once we measure the rate as a function of pH (or H^+), we can use the shape of the rate versus pH curve to identify the form of the rate law and perhaps observe the mechanism that may be operating (Gupta and Gupta 1984). We can also make predictions about how long the reaction will take during given situations. For example, we can calculate the half life of the chemical reaction or perhaps the time required for the reaction to be quantitative.

Collecting Fluorescence Spectra at Different pH Values

The rate law studies were done using a laboratory constructed stopped flow-fluorescence spectrometer (SFFS). Stopped-flow methods are designed to study fast reactions and they involve the rapid flowing together of separate solutions of the reactants. This rapid mixing can be coupled to a rapid-response method, such as fluorescence spectrometry, for monitoring the progress of the reaction. As mentioned earlier, the rate law studies are based on the dependence of rate of the reaction on H^+ concentration (Gordon et. al 2002, Gupta and Gupta 1984). The pH value of a solution containing individual THM4 or HAA5 species was varied from 3 to 13, and the formation of fluorescence intensity was recorded as a function of time at each pH value. In an ideal case and assuming that the rate law is first order in both NCA and DBP species, plotting the formation of fluorescence intensity versus time would result in a linear fit to one of the simple chemical kinetic models such as zero, first, or second order and the rate constant for the simple model would be obtained from the slope. However, the plots of fluorescence intensity versus time showed a non-liner relationship at each pH value of the study indicating an H^+ dependency on rate and a more complicated rate law than the simple models.

When the kinetic equation involves more than one reacting concentration, in this case, NCA and THM4 or HAA5 species, the reaction may often follow mixed-second order kinetics (Espenson 1995). The calculations for studying mixed second order reactions can be greatly simplified by adjusting their concentrations, so that one of them is present in considerable excess. Thus the concentration of the excess species will remain essentially constant during the course of the reaction, and the over-all order of the reaction will seem experimentally to be reduced (Wingrove and Caret 1981). Fortunately, this set of conditions corresponds well to the CMS-FIA system where NCA (~30% by weight) is present in excess to the THM4 or HAA5 species (μg/L level). Since the concentration of NCA is much larger than concentration of THM4 or HAA5 species, it follows pseudo first-order kinetics and the possible rate law for the reaction of NCA with DBPs would be of the mixed second-order form where k_{mix} refers to the mixed second order rate constant and DBP refers to the individual THM4 or HAA5 species in Equation 8.1.

$$Rate = k_{mix}\,[DBPs][NCA] \qquad (8.1)$$

where $Rate$ = rate of reaction

k_{mix} = mixed second order rate constant

DBP = individual THM4 or HAA5 species

NCA = nicotinamide reagent

To obtain k_{mix} it is necessary to solve the Equation 8.2 and plot it versus time. The further away the initial concentrations are to each other, the more accurately Δ_0 can be determined (Wingrove and Caret 1981).

$$\ln\left\{1+\frac{\Delta_0(Y_0-Y_\infty)}{[DBPs]_0(Y_t-Y_\infty)}\right\}=\ln\frac{[NCA]_0}{[DBPs]_0}+k_{mix}\Delta_0 t \qquad (8.2)$$

where k_{mix} = represents the mixed second order rate constant at a particular pH
Y_0 = represents initial instrument reading
Y_∞ = represents the final instrument reading when DBPs are limiting reagents
Y_t = represents instrument reading at any time and Δ_0 = (a [NCA] – b [DBPs]).
Δ_0 = difference in initial concentrations of reactants
DBPs = individual THM4 or HAA5 species
NCA = nicotinamide reagent

When the equation is plotted vs. time, a straight line should be obtained if the system follows mixed second-order kinetics. The k_{mix} can be calculated from the slope and is equal to the slope/Δ_0. Such plots were prepared for all DBPs when they reacted with NCA at pHs from 3 to 13 and interferents at the same pH values and experimental rate constants were calculated (Table 8.3).

Since k_{mix} has been calculated for the HAA5 and the THM4 reactions, half-lives and t = 99.99% can also be determined. It is more complicated to calculate half-lives for the mixed second-order reaction since the concentrations of two reagents are constantly changing. To calculate half-lives for the mixed-second order reaction, it is necessary to define half-life for NCA or DBPs. Since the concentration of NCA remains mostly constant during the reaction, the half-life for DBPs was defined and solved. The half-lives for DBPs show the time required to use up half of its initial concentration. Equation 8.3 was used to estimate half-lives:

$$t_{1/2}=\frac{a}{k_{mix}(bA_0-aB_0)}\times\ln\left(\frac{aB_0}{2aB_0-bA_0}\right) \qquad (8.3)$$

where k_{mix} = represents the mixed second order rate constant at a particular pH
A_0 = initial concentration of reactant A (NCA)
B_0 = initial concentration of reactant B (DBPs)

To calculate how much time it takes for a reaction to be complete, or quantitative, it is necessary to calculate time when t = 99.99%. The calculations showed that it takes about ~1 second for the reaction to be complete at pH values normally encountered during experimental conditions. Similar calculations were carried out for other HAA5 and THM4 species and the fluorescent products for all species also formed in approximately the same timescale at the conditions of the CMS-FIA and PC-IC flow cells. This indicates that formation of the fluorescent product is almost instantaneous and can be detected in real time during on-line monitoring of DBPs in drinking water.

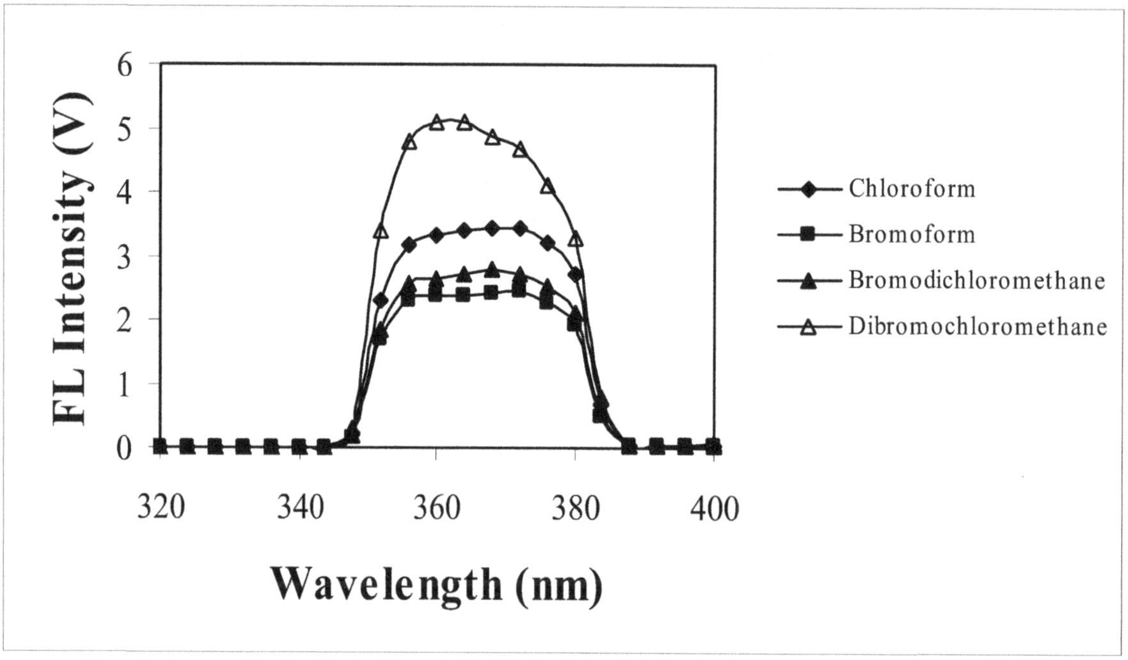

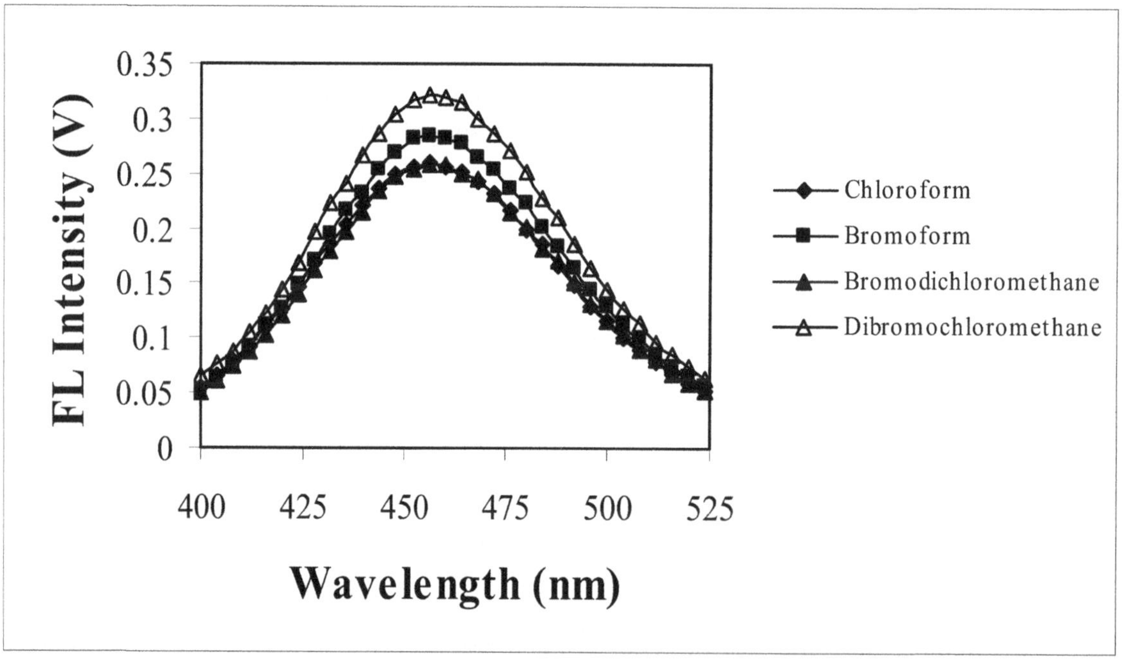

Figure 8.3 Fluorescence excitation spectra for THM4 species (top) and fluorescence emission spectra for THM4 species (bottom)

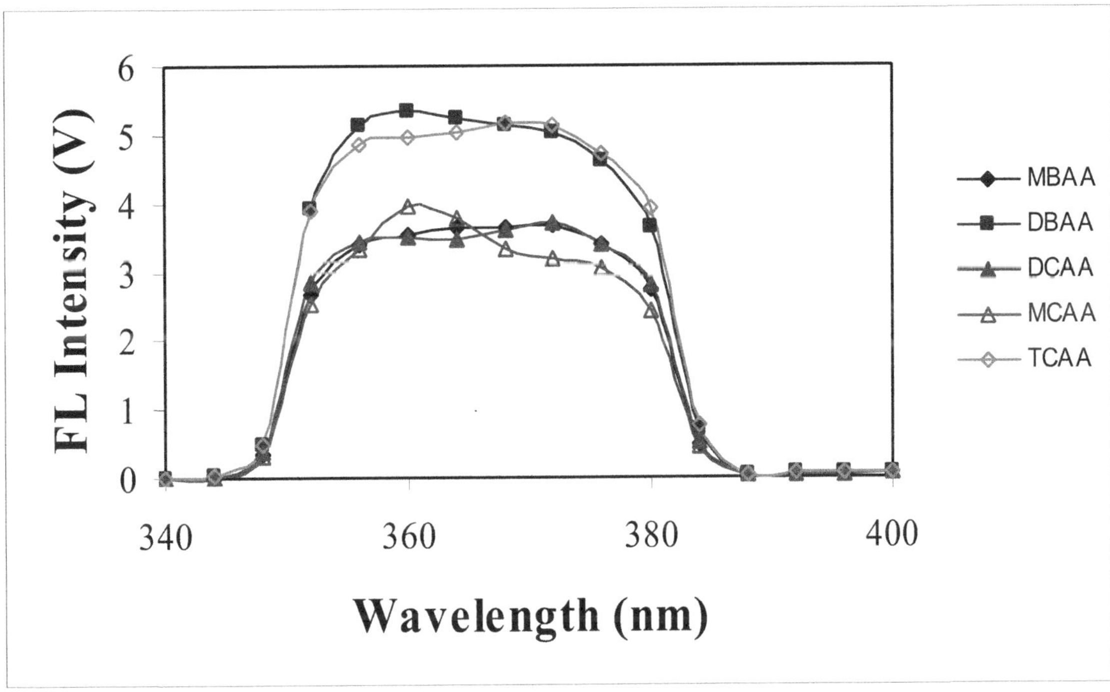

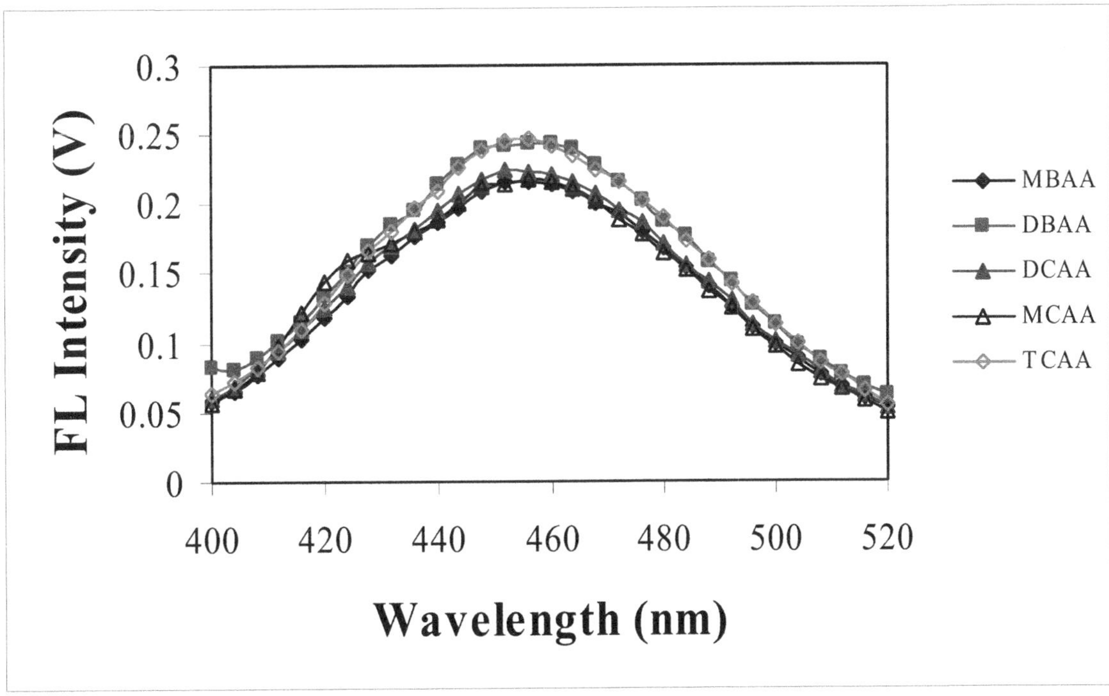

Figure 8.4 Fluorescence excitation spectra of HAA5 species (top) and fluorescence emission spectra of HAA5 species (bottom)

Preliminary law rate for DBPs

The mixed second-order rate constants (k_{mix}) shown in Table 8.3 indicate that the rate of the reaction between DBPs and NCA decreases as pH increases. In general there are three regions that can be observed: at low pH values (3-6); k_{mix} reaches a maximum value, at mid pH values (6-8), k_{mix} decreases; more sharply,; and at high pH values (8-12) k_{mix} is smallest.

Table 8.3
Experimental rate constants, half-lives, and t = 99.99% for TCAA at pH 3 to 13

pH	k_{mix} (M^{-1} s^{-1})	Half-life (s)	t = 99.99% (s)
3	4.62 ± 0.08	0.0611	0.2029
4	4.62 ± 0.05	0.0611	0.2029
5	4.87 ± 0.13	0.0580	0.1925
6	4.68 ± 0.05	0.0603	0.2003
7	4.40 ± 0.03	0.0641	0.2131
8	1.79 ± 0.09	0.1577	0.5237
8.5	1.00 ± 0.09	0.2822	0.9375
9	0.91 ± 0.10	0.3101	1.0302
10	0.89 ± 0.05	0.3171	1.0534
11	0.91 ± 0.06	0.3101	1.0302
12	0.86 ± 0.05	0.3282	1.0901
13	1.27 ± 0.03	0.2222	0.7382

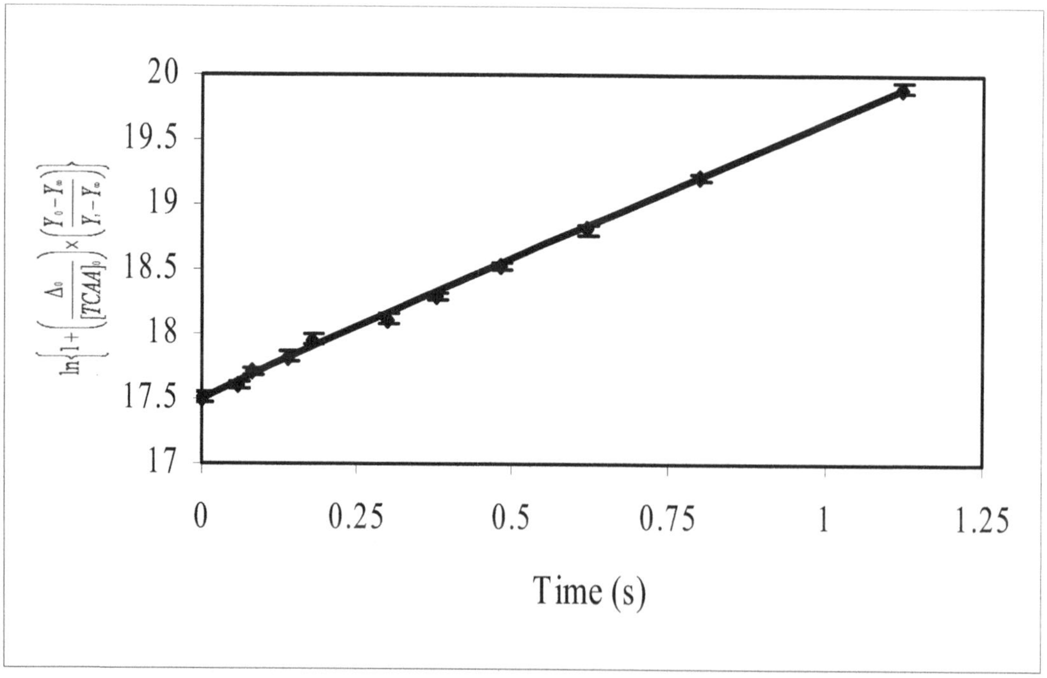

Figure 8.5 Mixed second-order kinetics for TCAA at pH = 13

All HAA5 species exhibited the three-region plot, though MBAA and MCAA features were somewhat less pronounced. Likewise, plots for the THM species showed similar trends to the HAA5 data . All four species have a similar decrease in the slope but $CHBr_3$ and $CHBr_2Cl$ do not plateau at low pH values. The data appeared as a plot with "upward bends" (Espenson 1995), which represents a transition with increasing order of $[H^+]$ from zero-order to first order in $[H^+]$ and signifying a mechanism with two independent pathways. This is characteristic of a rate law with Equation 8.4:

$$k_{mix} = k_1 + k_2 [H^+] \tag{8.4}$$

An implication of this rate law is that a pre-equilibrium between protonated and deprotonated forms of reactant (likely NCA) is rapid and that the protonated form is more reactive (Gupta and Gupta 1984). It is believed that at high pH values (8-12) the HAA species are completely ionized and the concentration of protonated HAA species is very low. The THM4 species do not exhibit acid/base characteristics, thus the two forms are likely the protonated and deprotonated NCA species ($HNCA^+$ and NCA, respectively). We assume that the heterocyclic –N is being protonated rather that –N from the amide group (Wingrove and Caret 1981). Thus we can write the following mechanisms consistent with this rate law. NCA undergoes base dissociation equilibrium in a very rapid reaction to establish equilibrium concentrations of NCA and $HNCA^+$. Then NCA reacts with the either the THM or HAA species (shown as DBP) to produce fluorescent products (products*).

$$NCA + H^+ = HNCA+ \qquad K_b = K_w/K_a$$
$$NCA + DBP \rightarrow products^* \qquad k_1$$
$$HNCA^+ + DBP \rightarrow products^* \qquad k_2$$

Considering TCAA as an example, it is possible to calculate the error of our rate law. The intercept and slope of the plot provided estimates of the rate constants and for TCAA $k_1 = 8.6 \times 10^{-1}\ M^{-1}\ s^{-1}$ and $k_2 = 5.0 \times 10^7\ M^{-1}\ s^{-1}$. Note that k_2 is large (10^7), so that even at high pH values when the concentration of $HNCA^+$ ($pK_a = 3.3$) and H^+ are low, the rate constant would still make the k_2 term significant compared to the k_1 term. Thus both terms must be considered.

The experimentally determined constants were used to calculate theoretical initial rate constants, $k_{mix\text{-}calc}$, for the reactions. Upon knowing the theoretical rate constants, it is possible to compare them to experimentally determined values and calculate the percent difference between the constants. Smaller percent difference shows a better agreement and one could conclude that the reaction fits the proposed rate law. Table 8.4 presents theoretically and experimentally determined values of k_{mix} for pH 7 – 13. Over the pH range 8-12, the average error between the experimentally and theoretically determined values average ~2.5%. At pH 7, the error rises to 25% and at lower pH values, the percent difference between the two constants was even larger (~100%). The rate law functions well in basic solution such as the conditions of our analytical measurements with CMS-FIA and PC-IC. However, at low pH values, the model fails. We would like to note that there is an exception in that the errors remained low for bromoform and dibromochloromethane over the entire pH region 3 to 13. For the remaining species, large deviations were observed at low pH values.

Table 8.4
Results from graphical fitting of the rate constant for TCAA

pH	k_{mix}	$k_{mic\text{-}calc}$	% Difference
7.0	4.40	5.86	25
8.1	1.27	1.31	3
8.5	1.00	1.02	2
9.0	0.91	0.91	<1
10.0	0.89	0.87	-3
11.0	0.91	0.86	-5
12.0	0.86	0.86	<1

Experimentally, we observe additional effects. While the fluorescent products form rapidly at lower pH values, they readily decompose. The analytical signal reaches greater magnitudes at higher pH values. We believe that fluorescent products are formed, but decompose at lower pH values and that this decomposition (for all practical purposes a fluorescent quenching) is accelerated by H^+. A more complete mechanism of what occurs might be shown by adding an additional step that shows the quenching of the fluorescent product and the production of a radiationless transition (Δ). We caution here that we have not yet been able to do more extensive fits to any revised models at this point. More work is being done at this time to further evaluate the implications of these observations.

$$NCA + H^+ = HNCA+ \qquad K_b = K_w/K_a$$
$$NCA + DBP \rightarrow Products^* \qquad k_1$$
$$HNCA^+ + DBP \rightarrow Products^* \qquad k_2$$
$$Products^* + H^+ \rightarrow Products + \Delta \qquad k_3$$

Completely different results were obtained for interferents. The disinfectants did not follow the same pH dependent profile as did the HAA5 and the THM4 species. In other words, while the disinfectant species will react with NCA to form fluorescent products, it does so by a different mechanism. The fact that the disinfectants react mean that they are possible interferents in the CMS-FIA and PC-IC methods. The interference effect would be expected to be much more problematic with CMS-FIA where no separation of the different chemical species occurs. With PC-IC, the HAA9 species can be likely be resolved from disinfectant species. It is possible that the disinfectants species could be minimized using a masking agent prior to analysis. The PI has experience with masking chemical disinfectants and we will likely use SFFS to evaluate the effects of various masking agents for use with these disinfectants.

Some trends were evident in the data. For example, the behavior of the free available chlorine, chlorine dioxide and monochloramine seem similar. The effect of dichloroamine seems more complex. It is possible that we might be able to learn some insight as to the mechanism of the disinfectants as we have done with the HAA5 and THM4 species. Theoretical considerations of this sort are being explored along with their significance in the analytical chemistry of the NCA reactions.

These results indicate that free available chlorine and chloramines will interfere in the nicotinamide (NCA) based CMS-FIA methods. Preliminary studies demonstrated that at the pH of drinking water (~pH 7), hypochlorous acid (HOCl), hypochlorite ion (OCl^-) and chloramines do not permeate the membrane readily compared to THM4 species. Thus, HOCl, OCl^- and chloramines did not appear to interfere in THM4 measurements. However, the concentration of the disinfectant species did act as interferents in measurements of HAA9 species. It was concluded that sodium thiosulfate ($Na_2S_2O_3$) was an effective masking agent for the disinfectant species. When $Na_2S_2O_3$ was introduced into the system, minor modifications were made to the instrumental setup. This revised CMS-FIA system is shown in Figure 8.6.

An additional peristaltic pump was needed to add the extra reagent. The sampling flow rate was 1.0 mL/min. The flow rates for other reagents were also changed. The flow rates of NaOH, NCA, $Na_2S_2O_3$ and reagent water were 0.9 mL/min, 0.6 mL/min, 0.6 mL/min, and 1.5 mL/min, respectively. An additional mixing coil (o.d. 1.6 mm, i.d. 0.5 mm, length 2000 mm) was used to mix NaOH and $Na_2S_2O_3$ solution (Figure 8.6). The optimum concentration of $Na_2S_2O_3$ at 0.5% (pH = 4.5) was chosen as the reducing agent for disinfectants and interference from these species was minimized in HAA9 measurements. Additional studies to further optimize the NaOH concentration were done before conducting additional monitoring studies. Studies were done over the range 0.2 - 3M NaOH solutions. It was concluded that in order to obtain the maximum analytical signal (peak heights), 3M NaOH was chosen for this extended monitoring study.

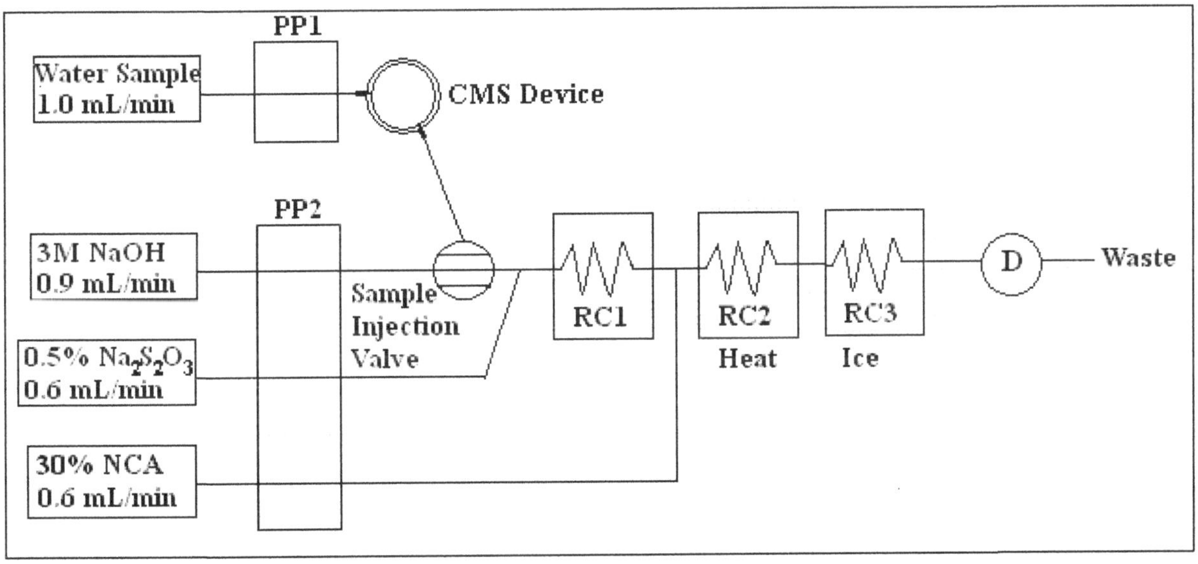

Figure 8.6 Schematic Diagram of CMS-FIA NCA-FL instrument for TTHM4 and THAA9 used in the extended studies. PP1 and PP2: peristaltic pumps; RC1: reaction coil; RC2: reaction coil in water bath (97°C); RC2: reaction coil in ice bath; D: fluorescence detector

Advances in the PC-IC Method and Pre-monitoring Surveys for PC-IC Method

The main advancement with the PC-IC method was the resolution of HAA9 species, which also helped to alleviate problems with interference. In earlier Chapters, we presented the results of studies aimed at combining the NCA-FL chemistry with post column reaction after HAA species are separated using ion chromatography. One disadvantage of this work was that we were not able to resolve the monohalogenated species while meeting the target sampling rate of one sample per hour. One goal has been to anticipate possible interfering species and to take them into account prior to carrying out on-line monitoring studies. For example, in the Houston monitoring studies of Phase 2, we demonstrated that the presence of bromochloroacetic acid (BCAA) complicated HAA5 analysis and interfered with HAA5 measurements (Figure 7.13). Specifically, BCAA interferes with DCAA and DBAA, but BDCAA, DBCAA, and TBAA are baseline resolved from TCAA and each other. Thus using the PC-IC chemistry, derivatization and detection of HAA9 is possible. It seemed that the best way to minimize interference from the HAA9 species was simply to alter the chromatography to allow for resolution of all HAA9 species. Then, these species will not interfere with HAA5 measurements, and likewise data on HAA9 can also be obtained.

Initial research focused on using one Dionex AS9-SC analytical column to separate the HAA9 species, but a single column was insufficient for this purpose. To fully resolve HAA9 species in complex matrices, it was necessary to connect two AS9-SC analytical columns in series. Figure 8.7 shows the separation of a 50 µg/L each HAA9 species. The total separation time is approximately 60 minutes with all HAA9 species near baseline resolved. Continuous flow PC-IC is used as opposed to earlier studies that used stopped-flow techniques. It is possible to do this in this case, because the separation time is longer than the 30 min optimum heating time needed for derivatization. Longer heating times could be used to improve MDL values, but the expense of the KOTs needed for trapping the chromatogram offset the need for longer heating times. The analysis time of 1.5 hours certainly exceeds the desired goal of one sample per hour, but the speciation of HAA9 (along with minimization of possible interferences) is an acceptable trade-off.

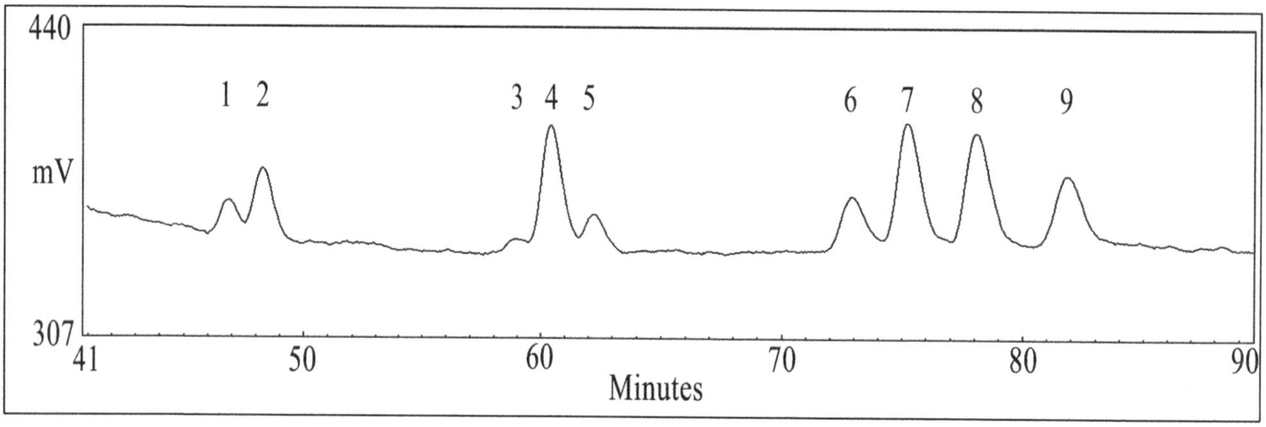

Figure 8.7 Preliminary PC-IC HAA9 chromatogram. **(1) MCAA (51.0 µg/L), (2) MBAA (51.0 µg/L), (3) DCAA (71.0 µg/L), (4) BCAA (48.2 µg/L), (5) DBAA (51.0 µg/L), (6) TCAA (52.0 µg/L), (7) BDCAA (46.8 µg/L), (8) DBCAA (46.8 µg/L), (8) TBAA (52.0 µg/L)**

Preliminary MDL, accuracy and precision studies with an older, less capable filter based fluoresence detector show MDLs ranging from 4.7 µg/L for BDCAA to 16.3 µg/L for DCAA, mean % recoveries ranging from 71.9 % for TBAA to 99.7 % for BCAA, and % RSD ranging from 10% for BDCAA to 21 % for DCAA. These preliminary results for MDL, accuracy and precision do not fall within the desired goals for the ideal analyzer, but prior research has shown that using a state-of-the-art detector should improve MDL values to within the desired goal of 0.5 - 5 µg/L for each HAA9 species. The City of Houston has made available to us a state-of-the-art fluorescence detector. With this detector, the MDLs ranged from 0.6 µg/L for BCAA to 10.1 µg/L for TBAA, mean % recoveries ranged from 58 % for DBAA to 161 % for TBAA, and % RSD values ranged from 3.5% for BCAA to 32 % for DBCAA (Table 8.5).

The main advantage of these revised methods is that they eliminate many interferences by quantifying all HAA9 species. However, the disadvantage of this approach is that it takes about 90 minutes to analyze a single sample on-line and the sampling is unrestricted. When conducting side-by-side with other methods, the PC-IC method can sample only about once each two hours.

Table 8.5
Results of MDL, accuracy, precision and linearity studies for the revised PC-IC method for HAA9 analysis with a relatively inexpensive filter based fluorescence detector and a higher end dual monochromator based fluorescence detector

PC-IC with Filter Based Fluorescence Detector					
HAA9 Species†	MDL (μg/L)	(Mean % Recovery)	% RSD	r^2	Conc. Range (μg/L)*
MCAA	7.3	84.4	13	0.969	15.3 - 76.5
MBAA	8.9	82.4	7	0.981	15.3 - 76.5
DCAA	16.3	88.1	21	0.983	21.3 - 106.5
BCAA	9.8	99.7	16	0.964	14.5 - 72.3
DBAA	7.7	93.9	13	0.897	15.3 - 76.5
TCAA	7.4	75.4	15	0.832	15.6 - 78.0
BDCAA	4.7	73.9	11	0.975	14.3 - 71.7
DBCAA	5.0	81.8	10	0.969	14.0 - 70.2
TBAA	7.9	71.9	17	0.955	15.6 - 78.0
PC-IC with Dual Monochromator Based Fluorescence Detector					
HAA9 Species†	MDL (μg/L)	(Mean % Recovery)	% RSD	r^2	Conc. Range (μg/L)*
MCAA	2.2	83	13	0.999	2.6 - 51.0
MBAA	2.1	94	11	0.999	2.6 – 51.0
DCAA	6.6	90	25	0.998	3.4 – 35.5
BCAA	0.6	86	3.5	0.999	2.4 – 48.2
DBAA	3.2	58	26	0.993	5.1 – 51.0
TCAA	2.1	79	12	0.992	2.6 – 52.0
BDCAA	2.8	82	17	0.998	2.4 – 47.8
DBCAA	5.6	91	32	0.996	2.3 – 46.8
TBAA	10.1	161	29	0.970	5.2 – 52.0

* Calibration based on five standards for all HAAs except DCAA, DBAA, and TBAA used four standards.
† Check standard concentrations were ~7.5 μg/L for all HAAs, except DCAA at 10.6 μg/L and were analyzed seven consecutive times.

Following this work, three pre-monitoring surveys were performed using EPA 300.0 (anions by IC-MSCD), EPA 552.3 and a Houston water sample spiked with approximately 10 μg/L of each HAA9 species. A Houston water sample was analyzed in duplicate using EPA 300.0 and found to have 25.5 mg/L chloride ion, 96.4 mg/L sulfate ion, 1.2 mg/L fluoride ion, and 0.8 mg/L nitrate ion as the primary common anions present in the matrix. This relatively high concentration of common anions could effect the separation of the HAA9 species on the IC columns. Another sample of Houston water was analyzed by EPA 552.3 to determine the identity and concentration of the HAA9 species present as well as the THAA9 concentration. The average concentration of MCAA, DCAA, TCAA, MBAA, DBAA, BCAA, BDCAA, DBCAA, TBAA and TTHM9 were 1.9 ± 0.5 μg/L, 10.4 ± 0.2 μg/L, 5.6 ± 0.2 μg/L, 1.3 ± 0.2 μg/L, 2.9 ± 0.3 μg/L, 7.0 ± 0.2 μg/L, 4.8 ± 0.3 μg/L, 1.1 ± 0.4 μg/L, 1.0 ± 0.2 μg/L, and 36.1 ± 1.7 μg/L, respectively. These concentrations were used to determine the range of the calibration curve to be used. Finally, Houston water spiked with approximately 10 μg/L of each HAA9 species was analyzed by PC-IC to confirm the chromatography would not be affected by the relatively high concentrations of chloride ions and sulfate ions present.

SUMMARY OF REVISED ANALYTICAL METHODS AND MONITORING STUDIES

In response to analytical issues identified in the first set of on-line monitoring studies in Memphis, TN and Houston, TX, there was significant effort devoted to revising the CMS-GC, CMS-FIA and PC-IC methods to eliminate these issues. Additionally, revised strategies were proposed for conducting on-line monitoring studies. The revised analyzers and associated methods were used to conduct a second round of real world studies in Memphis, TN and Houston, TX and these results are collected and discussed in Chapter 9.

CHAPTER 9
ADDITIONAL ON-LINE MONITORING STUDIES

ON-LINE MONITORING STUDIES USING THE REVISED METHODS

In this Chapter, the results of a second round of on-line monitoring studies using capillary membrane sampling-gas chromatography (CMS-GC), capillary membrane sampling flow injection analysis with nicotinamide-fluorescence (CMS-FIA NCA-FL), and post column-ion chromatography with NCA-FL chemistry (PC-IC) are presented. In addition, the on-line monitoring purge and trap gas chromatograph (OPTGC) method (Chapter 2) is also employed to monitor individual THM4 and total THM4 concentrations. These revised analyzers were subjected to further side-by-side comparisons to EPA 502.2 and 552.3 in Memphis, TN and Houston, TX. These results are summarized and compared to the EPA methods and to each other. Recommendations are made with respect to the analysers and their use.

CMS-GC STUDIES

Memphis EX On-Line Monitoring Study using two CMS-GC-ECD Systems

During the Memphis EX monitoring study, side-by-side comparisons were made to EPA Method 502.2 using two CMS-GC-ECD systems. This CMS-GC-ECD monitoring study was conducted from June 11 - 17, 2006 at the Memphis, TN sampling site one (MAMML laboratory). The CMS-GC-ECD was interfaced to the drinking water distribution system using on-line sampling configuration two as shown in Appendix A. In previous on-line monitoring studies, the drinking water was flowed to the CMS using pressure, yet this study employed a rugged peristaltic pump to flow the sample to the CMS-GC-ECD systems. THM4 species concentrations were monitored at ~1 sample/ hour (though the analyzer was typically taken off-line every 12 hours to run a check standard). TTHM4 is the sum of the four THM concentrations. The results along with details used in the analysis of the data are presented in Table 9.1. CMS-GC1 denotes the "portable" CMS-GC-ECD used in earlier studies and CMS-GC2 denotes the "benchtop" CMS-GC-ECD constructed for these additional studies. As mentioned in previous Chapters, the "percent recovery" (%REC_{Bias}) reported is calculated by assuming the EPA method (502.2) provides the "true value" of the analyte concentration and the CMS-GC-ECD provides the "experimental value."

The results for the Memphis EX monitoring study are summarized in Table 9.1. There were 83 values for EPA 502.2 with 72 direct comparisons to EPA 502.2 for CMS-GC1 and 62 direct comparisons for CMS-GC2. The concentration of $CHCl_3$ range from 0.1 to 0.4 µg/L for EPA 502.2, from 0.1 to 0.4 µg/L for CMS-GC1, and 0.0 to 0.2 µg/L for CMS-GC2. The data showed that the concentration for $CHCl_3$ is approximately the same for both Memphis 1 and Memphis EX studies which were over a year apart and conducted during a different time of the year. The average concentration measured by EPA 502.2 was 0.2 ± 0.1 µg/L, using CMS-GC1 was 0.1 ± 0.0 µg/L, and using CMS-GC2 was 0.1 ± 0.0 µg/L. The bias between the CMS-GC method EPA 502.2 ranged from –0.2 to 0.3 µg/L with an average of -0.1 ± 0.1 µg/L. The % REC_{Bias} values ranging from 29 to 292 % with an average of 65± 46 % for CMS-GC1. With the CMS-GC2 analyzer, the bias ranged from -0.3 to 0.0 µg/L with an average of -0.1 ± 0.1 µg/L.

The CMS-GC2 % REC_{Bias} values ranged from 5 to 150 % with an average of 41 ± 27 %). As with the Memphis 1 study the bias for Memphis EX was less than 1.0 μg/L for the CMS-GCECD analyzers, but was negative instead of positive. This is due to the low concentration of $CHCl_3$ when compared to the MDL of the CMS-GC analyzers. A small difference in concentration can swing the bias in the plus or minus direction; therefore, the recovery data is variable.

The $CHBrCl_2$ concentration ranges from 0.2 to 0.8 μg/L for EPA 502.2, 0.2 to 0.5 μg/L for CMS-GC1and 0.1 to 0.6 μg/L for CMS-GC2. The average concentration was 0.5 ± 0.2 μg/L for EPA 502.2, 0.3 ± 0.1 μg/L for CMS-GC1 and 0.4 ± 0.1 μg/L for CMS-GC2. The concentration for $CHBrCl_2$ observed in this study is also approximately the same as from the Memphis 1 study. The bias ranged from -0.5 to 0.2 μg/L (averaging -0.1 ± 0.2 μg/L) with % REC_{Bias} ranging from 33 to 206 % (averaging 87 ± 41 %) for CMS-GC1. The bias ranged from -0.3 to 0.3 μg/L (averaging -0.1 ± 0.1 μg/L) with % REC_{Bias} ranging from 32 to 224 % (averaging 94 ± 39 %) for CMS-GC2. Both of the CMS-GC analyzers were negatively biased, but with approximately the same magnitude as for the Memphis 1 study.

For $CHBr_2Cl$, the concentrations ranged from 0.2 to 1.5 μg/L for EPA 502.2, 0.3 to 0.9 μg/L for CMS-GC1, and 0.4 to 0.9 μg/L for CMS-GC2. These ranges are higher than with the Memphis 1 study. The average concentration was 0.8 ± 0.2 μg/L for EPA 502.2, 0.6 ± 0.1 μg/L for CMS-GC1 and 0.6 ± 0.1 μg/L for CMS-GC2. Using CMS-GC1, the bias ranged from -0.9 to 0.4 μg/L with an average bias of -0.2 ± 0.2 μg/L, and % REC_{Bias} ranged from 39 to 289 % and an average of 84 ± 36 %. Using CMS-GC2, the bias ranged -0.9 to 0.5 μg/L with an average of -0.1 ± 0.2 μg/L, the % REC_{Bias} ranged from 41 to 350 %, and the average % REC_{Bias} of 95 ± 47 %. Overall the average bias was the same for CMS-GC 1 analyzer when compared to results from Memphis 1; however, CMS-GC 2 had a larger negative bias. The recoveries for both CMS-GC analyzers did improve from previous results for Memphis 1 study.

The concentration of $CHBr_3$ ranged from 0.1 to 2.1 μg/L using EPA 502.2, 0.1 to 1.2 μg/L using CMS-GC1, and 0.1 to 1.3 μg/L using CMS-GC2. The average concentration was 0.5 ± 0.3 μg/L using EPA 502.2, 0.4 ± 0.2 μg/L using CMS-GC1, and 0.4 ± 0.3 μg/L using CMS-GC2. Using CMS-GC1, the bias ranged from -0.9 to 0.7 μg/L with an average of -0.2 ± 0.3 μg/L. The % REC_{Bias} values for CMS-GC1 ranged from 0 - 490 % with an average of 98 ± 86 %. The bias values using CMS-GC2 ranged from -0.8 to 0.7 μg/L with an average of -0.1 ± 0.3 μg/L. For CMS-GC2, the % REC_{Bias} ranged from 0 - 844 % with an average value of 126 ± 122 %. When compared to Memphis 1 the average bias for both CMS-GC analyzers for this study improved by at least 5.0 μg/L.

The concentrations of each individual THM4 species were summed to obtain total THM4. The total THM4 concentrations ranged from 0.7 to 4.1 μg/L using EPA 502.2, 0.7 to 2.2 μg/L using CMS-GC1, and 0.8 to 3.0 μg/L using CMS-GC2. The average concentration was 1.9 ± 0.7 μg/L using EPA 502.2, 1.4 ± 0.3 μg/L using CMS-GC1, and 1.5 ± 0.4 μg/L using CMS-GC2. Total THM4 concentrations were slightly lower for the Memphis EX study than for the Memphis 1 study. Using the CMS-GC1, the bias ranged from -2.4 to 0.9 μg/L with an average of -0.5 ± 0.6 μg/L and % REC_{Bias} ranged from 33 to 199 % with an average of 80 ± 32 %. The CMS-GC2 bias ranged from -1.6 to 0.8 μg/L with an average of -0.3 ± 0.5 μg/L. The CMS-GC2 % REC_{Bias} ranged from 47 to 185 % with an average of 88 ± 28 %. There is a larger bias value for this study for both analyzers when compared to Memphis 1 results, yet the bias range and recovery values have decreased. The most recognizable difference between the two studies is the change from positive bias to negative bias for the CMS-GC method. Since the single THM4 concentrations are all at the MDL for the method this difference is not uncommon.

Statistical analysis of the CMS-GC1 gives a $t_{critical}$ for 72 comparisons of 2.214. For $CHCl_3$, $CHBrCl_2$, $CHBr_2Cl$, $CHBr_3$ and TTHM4, $t_{calculated}$ is 8.189, 5.519, 7.110, 3.922, and 5.306 respectively. This indicates a statistical difference between the two methods for all analytes. The results for the CMS-GC2 data gives a $t_{calculated}$ for $CHCl_3$, $CHBrCl_2$, $CHBr_2Cl$, $CHBr_3$ and TTHM4 of 13.35, 3.426, 4.257, 0.771, and 4.961 respectively. There was not a statistical difference between the results reported for EPA 502.2 and CMS-GC2 for $CHBr_3$, however the remaining THM4 species and total THM4 did exhibit a statistical difference between the two sets of results. From the previous statistical results obtained during the Memphis 1 study, statistical agreement was observed for TTHM4 and for none of the single THM4 species. In contrast, the results above show that $CHBr_3$ agrees for CMS-GC2 while not for TTHM4.

Houston EX On-Line Monitoring Study using two CMS-GC-ECD Systems

The same two CMS-GC analyzers were used to conduct a similar study in Houston, TX. Side-by-side comparisons were made to EPA 502.2 from June 26 - 29, 2006. The THM4 species concentrations were monitored at ~1 sample/ hour with a check standard being analyzed approximately every 12 hours. The results along with details used in the analysis of the data are presented in Table 9.2.

There were 61 values for EPA 502.2 with 51 direct comparisons to EPA 502.2 for CMS-GC1 and 61 direct comparisons for CMS-GC2. The concentration for $CHCl_3$ ranged from 35.8 to 51.9 μg/L for the EPA Method, from 31.3 to 44.5 μg/L for CMS-GC1 and 31.7 to 46.7 μg/L for CMS-GC2. The average concentration for EPA 502.2 is 42.5 ± 3.4 μg/L, for CMS-GC1 is 40.4 ± 3.0 μg/L and for CMS-GC2 is 40.6 ± 3.1 μg/L. The average concentration for $CHCl_3$ for the Memphis EX study was ~13.0 μg/L lower than with this study. Using CMS-GC1, the bias ranged from –13.9 to 6.8 μg/L with an average of -1.5 ± 4.9 μg/L, and % REC_{Bias} values ranged from 71 to 119 % with an average of 97 ± 11 %. Using CMS-GC2, the bias ranged from -14.6 to 7.1 μg/L with an average of -2.0 ± 5.0 μg/L, and % REC_{Bias} values ranging from 72 to 119 % with and average of 96 ± 11 %. Compared to the previous Houston study the bias results for $CHCl_3$ are improved in both the average and range. Also the % REC_{Bias} values have improved especially the standard deviation, which dropped from 27% for the previous Houston study to 11% for the Memphis EX study. This improvement is most likely due to the further optimization of the CMS-GC-ECD analyzer.

The concentration of $CHBrCl_2$ ranged from 7.1 to 12.7 μg/L using EPA 502.2, 9.7 to 14.7 μg/L using CMS-GC1, and 9.3 to 13.4 μg/L using CMS-GC2. The average concentration was 9.2 ± 1.0 μg/L for EPA 502.2, 12.8 ± 1.5 μg/L for CMS-GC1, and 12.0 ± 0.8 μg/L for CMS-GC2. The average concentration for all methods is ~ 3.0 μg/L lower than see in the previous Houston study. The bias ranged from 0.0 to 7.2 μg/L with average of 3.7 ± 1.6 μg/L and % REC_{Bias} ranging from 100 to 201 %, and averaged 142 ± 20 % when the CMS-GC1 was used. Using the CMS-GC2, the bias ranged from 0.2 to 6.0 μg/L with average bias of 2.8 ± 1.2 μg/L. The % REC_{Bias} ranged from 101 to 184 % with and average of 131 ± 16 % using CMS-GC2. The average bias is noted to be slightly higher for this study for both the CMS-GC analyzers than with the previous Houston study; however, the bias range and its deviation are lower.

The $CHBr_2Cl$ concentrations ranged from 1.1 to 2.1 μg/L using EPA 502.2, 1.4 to 2.5 μg/L using CMS-GC1, and 0.9 to 2.2 μg/L using CMS-GC2. The average concentration was 1.5 ± 0.2 μg/L using EPA 502.2, 2.0 ± 0.3 μg/L using CMS-GC1, and 1.7 ± 0.3 μg/L using CMS-GC2. There was a lower average concentration for all methods for the Houston EX study than for the previous Houston study. Using CMS-GC1, the bias ranged from -0.3 to 1.2 μg/L with an average of 0.4 ± 0.3 μg/L. The % REC_{Bias} for CMS-GC1 ranged from 84 to 209 % with an average of 130 ± 24 %. With CMS-GC2, the bias ranged from -0.9 to 1.0 μg/L with an average of 0.2 ± 0.3 μg/L. The CMS-GC2 % REC_{Bias} ranged from 51 to 194 % with an average of 113 ± 25 %. There was similar average bias results obtained for both Houston studies, but the deviation of the bias was slightly lower for the Houston EX study.

None of the methods detected $CHBr_3$ in Houston drinking water, thus all the methods were in agreement with this regard. For THM4, the concentrations ranged from 45.6 to 62.0 μg/L using EPA 502.2, 42.5 to 61.4 μg/L using CMS-GC1, and 42.0 to 59.9 μg/L using CMS-GC2. The average concentration was 53.3 ± 3.8 μg/L using EPA 502.2, 55.1 ± 4.7 μg/L for CMS-GC1, and 54.3 ± 4.0 μg/L for CMS-GC2. THM4 concentrations for both the Houston EX and previous Houston studies were similar for the EPA 502.2. On the other hand, the CMS-GC analyzers were ~4.5 μg/L higher for this study. Using CMS-GC1, the bias ranged from -12.5 to 13.9 μg/L, with an average of 2.8 ± 6.1 μg/L. The CMS-GC1 % REC_{Bias} ranged from 80 to 131 % with an average of 106 ± 12 %. For CMS-GC2, the bias ranged from -11.8 to 12.0 μg/L with an average of 1.0 ± 5.9 μg/L. The CMS-GC2 % REC_{Bias} ranged from 81 to 126 % with an average of 102 ± 11 %. The average bias was lower in previous Houston study, yet had a deviation at least as twice a large as this study.

The result of statistical analysis for CMS-GC1 for 51 comparisons gives a $t_{critical}$ of 2.403. For $CHCl_3$, $CHBrCl_2$, $CHBr_2Cl$, and TTHM4, $t_{calculated}$ is 2.224, 16.832, 9.970, and 3.204 respectively. This indicates a statistical difference between the two methods for all analytes except for $CHCl_3$. The result of statistical analysis for CMS-GC2 for 61 comparisons gives a $t_{critical}$ of 2.390. The result for the CMS-GC2 data gives a $t_{calculated}$ for $CHCl_3$, $CHBrCl_2$, $CHBr_2Cl$, and TTHM4 of 3.087, 18.538, 3.839, and 1.288, respectively. There was not a statistical difference for TTHM4; however, there was a difference for each individual THM4 species. Both Houston studies showed TTHM4 to be statistically same for the methods.

Table 9.1

Memphis EX monitoring study: monitoring THM4 concentrations using CMS-GC1 and CMS-GC2 compared to EPA 502.2

	$CHCl_3$			$CHBrCl_2$			$CHBr_2Cl$			$CHBr_3$			TTHM4		
Statistic/ *Method*	*EPA*	CMS-GC1	CMS-GC2	*EPA*	CMS-GC1	CMS-GC2	*EPA*	CMS-GC1	CMS-GC2	*EPA*	CMS-GC1	CMS-GC2	*EPA*	CMS-GC1	CMS-GC2
N	83	72	62	83	72	62	83	72	62	83	72	62	83	72	62
Avg. Conc.	0.2	0.1	0.1	0.5	0.3	0.4	0.8	0.6	0.6	0.5	0.4	0.4	1.9	1.4	1.5
Std. Dev Conc.	0.1	0.0	0.0	0.2	0.1	0.1	0.2	0.1	0.1	0.3	0.2	0.3	0.7	0.3	0.4
Conc. Range	0.0 to 0.4	0.1 to 0.4	0.1 to 0.2	0.2 to 0.8	0.2 to 0.5	0.1 to 0.6	0.2 to 1.5	0.3 to 0.9	0.4 to 0.9	0.1 to 2.1	0.0 to 1.2	0.0 to 1.3	0.7 to 4.1	0.7 to 2.2	0.8 to 3.0
Avg. Bias	---	-0.1	-0.1	---	-0.1	-0.1	---	-0.2	-0.1	---	-0.2	-0.1	---	-0.5	-0.3
Std. Dev Bias	---	0.1	0.1	---	0.2	0.1	---	0.2	0.2	---	0.3	0.3	---	0.6	0.5
Bias Range	---	-0.2 to 0.3	-0.3 to 0.0	---	-0.5 to 0.2	-0.3 to 0.3	---	-0.9 to 0.4	-0.9 to 0.5	---	-0.9 to 0.7	-0.8 to 0.7	---	-2.4 to 0.9	-1.6 to 0.8
% REC_{Bias}	---	65	41	---	87	94	---	84	95	---	98	126	---	80	88
s.d. % REC_{Bias}	---	46	27	---	41	39	---	36	47	---	86	122	--	32	28
% REC_{Bias} Range	---	29 to 292	5 to 150	---	33 to 206	32 to 224	---	39 to 289	41 to 350	---	0 to 490	0 to 844	---	33 to 199	47 to 185
t-test* / $t_{critical}$ / $t_{calculated}$	---	Yes / 2.214 / 8.189	Yes / 2.222 / 13.35	---	Yes / 2.214 / 5.519	Yes / 2.222 / 3.426	---	Yes / 2.214 / 7.110	Yes / 2.222 / 4.257	---	Yes / 2.214 / 3.922	No / 2.222 / 0.771	---	Yes / 2.214 / 5.306	Yes / 2.222 / 4.961

* The $t_{critical}$ and $t_{calculated}$ values listed were calculated for the N listed using a two-tailed t-test with 98% confidence. Typically a $t_{calculated}$ value should be less than the $t_{critical}$. "Yes" means there exists a statistical difference between the two sets of results while "No" means there is no difference between the results.

Table 9.2

Houston EX monitoring study: monitoring THM4 concentrations using CMS-GC1 and CMS-GC2 compared to EPA 502.2

	$CHCl_3$			**$CHBrCl_2$**			**$CHBr_2Cl$**			**$CHBr_3$**			**TTHM4**		
Statistic/ *Method*	*EPA*	CMS-GC1	CMS-GC2	*EPA*	CMS-GC1	CMS-GC2	*EPA*	CMS-GC1	CMS-GC2	*EPA*	CMS-GC1	CMS-GC2	*EPA*	CMS-GC1	CMS-GC2
N	61	51	61	61	51	61	61	51	61	61	51	61	61	51	61
Avg. Conc.	42.5	40.4	40.6	9.2	12.8	12.0	1.5	2.0	1.7	n.d.	n.d.	n.d.	53.3	55.1	54.3
Std. Dev Conc.	3.4	3.0	3.1	1.0	1.5	0.8	0.2	0.3	0.3	n.d.	n.d.	n.d.	3.8	4.7	4.0
Conc. Range	35.8 to 51.9	31.3 to 44.5	31.7 to 46.7	7.1 to 12.7	9.7 to 14.7	9.3 to 13.4	1.1 to 2.1	1.4 to 2.5	0.9 to 2.2	---	---	---	45.6 to 62.0	42.5 to 61.4	42.0 to 59.9
Avg. Bias	---	-1.5	-2.0	---	3.7	2.8	---	0.4	0.2	---	n.d.	n.d.	---	2.8	1.0
Std. Dev Bias	---	4.9	5.0	---	1.6	1.2	---	0.3	0.3	---	n.d.	n.d.	---	6.1	5.9
Bias Range	---	-13.9 to 6.8	-14.6 to 7.1	---	0.0 to 7.2	0.2 to 6.0	---	-0.3 to 1.2	-0.9 to 1.0	---	---	---	---	-12.5 to 13.9	-11.8 to 12.0
% REC_{Bias}	---	97	96	---	142	131	---	130	113	---	n.d.	n.d.	---	106	102
s.d. % REC_{Bias}	---	11	11	---	20	16	---	24	25	---	n.d.	n.d.	--	12	11
% REC_{Bias} Range	---	71 to 119	72 to 119	---	100 to 201	101 to 184	---	84 to 209	51 to 194	---	---	---	---	80 to 131	81 to 126
t-test* $t_{critical}$ $t_{calculated}$	---	No 2.403 2.224	Yes 2.390 3.087	---	Yes 2.403 16.83	Yes 2.390 18.53	---	Yes 2.403 9.970	Yes 2.390 3.839	---	---	---	---	Yes 2.403 3.204	No 2.390 1.288

* The $t_{critical}$ and $t_{calculated}$ values listed were calculated for the N listed using a two-tailed t-test with 98% confidence. Typically a $t_{calculated}$ value should be less than the $t_{critical}$. "Yes" means there exists a statistical difference between the two sets of results while "No" means there is no difference between the results.

ADDITIONAL STUDIES WITH OPTGC

Memphis EX On-Line Monitoring Study using OPTGC

In past monitoring studies, we were not able to pursue work with the on-line monitoring purge and trap-gas chromatograph (OPTGC). We have been able to resolve certain logistic issues that prevented us using it in previous on-line monitoring studies in both Memphis and Houston. We present the results of additional studies in this Report. The Memphis EX monitoring study was conducted from June 11 to 17, 2006. The OPTGC was interfaced to the drinking water distribution system using on-line sampling configuration two as shown in Appendix A. The concentration of each THM4 species was monitored at a sample rate of ~1 sample/ hour. A check standard was analyzed approximately every 12 hours. The results for each THM4 species were summed and are reported as TTHM4. The results of these studies along with details used in the analysis of the data are presented in Table 9.3.

Table 9.3
Memphis EX monitoring study: monitoring THM4 concentrations using OPTGC

	$CHCl_3$		$CHBrCl_2$		$CHBr_2Cl$		$CHBr_3$		TTHM4	
Statistic/*Method*	*EPA*	*OPTGC*	*EPA*	*OPTGC*	*EPA*	*OPTGC*	*EPA*	*OPTGC*	*EPA*	*OPTGC*
N	83	83	83	83	83	83	83	83	83	83
Avg. Conc.	0.2	0.1	0.5	0.4	0.8	0.8	0.5	0.4	1.9	1.8
Std. Dev Conc.	0.1	0.0	0.2	0.1	0.2	0.3	0.3	0.1	0.7	0.5
Conc. Range	0.0 to 0.4	0.1 to 0.3	0.2 to 0.8	0.0 to 0.9	0.2 to 1.5	0.3 to 1.9	0.1 to 2.1	0.2 to 1.0	0.7 to 4.1	1.1 to 3.9
Avg. Bias	---	0.0	---	0.0	---	0.0	---	0.0	---	-0.1
Std. Dev Bias	---	0.1	---	0.2	---	0.2	---	0.3	---	0.7
Bias Range	---	-0.3 to 0.1	---	-0.4 to 0.3	---	-0.8 to 0.8	---	-1.0 to 0.5	---	-2.4 to 1.5
% REC_{Bias}	---	96	---	106	---	115	---	129	---	104
s.d. % REC_{Bias}	---	50	---	48	---	44	---	102	---	37
% REC_{Bias} Range	---	31 to 375	---	10 to 247	---	35 to 294	---	21 to 556	---	34 to 264
t-test* $t_{critical}$ $t_{calculated}$	---	Yes 2.209 3.882	---	No 2.209 1.782	---	No 2.209 1.710	---	No 2.209 1.922	---	No 2.209 1.278

* The $t_{critical}$ and $t_{calculated}$ values listed were calculated for the N listed using a two-tailed t-test with 98% confidence. Typically a $t_{calculated}$ value should be less than the $t_{critical}$. "Yes" means there exists a statistical difference between the two sets of results while "No" means there is no difference between the results.

There were 83 direct comparisons between EPA 502.2 and OPTGC. $CHCl_3$ concentrations ranged from 0.0 to 0.4 μg/L for the EPA 502.2, and from 0.1 to 0.3 μg/L for OPTGC. The average concentration from EPA 502.2 was 0.2 ± 0.1 μg/L and from OPTGC was 0.1 ± 0.0 μg/L. The OPTGC bias ranged from –0.3 to 0.1 μg/L with an average of 0.0 ± 0.1 μg/L, and the % REC_{Bias} ranged from 31 to 375 % with an average of 96 ± 50 %. Average bias and % REC_{Bias} values were slightly better than the CMS-GC method most likely due to the type of sampling that the GEC provides. It is not affected by sample temperature like the CMS; therefore, would generally be more stable and consistent.

The concentration of $CHBrCl_2$ ranged from 0.2 to 0.8 μg/L using EPA 502.2, and 0.0 to 0.9 μg/L using OPTGC. The average concentration was 0.5 ± 0.2 μg/L for EPA 502.2 and 0.4 ± 0.1 μg/L for OPTGC. The OPTGC bias ranged from -0.4 to 0.3 μg/L with an average of 0.0 ± 0.2 μg/L. The OPTGC % REC_{Bias} ranged from 10 to 247 % with an average of 106 ± 48 %. Again the bias data was slightly better than the CMS-GC and with lower % REC_{Bias} values.

$CHBr_2Cl$ concentrations ranged from 0.2 to 1.5 μg/L using EPA 502.2, and 0.3 to 1.9 μg/L using OPTGC. The average concentration was 0.8 ± 0.2 μg/L using EPA 502.2, and 0.8 ± 0.3 μg/L using OPTGC. The OPTGC bias ranged from -0.8 to 0.8 μg/L with an average of 0.0 ± 0.2 μg/L. The OPTGC % REC_{Bias} ranged from 35 to 294 % with an average 115 ± 44 %. The bias results for $CHBr_2Cl$ is similar to the previous THM4 species and to the CMS-GC results.

The concentration of $CHBr_3$ ranged from 0.1 to 2.1 μg/L using EPA 502.2, and 0.2 to 1.0 μg/L using OPTGC. The average concentration was 0.5 ± 0.3 μg/L using EPA 502.2, and 0.4 ± 0.1 μg/L using OPTGC. The OPTGC bias ranged from -1.0 to 0.5 μg/L with an average of 0.0 ± 0.3 μg/L. The OPTGC % REC_{Bias} ranged from 21 - 556 % with an average 129 ± 102 %. The error seems to be large due to the low concentrations. There was no calculated bias for all THM4 for the OPTGC though the deviation of bias and % REC_{Bias} increased with the more brominated THM4 species.

The concentration of total THM4 ranged from 0.7 to 4.1 μg/L using EPA 502.2, and 1.1 to 3.9 μg/L using OPTGC. The average concentration was 1.9 ± 0.7 μg/L using EPA 502.2, and 1.8 ± 0.5 μg/L for OPTGC. The OPTGC bias ranged from -2.4 to 1.5 μg/L with an average of -0.1 ± 0.7 μg/L. The OPTGC % REC_{Bias} ranged from 34.1 to 264 % with an average of 104 ± 37%. At these low concentrations the agreement between methods is often exaggerated in a positive or negative direction. For the OPTGC and CMS-GC analyzers, the bias to the EPA 502.2 is the determining factor. The OPTGC for this study shows a slight negative bias in terms of total THM4

The results of statistical analysis gives a $t_{critical}$ for 83 comparisons of 2.209. For $CHCl_3$, $CHBrCl_2$, $CHBr_2Cl$, $CHBr_3$ and TTHM4, 3.882, 1.782, 1.710, 1.922, and 1.278 $t_{calculated}$ is respectively. The results for $CHCl_3$ appear to be statistically different, while the remaining results for the THM4 species and total THM4 do not appear to be statistically different. Statitically, the results improved upon the CMS-GC results with only $CHCl_3$ showing statistical disagreement.

Houston EX On-Line Monitoring Study using OPTGC

The OPTGC on-line monitoring studies were conducted side-by-side with the other Houston studies from June 27 – 29, 2006. The OPTGC was interfaced to the drinking water distribution system as described earlier, and the concentrations of each THM4 species were monitored at a sample rate of ~1 sample/ hour. A check standard was analyzed every 12 hours.

As with other methods, the results for each THM4 species were summed and are reported as TTHM4. The results of these studies along with details used in the analysis of the data are presented in Table 9.4.

There were 61 values for EPA 502.2 with 41 direct comparisons for OPTGC with EPA 502.2 during the Houston EX study. The concentration of $CHCl_3$ ranged from 35.8 to 51.9 µg/L for the EPA Method and from 28.9 to 52.1 µg/L for OPTGC. The average concentration for EPA 502.2 was 42.5 ± 3.4 µg/L and for OPTGC was 36.5 ± 3.9 µg/L. The OPTGC bias values ranged from –10.9 to 12.7 µg/L with an average of -4.7 ± 3.9 µg/L. The OPTGC % REC_{Bias} values ranged from 74.2 to 132 % with an average of 88.7 ± 9.5 %. There is a negative bias for the OPTGC when analyzing Houston samples that is twice as large as the CMS-GC analyzers with also lower recovery.

The concentration of $CHBrCl_2$ ranged from 7.1 to 12.7 µg/L using EPA 502.2 and 6.4 to 9.9 µg/L using OPTGC. The average concentration was 9.2 ± 1.0 µg/L using EPA 502.2 and 8.0 ± 0.9 µg/L using OPTGC. The OPTGC bias ranged from -4.7 to 1.2 µg/L with an average of -1.3 ± 1.4 µg/L. The OPTGC % REC_{Bias} ranged from 59to 115 % with and average of 87 ± 13 %. In contrast to the CMS-GC analyzers the OPTGC had a negative bias and % REC_{Bias} values less than 100%. This is most likely due to the decrease in sensitivity at the upper section of a quadratic curve; therefore, a lower concentration would be closer to the EPA concentration.

The $CHBr_2Cl$ concentration ranged from 1.1 to 2.1 µg/L using EPA 502.2 and 0.7 to 1.4 µg/L using OPTGC. The average concentration was 1.5 ± 0.2 µg/L using EPA 502.2 and 1.0 ± 0.2 µg/L using OPTGC. The OPTGC bias ranged from -1.2 to -0.1 µg/L with an average of -0.6 ± 0.3 µg/L. The OPTGC % REC_{Bias} ranged from 39 to 93 % with an average of 63 ± 15 %. Again there is a negative average bias, yet lower for a more brominated (less quadratic) THM4.

The OPTGC was in agreement with other methods with respect to $CHBr_3$. None of the methods detected $CHBr_3$ in Houston drinking water. For total THM4, the concentrations ranged from 45.6 to 62.0 µg/L using EPA 502.2, and 36.0 to 63.3 µg/L using OPTGC. The average concentration was 53.3 ± 3.8 µg/L using EPA 502.2, and 45.5 ± 4.8 µg/L using OPTGC. The OPTGC bias ranged from -14.6 to 12.5µg/L with an average of -6.6 ± 4.9 µg/L. The OPTGC % REC_{Bias} ranged from 73.0 to 125 % with an average of 87.4 ± 9.1 %. The results of statistical analysis for OPTGC for 41 comparisons gave $t_{critical}$ of 2.423. For $CHCl_3$, $CHBrCl_2$, $CHBr_2Cl$, and TTHM4, $t_{calculated}$ is 7.764, 5.932, 13.192, and 8.696 respectively. This indicates a statistical difference between the two methods for all analytes, which is similar to the CMS-GC analyzers.

Table 9.4
Houston EX monitoring study: monitoring THM4 concentrations using OPTGC

	$CHCl_3$		$CHBrCl_2$		$CHBr_2Cl$		$CHBr_3$		TTHM4	
Statistic/*Method*	*EPA*	*OPTGC*	*EPA*	*OPTGC*	*EPA*	*OPTGC*	*EPA*	*OPTGC*	*EPA*	*OPTGC*
N	61	41	61	41	61	41	61	41	61	41
Avg. Conc.	42.5	36.5	9.2	8.0	1.5	1.0	n.d.	n.d.	53.3	45.5
Std. Dev Conc.	3.4	3.9	1.0	0.9	0.2	0.2	n.d.	n.d.	3.8	4.8
Conc. Range	35.8 to 51.9	28.9 to 52.1	7.1 to 12.7	6.4 to 9.9	1.1 to 2.1	0.7 to 1.4	---	---	45.6 to 62.0	36.0 to 63.3
Avg. Bias	---	-4.7	---	-1.3	---	-0.6	---	n.d	---	-6.6
Std. Dev Bias	---	3.9	---	1.4	---	0.3	---	n.d	---	4.9
Bias Range	---	-10.9 to 12.7	---	-4.7 to 1.2	---	-1.2 to -0.1	---	---	---	-14.6 to 12.5
% REC_{Bias}	---	88.7	---	87	---	63	---	n.d	---	87.4
s.d. % REC_{Bias}	---	9.5	---	13	---	15	---	n.d	---	9.1
% REC_{Bias} Range	---	74.2 to 132	---	59 to 115	---	39 to 93	---	---	---	73.0 to 125
t-test* $t_{critical}$ $t_{calculated}$	---	Yes 2.423 7.764	---	Yes 2.423 5.932	---	Yes 2.423 13.192	---	---	---	Yes 2.423 8.696

* The $t_{critical}$ and $t_{calculated}$ values listed were calculated for the N listed using a two-tailed t-test with 98% confidence. Typically a $t_{calculated}$ value should be less than the $t_{critical}$. . "Yes" means there exists a statistical difference between the two sets of results while "No" means there is no difference between the results.

ADDITIONAL STUDIES WITH CMS-FIA NCA-FL

Extended On-Line Monitoring Studies for TTHM4 using CMS-FIA NCA-FL

The first extended on-line monitoring study for CMS-FIA was conducted from June 11 to 17 at the University of Memphis sampling site (Memphis EX). The second extended on-line monitoring study was conducted from June 26 - 29 in Houston (Houston EX). The concentration of total THM4 species was monitored at a rate of 1 sample/ hour (though the analyzer was typically taken off-line every 12 hours to run a check standard). The results of these studies along with details used in the analysis of the data are presented in Table 9.5.

Table 9.5
Memphis EX (502.2) and Houston EX (502.2) monitoring studies: monitoring total THM4 concentrations using CMS-FIA NCA-FL

	Memphis EX (502.2)		**Houston EX (502.2)**	
Statistic/*Method*	*EPA*	*CMS-FIA*	*EPA*	*CMS-FIA*
N	79	79	52	52
Avg. Conc.	1.9	2.1	53.5	48.2
Std. Dev Conc.	0.7	0.9	4.1	3.4
Conc. Range	0.7 to 4.1	0.2 to 5.3	45.6 to 62.0	41.8 to 56.1
Avg. Bias	---	0.2	---	-5.3
Std. Dev Bias	---	1.1	---	3.7
Bias Range	---	-2.3 to 3.1	---	-11.4 to 4.2
% REC_{Bias}	---	128	---	90.3
s.d. % REC_{Bias}	---	78	---	6.5
% REC_{Bias} Range	---	11 to 523	---	80.4 to 108
t-test*		No		Yes
$t_{critical}$	---	2.375	---	2.402
$t_{calculated}$		1.887		10.461

* The $t_{critical}$ and $t_{calculated}$ values listed were calculated for the N listed using a two-tailed t-test with 98% confidence. Typically a $t_{calculated}$ value should be less than the $t_{critical}$. . "Yes" means there exists a statistical difference between the two sets of results while "No" means there is no difference between the results.

In the Memphis EX study, with 79 direct comparisons, the concentration for TTHM4 ranged from 0.7 to 4.1 μg/L for EPA 502.2 and from 0.2 to 5.3 μg/L for the CMS-FIA method. The average TTHM4 concentration from EPA 502.2 was 1.9 ± 0.7 μg/L and from CMS-FIA was 2.1 ± 0.9 μg/L. The CMS-FIA bias for total THM4 ranged from –2.3 to 3.1 μg/L with an average of 0.2 ± 1.1 μg/L. The CMS-FIA % REC_{Bias} for total THM4 ranged from 11 to 523 % with an average of 128 ± 78%. When compared to the Memphis 1 study the bias and % REC_{Bias} results for this study were very similar. The average bias for the Memphis 1 study was only 0.5 μg/L higher. The CMS-FIA results for this study compares well to the CMS-GC and OPTGC methods

For the Houston EX study, there were 52 direct comparisons with concentration ranges varying from 45.6 to 62.0 μg/L using EPA 502.2, and 41.8 to 56.1 μg/L using CMS-FIA. The average concentration was 53.5 ± 4.1 μg/L using EPA 502.2, and 48.2 ± 3.4 μg/L using CMS-FIA. The CMS-FIA bias for total THM4 ranged from -11.4 to 4.2 μg/L with an average of -5.3 ± 3.7 μg/L. The CMS-GC % REC_{Bias} for total THM4 varied from 80.4 to 108% with an average of 90.3 ± 6.5%. Despite being negative the average bias improved by at least 6.0 μg/L from the previous Houston study. This may be due to addition of the peristaltic pump, and the use of sodium thiosulfate. The negative bias is probably a result of the non-linearity of the CMS at these high concentrations of THM4. This is similar to the OPTGC bias results for TTHM4.

Statistically, for the Memphis EX study, the $t_{critical}$ for the 79 comparisons for TTHM4 is estimated to be 2.375. The $t_{calculated}$ is 1.887, which means that there is not a statistical difference between the methods. With the Houston EX study (502.2) for 52 comparisons for TTHM4, the $t_{critical}$ is 2.402 and $t_{calculated}$ is 10.461, which means the two methods are statistically different.

Extended On-Line Monitoring Studies for THAA9 using CMS-FIA NCA-FL

Recall that the CMS-FIA analyzer provide analysis of total THM4 and total HAA9 with each alternate injection. These studies were done at the same time as the total THM4 studies using the same analyzer. The concentration of total HAA5 species was monitored at a rate of 1 sample/ hour with the analyzer typically taken off-line every 12 hours to run a check standard. The results of these studies along with details used in the analysis of the data are presented in Table 9.6.

For 83 direct comparisons in Memphis drinking water, the concentration for Total HAA9 ranged from 1.9 to 7.6 μg/L using EPA 552.3 and from 0.6 to 6.3 μg/L using CMS-FIA. The average total HAA9 concentration using EPA 552.3 was 3.0 ± 0.8 μg/L, and 2.3 ± 1.2 μg/L using CMS-FIA. The CMS-FIA bias for total HAA9 ranged from –6.0 – 3.3 μg/L with an average of -0.7 ± 1.4 μg/L. The CMS-FIA % REC_{Bias} for total HAA9 ranged from 10 to 218% with an average value of 80 ± 42%. Comparing the average bias results of Memphis EX to Memphis 1, the CMS-FIA has a lower bias in the Memphis EX study. This is likely due to the addition of thiosulfate, which masks the FAC interference for THAA9.

The results of the Houston monitoring study with 44 direct comparisons give total HAA9 concentrations ranging from 68.6 to 83.2 μg/L using EPA 552.3, and from 44.2 to 54.6 μg/L using CMS-FIA. The average total HAA9 concentration was 77.1 ± 3.7 μg/L using EPA 552.3, and 50.9 ± 2.9 μg/L using CMS-FIA. The CMS-FIA bias for total HAA9 ranged from -33.5 to -15.1 μg/L with an average of -26.2 ± 3.9 μg/L. The CMS-FIA % REC_{Bias} for total HAA9 ranged from 58 to 78% with an average of 66.1 ± 4.2%. In the Houston EX study, there is a large negative bias, compared to Houston, where the bias was slightly positive after correcting for the presence of BCAA concentrations. This could be due to the preservation process in the EPA sampling procedures. Ammonium chloride is added to the sample producing chloramines, which are not as strong of a chlorinating agent compared to FAC. However, chloramines can still chlorinate and produce HAAs, leading to the negative bias seen by CMS-FIA and PC-IC.

Statistically, both studies give statistically different results. The Memphis study CMS-FIA results differ from EPA 552.3 results (t_{critical} = 2.373 versus $t_{\text{calculated}}$ = 4.430). The Houston results differ (t_{critical} = 2.416 versus $t_{\text{calculated}}$ = 44.001) as well.

Table 9.6
Memphis EX (552.3) and Houston EX (552.3) monitoring studies: monitoring total HAA9 concentrations using CMS-FIA NCA-FL

	Memphis EX (552.3)		Houston EX (552.3)	
Statistic/*Method*	*EPA*	*CMS-FIA*	*EPA*	*CMS-FIA*
N	83	83	44	44
Avg. Conc.	3.0	2.3	77.1	50.9
Std. Dev Conc.	0.8	1.2	3.7	2.9
Conc. Range	1.9 to 7.6	0.6 to 6.3	68.6 to 83.2	44.2 to 54.6
Avg. Bias	---	-0.7	---	-26.2
Std. Dev Bias	---	1.4	---	3.9
Bias Range	---	-6.0 to 3.3	---	-33.5 to -15.1
% REC_{Bias}	---	80	---	66.1
s.d. % REC_{Bias}	---	42	---	4.2
% REC_{Bias} Range	---	10 to 218	---	57.5 to 78.0
t-test* $t_{critical}$ $t_{calculated}$	---	Yes 2.373 4.430	---	Yes 2.412 44.001

* The t_{critical} and $t_{\text{calculated}}$ values listed were calculated for the N listed using a two-tailed t-test with 98% confidence. Typically a $t_{\text{calculated}}$ value should be less than the t_{critical}. "Yes" means there exists a statistical difference between the two sets of results while "No" means there is no difference between the results.

ADDITIONAL STUDIES WITH THE PC-IC ANALYZER

Memphis EX On-Line Monitoring Study using PC-IC

The Memphis EX monitoring studies were conducted June 11-17, 2006. Samples were analyzed at a rate of one samples/ two hours though the analyzer was typically taken off-line each 12 hours to analyze a check standard). The results for each HAA species were summed to determine the total HAA9 concentration. The results of these studies along with details used in the analysis of the data are presented in Table 9.7.

For 47 direct comparisons, the concentration of MCAA ranged from 0.3 - 1.3 μg/L with an average of 0.3 ± 0.1 μg/L using EPA 552.3. The PC-IC reported a concentration range from 1.7 to 13.7 for MCAA, and the average concentration of 2.7 ± 2.0 μg/L. The PC-IC bias ranged from 0.4 to 13.4 μg/L. The average bias was 2.4 ± 2.0 μg/L. The % REC_{Bias} ranged from 134 to 4727% with an average of 937 ± 696%. The Memphis EX bias for MCAA remains was slightly larger than the Memphis 2 study; however, the advantage here is the resolution of MCAA and MBAA.

The concentration of DCAA ranged 1.6 – 2.8 μg/L using EPA 552.3, and 6.6 – 11.1 μg/L using PC-IC. The average DCAA concentration was 1.9 ± 0.2 μg/L using the EPA 552.3, and 6.8 ± 0.9 μg/L using PC-IC. The PC-IC bias for DCAA ranged from 3.8 – 9.2 μg/L with an average of 4.9 ± 0.9 μg/L. The PC-IC % REC_{Bias} ranged from 232 - 592 % with an average of 361 ± 64 %. The average bias for DCAA has increased between the Memphis EX and Memphis 2 study. This is largely due to the increased DCAA MDL resulting from the loss of the stopped-flow techniques used in the Memphis 2 and Houston studies. However, the resolution of DCAA, BCAA and DBAA are a definite advantage in the Memphis EX and Houston EX studies.

TCAA concentrations ranged from 0.2 - 0.5 μg/L using EPA 552.3, and 2.1 – 4.1 μg/L using PC-IC. The average TCAA concentration was 0.2 ± 0.0 μg/L using EPA 552.3, and 2.2 ± 0.3 μg/L using PC-IC. The PC-IC bias for TCAA ranged from 1.6 to 3.9 μg/L with an average of 2.0 ± 0.3 μg/L. The PC-IC % REC_{Bias} for TCAA ranged from 420 - 2391 % with an average of 1249 ± 219 %. The average TCAA Memphis EX bias remained the same compared to the average Memphis 2 bias.

The concentration of MBAA ranged from 0.1 – 1.0 μg/L using EPA 552.3, and 2.1 – 4.8 μg/L using PC-IC. The average MBAA concentration was 0.5 ± 0.3 μg/L using EPA 552.3, and 2.2 ± 0.5 μg/L using PC-IC. The PCIC bias for MBAA ranged from 1.1 to 4.7 μg/L with an average of 1.7 ± 0.6 μg/L). The PC-IC % REC_{Bias} values for MBAA ranged from 203 - 4020 % with an average of 912 ± 858 %. The average bias for MBAA remained approximately the same in the Memphis EX study compared to Memphis 2, with a gain of resolving MCAA and MBAA.

DBAA concentrations ranged from 0.4 – 1.0 μg/L using EPA 552.3, and 2.6 – 4.8 μg/L using PC-IC. The average concentration reported by EPA 552.3 was 0.4 ± 0.1 μg/L, and for PC-IC was 2.7 ± 0.4 μg/L. The PC-IC bias for DBAA averaged 2.3 ± 0.4 μg/L and ranged from 1.6 – 4.4 μg/L. The PC-IC % REC_{Bias} for DBAA ranged from 258 - 1294 % with an average value of 715 ± 122 %). While the bias for DBAA increased slightly in the Memphis EX study compared to the Memphis 2 study, the bias still remains below 5 μg/L, and there is full resolution of the dihalogenated species.

The concentration of BCAA ranged from 0.4 – 0.9 μg/L for EPA 552.3, and 0.6 – 14.1 μg/L for PC-IC. The average BCAA concentration for EPA 552.3 and PC-IC was 0.5 ± 0.1 μg/L 1.4 ± 2.2 μg/L, respectively. The PC-IC bias for BCAA ranged from -0.3 – 13.7 μg/L with an average of 0.9 ± 2.2 µg/L. The PC-IC % REC_{Bias} for BCAA ranged from 64 -3294 % with an average of 296 ± 499 %. The average bias for BCAA does not have a comparison to the Memphis 2 study, but the average bias is still less than 5 µg/L, which is acceptable.

BDCAA was not detected and thus the concentrations were expected to be below the MDL for BDCAA. This was true for both EPA 552.3 and PC-IC. Thus, in this regard the two methods are in agreement. The MDL values for EPA 552.3 and PC-IC were 0.3 µg/L and 2.8 µg/L, respectively, for all 47 direct comparisons.

DBCAA concentrations were less than the MDL for DBCAA (0.4 µg/L) for 47 direct comparisons according to EPA 552.2. However, for PC-IC, the concentration of DBCAA ranged from 5.6 – 8.5 µg/L in an apparent disagreement. The average DBCAA using PC-IC was 5.7 ± 0.5 µg/L. The PC-IC bias, using the MDL value for EPA 552.3 as the "true value", ranged from 5.3 – 8.2 µg/L with an average of 5.3 ± 0.5 µg/L. The PC-IC % REC_{Bias} for DBCAA ranged from 1600 - 2443 % with an average of 1628 ± 139 %.

TBAA concentrations ranged from 0.5 – 0.6 µg/L using EPA 552.3, and 10.1 – 12.2 µg/L using PC-IC. The average concentration was 0.5 ± 0.0 µg/L using EPA 552.3, and 10.2 ± 0.4 µg/L using PC-IC. The PC-IC bias for TBAA ranged from 9.5 – 11.7 µg/L with an average of 9.7 ± 0.4 µg/L). The PC-IC % REC_{Bias} for TBAA ranged from 1592 - 2349 % with average of 1945 ± 113 %. The average bias for TBAA was somewhat high at 9.7 µg/L, but considering that the MDL for TBAA is 10.1 µg/L, the bias is not comparatively high.

As with the THM4 studies discussed earlier, the individual concentrations of HAA9 species were summed to give the total HAA9 concentration. Total HAA9 ranged from 2.1 - 7.6 µg/L using EPA 552.3, and 0.0 - 39.7 µg/L using PC-IC. The average total HAA9 concentration was 3.0 ± 0.8 µg/L using EPA 552.3, and 9.5 ± 9.7 µg/L using PC-IC. The PC-IC bias for total HAA9 ranged from -2.9 – 36.9 µg/L with and average of 6.5 ± 9.7 µg/L. The PC-IC % REC_{Bias} for total HAA9 ranged from 0 to 1415 % with an average of 318 ± 331 %. The average THAA9 Memphis EX bias is slightly less than the Memphis 2 bias, and all HAA9 species are being speciated and quantified, which is a substantial improvement over the Memphis 2 study.

Statistically, it is not suprising that the two methods do not give statistically the same results. The value of t_{critical} for the 47 comparisons is 2.410 and the $t_{calculated}$ for MCAA is 8.303; for DCAA, 35.709; for TCAA, 43.815; for MBAA, 19.331; for DBAA, 40.597; for BCAA, 2.690; for DBCAA, 75.174; for TBAA, 154.721; and for THAA9, 4.582. In each case, the results of the methods are statistically different.

Houston EX On-Line Monitoring Study using PC-IC

The Houston EX monitoring study was carried out along with other studies June 26 - 29, 2006. Samples were monitored at a rate of one samples/ two hours with the analyzer typically taken off-line each 12 hours to analyze a check standard. The results for each HAA9 species were summed to determine the total concentration of HAA9 species. The results of these studies along with details used in the analysis of the data are presented in Table 9.8.

For the MCAA concentrations, 23 direct comparisons showed that the concentration ranged from 2.1 to 6.1 µg/L with an average of 3.8 ± 1.4 µg/L using EPA 552.3. The PC-IC method measured concentrations ranging from 2.7 to 15.3 for MCAA with an average MCAA concentration of 8.6 ± 3.8 µg/L. The PC-IC bias for MCAA ranged from -2.9 to 13.2 µg/L for the PC-IC with an average bias for MCAA of 4.8 ± 4.8 µg/L. The PC-IC % REC_{Bias} for MCAA ranged from 53 to 735 % with an average of 276 ± 178 %. The average MCAA bias for Houston

EX decreased by a factor of seven when compared to the first Houston study. This large decrease is due to the resolution of MCAA from MBAA as well as another HAA-like species.

For 21 direct comparisons, DCAA concentrations ranged from 34.1 – 45.7 μg/L using EPA 552.3, and 9.8 – 46.8 μg/L for PC-IC. The average concentration of DCAA was 42.1 ± 2.7 μg/L for EPA 552.3, and 27.5 ± 9.8 μg/L for PC-IC. The PC-IC bias for DCAA ranged from -34.5 – 3.0 μg/L with an average of -14.7 ± 10.0 μg/L. The PC-IC % REC_{Bias} for DCAA ranged from 22 - 109 % with an average value of 66 ± 24 %. The DCAA average bias for Houston EX decreased from a positive 1.4 μg/L for the DCAA Corr. Values (chapter 7) to a negative 14.7 μg/L. This large increase in bias could be from the continued production of DCAA in the EPA sample vial, resulting from converting the FAC to chloramines.

The concentration of TCAA ranged from ranged from 16.5 – 28.4 μg/L using EPA 552.3, and 2.0 – 14.2 μg/L for PC-IC for 23 direct comparisons. The average TCAA concentration was 18.7 ± 2.4 μg/L using EPA 552.3, and 7.7 ± 4.2 μg/L using PC-IC. The PC-IC bias for TCAA ranged from -25.1 to -4.5 μg/L with an average of -11.0 ± 5.2 μg/L. The PC-IC % REC_{Bias} for TCAA ranged from 11 - 76 % with an average of 42 ± 23 %. The TCAA average bias remained consistently negative when comparing Houston EX to Houston. There seems to be a consistent underreporting for PC-IC compared to EPA 552.3 for DCAA and TCAA. This could be due to sampling problems or something more fundamental to the method. In either case, further study needs to be devoted to these problems.

For 23 direct comparisons, the MBAA concentration ranged from 0.1 – 2.3 μg/L using EPA 552.3, and 2.1 – 7.8 μg/L using PC-IC. The average MBAA concentration was 1.6 ± 0.7 μg/L using EPA 552.3, and 2.3 ± 1.2 μg/L using PC-IC. The PC-IC bias for MBAA ranged from -0.2 to 5.7 μg/L with an average of 0.7 ± 1.3 μg/L). The PC-IC % REC_{Bias} for MBAA ranged from 91 - 1750 % with an average of 346 ± 559 %. The MBAA average bias is greatly improved, again, this is largely due to the resolution of MCAA and MBAA, as well as the unknown HAA-like species.

The DBAA concentration for 21 direct comparisons were less than the MDL of 0.4 μg/L for all EPA 552.3 samples. For PC-IC, the concentration of DBAA ranged from 3.2 – 6.0 μg/L. with an average concentration of DBAA measured for PC-IC of 3.5 ± 0.8 μg/L. The PC-IC bias for DBAA using the EPA 552.3 MDL values (0.4 μg/L) as the "true values" ranged from 2.8 – 5.7 μg/L with an average of 3.1 ± 0.8 μg/L. The PC-IC % REC_{Bias} for DBAA ranged from 865 - 1629 % with an average of 936 ± 219 %. In this case for DBAA, the results are in direct disagreement.

With BCAA, there were 21 direct comparisons and the concentrations of BCAA ranged from 5.8 – 7.4 μg/L using EPA 552.3, and 0.6 – 3.1 μg/L using PC-IC. The average BCAA concentration was 6.5 ± 0.5 μg/L using EPA 552.3, and 1.9 ± 0.6 μg/L using PC-IC. The PC-IC bias ranged from -6.7 to -3.1 μg/L with an average of -4.6 ± 0.9 μg/L. The PC-IC % REC_{Bias} for BCAA ranged from 8.2 – 49.7 % with an average of 29.6 ± 9.9 %. While there is a negative average bias for BCAA, it is less than 5 μg/L and this is an improvement over the Houston study, in that BCAA is now quantifiable.

BDCAA concentrations ranged from 3.2 – 4.9 μg/L using EPA 552.3, and 2.8 – 3.8 μg/L using PC-IC for 23 direct comparisons. The average BDCAA concentration was 4.2 ± 0.4 μg/L using EPA 552.3, and 2.8 ± 0.2 μg/L using PC-IC. The PC-IC bias for BDCAA ranged from -2.1 to -0.1 μg/L with an average of -1.4 ± 0.5 μg/L. The PC-IC % REC_{Bias} for BDCAA ranged from 57.0 – 97.8 % with an average 67.9 ± 9.3 %. There is a slight negative bias for BDCAA, but it is less than 2 μg/L which is acceptable.

DBCAA was not detected using either EPA 552.3 or PC-IC methods. With DBCAA, the concentration was less than the MDL of 0.4 μg/L for EPA 552.3 and the MDL of PC-IC (5.6 μg/L) for all 23 direct comparisons. In this regard, the two methods are in agreement.

TBAA concentrations ranged from 0.5 – 0.7 μg/L using EPA 552.3, and were less than the MDL of 10.1 for TBAA using the PC-IC method for 23 direct comparisons. In this regard, the two methods agree. The average concentration of TBAA using EPA 552.3 was 0.6 ± 0.1 μg/L. The bias, using the MDL for PC-IC as the experimental value, ranged from 9.4 – 9.6 μg/L with an average of 9.5 ± 0.1 μg/L. The PC-IC % REC_{Bias} for TBAA ranged from 1366 - 1942 % with an average of 1736 ± 181 %.

The total HAA9 concentration ranged from 69.1 – 83.0 μg/L using EPA 552.3, and 33.6 – 67.8 μg/L using PC-IC for 21 direct comparisons. The average total HAA9 concentration was 77.4 ± 3.3 μg/L using EPA 552.3, and 49.7 ± 8.5 μg/L using PC-IC. The PC-IC bias for total HAA9 ranged from -42.3 to -8.7 μg/L with an average of -27.7 ± 8.1 μg/L. The PC-IC % REC $_{Bias}$ for total HAA9 concentration ranged from 45 to 89% with an average of 64 ± 10 %. The large negative THAA9 bias seems to be problematic. Taking a closer look at the average biases reveals that the majority of the negative bias results from DCAA and TCAA. Clearly there is an issue there; however, excepting TBAA, the remaining HAA9 species have biases are less than 5 μg/L.

Statistically, the two methods give statistically different results. The value of t_{critical} for the 23 direct comparisons of MCAA, TCAA, MBAA, BDCAA and TBAA is 2.508 and the $t_{calculated}$ for MCAA is 4.814; for TCAA, 10.226; for MBAA, 2.772; for BDCAA, 13.963; and for TBAA, 733.524. The value of $t_{critical}$ for the 21 direct comparisons of DCAA, DBAA, BCAA and THAA9 is 2.528. The $t_{calculated}$ for DCAA is 6.687; DBAA, 17.405; BCAA, 23.218; and THAA9, 15.776.

COMPARING THE ON-LINE METHODS IN TERMS OF ANALYTES

In previous discussions, the methods have been largely compared to EPA methods. In this section, we present a comparison of the methods in terms of the information they provide with respect to analyte concentrations when compared to each other. While these results still indicate that there are still analytical issues to be addressed with each method, they do show promise for further development.

Comparing the THM4 and Total THM4 Methods

In the revised monitoring studies, three methods were compared to EPA 502.2. Figure 9.1 presents two bar graphs with error bars comparing the average concentration predicted by each analytical method for each individual THM4 species and total THM4. The Memphis EX results are shown in the top plot, while the Houston EX results are presented in the bottom plot. A qualitative review of the data from studies these studies indicate that all three methods agree quite well at the very low levels of Memphis drinking water and in the more complex sample matrix of Houston drinking water. Note that the error bars in most cases overlap. The higher end chromatography based instruments provide data on all THM4 species that is in good agreement with the side-by-side EPA comparison. The same is true of these methods when used to predict total THM4 concentrations. Likewise, the results agree with the CMS-FIA method, though this method exhibits considerably more error. Each of these methods appears to provide reliable on-line monitoring data at low concentration levels (Memphis) and at moderate concentrations in a generally more complicated matrix (Houston).

Similar plots are shown comparing the bias of these analytical measurements assuming that the "true value" is reported by EPA 502.2 (Figure 9.2). All of the methods provide very low average bias values over the course of the study. In Memphis drinking water, the error on the average bias for each individual THM4 species is almost exclusively much less than 0.5 μg/L, while for total THM4 species, the error on the average bias is near the single μg/L level for all methods except CMS-FIA where the error is less than a few μg/L. In the more complicated Houston system, the errors are slightly larger (1-10 μg/ range), but reproducible.

Thus in summary, we believe all the proposed methods are useful for carrying out optimization studies in drinking water distribution systems as long as careful consideration is given to the historical data of the distribution system; coupled with careful premonitoring surveys. The proposed methods offer many advantages to EPA 502.2. It should be noted that while the on-line methods were largely operated in this side-by-side study at a rate of one sample per hour, each is capable of higher sample throughput. For example, the CMS-GC, OPTGC, and the CMS-FIA can provide sampling rates as high as two samples per hour, but certainly this is dependent on the matrix when chromatographic methods are considered. The MDL, accuracy and precision values are in acceptable ranges for on-line monitoring. Moreover, each method is accomplished with relatively low level instrumentation that is constructed from commercially available components. Once assembled and calibrated, the analyzers are largely automated and require minimal operator intervention. We believe our proposed methods offer considerable advantage over routine "shake and shoot methods" such as EPA standard methods EPA 551.

Comparing the HAA9 and Total HAA9 Methods

The revised HAA9 monitoring studies compared the CMS-FIA and the PC-IC methods to EPA 552.3. As shown earlier with the THM4 studies, Figure 9.3 presents two bar graphs with error bars comparing the average concentration predicted by each analytical method for each individual HAA9 species and total HAA9. The Memphis results are shown in the top plot while the Houston results are presented in the bottom plot. As before, considering the methods very broadly, the plots indicates that the results agree within reason at the very low levels of Memphis drinking water and in the more complex sample matrix of Houston drinking water. The error bars in most cases overlap, though there is considerably more error than exhibited with the THM4 species. However, one must keep in mind that the quantification of all HAA9 species and total HAA9 is a considerably more difficult analytical problem. The PC-IC analyzer provides data on all HAA9 species that is in acceptable agreement to a first approximation with the side-by-side EPA comparison. It is believed that this is the first ever on-line, post-column derivatization of all nine HAA9 species. There also appear to be significant differences in the individual results for each HAA9 species reported by EPA 552.3 and PC-IC. This observation will be discussed in more detail in Chapter 10. It is also important to note that the PC-IC and the CMS-FIA methods show excellent agreement. This means that while we have not been able to completely explain the differences between EPA 552.3 and the NCA based methods, they may differ by one or more similar reasons. More to the point, it is likely that the same factors are resulting in the discrepancy between the EPA and NCA based methods and that this factor is a positive bias in EPA 552.3. These two methods show promise for reliable on-line monitoring at the low concentration levels of Memphis and the moderate concentrations in the slightly more complicated matrix of Houston.

Similar plots are shown comparing the bias of the HAA9 analytical measurements assuming that the "true value" is reported by EPA 552.3 (Figure 9.4). All of the methods provide promising results for average bias values over the course of the study, though certainly it would be desirable to have smaller average bias values. The errors in these measurements are relatively small considering the difficulty of the analytical measurement. For the Memphis drinking water study, the average bias for each individual HAA9 species is almost exclusively much less than 5 μg/L . For total HAA9 concentration, the cumulative error of nine HAA species becomes quite large. There is a considerable, but consistent negative bias when the PC-IC and CMS-FIA methods are compared to EPA 552.3 that is mirrored in the average concentration results. There was an analytical complication in the Houston monitoring studies that is accounted for in Chapter 10. Regardless, the bias is relatively constant with low uncertainties, thus both NCA based methods could be used for optimization of treatment practices in the distribution system.

Thus in brief summary, both the PC-IC and CMS-FIA methods may provide reliable results for carrying out optimization studies in drinking water distribution systems. However, it is cautioned that careful consideration should be given to historical data of the distribution system and the detailed premonitoring surveys. Yet both methods, despite analytical concerns, have several advantages compared to EPA 552.3, such as both are largely automated. "Connect and automated sampling/analysis" is a lot less labor intensive than the "shake, extract, derivatize and shoot" approach of EPA 552.3. Sampling rate for PC-IC is an issue: we can provide speciation for all but the monohalogenated species as well as total HAA5 at a rate of one sample per hour. We can provide completely resolved HAA6 species and total HAA6 at a rate of about one sample every 90 minutes. For the HAA9 species and total HAA9, the sampling rate is about one sample every two hours. Slight improvements on the order of a few minutes could be possible. In these studies, the analysis times are coordinated with all other methods to achieve side-by-side comparisons. It is likely that the PC-IC method could provide slightly higher sample throughput when monitoring HAA concentrations alone. The CMS-FIA method might be able to provide sample rates of two or more sample per hour in certain circumstances. The MDL, accuracy and precision values are in acceptable ranges for on-line monitoring, but certainly improvements and further optimization is warranted. The PC-IC and CMS-FIA analyzers are relatively low level instrumentation that is constructed from commercially available components. As with the THM4 methods, once assembled and calibrated, the analyzers are largely automated and require minimal operator intervention. We believe our proposed methods offer considerable advantage over routine 551 and 552.3 methods, and should be further explored.

Table 9.7
Memphis EX monitoring study: monitoring HAA9 concentrations using PC-IC

	MCAA		DCAA		TCAA		MBAA		DBAA		BCAA		BDCAA		DBCAA		TBAA		THAA9	
Statistic/*Method*	*EPA*	*PCIC*	*EPA*	*PCIC*	*EPA*	*PCIC*	*EPA*	*PCIC*	*EPA*	*PCIC*	*EPA*	*PCIC*	*EPA*	*PCIC*	*EPA*	*PCIC*	*EPA*	*PCIC*	*EPA*	*PCIC*
N	47	47	47	47	47	47	47	47	47	47	47	47	47	47	47	47	47	47	47	47
Avg. Conc	0.3	2.7	1.9	6.8	0.2	2.2	0.5	2.2	0.4	2.7	0.5	1.4	n.d.	n.d.	n.d.	5.7	0.5	10.2	3.0	9.5
Std. Dev Conc.	0.1	2.0	0.2	0.9	0.0	0.3	0.3	0.5	0.1	0.4	0.1	2.2	n.d.	n.d.	n.d.	0.5	0.0	0.4	0.8	9.7
Conc. Range	0.3 to 1.3	1.7 to 13.7	1.6 to 2.8	6.6 to 11.1	0.2 to 0.5	2.1 to 4.1	0.1 to 1.0	2.1 to 4.8	0.4 to 1.0	2.6 to 4.8	0.4 to 0.9	0.6 to 14.1	--- --- ---	--- --- ---	--- --- ---	5.6 to 8.5	0.5 to 0.6	10.1 to 12.2	2.1 to 7.6	0.0 to 39.7
Avg. Bias	---	2.4	---	4.9	---	2.0	---	1.7	---	2.3	---	0.9	---	n.d.	---	5.3	---	9.7	---	6.5
Std. Dev Bias	---	2.0	---	0.9	---	0.3	---	0.6	---	0.4	---	2.2	---	n.d.	---	0.5	---	0.4	---	9.7
Bias Range	---	0.4 to 13.4		3.8 to 9.2	---	1.6 to 3.9	---	1.1 to 4.7	---	1.6 to 4.4	---	-0.3 to 13.7	---	---	---	5.3 to 8.2	---	9.5 to 11.7	---	-2.9 to 36.9
% REC_{Bias}	---	937	---	361	---	1249	---	912	---	715	---	296	---	n.d.	---	1628	---	1945	---	318
s.d. % REC_{Bias}	---	696	---	64	---	219	---	858	---	122	---	499	---	n.d.	---	139	---	113	---	331
% REC_{Bias} Range	---	134 to 4727	---	232 to 592	---	420 to 2391	---	203 to 4020	---	258 to 1294	---	64 to 3294	---	---	---	1600 to 2443	---	1592 to 2349	---	0.0 to 1415
t-test*	---	Yes	---	Yes	---	Yes	---	Yes	---	Yes	---	Yes	---	---	---	Yes	---	Yes	---	Yes
$t_{critical}$		2.410		2.410		2.410		2.410		2.410		2.410				2.410		2.410		2.410
$t_{calculated}$		8.303		35.71		43.82		19.33		40.60		2.690				75.17		154.7		4.582

* The $t_{critical}$ and $t_{calculated}$ values listed were calculated for the N listed using a two-tailed t-test with 98% confidence. Typically a $t_{calculated}$ value should be less than the $t_{critical}$. "Yes" means there exists a statistical difference between the two sets of results while "No" means there is no difference between the results.

Table 9.8
Houston EX monitoring study: monitoring HAA9 concentrations using PC-IC

	MCAA		DCAA		TCAA		MBAA		DBAA		BCAA		BDCAA		DBCAA		TBAA		THAA9	
Statistic/*Method*	*EPA*	*PCIC*	*EPA*	*PCIC*	*EPA*	*PCIC*	*EPA*	*PCIC*	*EPA*	*PCIC*	*EPA*	*PCIC*	*EPA*	*PCIC*	*EPA*	*PCIC*	*EPA*	*PCIC*	*EPA*	*PCIC*
N	23	23	21	21	23	23	23	23	21	21	21	21	23	23	23	23	23	23	21	21
Avg. Conc	3.8	8.6	42.1	27.5	18.7	7.7	1.6	2.3	n.d.	3.5	6.5	1.9	4.2	2.8	n.d.	n.d.	0.6	n.d.	77.4	49.7
Std. Dev Conc.	1.4	3.8	2.7	9.8	2.4	4.2	0.7	1.2	n.d.	0.8	0.5	0.6	0.4	0.2	n.d.	n.d.	0.1	n.d.	3.3	8.5
Conc. Range	2.1 to 6.1	2.7 to 15.3	34.1 to 45.7	9.8 to 46.8	16.5 to 28.4	2.0 to 14.2	0.1 to 2.3	2.1 to 7.8	--- --- ---	3.2 to 6.0	5.8 to 7.4	0.6 to 3.1	3.2 to 4.9	2.8 to 3.8	--- --- ---	--- --- ---	0.5 to 0.7	--- --- ---	69.1 to 83.0	33.6 to 67.8
Avg. Bias	---	4.8	---	-14.7	---	-11.0	---	0.7	---	3.1	---	-4.6	---	-1.4	---	n.d.	---	9.5	---	-27.7
Std. Dev Bias	---	4.8	---	10.0	---	5.2	---	1.3	---	0.8	---	0.9	---	0.5	---	n.d.	---	0.1	---	8.1
Bias Range	---	-2.9 to 13.2	---	-34.5 to 3.0	---	-25.1 to -4.5	---	-0.2 to 5.7	---	2.8 to 5.7	---	-6.7 to -3.1	---	-2.1 to -0.1	---	---	---	9.4 to 9.6	---	-42.3 to -8.7
% REC_{Bias}	---	276	---	66	---	42	---	346	---	936	---	29.6	---	67.9	---	n.d.	---	1736	---	64
s.d. % REC_{Bias}	---	178	---	24	---	23	---	559	---	219	---	9.9	---	9.3	---	n.d.	---	181	---	10
% REC_{Bias} Range	---	53 to 735	---	22 to 109	---	11 to 76	---	91 to 1750	---	865 to 1629	---	8.2 to 49.7	---	57.0 to 97.8	---	---	---	1366 to 1942	---	45 to 89
t-test*	---	Yes	---	Yes	---	Yes	---	Yes	---	Yes	---	Yes	---	Yes	---	---	---	Yes	---	Yes
$t_{critical}$		2.508		2.528		2.508		2.508		2.528		2.528		2.508				2.508		2.528
$t_{calculated}$		4.814		6.687		10.23		2.772		17.41		23.22		13.96				733.5		15.78

* The $t_{critical}$ and $t_{calculated}$ values listed were calculated for the N listed using a two-tailed t-test with 98% confidence. Typically a $t_{calculated}$ value should be less than the $t_{critical}$. "Yes" means there exists a statistical difference between the two sets of results while "No" means there is no difference between the results.

(Top plot) Memphis THM4 On-Line Monitoring Study—Average Concentration Results

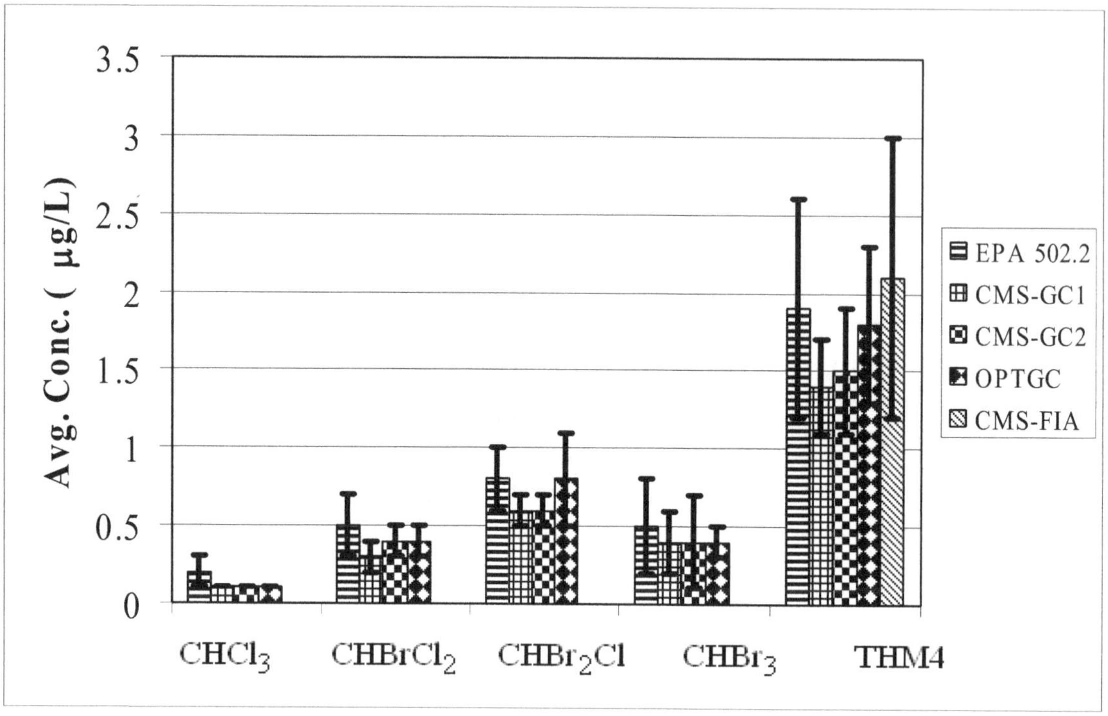

(Bottom Plot) Houston THM4 On-Line Monitoring Study—Average Concentration Results

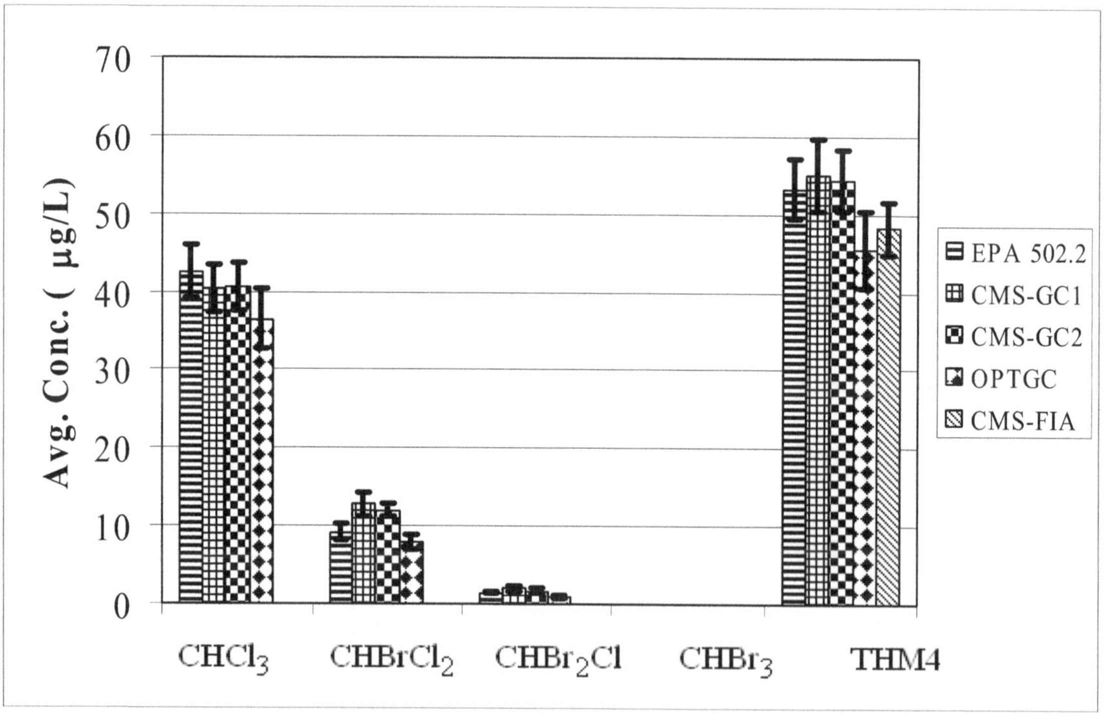

Figure 9.1 Plots comparing the average concentrations calculated for each THM4 species and total THM4 using EPA 502.2 and the candidate methods

(Top plot) Memphis THM4 On-Line Monitoring Study—Average Bias Results

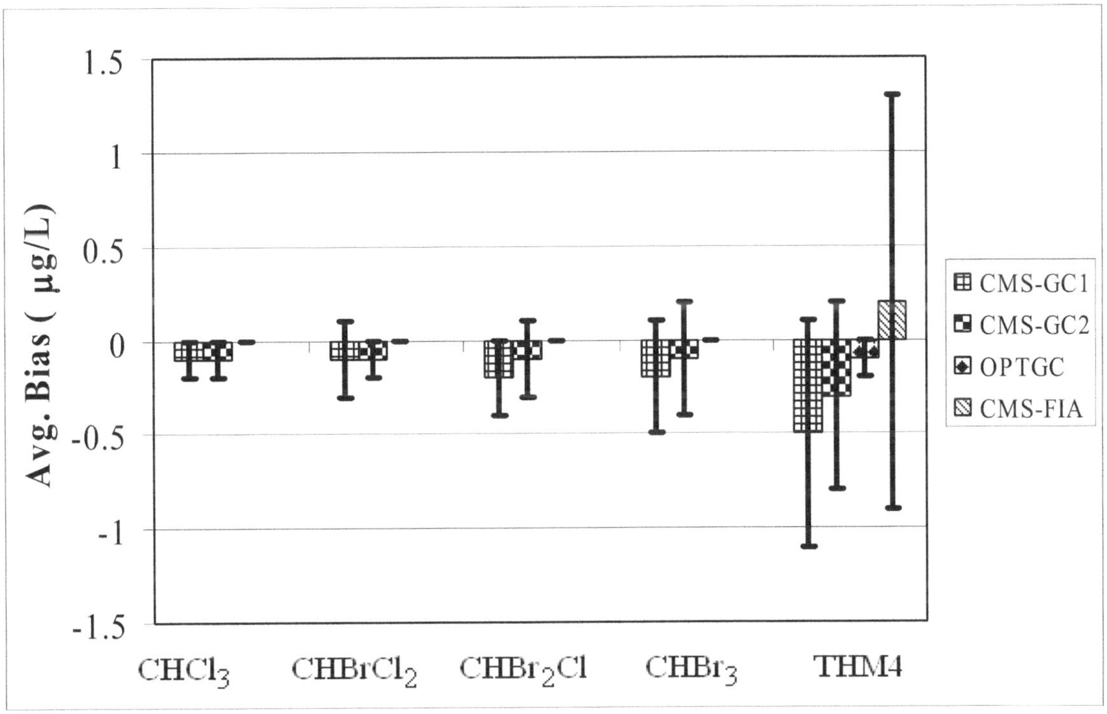

(Bottom Plot) Houston THM4 On-Line Monitoring Study—Average Bias Results

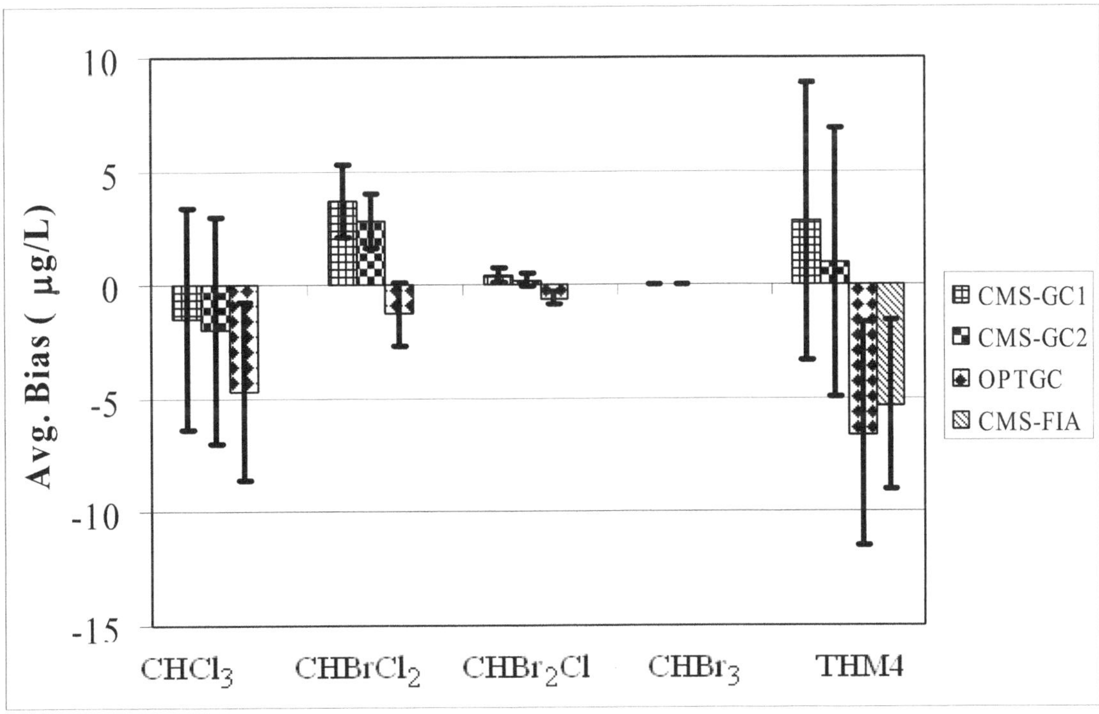

Figure 9.2 Plots comparing the average bias calculated for each THM4 species and total THM4 using EPA 502.2 and the candidate methods

(Top plot) Memphis HAA9 On-Line Monitoring Study—Average Concentration Results

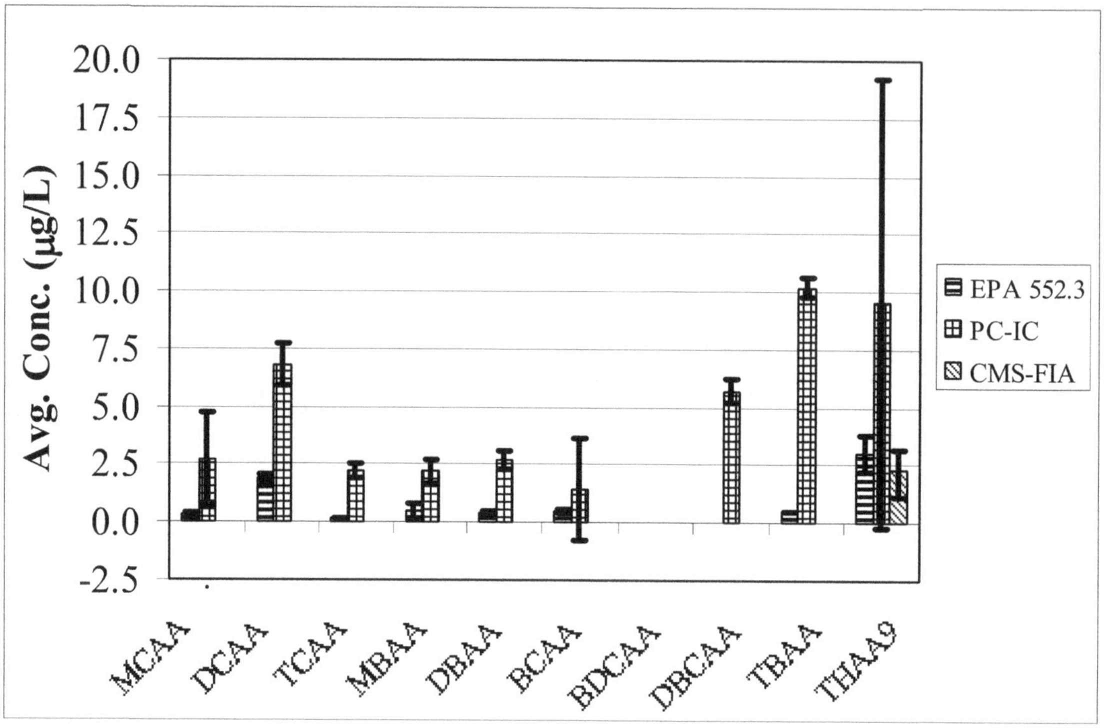

(Bottom Plot) Houston HAA9 On-Line Monitoring Study—Average Concentration Results

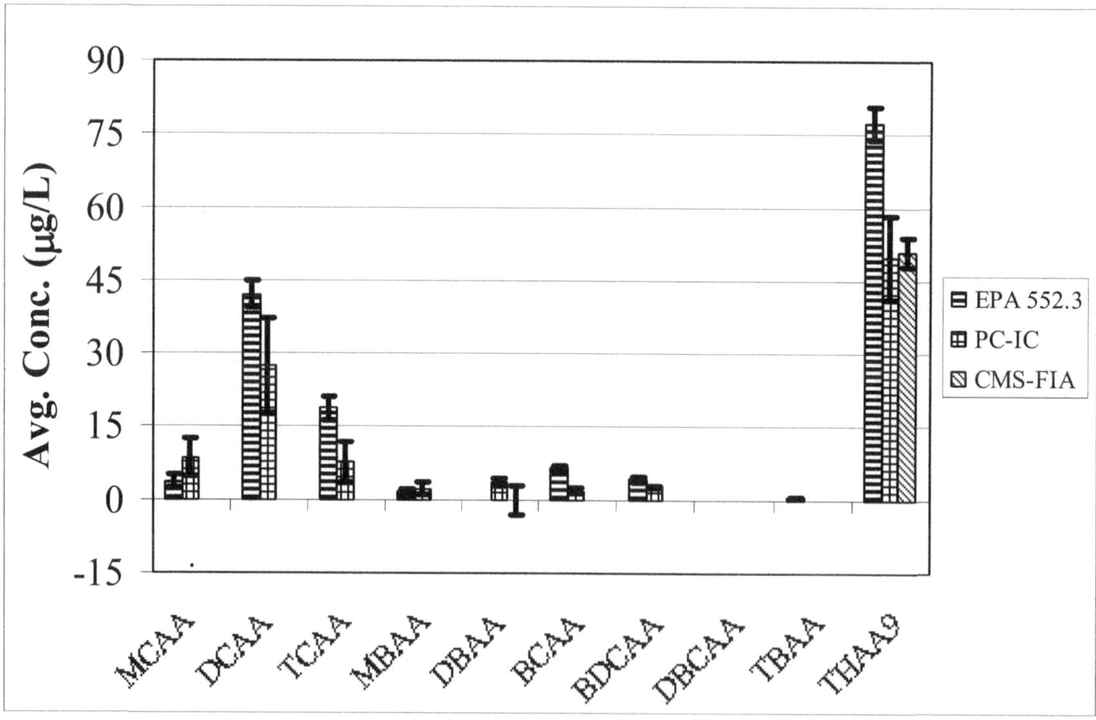

Figure 9.3 Plots comparing the average concentrations for each HAA9 species and total HAA9 calculated using EPA 552.3 and the candidate methods

(Top plot) Memphis HAA9 On-Line Monitoring Study—Average Bias Results

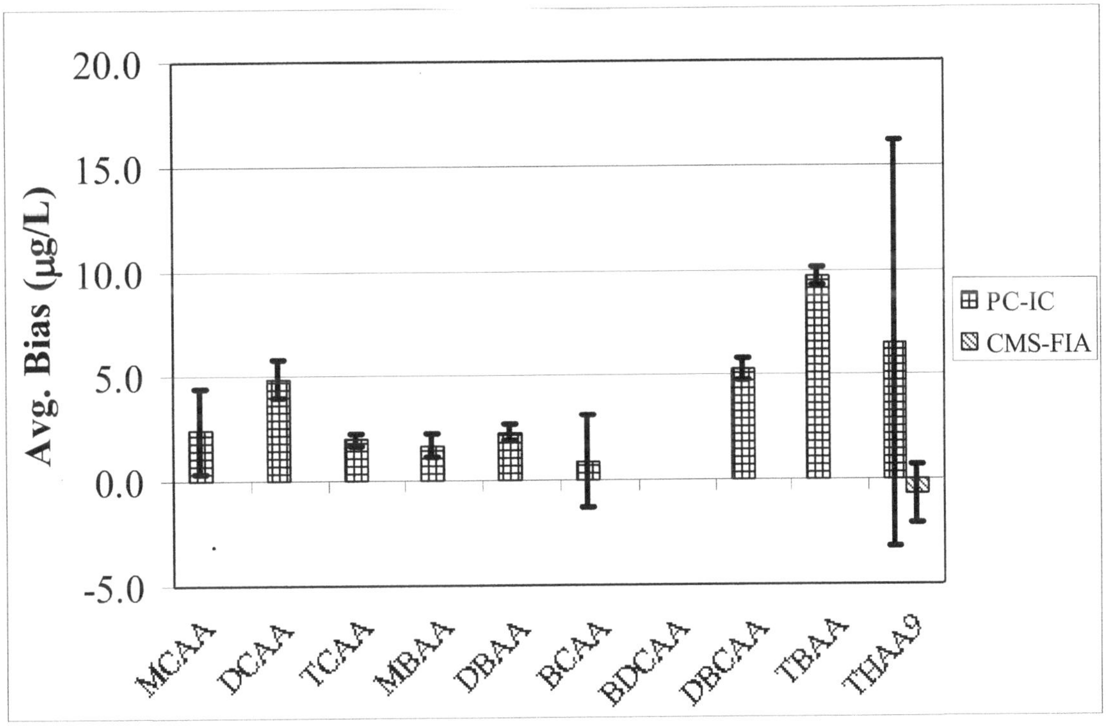

(Bottom Plot) Houston HAA9 On-Line Monitoring Study—Average Bias Results

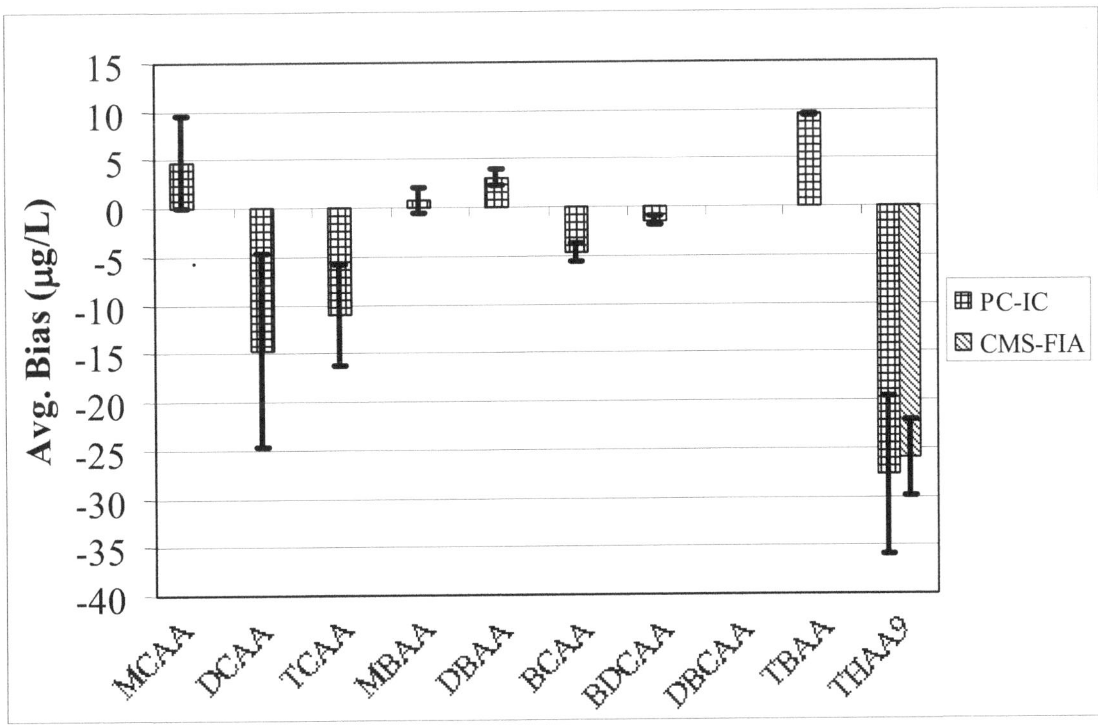

Figure 9.4 Plots comparing the average bias for each HAA9 species and total HAA9 calculated using EPA 552.3 and the candidate methods

CHAPTER 10
CONCLUSIONS AND RECOMMENDATIONS FOR FURTHER RESEARCH

PROJECT ACCOMPLISHMENTS

Over the course of this project, there was a great deal of progress toward developing methods for on-line monitoring of THMs and HAAs. Four types of monitoring methods were explored for THM4 monitoring: on-line monitoring purge and trap gas chromatograph (OPTGC), supported capillary membrane sampler (SCMS) -GC, capillary membrane sampling (CMS) -GC and capillary membrane sampling flow injection analysis with nicotinamide-fluorescence (CMS-FIA NCA-FL). Of these methods, three were subjected to extensive on-line monitoring studies as candidate methods. Three approaches were explored for HAA5 monitoring including ion chromatography-membrane suppressed conductivity detection (IC-MSCD), CMS-FIA and post column-ion chromatography with NCA-FL (PC-IC). Of these three approaches, CMS-FIA and PC-IC were subjected to extensive monitoring studies in Memphis, TN and Houston, TX. In Chapters 2-7, the details of the analytical methods were presented. In Chapter 8, each method was evaluated in terms of the target goals of this project with respect to the Ideal Method. The candidate methods were compared to one another and discussed with respect to their advantages and disadvantages. Strategies were developed for carrying out better on-line monitoring studies and instrument improvement. Specific recommendations for on-line monitoring of THM4 and HAA5 concentrations were reviewed and summarized. The revised analyzers and optimized methods were used with the revised strategies for on-line monitoring. Chapter 9 presents the results of additional on-line monitoring studies in Memphis, TN and Houston, TX. The results of the different analytical methods are discussed with respect to how well they agreed in real world studies. In Chapter 10, conclusions and recommendations for further research will be provided. Broader implications of the project will be discussed. An effort will be made to focus on practical considerations of a relatively small analytical laboratory with limited resources. Detailed descriptions of instrument construction and developing an on-line monitoring site are presented in the Appendices.

PRACTICAL CONSIDERATIONS: A LABORATORY PERSPECTIVE

Construction, Optimization and Maintenance of the On-Line Methods

The CMS-GC, CMS-FIA, and the PC-IC methods are all constructed from relatively inexpensive commercially available components. Like most laboratories, we do not have an unlimited budget. We have kept the cost of the analyzers low and the construction as simple as possible. Detailed procedures for constructing and optimizing the analytical methods are presented in this Report. It is likely that analysts interested in constructing and operating analyzers such as these might also need to further tailor the analyzers to have their own in-house methods for their own distribution system. This Report provides a good starting point for such studies. Specific details about the construction of each analyzer are presented in the Appendices. From a broader view, it should be noted that all of the analytical instruments that were used in this study were already available or were purchased used, for a considerable savings. For example, the filter-based fluorescence detectors that were used in this study were exclusively purchased on websites such as eBay (www.ebay.com) and LabX (www.labx.com) for a fraction of the cost of new equipment (typically in the price range \$25-\$250). The gradient pumps that were used for the PC-IC method were procured in the same way, typically costing approximately \$300 each. In practice, we purchased several of these components and used them as spares when a second device could be made functional or as parts when they could not be made functional. The portable GC used in CMS-GC1 was purchased on eBay for \$800 and parts units were bought in the range \$25-\$125. The GC used in CMS-GC2 was a bench-top GC that was procured through the campus surplus equipment program at the University of Memphis. The point is that functional on-line analyzers were constructed from low cost or no-cost (perhaps already available) instrumentation. This is an important aspect of the work, as it means that a significant cost savings can be realized with a little preplanning. However, the analyzers could likely give improved results if they were constructed with state-of-the art instrumentation. Most notably, the PC-IC method requires a high end (dual monochromator based) fluorescence detector for operation at the single μg/L concentrations levels and this cost is in the neighborhood of \$12-15K.

Analysts should be encouraged to use the EPA methods for comparison and confirmation studies. Historical analytical data is typically available to aid in taking into account site specific interferences issues. It has been demonstrated here, premonitoring surveys are an absolute necessity before conducting on-line monitoring studies. Additionally, analysts should be encouraged to use other available resources for constructing on-line analyzers. For example, many larger utilities may already have on hand, in storage or surplus, older analytical instruments that could be adapted into on-line monitoring analyzers. For example, a spare GC with an electrolytic conductivity detector could be used for THM4 monitoring.

Revisiting and Clarifying Sampling Rates for On-Line Monitoring Methods

The distinction needs to be made that the EPA methods actually limited our sampling rates when we conducted side-by-side studies. In every case, except the PC-IC method, the on-line method was operated for several days at sample rates of 1 sample per hour. This 1 sample per hour sampling rate is a result of the nature of the side-by-side comparison study with the EPA methods. In order to keep the number of EPA samples to a manageable size for our small laboratory (staffed for this project by the PI and three chemists), the on-line methods were coordinated to sample once an hour to coincide with the collection of a grab sample for EPA analysis. For the CMS-FIA methods, it would be possible to sample at a sampling rate of 3-4 samples per hour if it were allowed to operate at maximum sampling rate. For CMS-GC and OPTGC, the sampling rate could be increased to 2-3 samples per hour. The PC-IC method was sampling at a rate of 0.5 samples per hour (1 sample every two hours) for selectively measuring all HAA9 species. For HAA6, the sampling rate could be increased to ~1 sample per hour based on revised studies. While it is recommended that samples be analyzed periodically using EPA methods as a comparison, in routine studies focused less on comparison and more on evaluating the change in total THM4 or total HAA9 concentrations during a treatment process, the on-line analyzers can provide 3-4 times better time resolution of the concentration data, with much less manual labor.

Another inherent advantage of the on-line monitoring analyzers is that they are much gentler in sampling than traditional grab sampling. Sample preparation steps (collection, transfer, pre-treatment, etc) are eliminated because the sampling valve is connected directly to the drinking water distribution system. The sample is taken directly from the water distribution system in real time, and processed by the analyzer according to its own chemical basis. This on-line sampling eliminates instances for error and analyte loss. Moreover, sample timing is automated, regular, and documented by the analyzer computer system. Compared to the on-line methods, the EPA methods require a much higher degree of manual labor and were most susceptible to systematic errors. The proposed on-line monitoring methods are automated to different degrees. The CMS-GC and OPTGC analyzers are developed to the point of being able to operate and report data in near-real time. The CMS-FIA requires some degree of operator oversight in reporting the data, while the PC-IC requires even more operator intervention before reporting concentration data. All methods can ultimately be developed to full automation, but at this point all share the characteristic that once they are calibrated, the sampling and analysis process is done on-line in real time. The analyst, once a particular analyzer is calibrated and on-line, is free to do other tasks.

For all practical purposes, we have constructed simple robotic analyzers. They will not get tired and can, within reason, run continuously with minor maintenance. Once the analyzers are set up and operating, they require minimal user intervention. At present, the robotic analyzer is relatively dumb. While it is possible to set up various alarms (email, fax, phone, pager) to notify operators should the analyzer or treatment process go out of compliance, it should be noted that there are real limitations to the degree to which the analyzer can intervene. Ultimately, as the data handling becomes more complex, there is no reason why the chemical analyzer could not be made smarter. It would be possible to construct a smarter analyzer, an expert system, which could evaluate its own quality control on the fly, make decisions about treatment and take steps to intervene in the process. As perhaps an overly simple example, consider that the same computer that controls the analyzer is also in control of disinfectant addition. The analyzer tells the computer that DBP concentrations are on the rise and triggers the computer to slowly back off on the amount of disinfectant added to the raw water. The DBPs begin to fall, and the system is brought back in compliance.

Bias versus Instantaneous Concentration Changes

Extensive studies have been presented that measure the difference between the results reported by the on-line methods and the results of the appropriate EPA method. Assuming that the EPA method reports the true value, it is possible to calculate the bias between the two methods. The bias studies allow for estimation of the absolute error of the analytical methods. The small average bias values, measured to a high degree of precision, demonstrated that these proposed methods, on average, agree with one another. Certainly, there are some notable exceptions, like the large bias exhibited between the NCA based methods and EPA 552.3. Yet even in those cases, the bias is relatively constant and known very precisely. This means correction factors could be applied to bring the on-line monitoring method in line with EPA methods, if necessary. The possible cause of this large bias will be discussed in more detail later in this Chapter.

Plots of total THM4 (Figure 10.1) and total HAA9 (Figure 10.2) concentrations over time are shown to identify some general trends in the data. Ideally, all the data for a particular parameter, either total THM4 or total HAA9, should lie on top of one another. In the real world, this is seldom the case. Clearly there are differences between the on-line results and the EPA

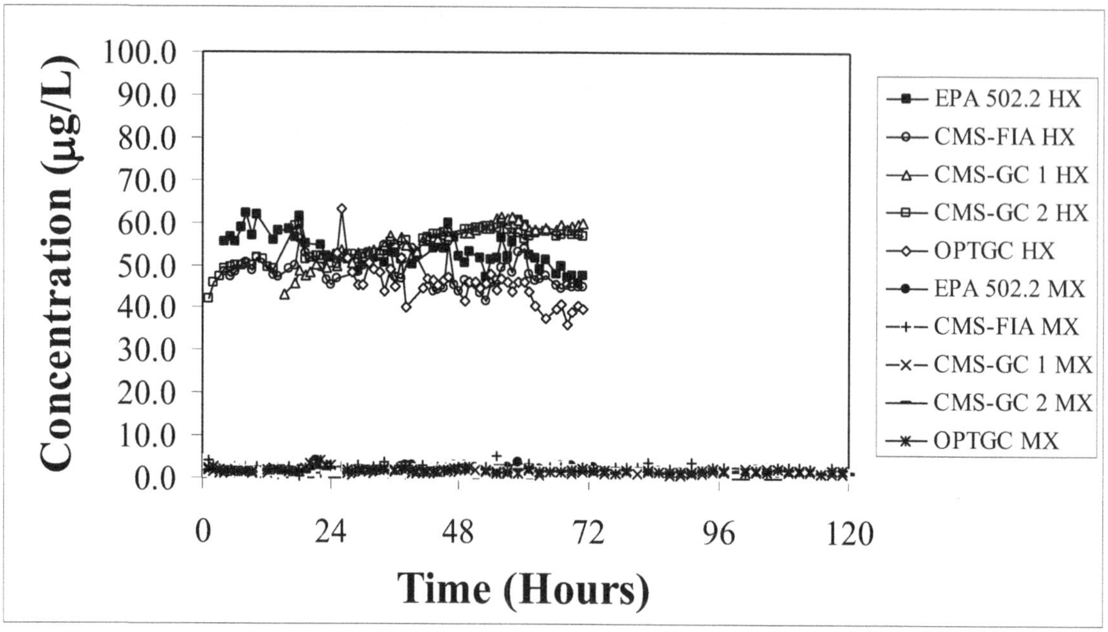

Figure 10.1 Combined plot of TTHM4 data in extended studies conducted in Memphis and in Houston

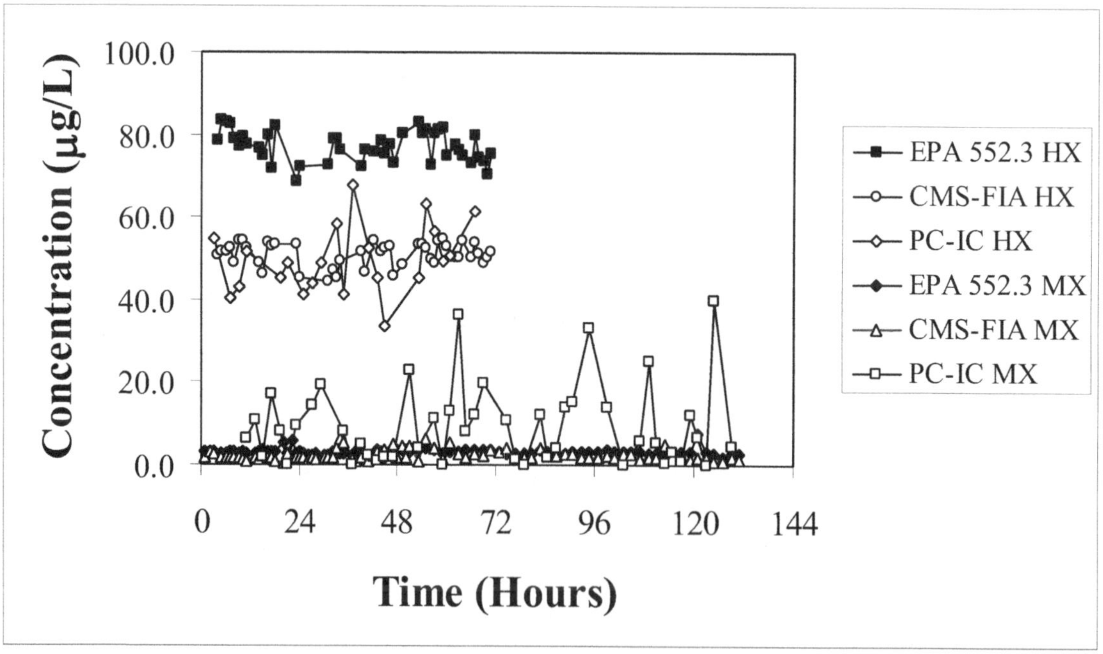

Figure 10.2 Combined plot of HAA9 data in extended studies conducted in Memphis and in Houston

results. The plots do not lie directly on top of one another as would be expected in the ideal case. However, the results do show some degree of agreement. This is reflected in the results of the bias studies discussed above.

Conducting studies in Memphis, TN offers the chance to evaluate very low levels of THM4 and HAA9 in a relatively simple ground water matrix where the concentrations of both THM4 and HAA5 species is expected to be quite low (~1-10 µg/L). This relatively simple sample matrix taxes the ability of the methods to operate at low concentration levels. This is in contrast to Houston, TX where the THM4 and HAA9 concentrations are significantly higher (near the MCL) and the sample matrix is typically more complex surface water. These online methods can be applied with some success over the concentration range of the old MCL (100 µg/L) down to the typical minimum reporting levels (MRL) of most methods (typically ~1 µg/L) based upon the MDL values reported in this work. This seems to be a reasonable concentration range that a professional would need to conduct monitoring and treatment optimization studies.

This work demonstrates that the proposed on-line methods can be used over the concentration range from single µg/L up to ~100 µg/L. Considering total THM4 data, operating at the MDLs in Memphis water, EPA 502.2, CMS-GC, OPTGC and CMS-FIA methods all are in relative agreement. Operating at the MCL level, the same methods remain in relative agreement, though certainly the amount of bias is generally higher.

The bias studies demonstrate the average difference between the two methods over the course of a monitoring study. This is an important statistic to consider, but ultimately the true test of the methods comes down to the question - How well do the methods follow one another, or "track" the concentrations of THM4 and HAA9 species? Earlier in this Report, the notion of "analyte tracking" was referring to how well the methods are in agreement at a particular point in time. This is in contrast to the bias where differences are averaged over the entire study. With tracking, we are referring to the instantaneous behavior of the two methods at a particular point in time.

As a simple analogy consider if two pH electrodes were placed in the same solution. If base is added, then both pH meters, if all is operating well, would exhibit an increase in pH. Likewise, if acid were added, one would expect both electrodes to register a decrease in the pH. Similarly, if there is a change in the concentration of THM4 or HAA9 concentration, we would expect both analyzers to respond accordingly. Both should exhibit either an increase or a decrease in the analytical signal. Thus we would expect that at each particular slice of time that the two methods would be responding in the same way. In other words, both analyzers (EPA and candidate method) should ideally be registering the same "instantaneous change"--either both positive, both negative, or both unchanged.

Considering "analyte tracking" in this way affords a way to take a more quantitative look at this phenomenon. The instantaneous concentration change for a particular sampling time slice (sampling time T_1 to sampling time T_2, where T_2 is greater than T_1 and the two times are in consecutive order in the study) was calculated by subtracting the concentration of analyte at T_1 from the concentration of analyte at T_2. The two methods agree if both exhibit a positive, negative or no change in concentration.

For the Memphis study, considering the instantaneous change in total THM4 concentration as a measure of analyte tracking, the CMS-GC1 agreed with EPA 502.2 62.0 % of the time. The CMS-GC2 showed agreement with EPA 502.2 54.1% of the time. The OPTGC and the CMS-FIA were in agreement with EPA 502.2 57.3% and 62.3 % of the time, respectively. For total HAA9, the CMS-FIA agreed with EPA 552.3 58.5% of the time and the PC-IC method agreed 47.8% of the time. Studies carried out in Houston gave similar results. For total THM4, CMS-GC1 and CMS-GC2 agreed with EPA 502.2 60.0% and 55.0% of the time, respectively. Likewise, OPTGC and CMS-FIA agreed with EPA 502.2 60.0% and 60.8% of the time, respectively. For total HAA9, the CMS-FIA agreed with EPA 552.3 62.7% of the time and the PC-IC agreed with EPA 552.3 65.0% of the time.

A more traditional way to look at analyte tracking would be to examine closely the concentration over time plots of the data. For completeness, the plots for total THM4 and total HAA9 species are collected at the end of this Chapter (Figures 10.11 – 10.22). While these plots are complicated, it is suggested that the way to read them is to simply locate the initial concentrations for each study (located at extreme left side of the plot) and follow both the EPA result and the on-line results through the progression of the study. In evaluating the plots in this manner, the on-line methods show good analyte tracking with EPA methods.

Accounting for the Houston EX Negative Bias of CMS-FIA and PC-IC

Figure 10.2 clearly indicates a large bias between the EPA method and the NCA based methods. In Memphis water, the methods agreed relatively well - the bias averaged ~1 to 7 μg/L for total HAA9 using CMS-FIA and PC-IC, respectively. However, the Houston study gave significantly larger biases (~30 μg/L). One thing to note is that while the NCA methods exhibit significant differences from the EPA method, the two NCA based methods tend to be in agreement. This is in contrast to earlier studies with the two methods. It is important to explore the significant negative bias for CMS-FIA and PC-IC and understand the chemical basis for it.

There are two immediate possibilities. Either the NCA based methods underestimate the THAA9 concentrations or EPA 552.3 over estimates the THAA9 concentrations. Extensive work has gone into identifying and minimizing possible interferences for the CMS-FIA and PC-IC methods. One possible explanation is that the NCA chemistries underestimate the THAA9 concentrations, which could result from sample loss. However, this seems unlikely as the sample is taken directly from the distribution system, loaded into either the CMS-FIA or PC-IC analyzer, processed and analyzed immediately. It seems unlikely that there is a logical place for sample loss to occur.

Another possible explanation for the difference might be that the HAA9 species continue to be produced in the sample vial after sample collection. This explanation makes even more sense when one considers the logistics of the Memphis and Houston studies. During both the Memphis and Houston studies, 40 mL of sample was collected in 60 mL extraction vials with 4 mg of added ammonium chloride crystals (NH_4Cl), as per sampling guidelines detailed in EPA 552.3 (USEPA 2003a). The analysis of the Memphis samples was typically done within 24-36 hours after collection. This was not possible with the Houston study as samples were collected, stored and then analyzed upon return to the University of Memphis laboratory. This meant that the Houston samples were analyzed within seven to ten days after being collected, well within the 14 day holding time specified by the EPA method.

In EPA 552.3, ammonium chloride crystals are added to the sample vial during sampling to quench free available chlorine (FAC) by converting FAC to chloramines. Chloramines were shown to exhibit decreased HAA formation, compared to FAC, in processed (but yet to be chlorinated) waters. These waters contained above average TOC concentrations (~4 mg/L) in the presence of above average FAC concentrations (~6 mg/L) (Pepich et al. 2004). Samples quenched with ammonium chloride crystals (Pepich et al. 2004) produced an additional 5 μg/L of HAA9 species, the majority being 4 µg/L of DCAA, 1.7 µg/L of BCAA and 0.7 µg/L of DBAA over a 28-day study period. Seemingly, the mass balances do not seem to hold, but the trend is clear. The majority of additional formation occurred within the first two holding days. The concentration of MCAA varied over the course of the 28-day study, but ultimately exhibited a net decrease of 1.5 µg/L. This indicated that MCAA was being further chlorinated in the sample vials. In contrast, unquenched samples produced an additional 58 µg/L with the primary constituents being DCAA (26 µg/L), TCAA (14 µg/L), BCAA (11 µg/L), BDCAA (15 µg/L) and small increases of MCAA (3.3 µg/L) and DBAA (3.6 µg/L). The concentrations of DCAA, TCAA and BCAA significantly increased within the first two days, while BDCAA increased steadily over the first 21 days. In addition, the conversion of FAC to chloramines still retained the antimicrobial activity such that HAAs were not significantly degraded by microbial processes. These results reported by Pepich et al. (2004) are the primary reason ammonium chloride crystals were added to EPA 552.3 as the sample preservation procedure. These results show the need for quenching the samples to minimize continued HAA9 formation. But they also demonstrate that HAA9 formation is not really stopped, only slowed upon addition of ammonium chloride crystals. Thus, EPA 552.3 has a clear and documented positive bias that becomes more evident in the first 48 hours following sampling, despite following recommended protocols for sample quenching. It is suggested that additional steps may be needed to quench the HAA9 formation reactions in more complicated systems, however no detail is provided as to what those steps should be.

While Pepich et al. studied systems using FAC as the primary disinfectant, no results are reported for systems where chloramines are used as the primary residual disinfectant. This is important because the City of Houston uses a combination of FAC and chloramines as the primary residual disinfectant, 1.3 mg/L and 2.1 mg/L respectively (Houston Water Quality Report 2005). In a system such as Houston's, the addition of the ammonium chloride crystals will slow the formation of additional HAA9 species, but even the slowed formation might account for the large bias between the two methods when the samples were analyzed approximately seven to ten days later. If the assumption is made that both PC-IC and CMS-FIA are providing the concentration value of THAA9 at the specific sampling time, which is warranted as both methods agree quite well, then the EPA 552.3 concentration value can be explained as the continued production of HAA9 species in the sampling vial as reported by Pepich et al. (2004).

This large bias is supported further by close examination of the individual HAA9 species from the PC-IC studies. The PC-IC average biases for DCAA, TCAA, BCAA and BDCAA were -14.7 µg/L, -11.0 µg/L, -4.6 µg/L and -1.4 µg/L respectively, indicating that concentration increased for these HAA9 species in the EPA sample vials. These are the same HAA9 species that Pepich et al. (2004) showed increasing over time in quenching studies. Thus there is a clear, documented, expected positive bias in EPA method 552.3 in chloraminated systems when the samples are analyzed within the specified holding time of 14 days. While immediate analysis may minimize this bias, it was not possible to do this in the on-line monitoring studies conducted in Houston.

Additional information is revealed by taking an even closer look at the data. After a seven-day holding time and analysis using EPA 552.3, MCAA and TCAA do not show the trends that are seen in the PC-IC concentration plots (Figures 10.3 and 10.4). For MCAA, PC-IC shows an upward trend in concentration after the first 24 hours, whereas EPA 552.3 shows a flat trend up until 24 hours, followed by a sharp decrease, and then another flat trend (Figure 10.3).

For TCAA, PC-IC shows a downward trend in concentration after 24 hours, while EPA 552.3 shows a flat trend until the last sample, which increases by ~10 μg/L in concentration (Figure 10.4). Figure 10.5 shows the plot of concentration for PC-IC of MCAA and TCAA superimposed on each other, and demonstrates that after 24 hours, the concentration of TCAA shows a steady decrease, while MCAA shows a steady increase, with both trends occurring simultaneously; and neither of these trends is shown to be occurring by EPA 552.3. Additionally, the THAA9 plots for EPA 552.3, CMS-FIA and PC-IC do not show these trends (Figure 10.6).

The MCAA and TCAA concentrations are essentially a mirror image of each other in the concentration vs. time plot (Figure 10.5), and because of the mirror imaging, THAA9 concentrations would not reflect a change over time as is shown by PC-IC for the individual species. These trends shown by PC-IC for the individual species indicate a time-based phenomenon which, from a realistic time and labor perspective, is not easy to study with EPA 552.3. Samples would need to be processed immediately upon collection, likely including the derivatization process, as preservation and holding of the sample would lead to the same positive bias already observed in the final Houston study.

The analysis of HAA9 species by EPA 552.3 is not a simple or quick method, as it was designed from a compliance monitoring perspective, where simple and fast analysis is not a requirement, but desired. However, the PC-IC method is able to provide these kinds of time resolution results because the method was designed from the beginning as an automated method capable of analysis without operator intervention or sample preparation. Further research should be devoted to improving the PC-IC method for improved MDLs with a goal of providing highly reliable time resolved data. Moreover, additional side-by-side studies with the PC-IC and EPA 552.3 methods could provide additional insight in the chemistry of these two methods.

Temporal and Seasonal Variation in THM4 and HAA9 Concentrations

One of the goals of developing on-line monitoring methods was to be able to identify any seasonal or temporal (perhaps diurnal) changes that might occur in THM4 and HAA9 concentrations. For example, formation of the DBPs might be expected to be slower in winter months than in summer months simply because lower temperatures would decrease the rate of formation. Another possibility is that sunlight induced photolysis might decrease DBPs during the day and the lack of photolysis might allow them to build up overnight.

The data presented in this study was aimed mostly at direct comparison to EPA methods. However, the Memphis 1 monitoring study and the Memphis EX monitoring study were carried out in the same location, but Memphis 1 was conducted in December, while Memphis EX was done in June. There did not appear to be any significant variation in the DBP concentrations between the two studies indicating a seasonal variation. However, it should be noted that Memphis experiences relatively mild winters. More marked effects might be visible in other systems, and research needs to be done to thoroughly investigate this variation. These measurements are also complicated since they are done at very low levels of DBPs.

Figures 10.7-10.10 show the same data as shown earlier (Figures 10.1 and 10.2) for total THM4 and total HAA9, except that superimposed on top of each figure in a series of vertical lines to indicate approximated sunrise and sunset times—identified on the plots with a "N" for night and a "D" for day. Therefore, each alternating section refers to data collected at either night time or during the day. While again the data is very limited and perhaps more detail should be given than our approximate time demarcations, it is possible that there are diurnal variations in the THM4 and HAA9 data, that is especially visible in the Houston EX study. Again, the low levels of DBPs in Memphis water complicate the identification of any trends. More detailed studies need to be done to further investigate this phenomenon.

Ultimately, while some very simple trends may be visible, at this stage, these studies are inconclusive as to whether diurnal and seasonal varations are indeed occurring. More detailed studies in the laboratory and in real world systems would help improve our understanding of these processes. The on-line monitoring methods proposed in this work should be useful in carrying out such studies in the future.

Need for Miniaturization and Next Generation Lab-On-A-Chip Devices

It is important to contrast where we really are across the board in water analysis and where we would like to be. This research has developed relatively simple and inexpensive approaches to on-line monitoring that are bench-top, but portable, chemical analyzers. These analyzers have been modified to facilitate on-line monitoring directly from drinking water distribution systems, so that operators could collect data and optimize treatment practices. We provide useful approaches for drinking water professionals, and have largely met the initial and subsequent goals of this project. However, these instruments are operating at their limits—it is doubtful that much more utility can be squeezed out of these technologies.

There is a need to take a 21st century approach to developing these types of chemical analyzers. Despite all of our advances in the past 30 years, many of the techniques we use for chemical analysis of drinking water are still largely based upon 1970's technology. This is true of the work we report here. The notable exception to 1970's era technology would be advances made in using LC-MS techniques, and LC-MS is generally agreed to be beyond the cost limitations established for an Ideal Method for this research.

Studies need to be done to understand more about the basic chemical kinetics of the NCA based reactions with THM4 and HAA9 species. This is an important chemistry in that it opens the possibility of developing state-of-the-art; integrated lab-on-a-chip (LOC) styled microdevices.

One of the current trends in analytical chemistry is to develop clever LOC devices and this approach needs to be explored more for drinking water analysis. The LOC systems reduce both size and cost of the instrumentation. These devices might be used alone as an inexpensive approach for on-line monitoring and optimization for small distribution systems. As miniaturized devices decrease in both size and cost, it becomes practical to construct arrays of miniaturized devices that can be placed throughout the distribution system and operated simultaneously. It will likely never be feasible to assemble arrays of GC-MS and LC-MS devices to the degree that we could assemble such an array of LOC microdevices. Such miniaturized systems can provide chemical, spatial, and temporal information about THM4 and HAA9 concentrations over time throughout the distribution system. Such a sensor array could allow for the construction of detailed chemical kinetic models describing distribution systems. In particular, the CMS-FIA may be uniquely suited for miniaturization, but CMS-GC and PC-IC methods could be miniaturized as well.

SUMMARY OF REPORT

Several on-line monitoring methods were explored, along with a detailed report on construction, operation and maintenance of the promising analyzers. These methods included on-line monitoring purge and trap gas chromatograph (OPTGC), capillary membrane sampling gas chromatograph (CMS-GC), capillary membrane sampling flow injection analysis with nicotinamide-fluorescence (CMS-FIA NCA-FL), and post column-ion chromatography with NCA-FL (PC-IC). In each case, these methods were subjected to detailed comparison studies to EPA methods in the laboratory as well as diverse and challenging real world situations.

The OPTGC is a commercially available approach to purge and trap sampling that was developed in conjunction with SRI instruments and was applied to individual and total THM4 analysis. The OPTGC approach combined membrane sampling and purge and trap sampling to produce a relatively simple approach for on-line monitoring of THM4. The MDLs for each THM4 species using OPTGC are all less than 0.5 μg/L, with accuracies, measured as mean % recoveries ranging from 103 to 127%. The OPTGC also provides for relatively precise data exhibiting % RSD values less than 5%, typically ranging from 2.1-4.9%. The OPTGC was developed based on EPA methods, being similar to purge and trap methods such as EPA 502.2 and EPA 524.2. The OPTGC instrument is relatively inexpensive with an estimated price in the neighborhood of $15-20K. It can achieve sampling rates of 2-3 samples per hour depending on the specific sample matrix and it can be easily automated to report data in near-real time from remote locations via wired or wireless internet.

The CMS sampling device developed for FIA and GC can distinguish between volatile and non-volatile species. This device is simple and inexpensive to construct and can be made from commercially available components for less than $200.

The CMS device was interfaced to a GC to develop a new approach to on-line monitoring of THM4 that provides individual and total THM4 concentrations. The CMS-GC can be interfaced to an existing GC with an electron capture detector (ECD), thus offers alternatives to laboratories that may already have such an instrument available. The CMS-GC method with ECD provides MDL values for the THM4 species in the range 0.3 to 0.5 μg/L, with accuracies ranging from 100 to 125% and precision estimates ranging from 3.5 to 8.6%. If a utility already has a GC with ECD, then this approach is very economical. Even if it is necessary to purchase a low end GC-ECD, the cost of this approach is still low, in the neighborhood of $8-12K. The CMS-GC can reasonably be expected to achieve sampling rates of 2-3 samples per hour and exhibits all the same reporting and communications options discussed with OPTGC.

A new approach to simplifying GC analysis, called gas chromatography on a valve (GCOV), was detailed. GCOV is a practical starting point for miniaturizing analytical instrumentation useful for on-line monitoring. GCOV offers a low cost approach to analysis that is perhaps a bit more challenging to construct compared to CMS-GC. GCOV is still a relatively new approach, but may offer unique advantages, and uses a commercially available pulsed discharge photoionization detector (PDPID) detector to detect THM4 species. It is possible to achieve sampling rates of 3-4 samples per hour with this device, but the operator skill level is slightly higher than other GC based methods. Preliminary studies gave MDL estimates in the range 1.1-1.6 μg/L with mean % recoveries ranging from 91-119% and % RSD values ranging from 8.6 to 16%.

The analytically important reaction between NCA and THM4 and HAA9 species was examined in detail. NCA reacts with THM4 and HAA9 species to form fluorescent products.

Two analytical methods, CMS-FIA and PC-IC, were developed based on this chemistry that allowed for on-line monitoring in near-real time.

The CMS-FIA approach is, to the best of our knowledge, the only method that allows for selective on-line monitoring of total THM4 and total HAA9 in near-real time. This method is automated and allows for direct on-line monitoring of both total THM4 and total HAA9 concentrations directly from the drinking water distribution system. The MDL estimates for the CMS-FIA method are 1.1 µg/L for total THM4 and 1.6 µg/L for total HAA9. The CMS-FIA method gave accuracies of 111-115% and precision estimates of 2-3%. Perhaps the most important aspect to the CMS-FIA method is that it is based on florescence detection, which simplifies on-line low level detection. The operator skill needed to operate the CMS-FIA is relatively low compared to the other methods in this project. Correspondingly, the simplicity has an analytical cost in that CMS-FIA cannot be used to speciate THM4 and HAA9 components, but thanks to the CMS device it can distinguish between total THM4 and total HAA9. The CMS-FIA method can provide sampling rates up to four samples per hour. The fluorescence detector needed to construct the CMS-FIA can be a low end filter based unit that can greatly reduce expense compared to high end monochromator based detectors. The entire CMS-FIA analyzer could be constructed from commercially available components for ~$10K, or significantly cheaper if the laboratory already has a fluorescence detector available.

The PC-IC method allows for on-line monitoring of individual HAA5 species as well as total HAA5. The initial PC-IC analyzer was developed to quantify HAA5 and could not resolve the monohalogenated species. The method exhibits good selectivity for HAA5 species and offers a much simpler approach to sample collection; processing and analysis compared to common derivatization procedures that employ liquid-liquid extraction. The PC-IC method separates the HAA5 species using an anion exchange column followed by post column reaction with NCA and fluorescence detection. The MDL estimates using a state-of-the-art dual monochromator based fluorescence detector range from 1.2 to 2.1 µg/L with accuracies of 79.4 to 105% and precision estimates of 5.9-11%. This PC-IC method achieved sampling rates of 1 sample per hour and requires a higher level of operator skill and instrumentation compared to other methods.

Three on-line monitoring methods (CMS-GC, CMS-FIA, and PC-IC) were used in side-by-side studies with appropriate EPA methods in Memphis, TN and Houston, TX. A systematic approach for evaluating the data analytically as well as strategies for carrying out on-line monitoring studies was proposed. The analyzers and methods were revised based on limitations identified in real world monitoring studies. The specific analytical shortcomings that were identified were related mostly to chromatography and interference issues. It was proposed that prior to conducting on-line monitoring studies; analysts should conduct premonitoring surveys using both standard EPA methods as well as the proposed on-line monitoring methods. Chromatographic issues were modified to allow for better resolution with interferents. Major advances were made with respect to the NCA based methods for HAAs as the methods were extended to analyze all HAA9 species and thereby eliminate four additional chemically similar species that would certainly interfere in measurements. A revised PC-IC method was developed that allows for measuring all HAA9 species. The MDL estimates range from 5-16 µg/L using an inexpensive filter based fluorescence detector and 0.6 to 10 µg/L using a state-of-the-art dual monochromator based fluorescence detector. The filter based PC-IC reports accuracies in the range 73.8 to 99.7% with precision estimates of 7-21 %, close to the MDLs of the method. The higher-end dual monochromator based PC-IC gave accuracies of 79-161 percent with precision estimates of 3.5 to 32%, again operating near the MDLs of the proposed method. Adding the additional four HAA species did result in a significant decrease in the sampling rate for the PC-IC method, from 1 to 0.5 samples per hour. However, the additional information provided by this revised method makes the decrease in sampling rate worthwhile.

Four revised chemical analyzers and associated methods (OPTGC, CMS-GC, CMS-FIA and PC-IC) were used in additional side-by-side studies with appropriate EPA methods in Memphis, TN and Houston, TX. A detailed statistical analysis of the results of the monitoring studies was done to compare the methods of choice. As long as one keeps in mind these analyzers are designed for treatment practice optimization studies rather than compliance monitoring, all four methods provided acceptable results when used in on-line monitoring studies in water distribution systems. It is possible to monitor individual THM4 and HAA9 concentrations using either OPTGC or CMS-GC for THM4 and PC-IC for HAA9. The CMS-FIA provides for a mass balance with the other methods as it can provide total THM4 and total HAA9. Used in concert, these three analyzers provide detailed on-line monitoring data directly from drinking water distribution systems over the concentration range ~1 to 100 μg/L with sampling rates of up to 3 - 4 samples per hour for all methods with the notable exception of PC-IC. For PC-IC, sampling rates are restricted to ~0.5-1 sample per hour with the current instrument.

A broader overview, a laboratory perspective, was presented that describes the implications of the results reported here, with specific challenges faced by relatively small water quality laboratories with limited resources. Specific analytical issues with EPA 552.3 were identified and discussed. Broadly, this research provides a starting point for on-line monitoring with other species of interest such as emerging DBPs, pesticides, and other possible contaminants in drinking and wastewater. This is especially true of other VOCs and halogenated species, but also of compounds of interest to security monitoring. These methods also provide a fundamental chemical basis for the development of miniaturized analytical devices such as lab-on-a-chip technologies.

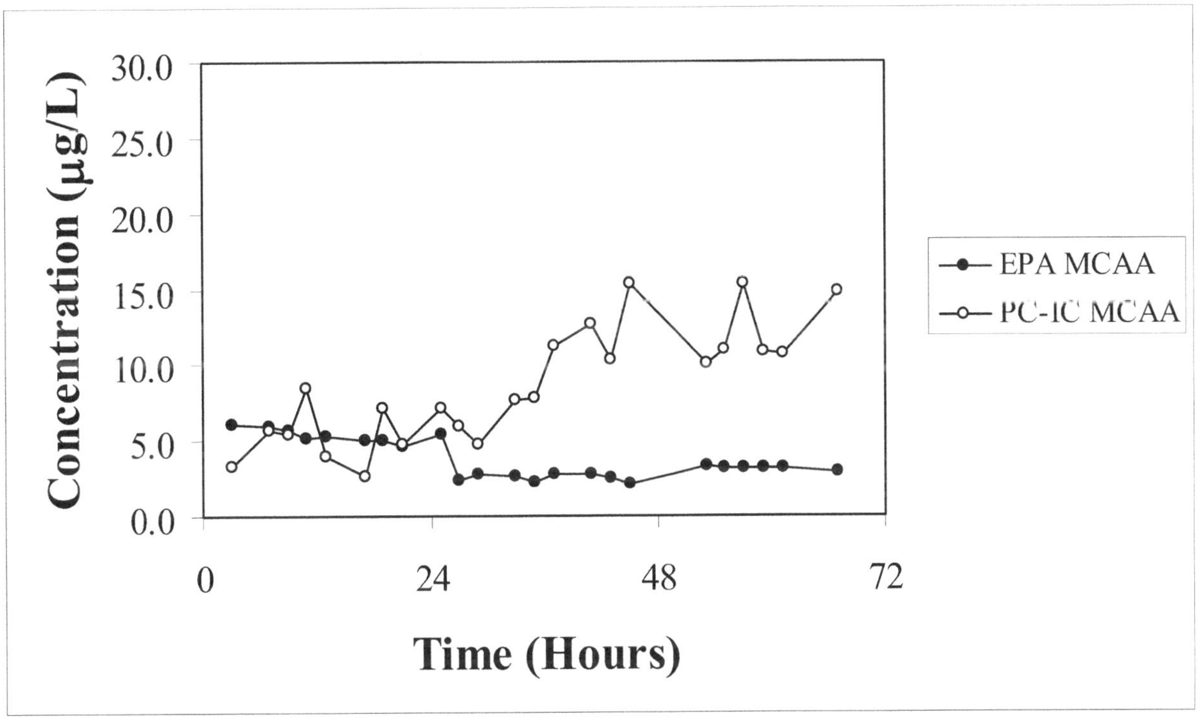

Figure 10.3 Houston EX concentration plot of MCAA for EPA 552.3 and PC-IC

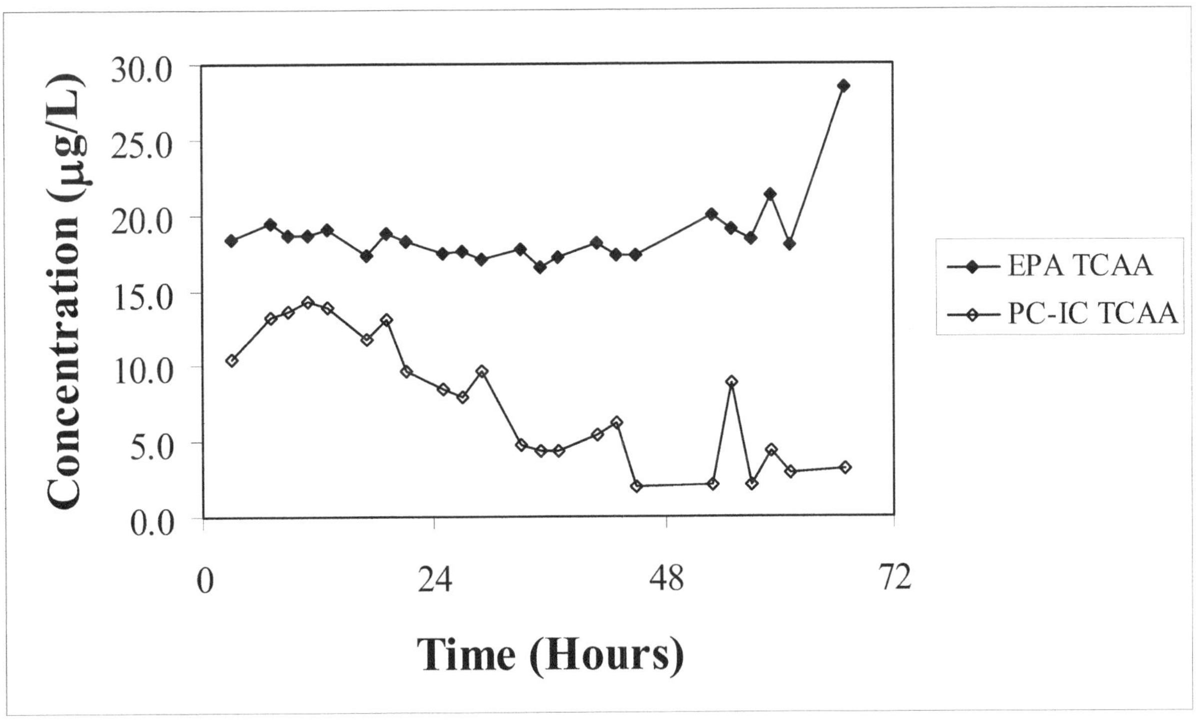

Figure 10.4 Houston EX concentration plot of TCAA for EPA 552.3 and PC-IC

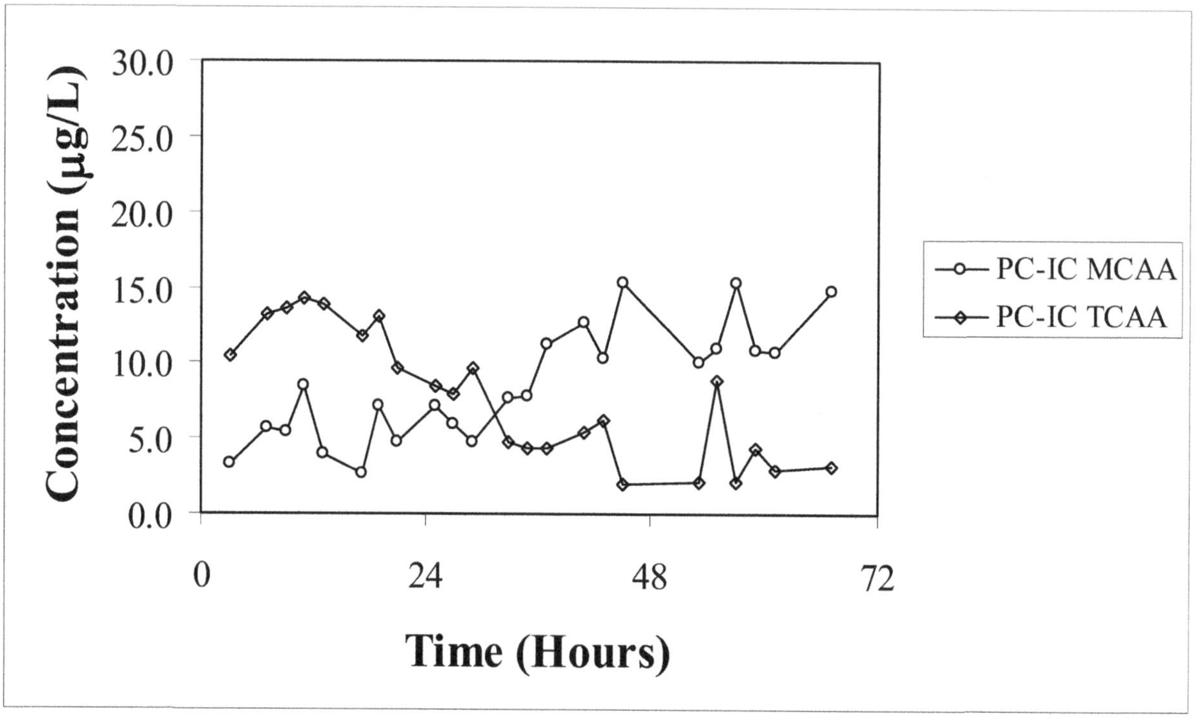

Figure 10.5 Houston EX concentration plot of MCAA and TCAA for PC-IC

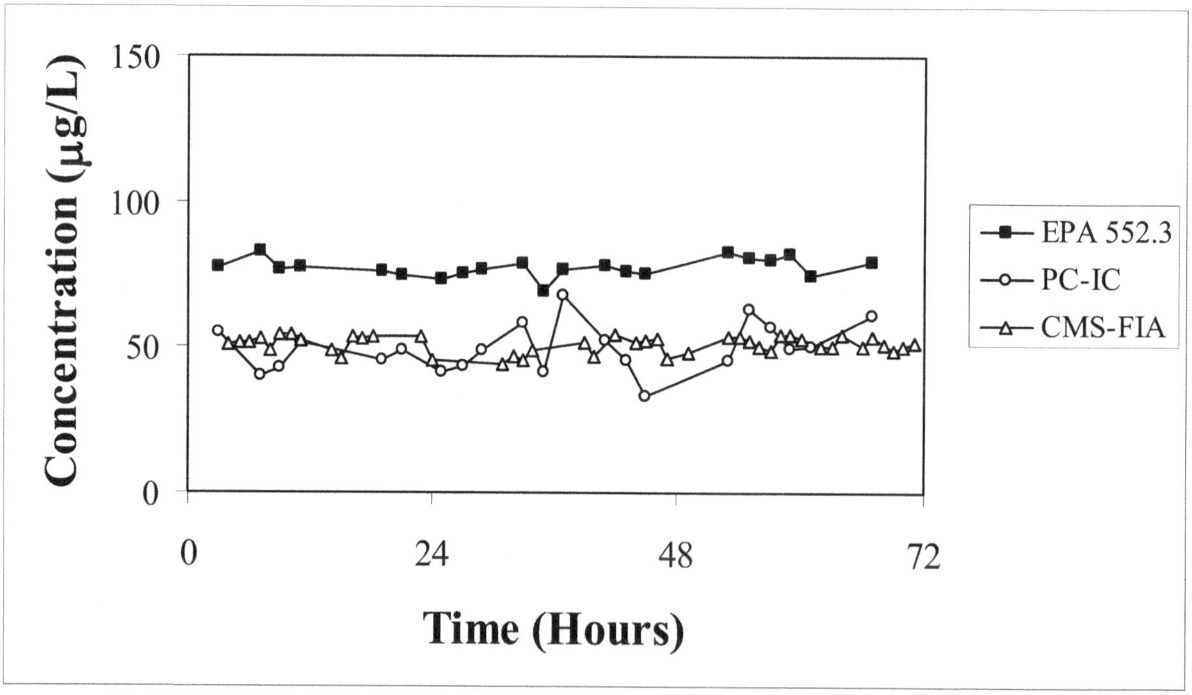

Figure 10.6 Houston EX concentration plot of THAA9 for EPA 552.3, PC-IC and CMS-FIA

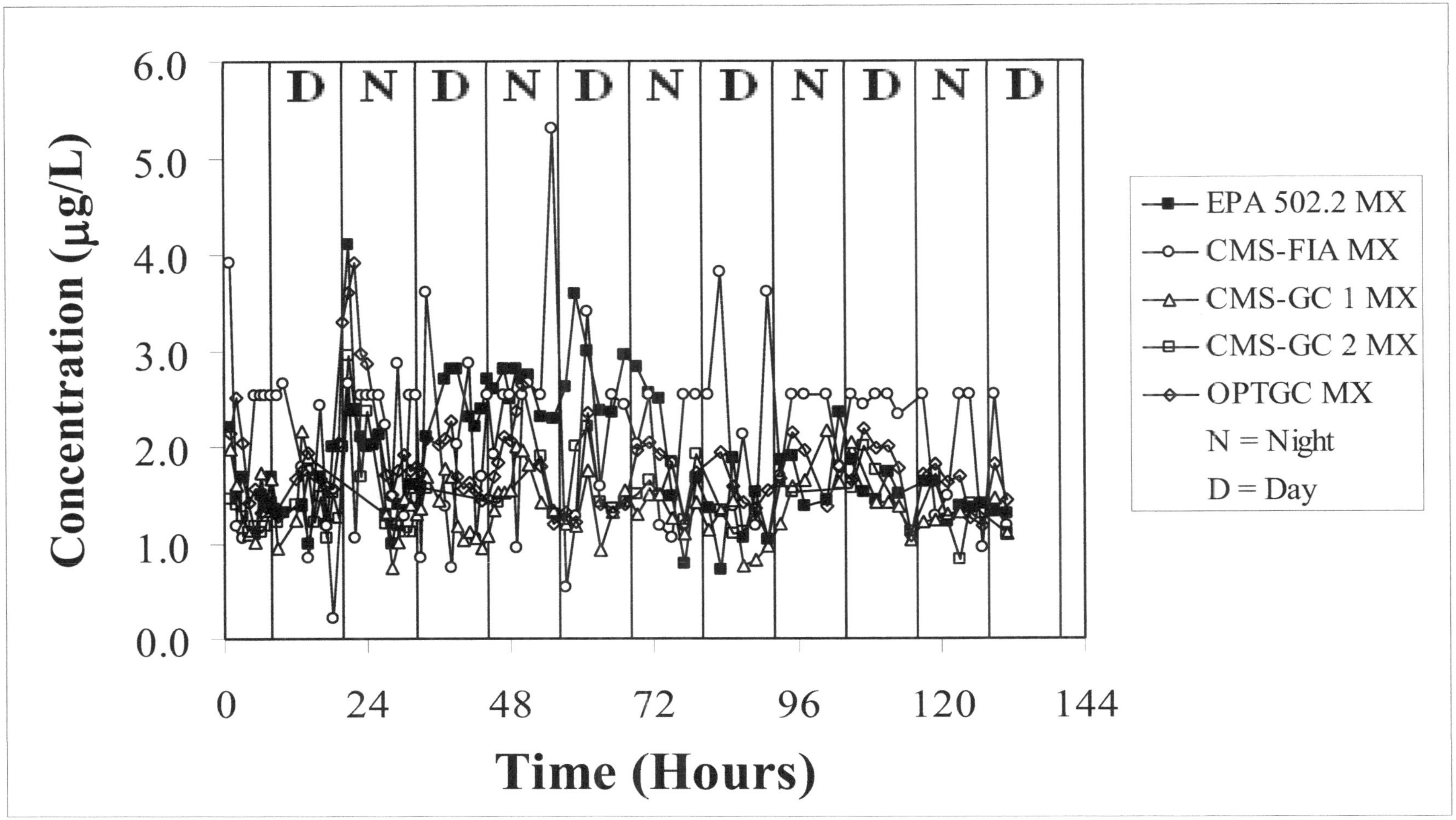

Figure 10.7 Combined plot of TTHM4 for Memphis EX monitoring data with labelled night and day periods

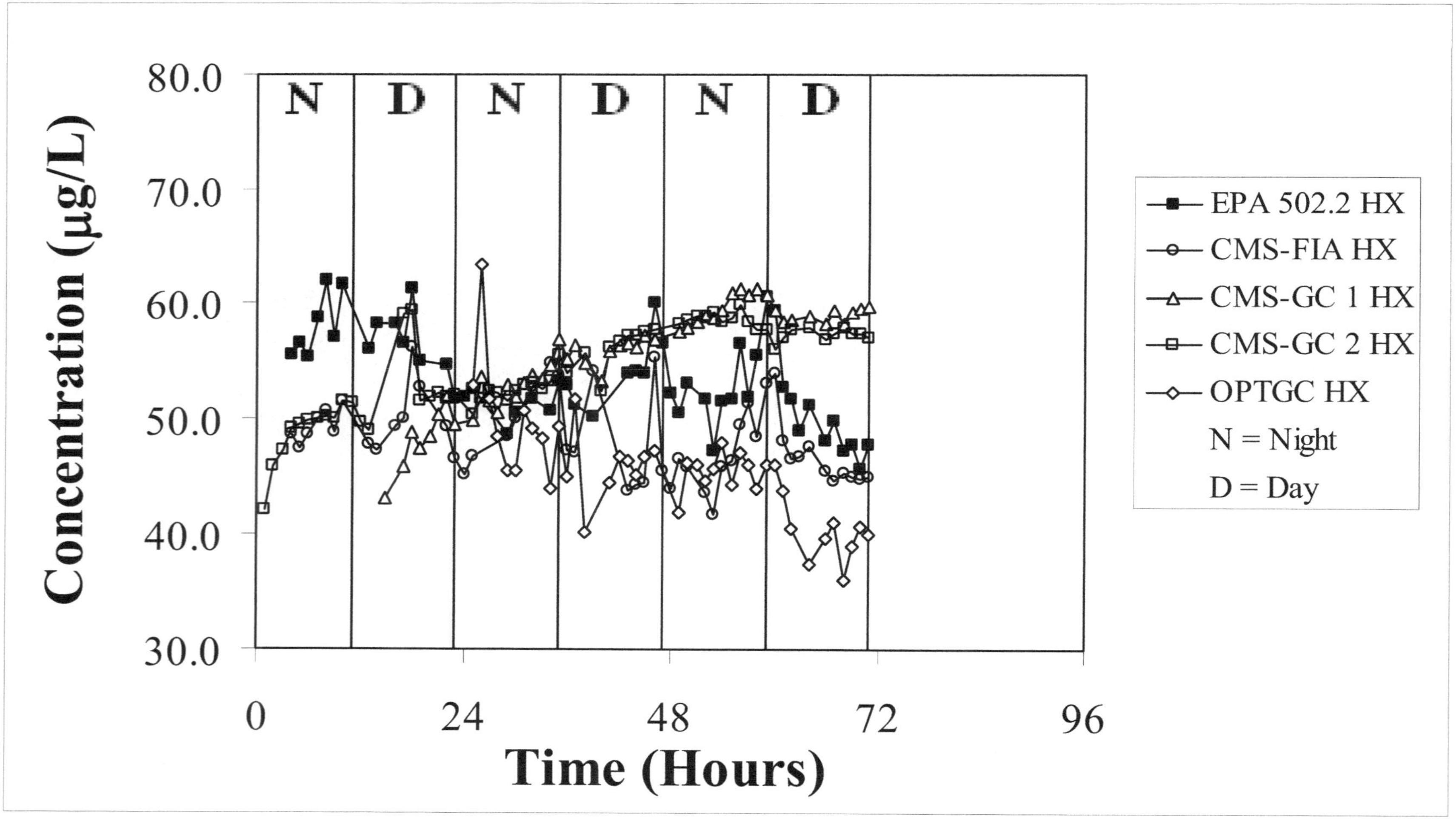

Figure 10.8 Combined plot of TTHM4 for Houston EX monitoring data with labelled night and day periods

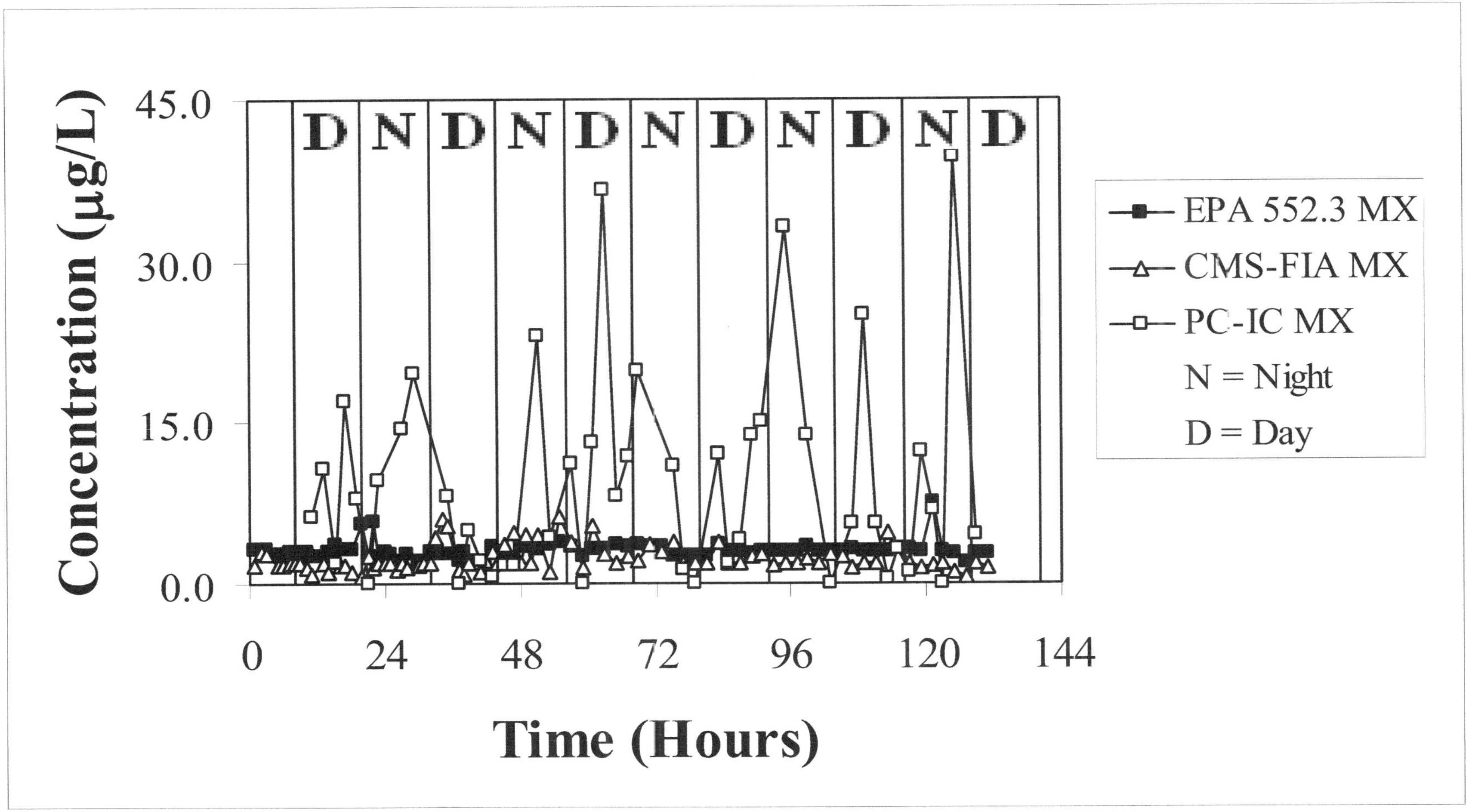

Figure 10.9 Combined plot of THAA9 for Memphis EX monitoring data with labelled night and day periods

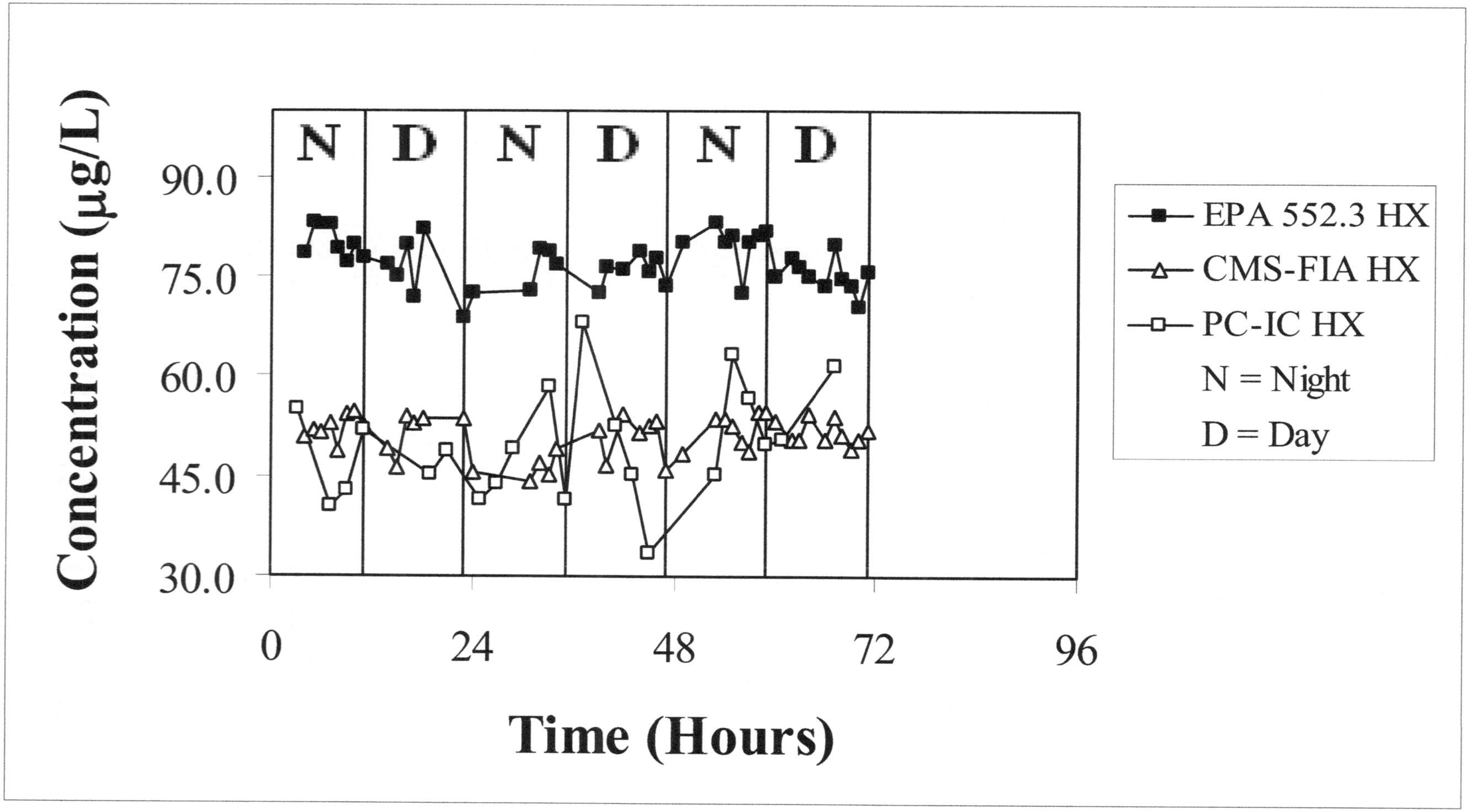

Figure 10.10 Combined plot of THAA9 for Houston EX monitoring data with labelled night and day periods

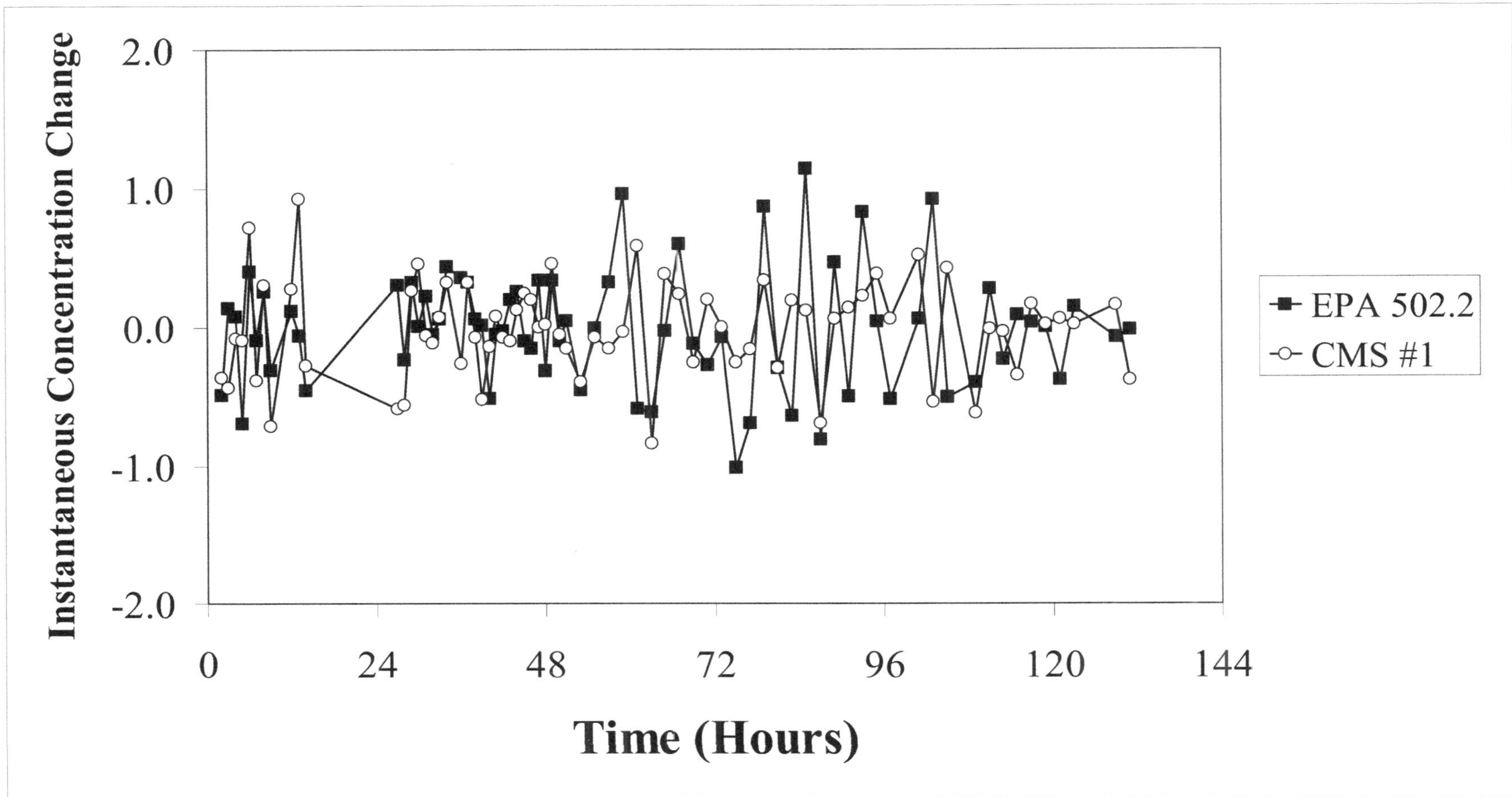

Figure 10.11 CMS-GC1 for TTHM4 (Memphis EX) ---62.0% of the time, the instantaneous concentration changes of the two methods agree

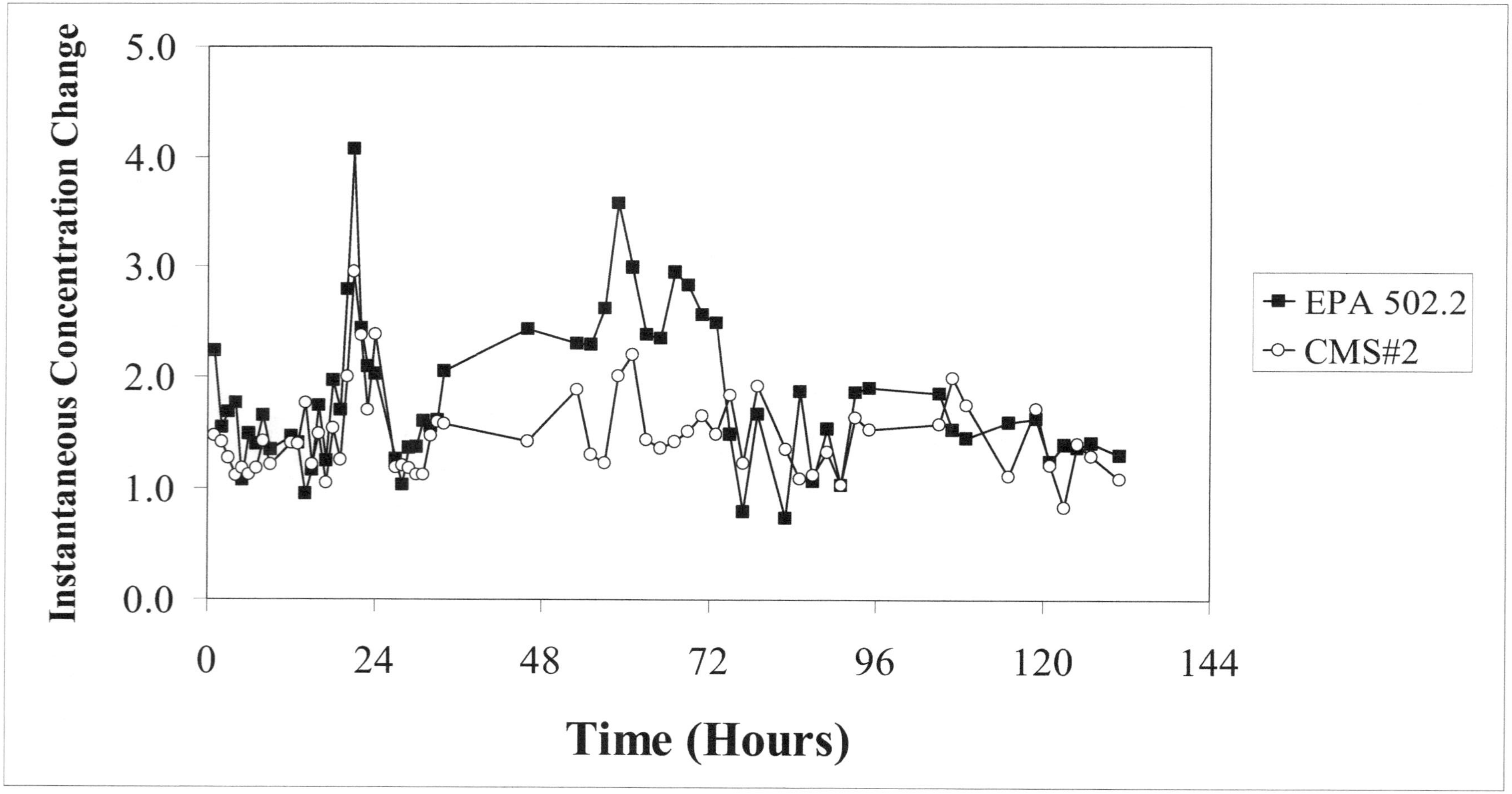

Figure 10.12 CMS-GC2 for TTHM4 (Memphis EX) ---54.1% of the time, the instantaneous concentration changes of the two methods agree

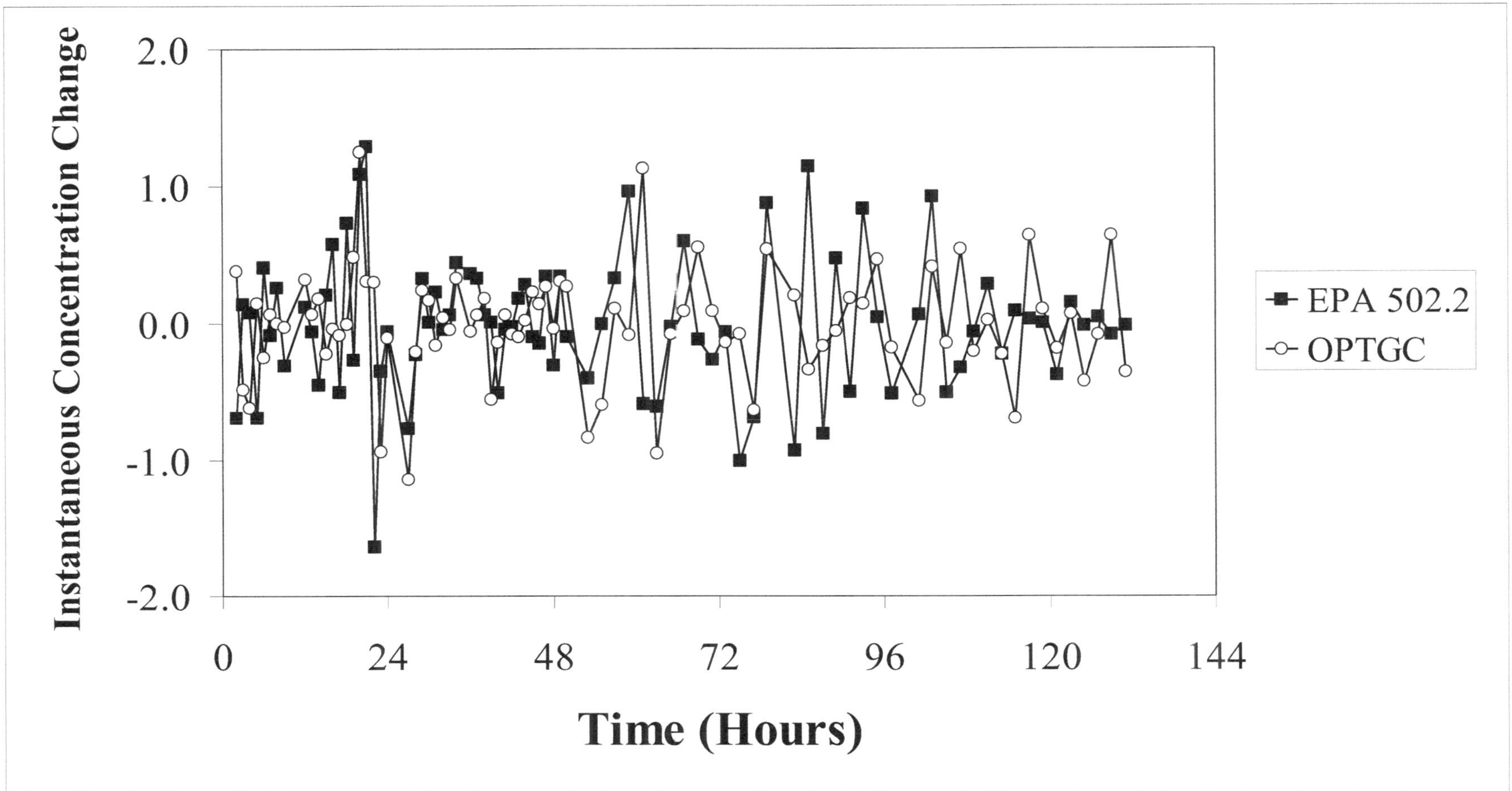

Figure 10.13 OPTGC for TTHM4 (Memphis EX) ---57.3% of the time, the instantaneous concentration changes of the two methods agree

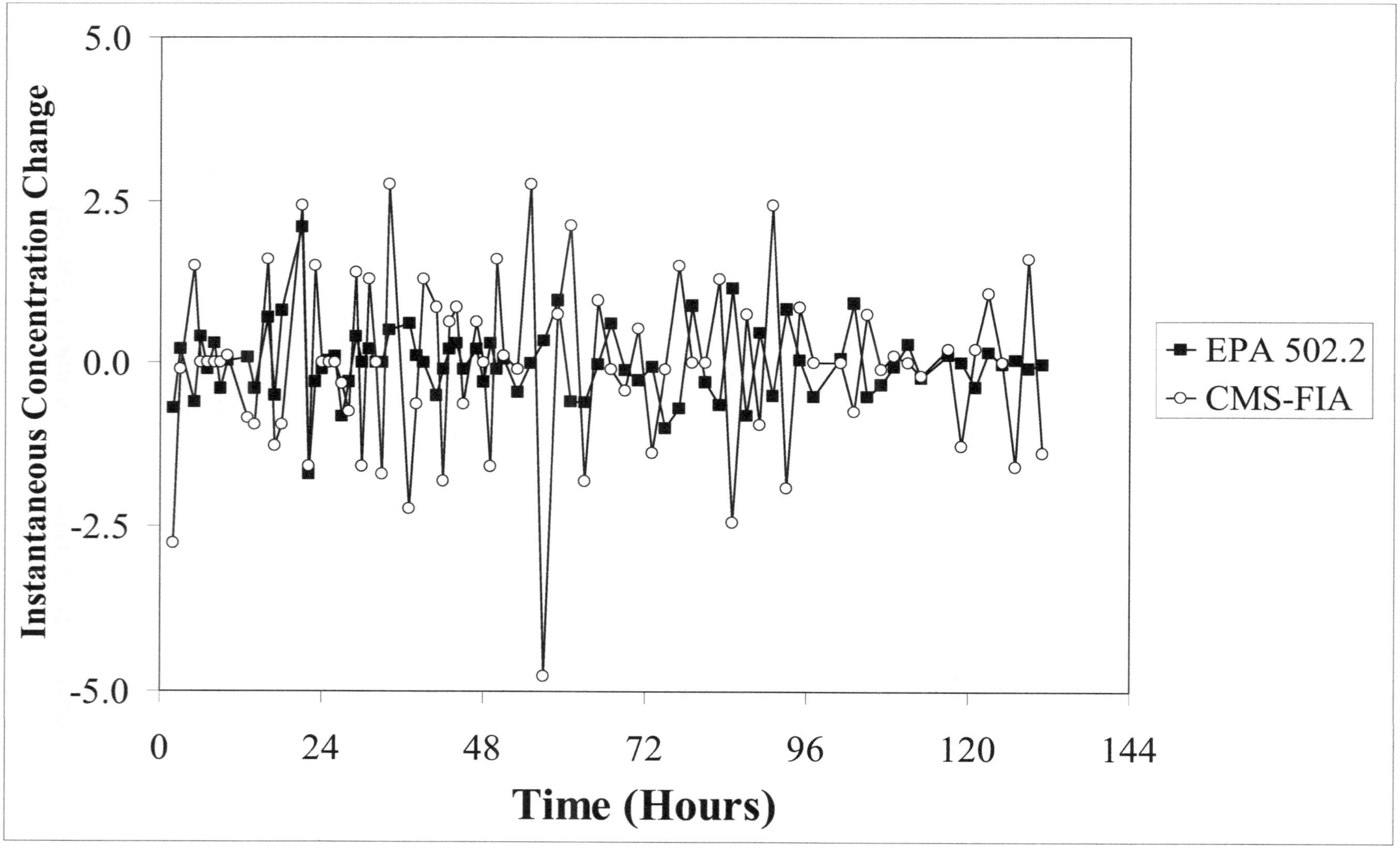

Figure 10.14 CMS-FIA for TTHM4 (Memphis EX)-- 62.8% of the time, the instantaneous concentration changes of the two methods agree

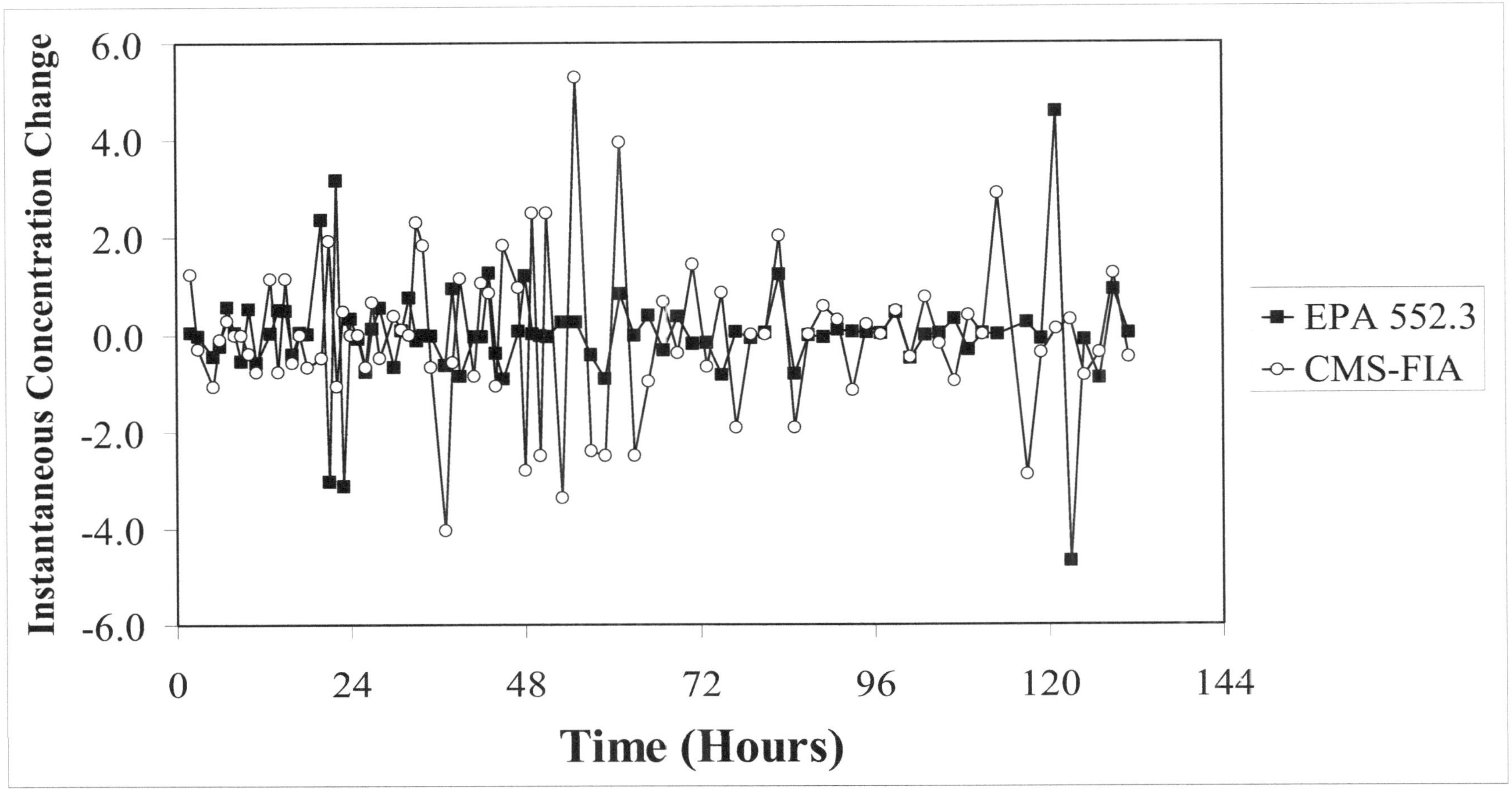

Figure 10.15 CMS-FIA for THAA9 (Memphis EX) – 58.5% of the time, the instantaneous concentration changes of the two methods agree

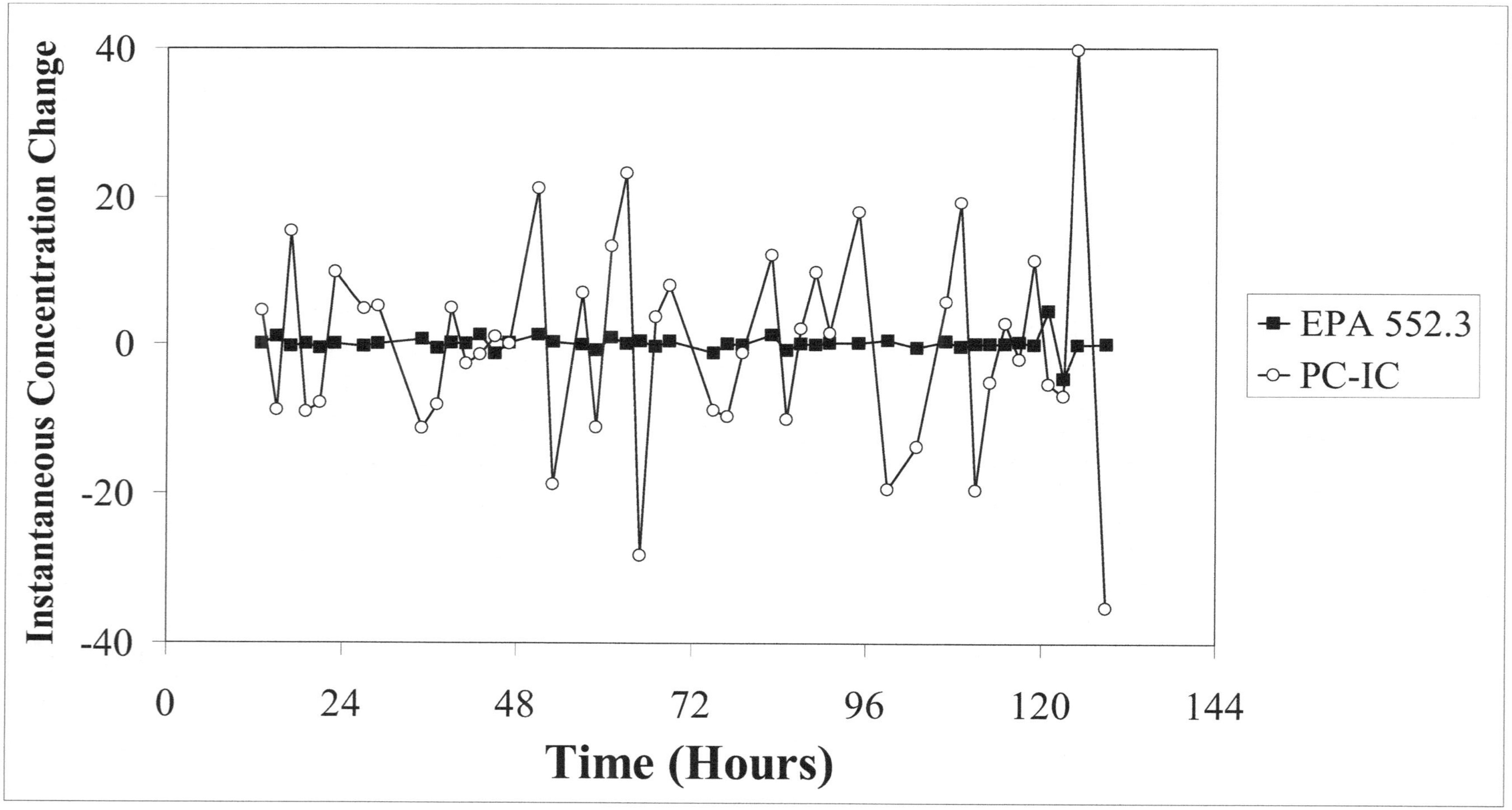

Figure 10.16 PC-IC for THAA9 (Memphis EX) – 47.8% of the time, the instantaneous concentration changes of the two methods agree

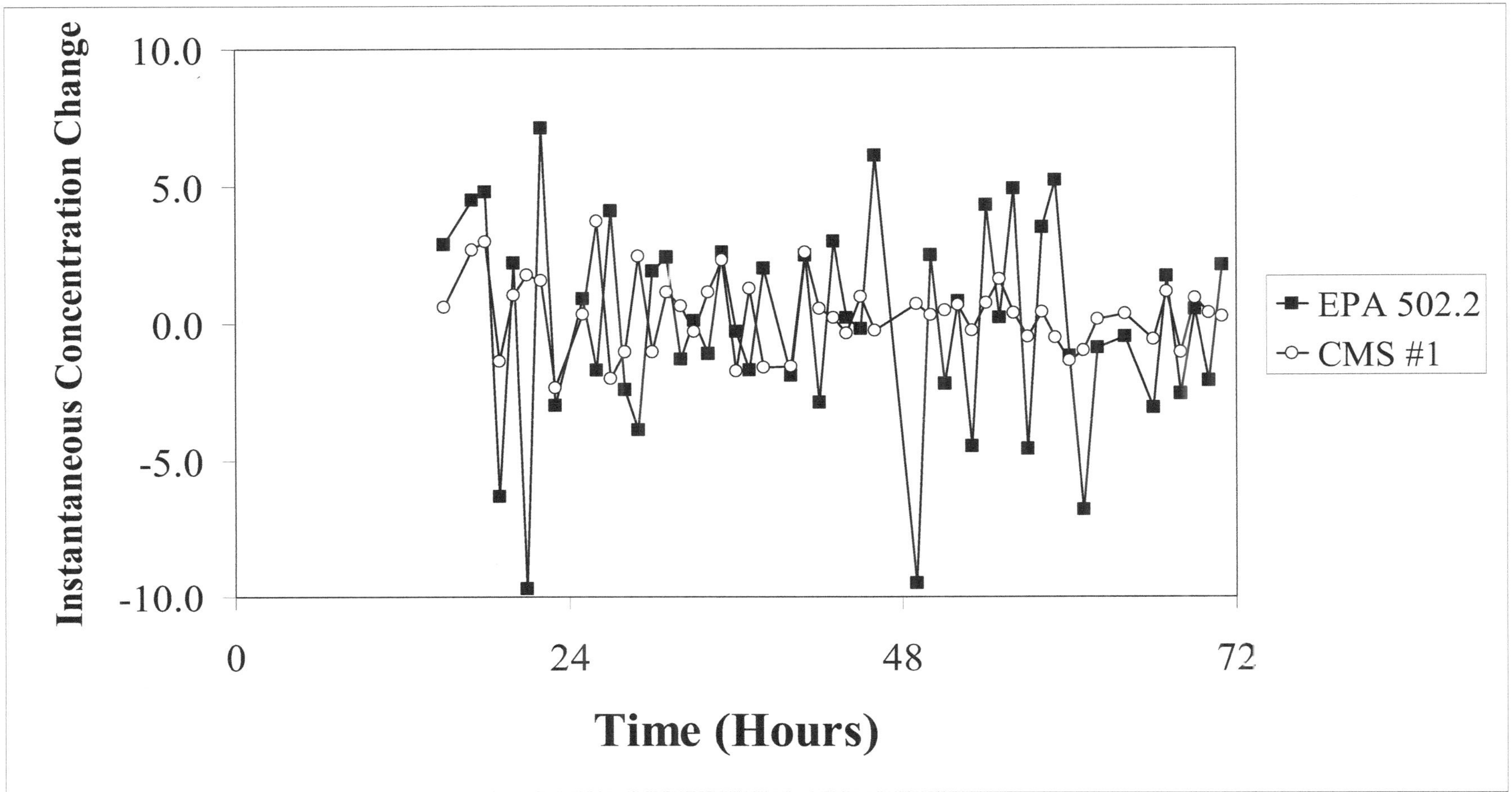

Figure 10.17 CMS-GC1 for TTHM4 (Houston EX) ---60.0% of the time, the instantaneous concentration changes of the two methods agree

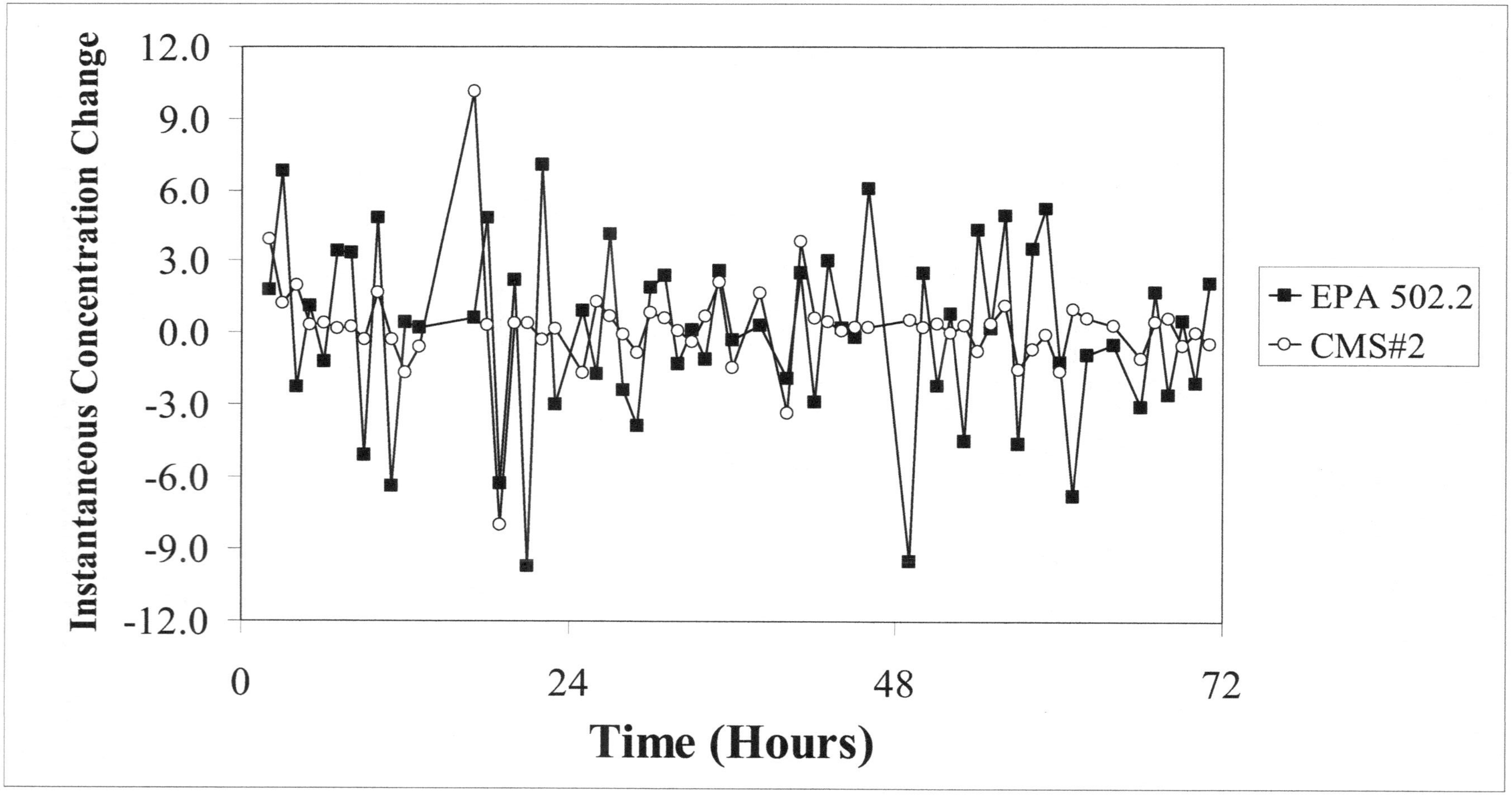

Figure 10.18 CMS-GC2 for TTHM4 (Houston EX) ---55.0% of the time, the instantaneous concentration changes of the two methods agree

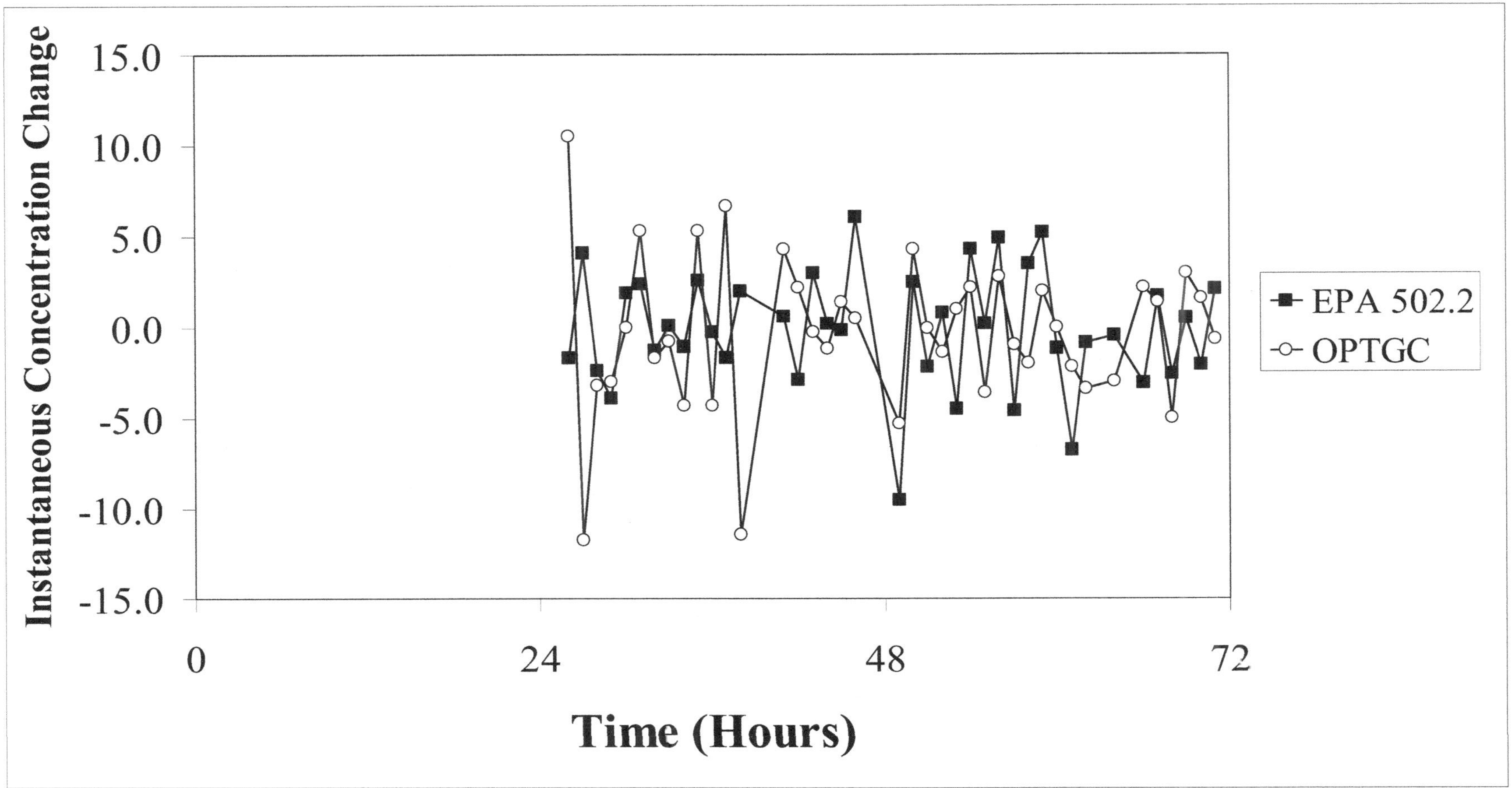

Figure 10.19 OPTGC for TTHM4 (Houston EX) ---60.0% of the time, the instantaneous concentration changes of the two methods agree

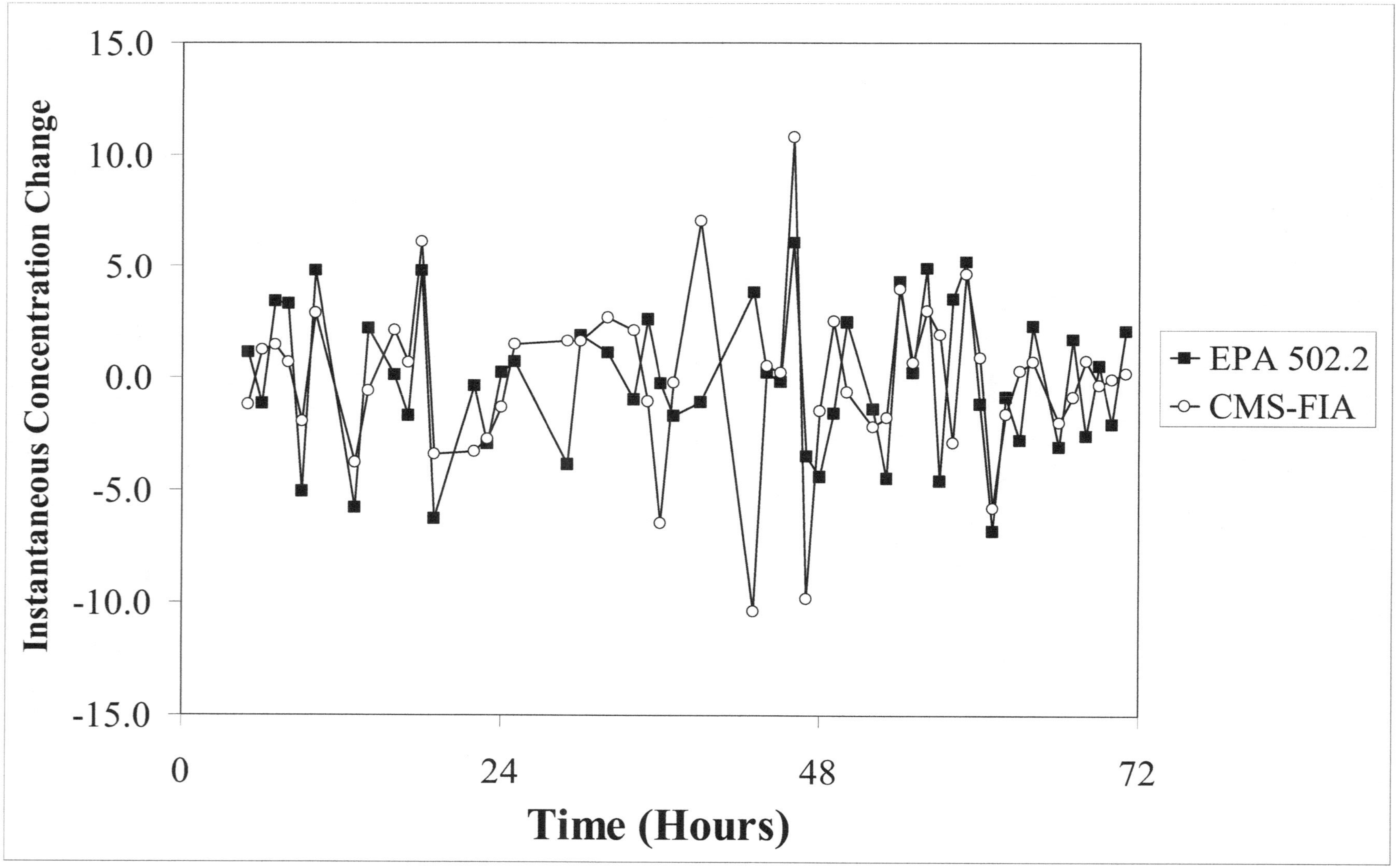

Figure 10.20 CMS-FIA for TTHM4 (Houston EX)-- 60.8% of the time, the instantaneous concentration changes of the two methods agree

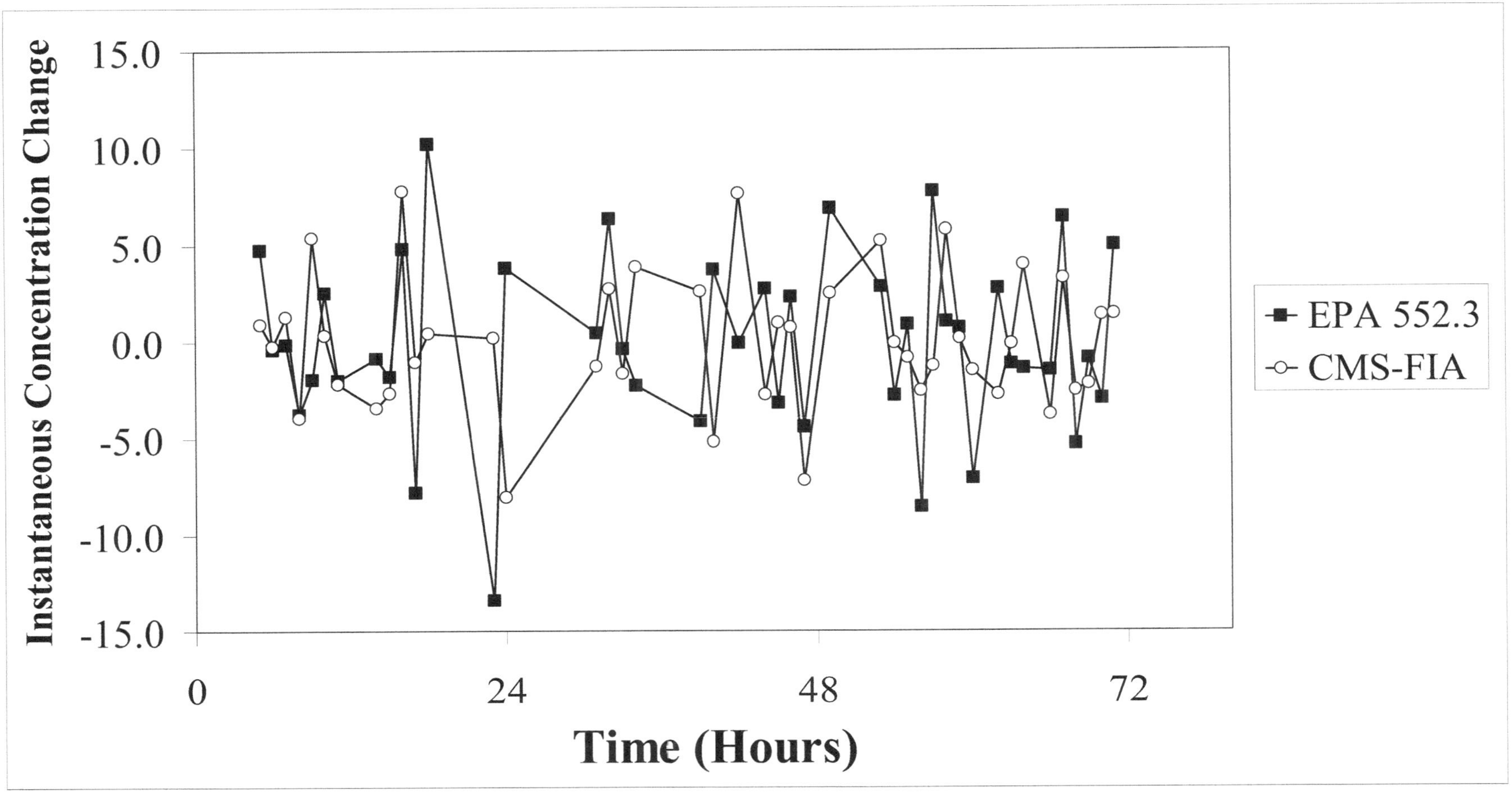

Figure 10.21 CMS-FIA for THAA9 (Houston EX) – 62.7 % of the time, the instantaneous concentration changes of the two methods agree

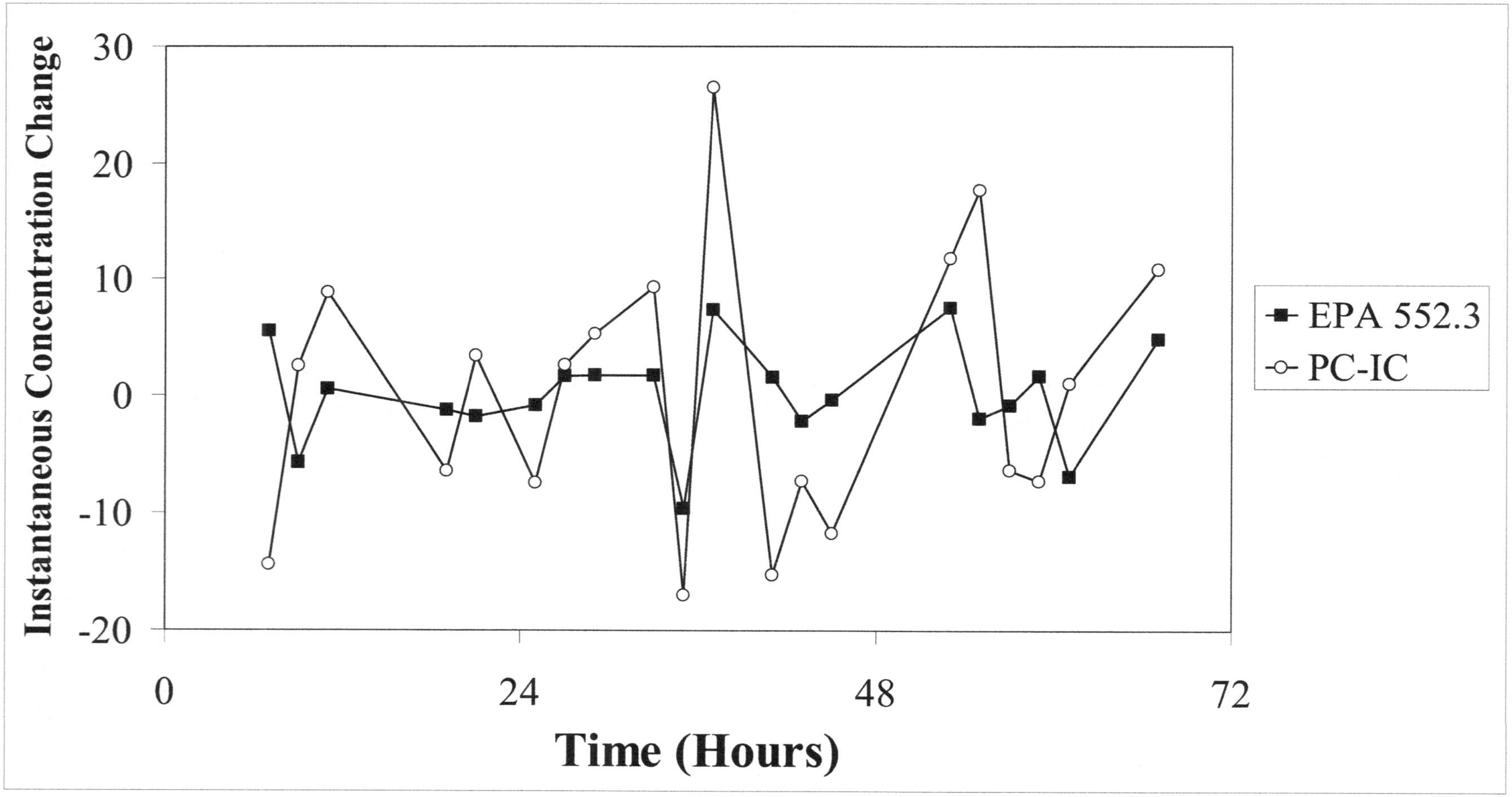

Figure 10.22 PC-IC for THAA9 (Houston EX) – 65.0% of the time, the instantaneous concentration changes of the two methods agree.

APPENDICES

INSTRUMENT CONSTRUCTION AND EXPERIMENTAL PROCEDURES FOR SELECTED CANDIDATE METHODS FOR THM4 AND HAA5 ANALYSIS IN DRINKING WATER DISTRIBUTION SYSTEMS

INSTRUMENT CONSTRUCTION

Appendix A. CMS-GC-ECD

Appendix B. CMS-FIA NCA-FL

Appendix C. PC-IC

EXPERIMENTAL PROCEDURES

Appendix D. CMS-GC-ECD

Appendix E. CMS-FIA NCA-FL

Appendix F. PC-IC

APPENDIX A
CONSTRUCTION OF CMS-GC-ECD INSTRUMENT

A.1 CMS-GC-ECD Component Lists

Note: All prices given in USD and reflect a minimum order

Table A.1
Capillary Membrane Sampler (CMS) component list

Component Description	Vendor	Amt. Used	U of Memphis	Vendor Price
1/8" Reaction Tee (1/16"to1/8")	Valco	2	$80.00	Comm.
Tefzel® Tubing (1/8")	Valco-Jour	120 cm	$5.50/m	Avail.
FEP Tubing (1/16")	Valco	~85 cm	$1.40/m	From
Silicone Tubing	Global FIA	124 cm	~$2.25/m	Global FIA
SS Hypodermic Tubing Segments	Global FIA	2	~$10	
		Total	**~$100**	**~$490**

Table A.2
Desiccator vessel component list

Component Description	Vendor	Amt. Used	U of Memphis	Vendor Price
500 mL Nalgene® Container	Fisher Sci.	1	$3.40	$3.40
Nafion tubing	Global FIA	18 cm	~$15.00/m	~$15.00/m
1/16" SS tubing	Valco	5 cm	$4.00/m	$4.00/m
1/16" Bulkhead fitting	Valco	2	$38.00	$38.00
Desiccant (Drierite)	Fisher Sci.	1	$43.02	$43.02
		Total	**~$104**	**~$104**

Table A.3
Injection valve system component list

Component Description	Vendor	Amt. Used	U of Memphis	Vendor Price
10-Port Injection Valve	Valco	1	$1,235.00	$1,235.00
Heated Valve Enclosure	Valco	1	$200.00	$210.00
1/16" SS tubing	Valco	~2.5 m	$12.00	$12.00
100 µL SS Sample Loop	Valco	2	$44.00	$44.00
Instrument Temp. Controller	Valco	1	$335.00	$350.00
Heating Tape	Fisher Sci.	1	$60.00	$60.00
Transformer	Fisher Sci.	1	$50.00	$70.00
		Total	**$1,937.00**	**$1,982.00**

Table A.4
Gas chromatograph component list

Component Description	Vendor	Amt. Used	U of Memphis	Vendor Price
Main Instrument (w/data board)	Buck Sci.	1	~$1,600	~$10,100
Analytical Column (VB-5)	Valco	1	~$220	~$220
		Total	**~$1,820**	**~$10,320**

Table A.5
Gas delivery system component list

Component Description	Vendor	Amt. Used	U of Memphis	Vendor Price
Brass Tee (1/8")	Agilent	1	$13.00	$13.00
1/8" Brass Fitting with ferrules	Agilent	6	$20.00	$20.00
CMS Gas flow controller	Brooks	1	$50.00	$50.00
1/8" Tefzel® Tubing	Valco Jour	~0.3 m	$5.50/m	$5.50/m
1/8" Copper Tubing	OhioValley	~1.2 m	$3.75/m	$3.75/m
1/16" SS tubing	Valco	~0.7 m	$4.00/m	$4.00/m
Helium Purifier	Valco	1	$750.00	$750.00
1/8" Ext / 1/16" Int Reducing Union	Valco	2	$28.00	$28.00
1/8" to 1/16" Int Reducing Union	Valco	1	$13.00	$13.00
1/4"Int/1/8"Ext Brass Union	Agilent	2	$7.00	$7.00
Gas Regulator	Fisher Sci.	2	$349.94	$349.94
Tank Anchor	Fisher Sci.	2	$85.14	$85.14
99.9995% Helium	Airgas	1	~$250	~$250
99.999% Nitrogen	Airgas	1	~$250	~$250
		Total	**~$2,053**	**~$2,053**

Table A.6
Software component list

Component Description	Vendor	Amt. Used	U of Memphis	Vendor Price
PeakSimple	SRI	1	free	free

Table A.7
Computer component list

Component Description	Vendor	Amt. Used	U of Memphis	Vendor Price
Laptop Computer	Dell (refurb.)	1	~$450	~$900
		Total	**~ $450**	**~$900**

Table A.8
On-line monitoring adaptors component list

Component Description	Vendor	Amt. Used	U of Memphis	Vendor Price
Tubing Clamp	IDEAL	2	\$7.00	\$7.00
Tubing Clamp	IDEAL	1	\$3.50	\$3.50
Tygon Tubing o.d. 2.5cm, i.d. 1.9cm	Fisher Sci.	8 cm	\$17.69	\$17.69
Tygon Tubing o.d. 1.8cm, i.d. 1.3cm	VWR	4 cm	~\$19	~\$19
Tygon Tubing o.d. 1.1cm, i.d. 0.5cm	VWR	135 cm	~\$35	~\$35
Liquid Flow Restrictor	Herbach&R..	1	~\$15	~\$15
PEEK Tee	Upchurch	1	\$28.00	\$28.00
PEEK 1/16" Int Union	Valco	1-3	\$45.00	\$45.00
		Total	**~\$170**	**~\$170**

Table A.9
Chemicals list

Component Description	Vendor	Amt. Used	U of Memphis	Vendor Price
$CHCl_3$ Std. (≥99.8%, 500mL)	Aldrich	1 mL	\$20.20	\$20.20
$CHBrCl_2$ Std. (≥98%, 50 g)	Aldrich	1 mL	\$56.20	\$56.20
$CHBr_2Cl$ Std. (98%, 10 g)	Aldrich	1 mL	\$35.30	\$35.30
$CHBr_3$ Std. (≥99%, 50 g)	Aldrich	1 mL	\$29.10	\$29.10
Methanol (HPLC Grade, 4L)	Fisher Sci.	40 mL	\$63.70	\$63.70
		Total	**\$204.50**	**\$204.50**

Table A.10
Additional supplies list

Component Description	Vendor	Amt. Used	U of Memphis	Vendor Price
Peristaltic Pump	AUTOCLUDE	1	~\$300	~\$300
Surge Protector (8-outlet)	Radio Shack	1	\$24.00	\$24.00
Digital Flowmeter	Agilent	1	\$704.00	\$704.00
SS tubing cutters	Valco	1	\$12.95	\$12.95
Thermometer	Fisher Sci.	1	\$9.11	\$9.11
10 uL Gas tight Syringes	Fisher Sci.	6	\$110.83	\$110.83
50 ul Gas tight Syringe	Fisher Sci.	1	\$28.81	\$28.81
10 mL Volumetric Flasks	Fisher Sci.	2	\$57.94	\$57.94
100 mL Volumetric Flasks	Fisher Sci.	13	\$321.75	\$321.75
5 mL Amber Vials	Supelco	10	\$77.40	\$77.40
		Total	**~\$1,647**	**~\$1,647**
	GRAND TOTAL		**~\$8,485**	**~\$17,870**

A.2 Capillary Membrane Sampler (CMS)

Note: Wear Latex gloves and work on a lint free surface to reduce contamination of silicone.

A.2.1 A length of 1/8" Tefzel® tubing was cut to a desired length (ex. 120 cm) to serve as the body for the CMS device. A piece of silicone rubber tubing (Global FIA) was cut, so that it was 3 to 4 cm longer than the length of Tefzel® tubing from the previous section (~124 cm). This extra length allows the ends of the silicone tubing to extend through the entire reaction tee for attachment.

A.2.2 The length of silicone tubing was then inserted through the entire length of 1/8" Tefzel® by using a thin electrical wire (ex. copper) to pull the silicone tubing through. Approximately 1.5 cm of the silicone tubing was left exposed at each end of the Tefzel® tubing.

A.2.3 Each end of the silicone tubing was then attached to a reaction tee using the following procedure. First, a SS hypodermic tubing segment (SS-TUB) was inserted halfway into the end of the silicone tubing. This requires some patience as the tubing i.d. was quite small. It is recommended that no gloves be worn during this step, also wetting the SS segment with reagent water helps with insertion.

A.2.4 After the SS hypodermic segment was inserted into the silicone tubing, a 1/8" SS fitting with ferrule was placed over the silicone tubing and the 1/8" Tefzel® tubing. The silicone tubing end was inserted into a reaction tee (TCEF211T) at the 1/8" size end and out of the 1/16" end (Figure A.1).

A.2.5 1/16" FEP tubing was cut into four segments (two, 7.5 cm segments and two, 35 cm segments). A 1/16" SS fitting and ferrule was placed on a 35 cm segment of 1/16" FEP tubing. The piece of SS hypodermic tubing exposed from the silicone tubing was then inserted into the end of the 1/16" FEP tubing. The 1/16" SS fitting with ferrule were then clamped to the FEP/silicone tubing assembly. This completes the attachment of one end of the silicone tubing to the gas inlet/outlet of the reaction tee. Repeat for the other end of the silicone tubing. Note it will require that the silicone tubing be pulled and cut to appropriate length once through the other reaction tee. It was important not to pull too hard as the membrane may tear. Once through the tee hold on to the silicone tubing until FEP tubing is attached (Figure A.1).

A.2.6 Once the silicone tubing is secured, the two 1/8" SS fittings with ferrules were tightened, thus clamping to the 1/8" Tefzel® tubing. Finally, the 7.5 cm lengths of 1/16" FEP tubing were combined with a 1/16" SS fitting and ferrule and installed to the sample inlet/outlet locations on the reaction tees (Figure A.2).

A.2.7 The CMS was attached to a flow controller (Brooks) at the Carrier IN (Figure A.8) by using an internal reducing union (1/8" to 1/16"). To test for liquid leaks place the device on a dry paper towel (preferably brown) and watch for the formation of wet areas when pumping reagent water (18.2 MΩcm) through using a peristaltic pump. If there is a liquid leak the fittings may be too loose or too tight. To test the device for gas leaks, submerge the device underwater and watch for air bubbles and monitor the gas flow rate using a digital flowmeter. A gas leak is either the result of a torn or pinched membrane, or there might be a loose connection at the membrane/fitting connection. After the leak checks, several mL of reagent water was pumped through the device. The CMS was then connected to the Desicator Vessel (Figure A.4) by using a 1/16" SS fitting and ferrule, which was clamped onto the 1/16" FEP tubing at the Carrier Out side of the CMS.

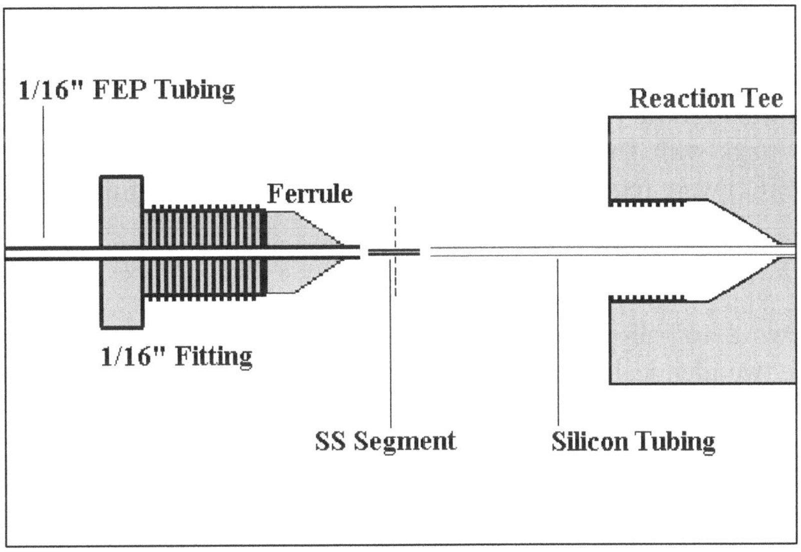

Figure A.1 Diagram of the silicone tubing membrane attachment

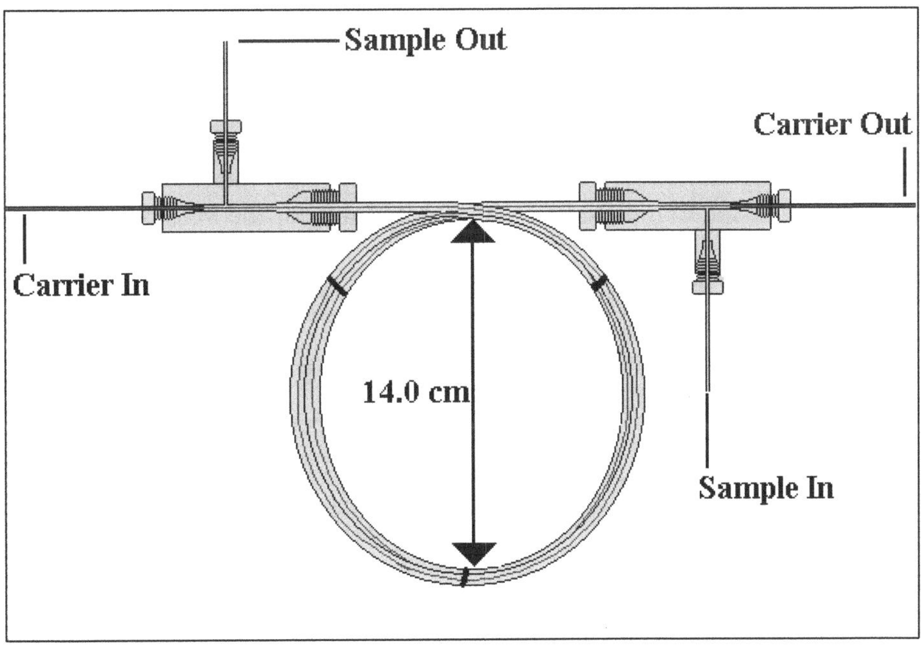

Figure A.2 Diagram of completed the CMS device

A.3 Desiccator Vessel
Note: Wear protective eyewear, gloves, and have adequate ventilation during construction.

A.3.1 A 500 mL Nalgene container serves as the body for this apparatus. First the cap was removed and the bottom of the container was prepared for drilling two holes in it by tapeing around the edge to prevent possible cracking (just as a precautionary step). Two marks are placed centered on the bottom of the container ~ 2.5 cm apart and ~ 2.0 cm from the edge (w/diameters ~0.8 mm). A Dremel® tool with appropriate fittings was used to drill holes.

A.3.2 Nafion tubing (NT-060, Global FIA), two SS tubing segments (L ~2.5 cm), four 1/16" internal fittings with ferrules, one 1/16" external fitting with ferrule, and two bulkhead unions (ZBU1, VICI) were required to construct the vessel using the following steps.

A.3.3 An 18 cm length of Nafion was cut using a razor blade.

A.3.4 Two 1/16" SS fittings, ferrules, and two SS segments (length ~2.5 cm) were combined to give two identical pieces.

A.3.5 Each piece was screwed into a SS bulkhead union's hex end. Each end of the Nafion tubing was slipped onto one of the SS tubing segments sticking from a bulkhead union.

A.3.6 The assembly was lowered into the top of the container and the two ends were placed through the two holes at the bottom of the container. SS nuts were used to fasten the bulkhead unions to the container. The sample transfer line from the valve was directed through a factory made hole and to one of the bulkhead unions on the vessel by using a 1/16' SS internal fitting with ferrule (Figure A.1).

A.3.7 The other union was connected to the CMS outlet by using a 7 cm SS tubing segment, a 1/16" SS internal fitting with ferrule, and a 1/16" SS internal union. Figure A.4 shows the main components of the completed vessel. Once finished, the vessel was slowly filled with desiccant (Drierite®).

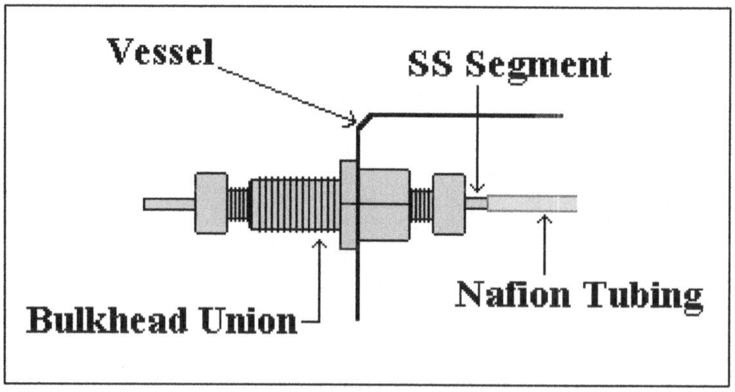

Figure A.3 Diagram of the desiccator vessel union

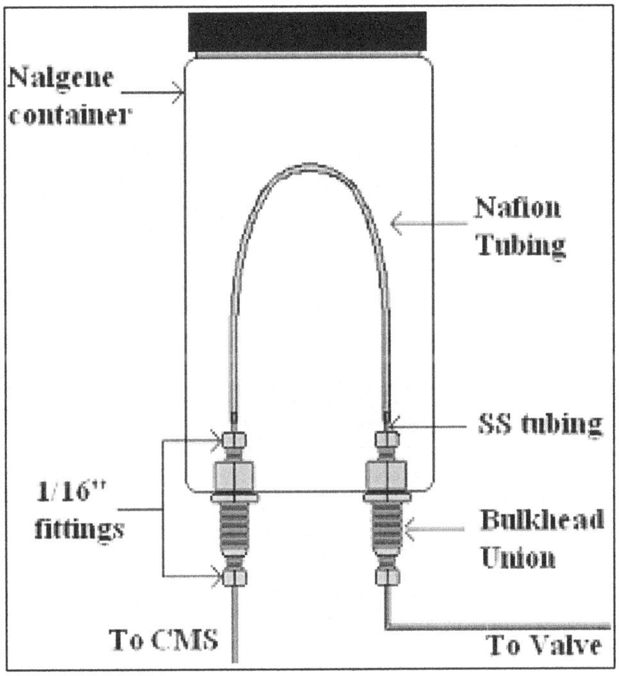

Figure A.4 Diagram of the completed desiccator vessel

A.4 Injection Valve System

A.4.1 A ten-port electrostatically actuated injection valve (Valco) was mounted on the heated valve enclosure (Valco) by removing the valve from the actuator and reattaching it on the other side. The valve clamp on the top of the enclosure was tightened in order stabilize the valve. The instrument temperature controller (ITC), heater block and thermocouple will be discussed in section A.4.4 (See Figure A.5 for finished system).

A.4.2 Two 100 μL SS sample loops (SL100UW, VICI) were attached to ports 7, 10 and 2, 5. A jumper loop was connected to ports 4, 9 to allow for continuous loading of sample onto one loop at a time with alternating injections (Figure A.6).

A.4.3 Three lengths (~40 to 50 cm) of SS tubing (1.6 mm o.d., 0.75 mm i.d., 625-30, Valco) were used to make a sample in, a sample out, and a GC carrier gas transfer line. These lines were then connected to the specified valve ports as depicted in Figure A.6.

A.4.4 The heater block for the valve enclosure was attached to the valve head as described in the included instrument manual. The instrument temperature controller (ITC) has a plug-in for the heater block and for a thermocouple. The thermocouple was carefully inserted into the enclosure before closing it. All operation procedures for the ITC (ITC10399, Valco), controls the valve heater, were followed directly from the manual. The ITC was set to 65 °C.

A.4.5 The two-position actuator control module for the valve was connected to TTL output "A" on the data acquisition board inside the GC. The first wire was inserted into ground and the fifth wire was inserted into TTL output "A". The valve was tested by going to the relay window found in PeakSimple software and pressing the (A) button. The GC has to be on for this test to work.

A.4.6 Next the sample transfer lines are wrapped with the heating tape (11-463-50C, Fisher) and then with two layers of aluminum foil. The tape was then connected to the transformer (2PF1010, Fisher). The transformer was adjusted until a thermometer measured a temperature of ~ 65°C underneath the aluminum foil (setting 27). The procedure for the connection of the GC carrier gas to the valve will be described in section A.6.

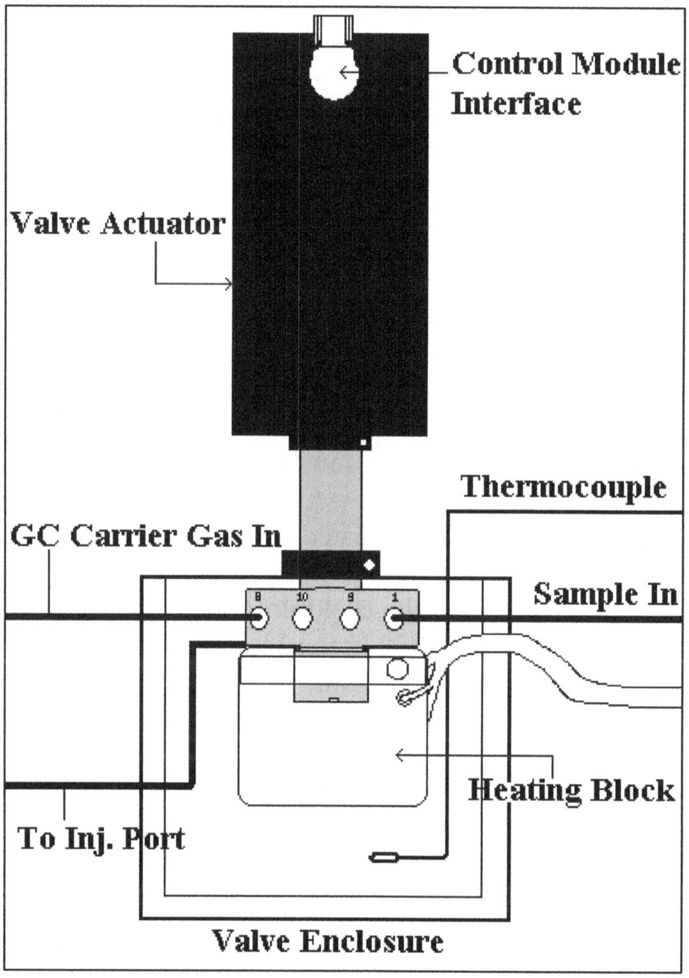

Figure A.5 Diagram of the assembled ten-port injection valve in the heated valve enclosure

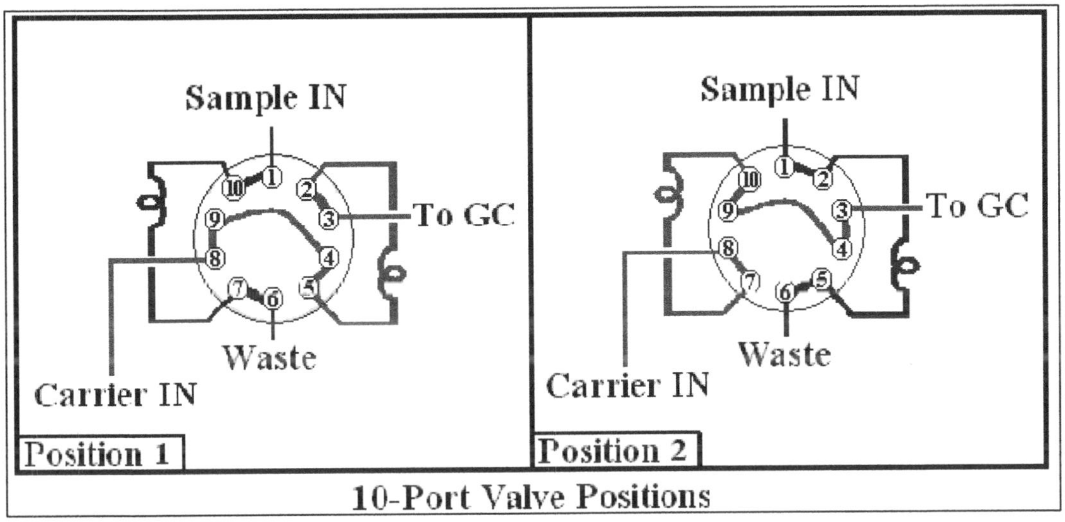

Figure A.6 Diagram of the ten-port injection valve port connections

A.5 Gas Chromatograph

A.5.1 The Buck Scientific 610 GC has its own data acquisition board used to automate the valve assembly, control the temperature of the oven, and to receive the signal from the ECD. To make the appropriate connections, the GC must be turned off and unplugged. Open the bottom of the GC. Touch bare metal to eliminate any residual static electricity. A typical precision flathead screwdriver was needed to loosen and tighten the board's terminals.

A.5.2 The first connection made was with the board and the ECD cable to the plus (white wire) and minus (black wire) terminals and ground (orange wire) in the channel 1 slot on the board. Next the temperature programming wires were connected to the board (white wire- to TP1 terminal, black wire - to ground terminal).

A.5.3 A connection was made between the control module for the injection valve and the TTL output "A" on the data board. Ignore this step if already completed. Only the first (red is one) and fifth wires were used. Wire one went to ground and the fifth wire went to TTL output "A". The wires were inserted into a small hole through a small opening on the GC to allow the bottom to close completely (Figure A.7).

A.5.4 Finally the serial out (RS-232) from the data acquisition board was connected to a computer by using a male/female serial cable. Com Port 1 on the PC was used to communicate. To test the GC, the bottom was closed; the GC was plugged in, and turned on. The temperature set-point for the ECD was set to ambient and the set-point for the oven was set to zero. No gases were needed for this procedure since the ECD and oven were not heated.

A.5.5 With the GC initiated the PeakSimple software was loaded from the interfaced PC. There should be a connection sequence between the data acquisition board and PC that will verify a successful initialization.

A.5.6 First the valve automation was tested by opening the relay/pump window through the View menu. The valve control module needs to be powered up and connected to the valve. By pressing the (A) button, the valve control module should switch positions if working.

A.5.7 Second the temperature control was checked by going to the **Channel 1, 2, 3, 4** box. The temperature box of Channel 1 was pressed. By changing the initial temperature to 30°C, the reading on the GC Oven Setpoint LCD should change instantly to this temperature if working properly. The Column Oven Heater light on the GC will come on. To prevent further heating the initial temperature was set back to zero.

A.5.8 To test the ECD signal output to the data acquisition board, the attenuation, standing current and zero knobs for the ECD were turned to observe different affects on the signal. Only with the ECD heated and fully operating will the signal be assessed. The installation of the capillary column will be discussed in the section A.6.5.

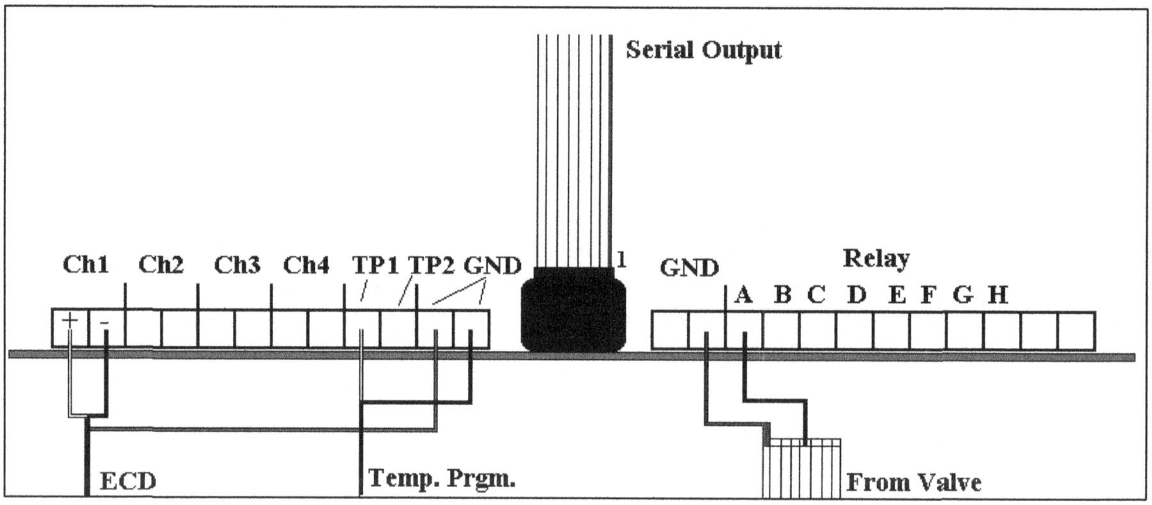

Figure A.7 Diagram of the PeakSimple data acquisition board connections

A.6 Gas Delivery System

Note: Proper handling of gas tanks including anchors and regulators were followed.

A.6.1 The He and N_2 tanks were installed next to the CMS-GC-ECD using the appropriate anchors. Regulators (Fisher Scientific) were installed and purged of all air according to regulator manuals (Teflon tape was required). The pressure was set to 108 psi for the N_2 tank and 52 psi for the He tank regulators. The regulators were each fitted with a 1/4" to 1/8" internal/external brass union to attach gas transfer lines.

A.6.2 A He purifier (HP2) was interfaced to the regulator by use of a 1/16" SS internal/external reducing union. The purifier was initialized and operated according to the instrument manual. The 1/16" SS outlet fitting from the purifier was connected to the 1/8" SS carrier gas inlet on the GC by use of another 1/16" to 1/8" SS internal/external reducing union and a ~ 25 cm length of 1/8" copper tubing coupled to two 1/8" brass fittings with ferrules (Figure A.8).

A.6.3 A ~ 1.2 m length of 1/8" copper tubing was connected to the N_2 regulator using a 1/8" brass fitting with ferrule. This tubing was directed to a brass tee equipped with three 1/8" brass ferrules/ferrule assemblies. The tee was needed in order to provide N_2 for both the ECD and the CMS device. A ~ 25 cm length of 1/8" copper tubing was used to connect the tee to the 1/8" ECD make-up gas inlet on the GC. An additional 1/8" brass fitting and ferrule was needed. To connect the tee to the CMS carrier gas flow controller a length of 1/8" Tefzel® tubing was required. This tubing allows for more movement than copper tubing if the CMS configuration should be moved. A shorter length of 1/8" Tefzel® tubing was used to connect the flow controller (Brooks, 8744) to the 1/8" to 1/16" SS internal reducing union. The union was equipped with a 1/8" SS and 1/16" SS fitting/ferrule assemblies. The end of the CMS carrier gas inlet (1/16" FEP tubing) was inserted into this union.

A.6.4 The outlet (1/16" FEP tubing) of the CMS was connected to the desiccator vessel using the 1/16" SS fitting/ferrule assembly. Next a length of 1/16" SS was used to connect the injection valve enclosure mentioned in section A.4 with the GC carrier gas outlet on the GC. No heating tape was required as this is just carrier gas.

A.6.5 The analytical capillary column (Valco, VB-5) was connected to the GC injection port by using the supplied 1/8" SS fitting with graphite ferrule and inlet liner ~2.5 cm. The opposite end of the column was connected to the ECD inlet by using the supplied 1/8" SS fitting and ferrule. The column was conditioned for at least 2 hrs at 230°C with 5.0 mL/min of helium before initial use.

A.6.6 The heated sample transfer line exiting from the injection valve assembly was adapted to GC injection port by using two 1/8" SS fittings and ferrules, a ~8 cm length of 1/8" SS tubing, and a 1/16" to 1/8" internal/external reducing union (Figure A.9) also heated.

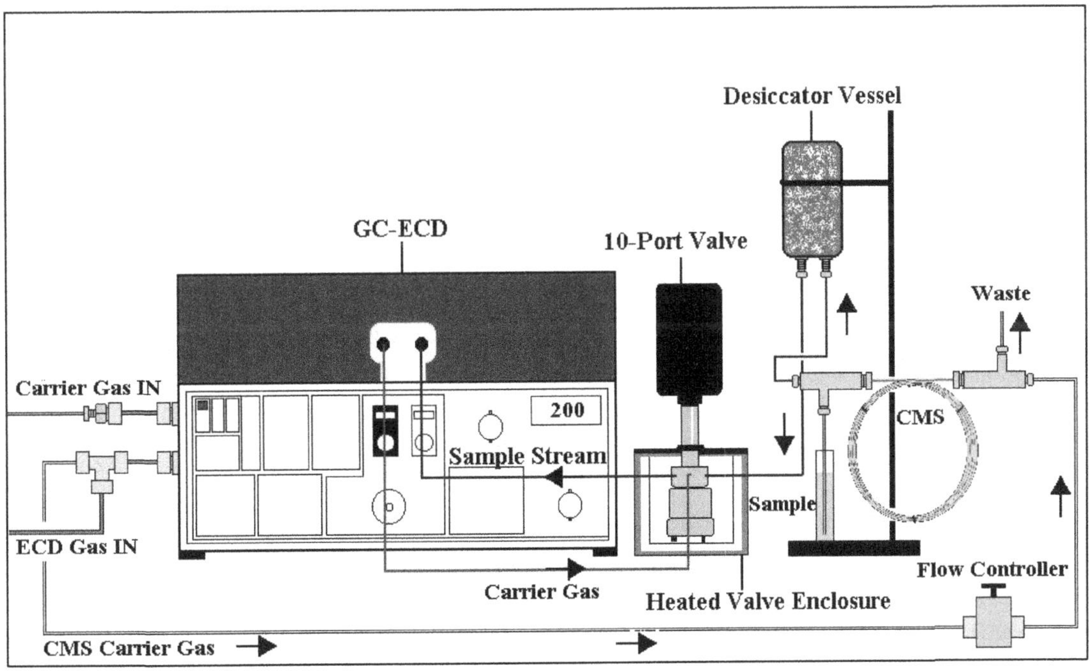

Figure A.8 Diagram of the gas delivery configuration

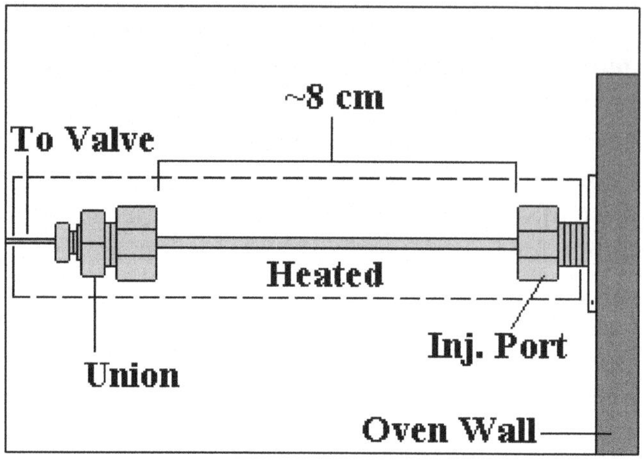

Figure A.9 Diagram of the GC injection port adapter assembly

A.7 Software

A.7.1 PeakSimple v3.20 was installed onto a Dell Latitude Pentium II laptop computer with the Windows 98SE Operating System. It communicated by using COM Port 1 of the PC.

A.7.2 All files were saved to the PeakSimple main software folder with documented names and numbers. All collected data was backed up to at least three locations (Main instrument PC, detachable USB storage device(s), and server PC).

A.8 Data Acquisition Hardware

A.8.1 Internal 4-channel Serial Data Acquisition System Interface Board (Lawson Type Board) was pre-installed by manufacturer, so no further installation was necessary.

A.9 On-line Sampling Device 1 Construction

A.9.1 Preparation of the water tap only requires removal of the aerator if present. A typical water tap pipe has a ~2.4-2.5 cm o.d so appropriate tubing was selected. The tap was purged for 30 minutes at 500 mL/min before sampling.

A.9.2 The first Tygon® tubing segment had the dimensions L~8.0 cm, o.d. 2.5 cm, i.d. 1.9 cm. This segment was used to connect directly to the water tap, so the exact dimension may vary depending on the size of the water tap. It is recommended that the Tygon® tubing be thick walled to minimize swelling due to pressure of water tap (Figure A.10).

A.9.3 Tygon® Tubing Segment #2 (L~4.0 cm, o.d. 1.8 cm, i.d. 1.3 cm) served as a reducing union for the third Tygon® tubing segment. Tygon® Tubing Segment #3 (L~40.0cm, OD 1.1cm, 0.5cm) was attached was inserted into segment #1. This assembly was then inserted into the first Tygon® tubing segment. A 1"-1"1/8 tubing clamp was used to secure all the tubing segments, together. The end of segment #3 was adapted directly to a liquid flow restrictor on the right side by use of a 3/8"-3/4" tubing clamp.

A.9.4 The left side of the liquid flow restrictor was fitted with a 1/16" to 1/4" external/internal SS bulk union. A ~25 cm length of 1/16" FEP was cut and attached to the bulk union by using a 1/16" external fitting with ferrule. A 1/16" PEEK tee was attached to the end of the 1/16" FEP tubing. An additional length of 1/16" FEP tubing was cut and attached to one of the tees outlet. This serves as the transfer line to the CMS sample inlet. The other outlet on the tee goes to waste.

A.9.5 This configuration was attached to the water tap by placing the Tygon® tubing over the tap and securing it using another 1"-1" 1/8" tubing clamp. The CMS was attached to the 1/16" FEP tubing from the configuration by using a small segment of peristaltic tubing. A pressure reading between 6.5 to 8.0 psi gave an approximate flow through the CMS of ~2.6 mL/min. Figure 5.3 in Chapter 5 depicts both of the completed configurations for the CMS-GC-ECD.

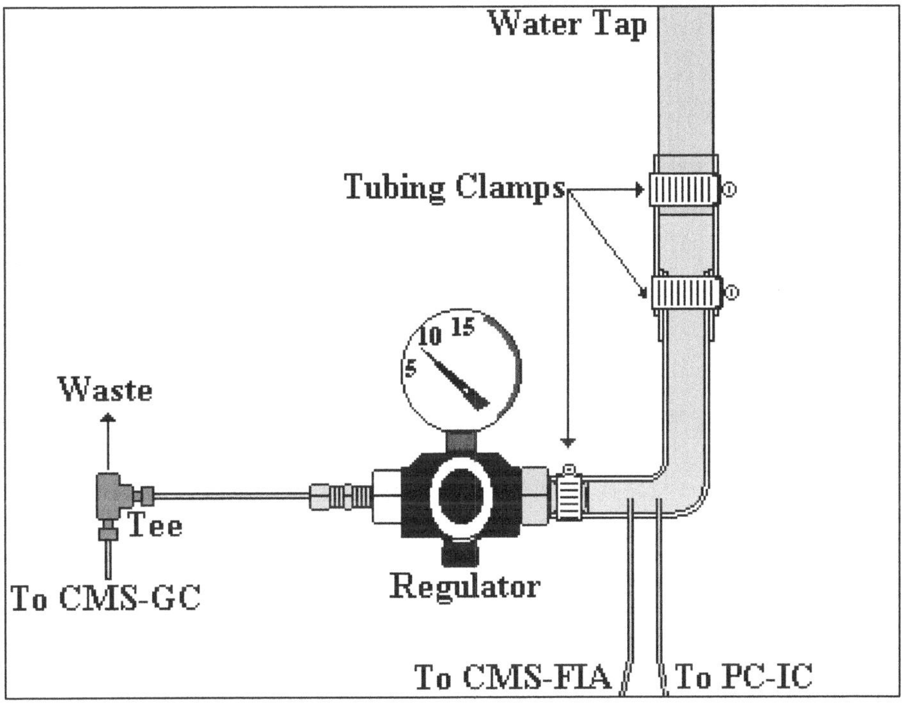

Figure A.10 Diagram of the on-line sampling liquid flow restrictor configuration

A.10 On-line Sampling Device 2 Construction

A.10.1 Preparation of the water tap only requires removal of the aerator if present. A typical water tap pipe has a ~2.4-2.5 cm o.d so appropriate tubing or adapter was selected. The Tygon® tubing was secured using a 3/8"-3/4" tubing clamp. The tap was typically purged for ~30 minutes at 500 mL/min before sampling.

A.10.2 A Tygon® tubing segment with the dimensions (L~ 95 cm, o.d. 1.2 cm, i.d .0.8 cm). This segment was connected directly to the water tap, so the exact dimensions chosen may vary depending on the size of the water tap. The tubing was configured in a ~25 cm diameter loop to allow for reduction of possible siphon effect from straight linear flow. (Figure A.11).

A.10.3 For a single instrument using a membrane sampler, a ~7 cm length of 1/16th PEEK tubing (0.75 i.d.) and a 1/16" PEEK union was used to adapt the 1/16" PTFE tubing to to the bottom of the Tygon® loop assembly This PEEK is inserted into the Tygon® tubing.

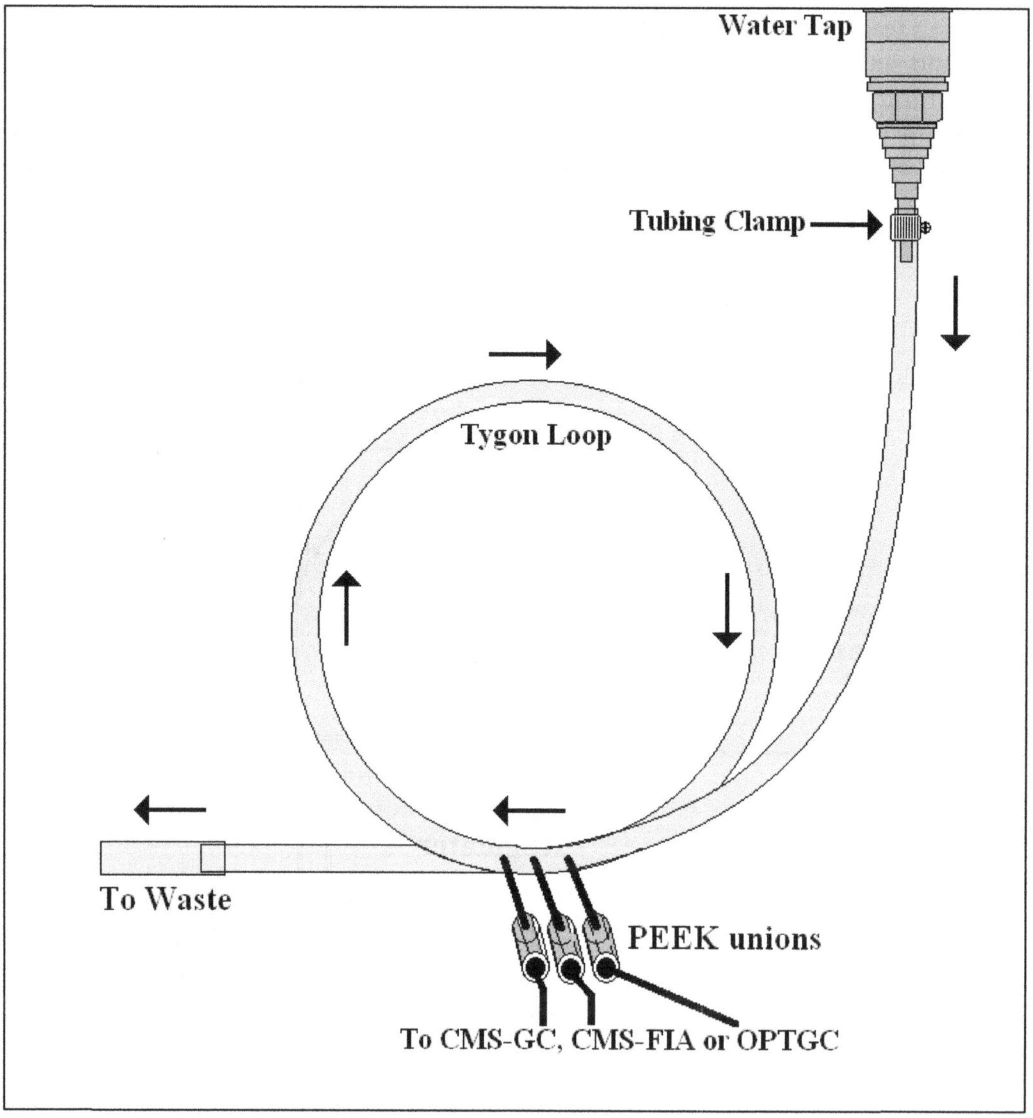

Figure A.11 Diagram of the on-line sampling loop configuration

APPENDIX B
CONSTRUCTION OF CMS-FIA NCA-FL INSTRUMENT

B.1 CMS-FIA NCA-FL Component Lists

Table B.1
Capillary Membrane Sampler (CMS) component list

Component Description	Vendor	Amt. Used	U of Memphis	Vendor Price
1/8" Reaction Tee (1/16"to1/8")	Valco	2	$80.00	Comm.
Tefzel® Tubing (1/8")	Valco-Jour	120 cm	$5.50/m	Avail.
FEP Tubing (1/16")	Valco	~85 cm	$1.40/m	From
Silicone Tubing	Global FIA	124 cm	~$2.25/m	Global FIA
SS Hypodermic Tubing Segments	Global FIA	2	~$10	
		Total	**~$100**	**~$490**

Table B.2
Flow injection analyzer system component list

Component Description	Vendor	Amt. Used	U of Memphis	Vendor Price
FIALab-2000	Alitea	1	~$3,500	~$3,500
Pump tubing (0.040")	Fisher	1	$22.90	$22.90
Pump tubing (0.040")	Fisher	1	$25.65	$22.65
Pump Tubing (0.025")	Fisher	1	$23.60	$23.60
Pump Tubing (0.020")	Fisher	1	$23.60	$23.60
10m Knitted Open Tubular Reaction Coil (0.75 mm i.d., 1/16" o.d.)	SeQuant	1	$400.00	$400.00
4 m Knitted Open Tubular Reaction Coil (0.75 mm i.d., 1/16" o.d.)	SeQuant	1	$185.00	$185.00
Heated Water Bath	Fisher	1	$556.43	$556.43
Thermometer	Fisher	1	$28.50	$28.5
Filter Fluorescence Detector	Varian	1	$200.00	N/A
Filter Fluorescence Detector	Ruhle	1	N/A	$6,945.00
Filter Band U-360, 2" square	Edmund Optics	1	N/A	$71.00
Filter Band BG-38, 2" square	Edm. Opt.	1	N/A	$48.50
Extension Cord	Belkin	2	$24.00	$24.00
		Total	**~$4,990**	**~$12,110**

Table B.3
Injection valve system component list

Component Description	Vendor	Amt. Used	U of Memphis	Vendor Price
Injection Valve	Valco	1	$1,165	$1,165
FEP Tubing (1/16")	Valco	1 m	$12.00	$12.00
FEP Sample Loop (60 μL)	FIALab	60 cm	$24.00	$24.00
		Total	**~$1,200**	**~$1,200**

Table B.4
Reagents list

Component Description	Vendor	Amt. Used	U of Memphis	Vendor Price
Nicotinamide (98%) (1 kg)	Sigma-Aldrich	75g	$45.40	$45.40
Sodium Hydroxide (ACS) (1 kg)	Fisher Sci.	40g	$70.18	$70.18
$CHCl_3$ Std. (≥99.8%, 500mL)	Aldrich	0.1 g	$20.20	$20.20
$CHBrCl_2$ Std. (≥98%, 50 g)	Aldrich	0.1 g	$56.20	$23.70
$CHBr_2Cl$ Std. (98%, 10 g)	Aldrich	0.1 g	$35.30	$19.60
$CHBr_3$ Std. (≥99%, 50 g)	Aldrich	0.1 g	$29.10	$29.10
MCAA (99%+) (100 g)	Aldrich	0.1 g	$31.10	$31.10
DCAA (99%+)(100 g)	Aldrich	0.1 g	$20.10	$20.10
TCAA (99%+) (250 g)	Aldrich	0.1 g	$48.30	$48.30
MBAA (99%+)(100 g)	Aldrich	0.1 g	$69.50	$69.50
DBAA (97%+) (25 g)	Aldrich	0.1 g	$106.50	$106.50
Methanol (HPLC grade, 4L)	Fisher	500 mL	$63.70	$63.70
		Total	**~$514.70**	**~$514.70**

Table B.5
Software component list

Component Description	Vendor	Amt. Used	U of Memphis	Vendor Price
PeakSimple	SRI	1	free	free

Table B.6
Computer component list

Component Description	Vendor	Amt. Used	U of Memphis	Vendor Price
Laptop Computer	Dell	1	~$450	~$900
		Total	**~$450**	**~$900**

Table B.7
Additional supplies list

Component Description	Vendor	Amt. Used	U of Memphis	Vendor Price
Pipettes (1 mL)	Fisher	2	$85.92	$85.92
Pipettes (10 mL)	Fisher	2	$97.57	$97.57
Bulbs for Pipettes	Fisher	2	$13.00	$13.00
Volumetric Flasks (100 mL)	Fisher	12	$297.50	$297.50
Volumetric Flasks (250 mL)	Fisher	2	$254.00	$254.00
Syringes (10 μL)	Fisher	6	$113.00	$113.00
Nalgene Narrow-Mouth Bottles (500 mL)	Fisher	2	$101.01	$101.01
Amber HDPE Bottles (125 mL)	Fisher	6	$116.07	$116.07
		Total	**~$1,078**	**~$1,078**
		GRAND TOTAL	**~$8,333**	**~$16,293**

B.2 Capillary Membrane Sampler (CMS) Prototype
Note: Wear Latex gloves and work on a lint free surface to reduce contamination of silicone.

B.2.1 A length of 1/8" Tefzel® tubing was cut to a desired length (ex. 94 cm) to serve as the body for the CMS device. A piece of silicone rubber tubing (Global FIA) was cut, so that it was 3 to 4 cm longer than the length of Tefzel® tubing from the previous section (~98cm). This extra length allows the ends of the silicone tubing to extend through the entire reaction tee for attachment.

B.2.2 The length of silicone tubing was then inserted through the entire length of 1/8" Tefzel® by using a thin electrical wire (ex. copper). Approximately 1.5 cm of the silicone tubing was left exposed at each end of the Tefzel® tubing to pull the silicone tubing through.

B.2.3 Each end of the silicone tubing was then attached to a reaction tee using the following procedure. First, a SS hypodermic tubing segment (SS-TUB) was inserted halfway into the end of the silicone tubing. This requires some patience as the tubing i.d. was quite small. It is recommended that no gloves be worn during this step, also wetting the SS segment with reagent water helps with insertion.

B.2.4 After the SS hypodermic segment was inserted into the silicone tubing, a 1/8" SS fitting with ferrule was placed over the silicone tubing and the 1/8" Tefzel® tubing. The silicone tubing end was inserted into a reaction tee (TCEF211T) at the 1/8" size end and out of the 1/16" end (Figure B.1).

B.2.5 1/16" FEP tubing was cut into four segments (2, 7.5 cm segments and 2, 35 cm segments). A 1/16" SS fitting and ferrule was placed on a 35 cm segment of 1/16" FEP tubing. The piece of SS hypodermic tubing exposed from the silicone tubing was then inserted into the end of the 1/16" FEP tubing. The 1/16" SS fitting with ferrule was then clamped to the

FEP/silicone tubing assembly. This completes the attachment of one end of the silicone tubing to the gas inlet/outlet of the reaction tee. Repeat for the other end of the silicone tubing. Note it will require that the silicone tubing be pulled and cut to appropriate length once through the other reaction tee. It was important not to pull too hard as the membrane may tear. Once through the tee hold on to the silicone tubing until FEP tubing is attached (Figure B.1).

B.2.6 Once the silicone tubing is secured, the two 1/8" SS fittings with ferrules were tightened, thus clamping to the 1/8" Tefzel® tubing. Finally, the 7.5 cm lengths of 1/16" FEP tubing were combined with a 1/16" fitting and ferrule and installed to the sample inlet/outlet locations on the reaction tees (Figure B.2).

B.2.7 The CMS was attached to a flow controller (Brooks) at the Carrier IN (Figure A.8) by using an internal reducing union (1/8" to 1/16"). To test for liquid leaks place the device on a dry paper towel (preferably brown) and watch for the formation of wet areas when pumping reagent water (18.2 MΩcm) using a peristaltic pump through. If there is a liquid leak the fittings may be too loose or too tight. To test the device for membrane leaks, submerge the device underwater and watch for air bubbles and monitor the gas flow rate using a digital flowmeter. A gas leak is either the result of a torn or pinched membrane, or there might be a loose connection at the membrane/fitting connection. After the leak checks, several mL of reagent water was pumped through the device.

B.2.8 The CMS was then connected to the valve. The CMS Carrier Out connects to port 8; CMS Sample Out connects to port 4, both using 1/16" 10-32 PEEK fittings and ferrules. CMS Carrier In and CMS Sample In are connected to the peristaltic pump tubing. (Figure B.2)

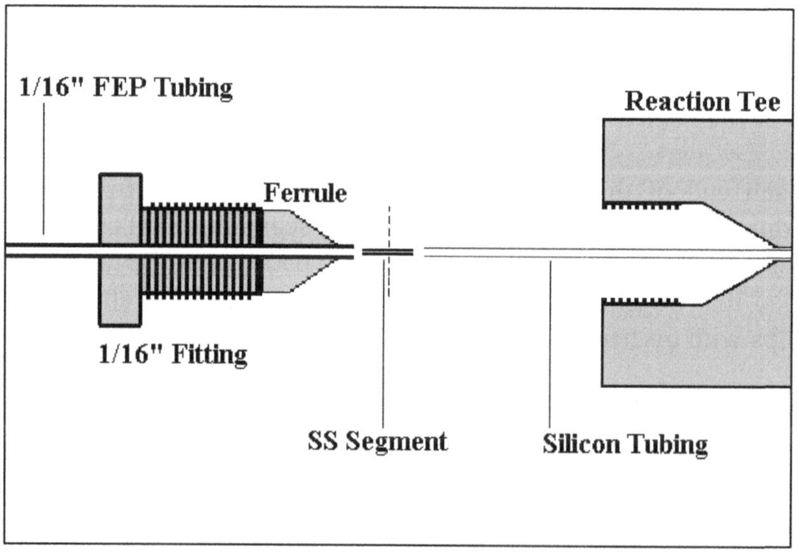

Figure B.1 Diagram of the silicone tubing membrane attachment

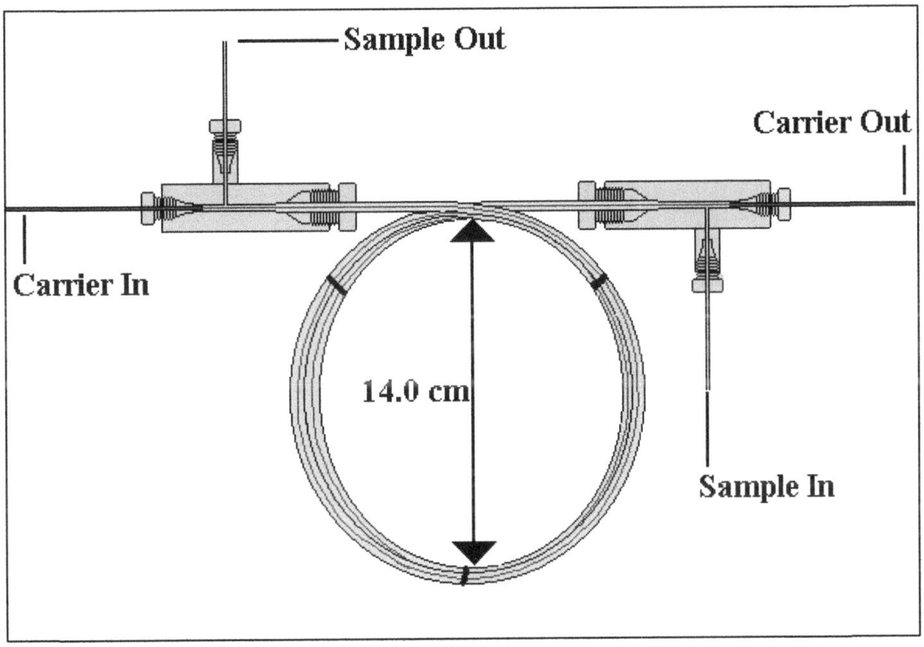

Figure B.2 Diagram of the completed CMS device

B.3 Injection Valve System

B.3.1 A ten-port electrostatically actuated injection valve (Valco) was mounted on the FIA Lab-2000 instrument by attaching it to right-hand side of the metal panel using two sheet metal screws (Figure B.3).

B.3.2 Two 30 cm long (60uL) PTFE sample loops (TTF 120, Valco) were attached to ports 2, 5 and 7, 10. Ports 8 and 4 were connected to the CMS device by FEP tubing. Approximately 30 cm long PTFE tubing were inserted into the ports 3 and 9 and were used for solutions exit to waste. Port 6 was connected to RC1 by 1/16" FEP tubing.

B.3.3 The two-position actuator control module was placed inside the FIA Lab-2000 unit and the rectangular hole was cut in the backside of the instrument. The power cord and the wires to be connected to the data acquisition board were pulled through the hole (Figure B.4).

B.3.4 The two-position actuator control module for the valve was connected to TTL output "C" on the data acquisition board that was placed next to the instrument. Pin one (red wire) was inserted into 'ground' terminal, and Pin 5 was inserted into TTL output "C" (Figure B.5). The valve was tested by going to the relay window found in PeakSimple (Model 203) software and clicking on the (C) relay button twice. The PeakSimple board has to be operating to perform this test. The valve should switch positions if it is working properly.

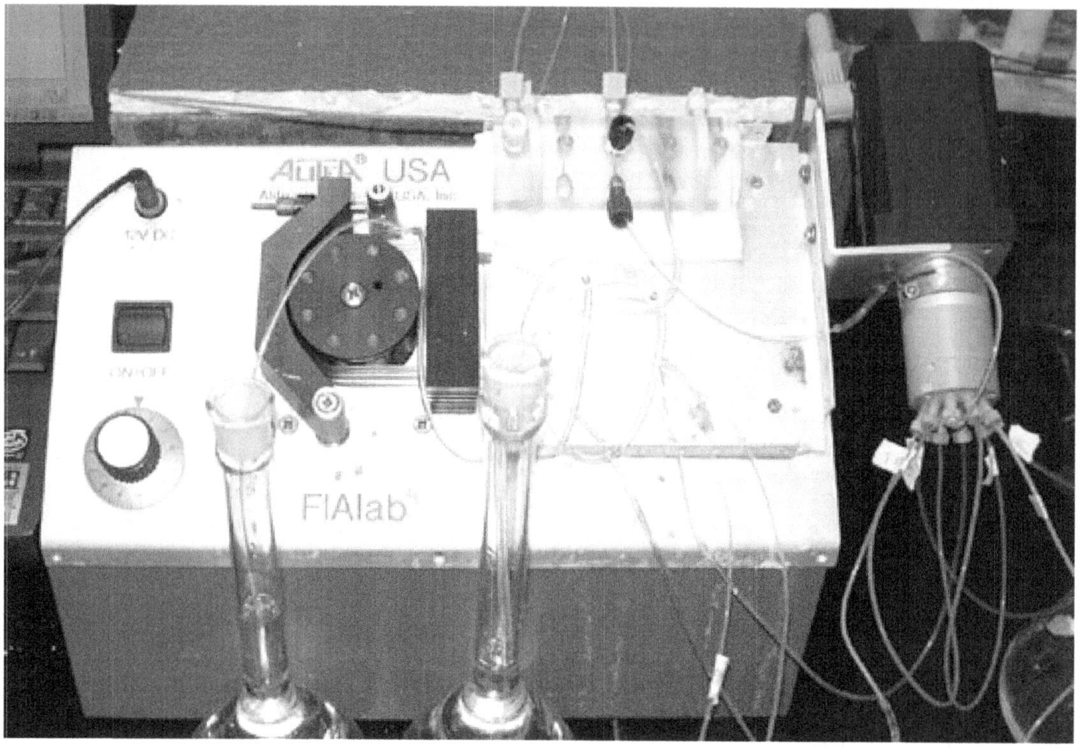

Figure B.3 Picture of the ten-port injection valve mounted on the side of the FIA Lab-2000 metal panel

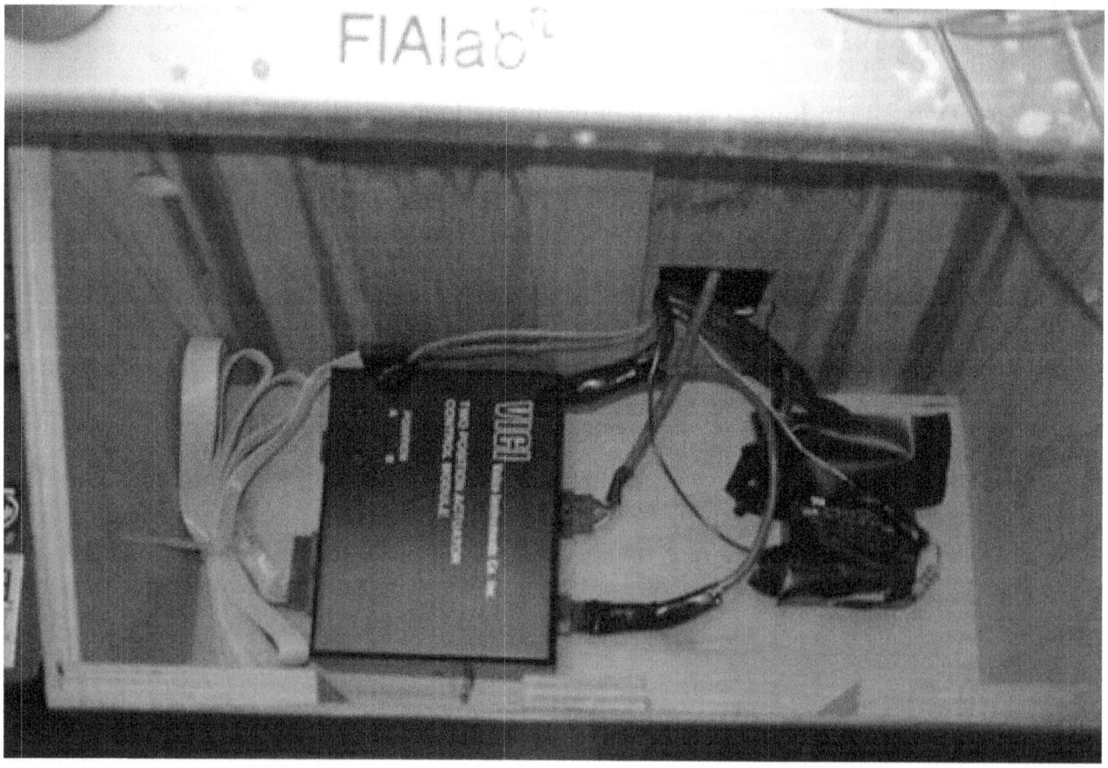

Figure B.4 Picture of the inside view of the modified FIA Lab-2000 instrument

B.4 FIA System

B.4.1 A single channel serial port data acquisition system (PeakSimple) was used to automate the valve assembly and to receive the signal from the fluorescence detector. To make the appropriate connections the data acquisition system was unplugged. The lid was removed to perform necessary connections. A typical precision flathead screwdriver was needed to loosen and tighten the board's terminals.

B.4.2 The first connection was made with the fluorescence detector cable. The (black wire) was inserted into the 'minus' terminal, (gray wire) was inserted in the 'plus' terminal and (bare wire) was placed in the ground terminal and they were tightened using a small flathead screwdriver (Figure B.5).

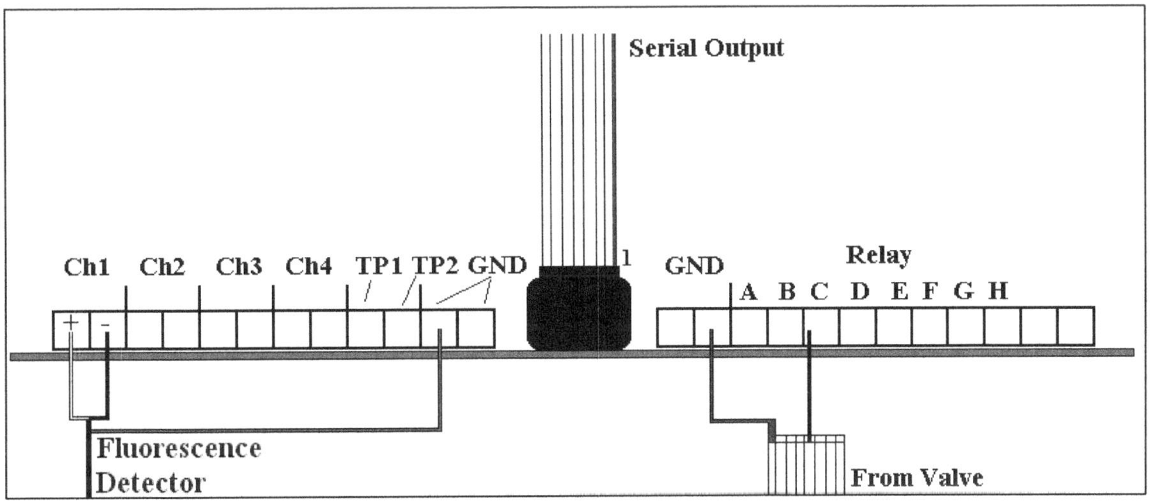

Figure B.5 Diagram of the PeakSimple data acquisition board connections

B.4.3 Finally, the data acquisition board was connected to a computer by using a RS-232 serial cable. Com Port 1 on the PC was used to communicate.

B.4.4 The PeakSimple software was loaded from the interfaced PC. There should be a connection sequence between the data acquisition board and PC that will verify a successful initialization.

B.4.5 First the valve automation was tested by opening the relay/pump window through the View menu. By clicking on the (C) button couple of times, the valve has to switch between two positions.

B.4.6 To test connections between fluorescence detector and the data acquisition board, fluorescence detector was turned on. 30% NCA solution can be injected through the flow cell to observe increase or decrease in the signal.

B.4.7 To drive solutions from the reagent bottles, necessary size pump tubing were placed over the four channel pump, tightened and connected to the injection valve and manifold by the 1/16" FEP tubing. Reagent water was pumped through the tubing and the flow rate was measured for the system. Note that pump tubing wears out and the flow rate has to be checked regularly and if it is needed they have to be replaced. If there is backpressure in the system, PEEK nuts should be checked, as well as the fluorescence detector flow cell. The flow cell can be cleaned by injecting 0.05M sulfuric acid through it. The sulfuric acid will dissolve any NCA deposits formed during the normal operation. After injecting the sulfuric acid through the detector, be sure to inject reagent water using at least twice the volume of sulfuric acid used.

B.4.8 The reaction manifold was prepared by placing the FEP tubing exiting from the injection valve into a ¼-28 flangeless fitting and ferrule, and then it was screwed into the first hole on the manifold. The same was done with FEP tubing that was delivering 30% NCA solution but this time it was screwed into the middle hole on the manifold. Finally, the last hole in that row was used for exiting the solutions to RC.

B.4.9 The reaction coils have to be chosen so that the maximum mixing of solutions is provided. Different lengths of reaction coils can be added in a chain to improve the mixing of solutions.

B.4.10 The reaction coils that were used to heat solution were completely submerged in the heat bath. The reaction coil that was used to cool down solution was placed in an ice bath.

B.5 Reagents

B.5.1 All water used was distilled and then deionized by a Millipore Mill-Q Plus/Mill-RO® 10 Plus Water Purification System Glassware and sample containers were meticulously cleaned with 1:1 nitric acid soaked for 24 hours and then rinsed with reagent grade water six times. All chemicals were HPLC grade or better and were purchased from Fischer Scientific Inc. Stock standard and NCA solutions were purchased from Aldrich Chemical Company.

B.5.2 Sodium hydroxide was prepared by dissolving appropriate amount of NaOH pellets in reagent water. Plastic 500 mL bottles were used to store the solution.

B.5.3 Nicotinamide solution was prepared by diluting 75 g of NCA to a total volume of 250 mL in a 250 mL volumetric flask. Solution was kept wrapped in aluminum foil. A solution was prepared immediately prior to use.

B.5.4 Stock solutions of the individual THM4 at a concentration of 1000 mg/L were prepared by dissolving each THM in 100 mL of methanol. Different aliquots of each stock standard solution were added with a separate gas tight 10 μL syringe without contacting the sides of the flasks. Appropriate amounts of the stock standard solution were added to a 100 mL volumetric flask and diluted to volume using reagent water to give various concentrations of the working standard. This standard lasts approximately three to four weeks.

B.5.5 Stock solutions of the individual HAA5 were prepared by dissolving each HAA5 in 100 mL of deionized water resulting in a concentration of 1000 mg/L. The working standards were prepared by diluting stock solution in reagent water. To preserve the standard, 10 mg of ammonium chloride was added to the working standard. The standard was transferred to an amber HDPE bottle and kept refrigerated. This standard lasts approximately two weeks.

B.6 Software

B.6.1 PeakSimple version 3.20 was installed on Dell Latitude Pentium III laptop Computer with the Windows 98SE Operating System.

B.6.2 All files were saved to the main software folder with documented names and numbers. All collected data was backed up at least three locations (Main instrument PC, detachable USB storage device(s), and server PC).

B.7 On-line Sampling Configuration

B.7.1 Water was sampled from the tap as described in Section A.9.

B.7.2. Approximately 40 cm long 1/16" FEP tubing was inserted into a small hole made in side of the Tygon® tubing. The water tap flow rate was kept at ~ 500 mL/min throughout the Tygon® tubing.

B.7.3 Approximately, every 12 hours, the 1/16" FEP tubing connected to Tygon® tubing at the water tap was taken off and inserted into a volumetric flask with a check standard that contained both THM4 and HAA5 species to assure calibration of the system. After the injection of THM4 and HAA5 standard into the FIA system, the 1/16" FEP tubing was then inserted back into the Tygon® tubing for approximately 25 minutes.

APPENDIX C
CONSTRUCTION OF THE PC-IC INSTRUMENT

C.1 PC-IC Component Lists

Table C.1
IC system component list

Component Description	Vendor	Amt. Used	U of Memphis	Vendor Price
Gradient IC Pump (GPM-1)	Dionex	1	$400.00	N/A
Gradient IC Pump (GP-50)	Dionex	0	N/A	$12,450.00
IonPac AG9-SC (50 x 4mm)	Dionex	1	$250.00	$250.00
IonPac AS9-SC (250 x4mm)	Dionex	1	$795.00	$795.00
EPC10W 10- Port Injection Valve	Valco	1	$1,235.00	$1,235.00
1/16' o.d. PEEK nuts (10 pk.)	Valco	5	$25.00	$25.00
1/16" o.d. PEEK ferrules (10 pk.)	Valco	5	$30.00	$30.00
Female Luer Adaptor to 10-32 male fitting	Valco	1	$20.00	$20.00
		Total	**$2,765.00**	**$14,815.00**

Table C.2
Post column reaction system component list

Component Description	Vendor	Amt. Used	U of Memphis	Vendor Price
FIAlab 2500 Peristaltic Pump	Alitea	1	$8,000.00	$5,245.00
Peristaltic Pump tubing (0.010" i.d.) (orange-blue) (12 pk)	Fisher	1	$25.53	$25.53
Mixing Tee 1/16" o.d. tubing, 0.75mm Bore (with flangeless ¼-28 fittings)	Valco	2	$120.00	$120.00
10m KOT PTFE 0.75mm i.d.	SeQuant	3	$1,110.00	$1,110.00
5m KOT PTFE 0.75mm i.d.	SeQuant	1	$170.00	$170.00
2m KOT PTFE 0.75mm i.d.	SeQuant	2	$210.00	$210.00
Internal unions – PEEK (1/16" o.d. tubing, 0.75mm bore) (includes 10-32 fittings)	Valco	8	$105.00	$105.00
Heated Water Bath (minimum 11" wide)	Fisher	1	$556.43	$556.43
Ice Bucket	Fisher	1	$47.27	$47.27
Thermometer (0 to 100°C)	Fisher	1	$9.11	$9.11
		Total	**$10,353.34**	**$7,598.34**

Table C.3
Detector component list

Component Description	Vendor	Amt. Used	U of Memphis	Vendor Price
Waters 2475 Dual Wavelength Fluorescent Detector	Waters Corp.	1	$13,225.00	$15,100.00
		Total	**$13,225.00**	**$15,100.00**

Table C.4
Tubing list

Component Description	Vendor	Amt. Used	U of Memphis	Vendor Price
40 ft PEEK (0.02" i.d., 1/16" o.d.)	Valco	40 ft	$112.00	$112.00
10ft PEEK (0.01" i.d., 1/16" o.d.)	Valco	~ 4 ft	$14.00	$14.00
10 ft FEP (0.03" i.d., 1/16" o.d.)	Valco	~ 5 ft	$14.00	$14.00
10 ft Tefzel® (1/16" i.d., 1/8" o.d.)	Valco	10 ft	$12.00	$12.00
2 ft Tygon® R-3603 (4.8mm i.d., 7.9mm o.d.) (10 ft)	Fisher	2 ft	$11.46	$11.46
		Total	**$163.46**	**$163.46**

Table C.5
PeakSimple hardware and software component list

Component Description	Vendor	Amt. Used	U of Memphis	Vendor Price
Peak Simple A/D Box Model 203	SRI	1	$1,395.00	$1,395.00
Peak Simple Software	SRI	1	free	free
		Total	**$1,395.00**	**$1,395.00**

Table C.6
Computer component list

Component Description	Vendor	Amt. Used	U of Memphis	Vendor Price
Laptop Computer	Dell (TigerDirect)	1	~$450	~$600
		Total	**~ $450**	**~ $600**

Table C.7
On-line monitoring adaptor component list

Component Description	Vendor	Amt. Used	U of Memphis	Vendor Price
¼-28 Flangless Fitting (PEEK, 1/8" O.D. tubing)	Valco	1	$10.00	$10.00
10-32 to ¼-28 internal union	Valco	1	$18.00	$18.00
Swagelok Stainless steel external union (1/8" OD tubing)	Agilent	1	$60.00	$60.00
		Total	**$88.00**	**$88.00**

Table C.8
Additional Supplies, Reagents and Standards lists

Component Description	Vendor	Amt. Used	U of Memphis	Vendor Price
25 ft electrical wire	Lowe's	10 ft	$6.95	$6.95
Surge Protector Strip	Radio Shack	1	$24.00	$24.00
1 mL Mohr Pipette (Graduated)	Fisher	1	$92.22	$92.22
10ml Mohr Pipette (Graduated)	Fisher	2	$152.52	$152.52
100 mL Volumetric Flasks	Fisher	9	$345.42	$345.42
Ammonium Chloride (500g)	Fisher	10 g	$45.34	$45.34
MCAA (99%+) (100 g)	Sigma-Aldrich	10 mg	$31.10	$31.10
DCAA (99%+)(100 g)	Sigma-Aldrich	10 mg	$20.10	$20.10
TCAA (99%+) (250 g)	Sigma-Aldrich	10 mg	$48.30	$48.30
MBAA (99%+)(100 g)	Sigma-Aldrich	10 mg	$69.50	$69.50
DBAA (97%+) (25 g)	Sigma-Aldrich	10 mg	$106.50	$106.50
Nicotinamide (98%) (1 kg)	Sigma-Aldrich	75g	$45.40	$45.40
Sodium Hydroxide (ACS) (1 kg)	Fisher Scientific	22g	$70.18	$70.18
Boric Acid (ACS) (500 g)	Fisher Scientific	50g	$56.38	$56.38
		Total	**$1,113.91**	**$1,113.91**
	GRAND TOTAL		**$29,553.71**	**$40,873.71**

C.2 Reagents

IC eluent - 400mM Boric Acid (H_3BO_3), 80mM Sodium Hydroxide

C.2.1 Dissolve 49.5 g H_3BO_3 and 6.4 g NaOH into 2 L of reagent water, stir until all solids are dissolved. Pour into eluent container, then bubble nitrogen through it for 20 min.

C.2.2 In another eluent container, pour reagent water into it, and bubble nitrogen through it for another 20 min.

Post-column reagents

C.2.3 2M NaOH - Dissolve 16 g NaOH into 198 mL of reagent water, stir until cool to the touch, and pour into a 250 mL bottle. (this will last ~24 hours).

C.2.4 37.5% NCA – Dissolve 75 g NCA into 140 mL of reagent water. Stir and slightly heat to dissolve faster. Pour into 250 mL bottle (this will last ~24 hours).

C.3 Ion Chromatograph Setup

C.3.1 Connect the IC pump-out to port 1 of the 10 port valve using the 0.25 mm i.d. PEEK tubing. Use the stainless steel fittings with the PEEK tubing.

C.3.2 The sample loop is 990 cm of 0.50 mm i.d. PEEK tubing, connect this section to ports 2 and 5 on the valve.

C.3.3 Connect 2 feet of 0.50 mm i.d. PEEK tubing to port 3 and 4, port 4 is sample in, and port 3 goes to waste. To the PEEK tubing connected to port 4, attach the female luer to 10-32 adaptor using a PEEK fitting (Figure C.1).

C.3.4 For on-line monitoring, remove the luer adaptor, and replace with a 10-32 to ¼-28 union. Using a 2 ft piece of 1/8" Tefzel® tubing place a ¼-28 flangeless fitting on one end and attach to the PEEK tubing using the internal union. On the other end, place a stainless steel fitting, and attach to the stainless steel external union. On the other side of the external union, screw on the 5 mm i.d. Tygon® tubing. Attach the Tygon® tubing to the faucet.

C.3.5 Connect 8 in. of 0.25 mm i.d. PEEK to ports 10 and 7, and another short section of 0.25 mm PEEK to ports 8 and 9 (Figure C.1).

C.3.6 Connect a short section (6 in.) from port 6 using the stainless steel fitting connected to the valve, and a 10-32 PEEK fitting connected to the AG9-SC. Using another short piece of PEEK (3 in.) connect the AG9-SC to the AS9-SC using the 10-32 PEEK fittings (Figure C.1).

C.3.7 Connect the AS9-SC out to one side of Mixing Tee 2 (MT2), using a 10-32 PEEK fitting on the column, and a ¼-28 flangeless fitting on MT2 with 0.25 mm i.d. PEEK (Figure C.1).

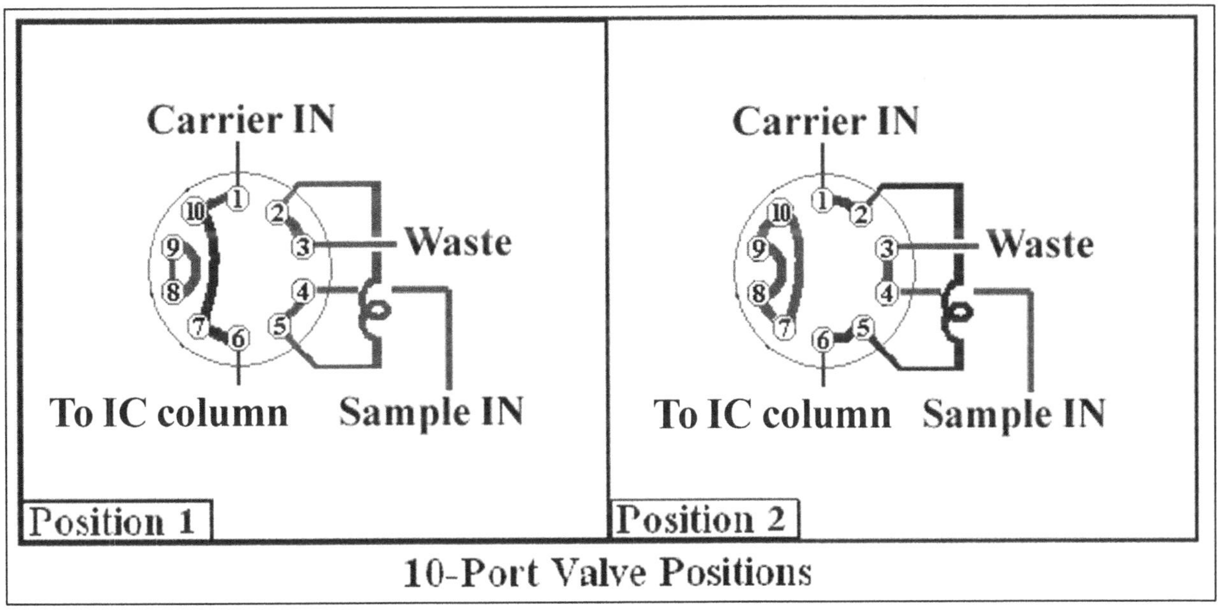

Figure C.1 Diagram of the ten-port valve showing port connections

C.4 Post Column Reaction System (Figure C.2)

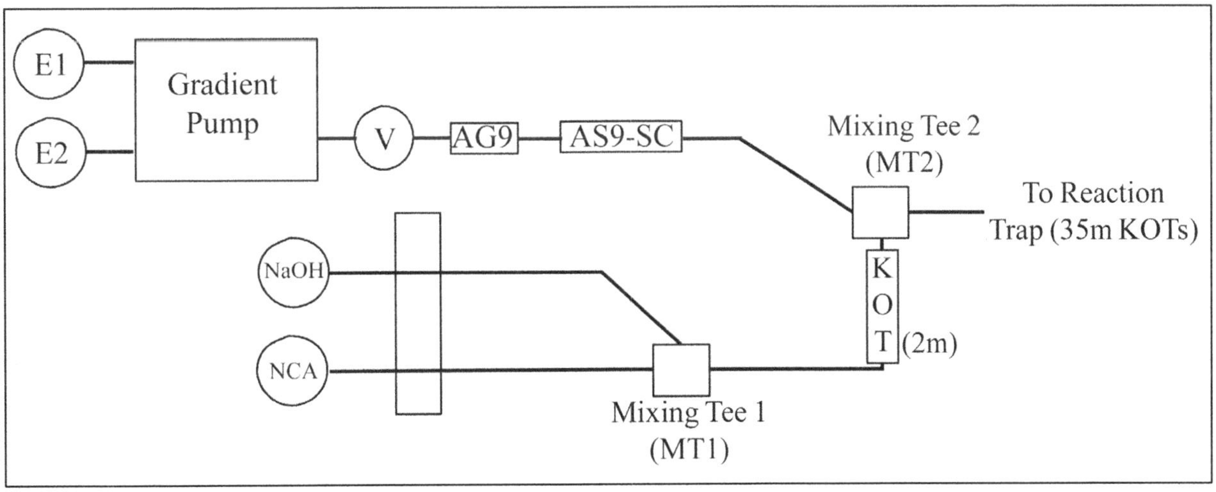

Figure C.2 Detailed diagram of the post column manifold and KOT connections

C.4.1 Using a razor blade, cut 3 short sections (10 cm) of 0.50 mm i.d. PEEK tubing, and 2 long sections (40 cm) of 0.25 mm I.D. PEEK tubing.

C.4.2 Using a razor blade, bring to a dull point one end of both sections of 0.25 mm i.d. PEEK, and one end of two sections of the 0.50 mm i.d. PEEK tubing, and both ends of the third section.

C.4.3 In a heat bath, at 94°C, place one end of the Tygon® peristaltic pump tubing in for ~5 sec to make the Tygon® more pliable. Then immediately insert the pointed end of the 0.25 mm i.d. PEEK tubing into the heated end of the peristaltic pump tubing. Do this once more. The 0.25 mm i.d. PEEK is placed into each reagent bottle to transfer sample more quickly to the pump, due to the slow flow rate. PEEK is used here because inserting FEP tubing into the small bore pump tubing is difficult at best; PEEK is a much stiffer material, allowing for easier insertion.

C.4.4 In a heat bath, at 94°C, place the other end of the Tygon® peristaltic pump tubing in for ~5 sec to make the Tygon® more pliable. Then immediately insert the sharpened end of the 0.50 mm i.d. PEEK tubing into the heated end of the peristaltic pump tubing. Do this once more with the double sharpened section.

C.4.5 With the double sharpened section, using a small piece of 0.50mm i.d. Tygon® tubing, connect the double sharpened section of PEEK to the other sharpened section of PEEK.

C.4.6 The two unsharpened ends of the 0.50 mm i.d. PEEK tubing are connected to MT1 using the 1/4-28 flangeless fittings. One section is connected to one side of the tee, and the other is connected to the top.

C.4.7 To the remaining port in Mixing Tee 1, connect a 2 m KOT to Mixing Tee 1 using the ¼-28 flangeless fitting. Connect the other end of the KOT to the top port of Mixing Tee 2 using a ¼-28 flangeless fitting.

C.4.8 On MT2, connect a ~40 cm section of 0.75 mm i.d. FEP tubing to the remaining side port using the ¼-28 flangeless fitting.

C.4.9 Connect the first 10 m KOT to the FEP tubing by using one of the PEEK internal unions (0.75 mm bore).

C.4.10 Connect the remaining two 10 m KOTs and the 5 m KOT in series using the PEEK internal unions (0.75 mm bore). The three 10 m KOTs and 5 m KOT make up the post column reaction trap, and are placed in the heat bath before analysis begins.

C.4.11 Connect another ~40 cm section of FEP tubing to the last KOT (the 5 m section) using the PEEK internal unions, and on the other end of the FEP tubing, connect the remaining 2 m KOT to the 40 cm section of FEP using the PEEK internal unions. This 2 m KOT goes into the ice bath.

C.4.12 Connect another small section (10 to 15cm) of FEP to the end of the 2 m KOT using another PEEK internal union, then connect the end of the FEP tubing to the detector in stainless steel tubing using the remaining PEEK internal union.

C.4.13 Send the "waste out" of the detector to a waste container.

Electrical / Signaling connections
Note: Strip all signaling wire ~0.25 inches down to reveal bare wire.

C.5 Valve Signaling

C.5.1 Connect pin 1 of eight conductor ribbon cable to the ground on the PeakSimple board. Pin 1 cable is marked with a red stripe. Ground on the PeakSimple board is marked as GD (Figure C.4).

C.5.2 Connect Pin 5 to TTL out "C". This will actuate the valve between position A and B. Turn the TTL out "on" to prime the valve, turn the TTL out "off" to make the valve switch positions (Figure C.4).

C.5.3 Turn the valve on by plugging it into the power system.

C.6 Ion Chromatograph Signaling

Note: When facing the rear of the GPM-1 pump, pin 1 of the telephone style connector is on the right, pin 6 on the left. (The GP-50 has TTL inputs as well, refer to the manual for connecting the PeakSimple board to the pump.)

C.6.1 Cut the six conductor telephone wire in half, and strip away ~6 inches of the outer protective sheath on the cut end. Then plug in the telephone connector into the back of the IC pump.

C.6.2 Connect Pin 2 to TTL output "A", connect pin 4 to TTL output B, and pins 5 and 6 to ground (GD). Pins 1 and 3 are not used (Figure C.4).

C.7 FIAlab 2500 signaling and Fluorescence detector

C.7.1 Inside the FIAlab 2500, connect the wire leading to "CW" (or "CCW") on the underside of the mounting plate to TTL out "E" on the PeakSimple board, using a piece of electrical wire as an extension (Figures C.3 and C.4).

C.7.2 Inside the FIAlab 2500, connect the wire leading to "EXT" on the underside of the mounting plate to TTL out "F" on the PeakSimple board, using a piece of electrical wire as an extension (Figures C.3 and C.4).

C.7.3 Inside the FIAlab 2500, connect the wire leading to "INP" on the underside of the mounting plate to the "TP1" output on the PeakSimple board, using a piece of electrical wire as an extension (Figures C.3 and C.4).

C.7.4 Inside the FIAlab 2500, connect the wire leading to "GND" on the underside of the mounting plate to the ground "GD" on the PeakSimple board, using a piece of electrical wire as an extension (Figures C.3 and C.4).

C.7.5 Plug the FIAlab 2500 into the power system.

C.7.6 Connect the (+) out on the Waters 2475 to channel 1 (+) on the PeakSimple board. Also, connect (-) to (-), and ground to ground, from detector to the PeakSimple board connections (Figure C.4).

Figure C.3 Picture of the FIAlab 2500 pump PCB for signaling connections

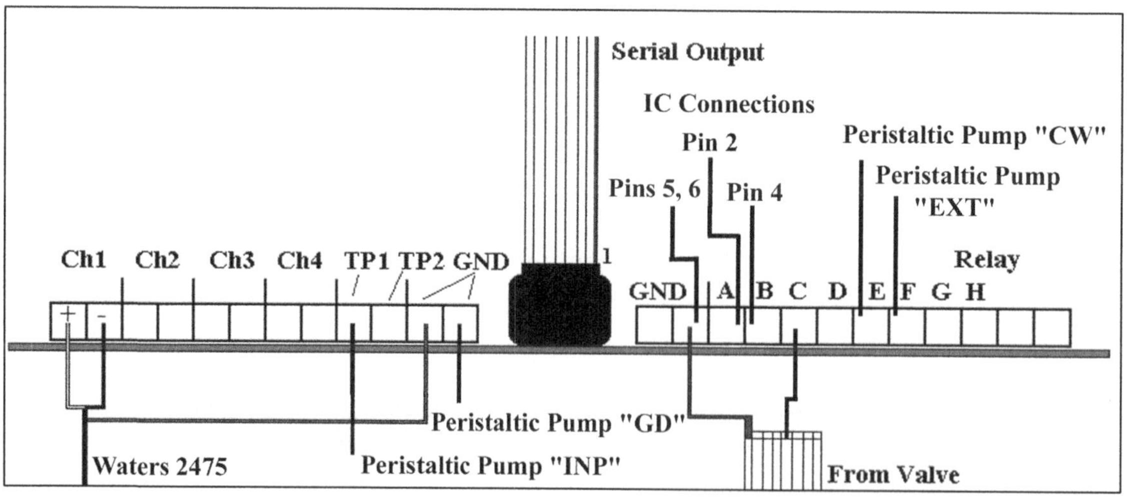

Figure C.4 Diagram of the PeakSimple Data Acquisition Board Connections

C.8 PeakSimple interface with the computer

C.8.1 Connect the PeakSimple hardware A/D box to serial port 1 (usually COM 1) on the computer with an RS-232 cable.

C.9 Software

C.9.1 PeakSimple v3.12 was installed to Dell Optiplex Pentium III Computer with Windows 98SE Operating System. Communication occurs over COM Port 1 of the PC (by default).

C.9.2 All files were saved to the main software folder with documented names and numbers. Additional copies of the software were also loaded onto notebook computers for on-line studies. All data was backed up at three locations (instrument PC, detachable USB storage device(s), and server PC).

APPENDIX D
CMS-GC-ECD EXPERIMENTAL PROCEDURE

D.1 CMS-GC-ECD Setup

D.1.1 It is assumed that all installation steps for constructing the instrument have been completed prior to arriving at this section. All devices (total of 6) that require power were plugged into an 8-outlet surge protector. The peristaltic pump was not plugged in at this time.

D.1.2 The surge protector was then turned on. All devices were powered up except the transformer, ITC, and the GC, as these all have power switches.

D.1.3 The helium purifier was purged and heated (at 25 mL/min) 30 minutes prior to being directed to the GC carrier gas inlet fitting for the first time.

D.1.4 The GC was turned on.

D.1.5 The manual set point values for the column heater and ECD heater were set manually to zero on the GC. For initial usage, the ECD standing current was set to 0 nA. The ECD make-up gas (N_2) was turned on and set to 108 p.s.i.. The flow rate for the make-up gas should be ~40.0 mL/min. Once optimum flow was achieved the ECD temperature was manually set to 300°C and the standing current was set to 3.0 nA. It takes approximately 1.5-2 hours for the ECD to heat up and equilibrate before connection of the analytical column and initial use.

D.1.6 The transfer lines were heated by plugging the heating tape terminal into the transformer. The transformer was placed on a setting between 27 and 28 and turned on (120 VAC). The GC carrier gas flow was also turned on and flowed through the flow controller and transfer lines for 15 minutes before being flowed to the GC column adapter. Once the GC carrier gas was switched to flow to the column, the flow controller was set ~5.0 mL/min and measured using the digital flowmeter. Table D.1 lists additional parameters that were set as well as the settings used for the on-line studies. The column was connected to the ECD. The total flow rate through the ECD vent should be ~ 45.0 mL/min The GC oven's initial settings will be reached after the PeakSimple software is initiated discussed in the next section.

Table D.1
Typical Settings for the CMS-GC-ECD (peristaltic pump used during calibrating).

Parameter	Setting
GC Column	Valco VB-5, 15.0m x 0.53 mm x 1.0 μm f.t.
Carrier Gas	Helium (99.999% purity) at 5.0 mL/min
CMS Carrier Gas	Nitrogen at 6.0 mL/min
ECD Make-up Gas	Nitrogen at 40.0 mL/min
Transfer Line Temperature	65 °C
Valve Enclosure Temperature	65 °C
ECD Temperature	300 °C
CMS Length	120-cm
Process Line Flow Rate to Regulator	~300 mL/min (~7-8 psi on regulator)
Flow Rate From Regulator	~2.6 mL/min
Initial Pump Time	15.0 min
Initial GC Oven Temperature	65 °C for 1.00 minutes
Temperature Ramp	65 °C to 225 °C at 60°C/min
Final GC Oven Temperature	220°C for 3.334 minutes
Total Chromatographic Time	7.00 min
Return to Initial Conditions	8.00min
Total Run Time	15.00 min

D.2 PeakSimple Software Setup

D.2.1 The PeakSimple software (previously installed) was loaded. An initialization sequence took place between the software and the data acquisition board in the GC.

D.2.2 A control file (*.con) must be created for a particular method in order to save initial conditions and operating conditions for the GC. The control file **CMS-GC-Control.con** was created and loaded by going to the File menu in the software toolbar and selecting open control file (Figure D.1). Three sub-files were made first, these are the ***.tem, *.evt, and *.cpt**.

D.2.3 Once the control file was loaded the Overall controls menu was accessed by selecting Edit on the main toolbar. The only settings changed were the Com Port (set to Com Port 2) and the default display was changed to 4.00 min. All other settings were left on default (Figure D.2).

D.2.4 The Channels menu was accessed next by selecting the **1234 Channels box** in the main toolbar (Figure D.3). First the Details box was selected and the settings were inputted as shown in Figure D.4.

D.2.5 The Temperature programming menu was opened and the temperature programming ramps given in Table D.1 were inputted (Figure D.5). It should be noted that the actual temperature of the GC heater was several °C behind the software, so newer instruments may vary. Once the ramps were inputted a temperature file was saved as **CMS-GC-TEMP.tem.**

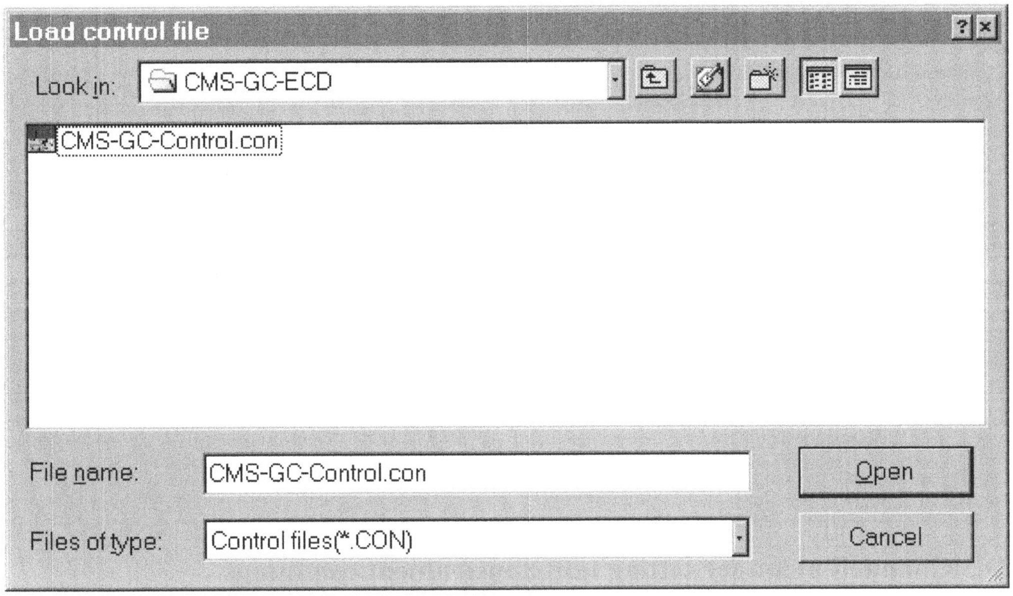

Figure D.1 Loading CMS-GC-Control.con file

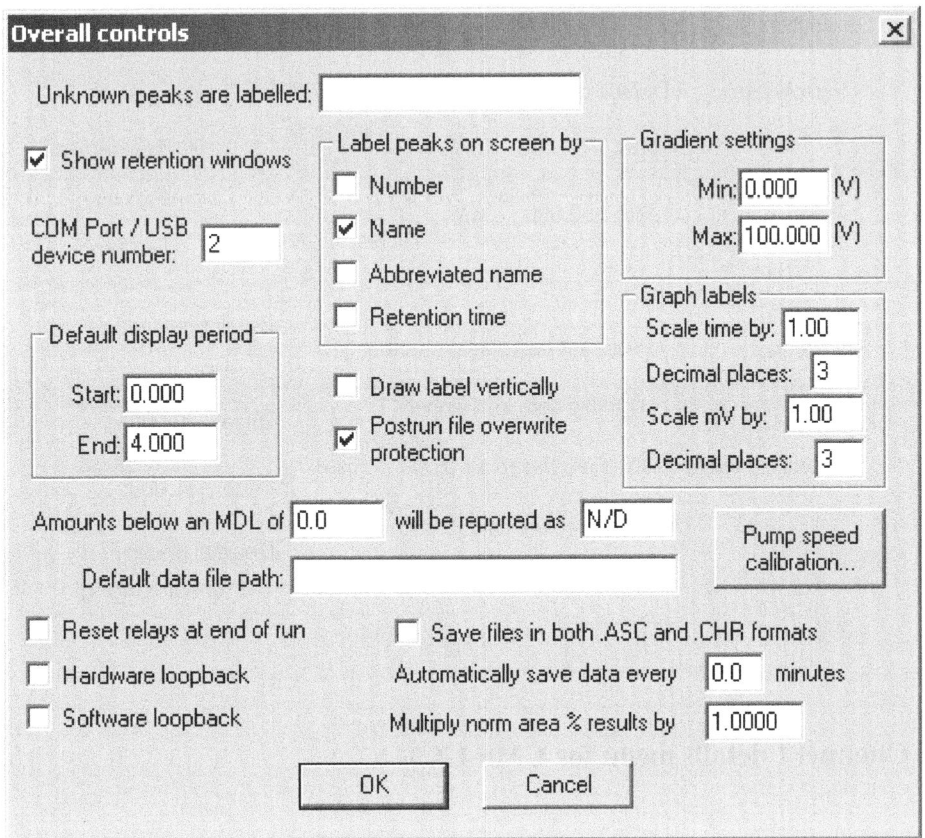

Figure D.2 Overall controls settings

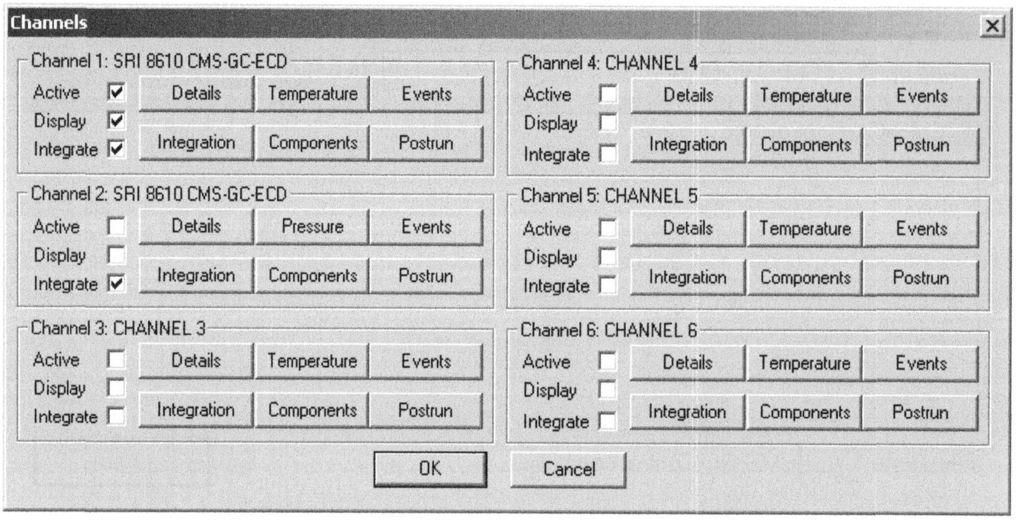

Figure D.3 Channels menu for setting initial instrument conditions

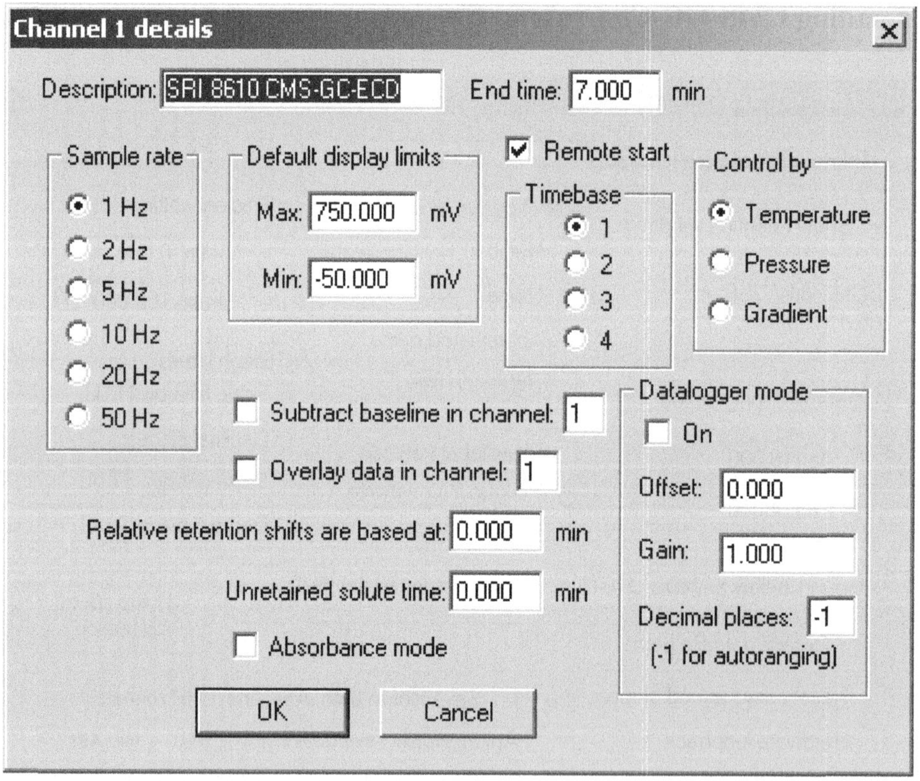

Figure D.4 Channel 1 details menu for CMS-GC

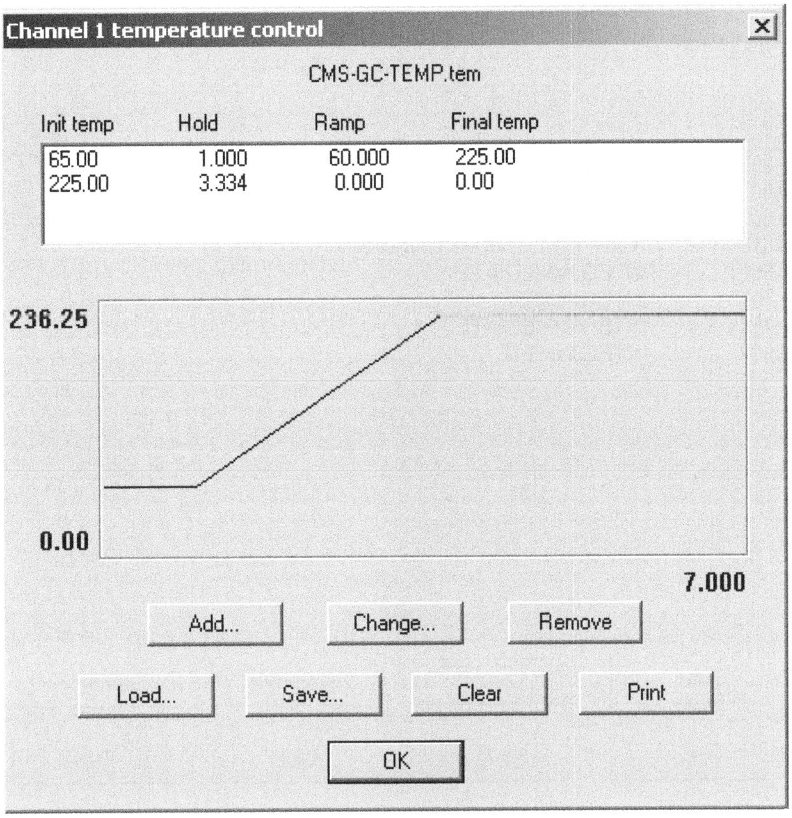

Figure D.5 Temperature programming for CMS-GC

D.2.6 The timed events for the CMS-GC were inputted in the Events menu. Figure D.6 lists the final events that were added at each time depicted. At time 0.000 min there is a zeroing of the analytical signal, relay (A) is turned on to prepare for injection valve actuation, and there is inhibition of integration to prevent a drop in the integrated baseline near due to the slight drop that occurs with injection valve rotation. At time 0.010 min relay (A) is turned off, thus actuating the injection valve at the beginning of an analysis. At time 0.850 min, there is an integration-based immediate function that is applied to correct for the negative peak that occurred several seconds earlier in the chromatogram. Next the integration is set to automatic at 1.100 min prior to the first THM4 peak elution. Since the baseline does tend to shift after the second THM4 elutes another integration based immediate was applied at 1.170 min. Finally, an integration inhibition was applied at 3.000 min after all of the THM4 have eluted to simplify the results to be transferred to an Excel spreadsheet. These settings suit the instrument and column combination for this research, so variations are likely meaning each event file might be completely different than the ones listed. The Event file was saved as **CMS-GC-Event.evt.**

D.2.7 The Integration menu was opened and the following settings were changed from default (Figure D.7). The Peak sensitivity was set at 100% and the Baseline sensitivity was set at 100%. Using 100% is only recommended for simple separation such as the THM4 and also if the detector noise is relatively low as with this system. Default is 95% for Peak sensitivity and 60 % for base sensitivity. It was determined that for the fast resolution of the THM4 these settings would not work.

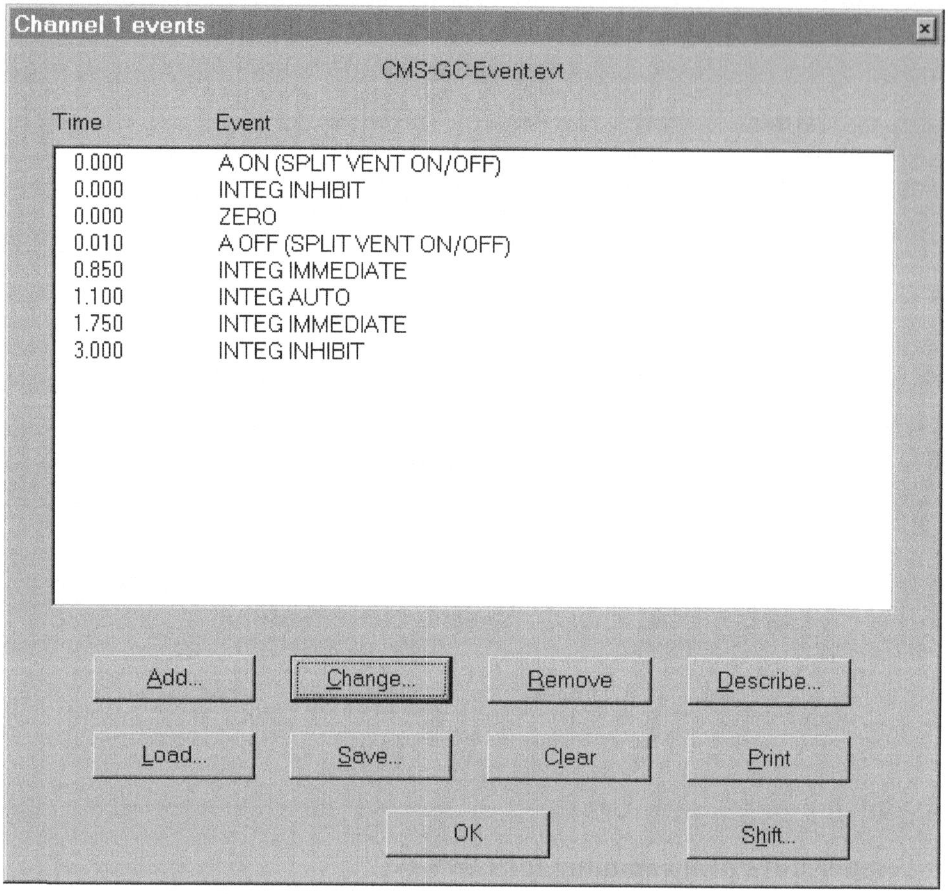

Figure D.6 CMS-GC-Event file

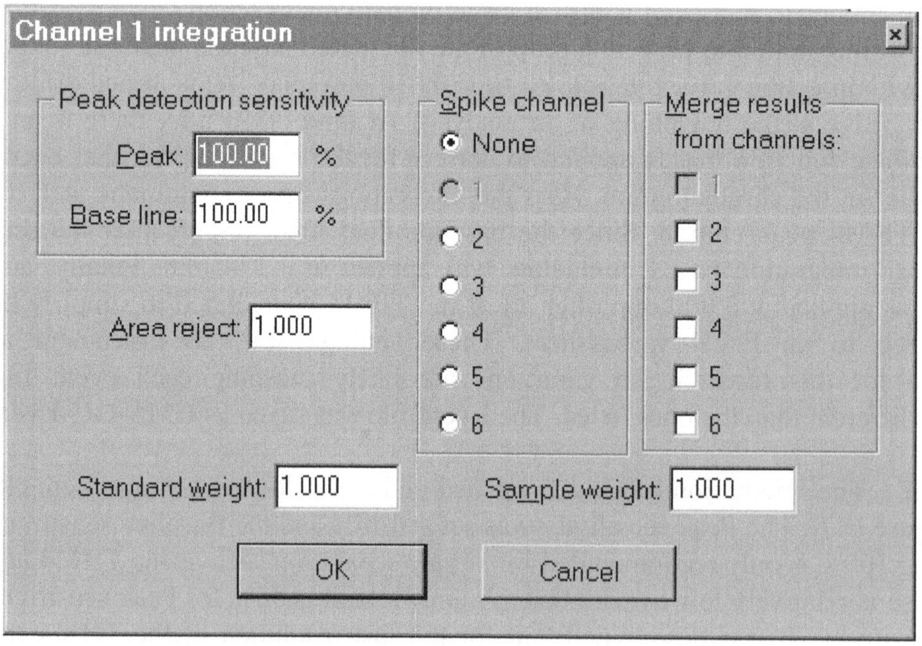

Figure D.7. Integration parameters for CMS-GC-ECD

D.2.8 Fifth, the Component menu is opened and the THM4 species are added according to a previously analyzed standard where retention times were recorded. An actual chromatogram was needed to add a retention window range for each component (Figure D.8). Each analyte has a separate edit menu as shown in Figure D.8 for chloroform. Chloroform will be used as an example for the set up of each component. Figure D.9 depicts a typical chromatogram segment showing just chloroform (~20 µg/L). Only the chloroform peak and the Component details menu for chloroform are shown in detail to exemplify typical component setup. Chloroform was labeled peak #1 and a general retention window where the maxima of the peak will reproducibly occur was given. The units of peak measurement were set to area and only the peak within the retention window that has the highest area will be measured. Other much smaller peaks are either skimmed off or separated by the integration functions. This was done for each of the THM4. Once this is completed the component was saved as **CMS-THM4.cpt.**

D.2.9 After the component file has been saved each THM was given a calibration file to be completed as calibration standards are analyzed. A different file must be named and saved for each THM if subsequent calibrations of the CMS-GC-ECD are performed. For example, Figure D.10 depicts the calibration file **062005-Houston-CMS-Cl3.cal** used during the Houston on-line study. The three additional files were named similarly. A quadratic fit is being used for chloroform, since the CMS has a non-linear trend for chloroform past the 25 to 30 µg/L range. Because Houston had higher levels of THM4 than Memphis a broader and higher calibration range was selected for the THM4 as noted by the fifth calibration standard ~120 µg/L of TTHM4. The $Ax^2 + Bx + C$ method for calibration was selected. The "Injected" column is the concentration of the THM in the calibration standard. The Area/height column is the corresponding area for a THM for a particular concentration in the calibration standard. The calibration files enable for immediate calculation of concentration for each THM which is then sent to an Excel spreadsheet by using a DDE link (section D.3).

D.2.10 In the Post-run actions menu (Figure D.11), a general chromatogram file name was inputted with **_01.chr** placed at the end for the first analysis. The auto increment box was selected to allow subsequent analysis to be named accordingly (ex. Houston study for number of hours). The Save results and the Add to results boxes were also selected to allow the chromatogram results to be transferred to the Excel spreadsheet. The Update DDE link box was selected and the Restart run after box. The Restart run after box was only selected for the on-line monitoring studies. The CMS-GC was set to restart after 53.00 minutes from each previous run, thus one analysis per hour during the on-line monitoring studies. The number of times to restart can either be set at 99 times or infinite by inputting zero. All other settings were left at default.

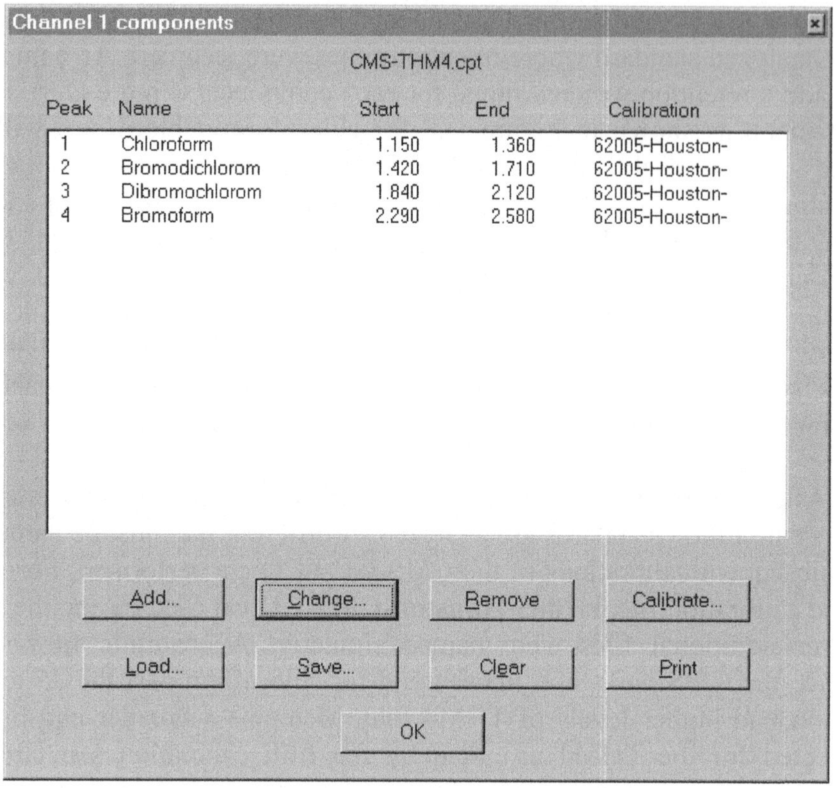

Figure D.8 Component list for CMS-GC-ECD

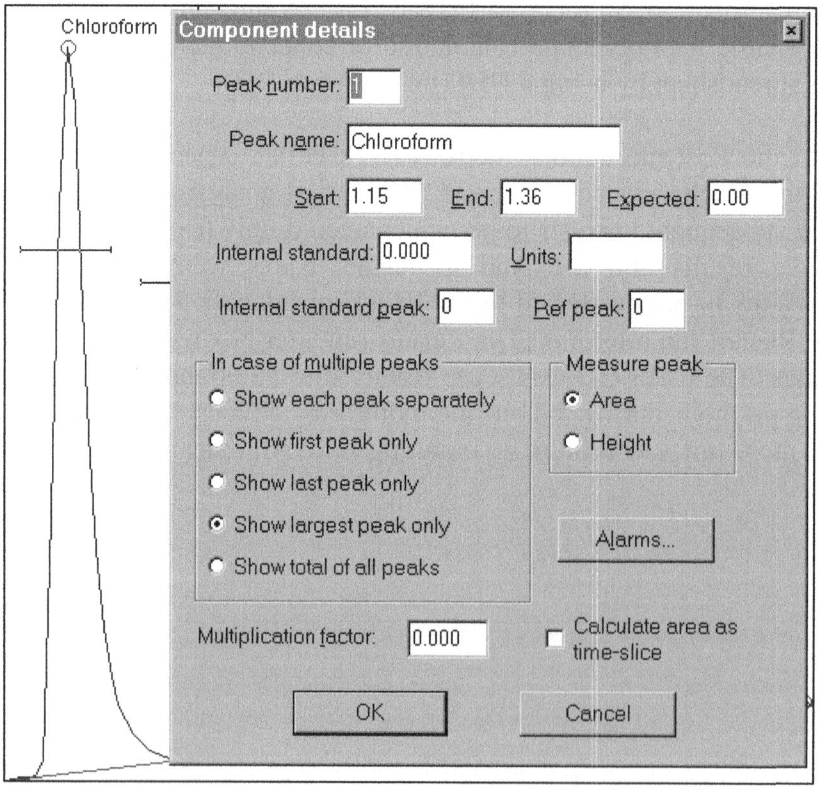

Figure D.9 Chloroform peak and component details menu

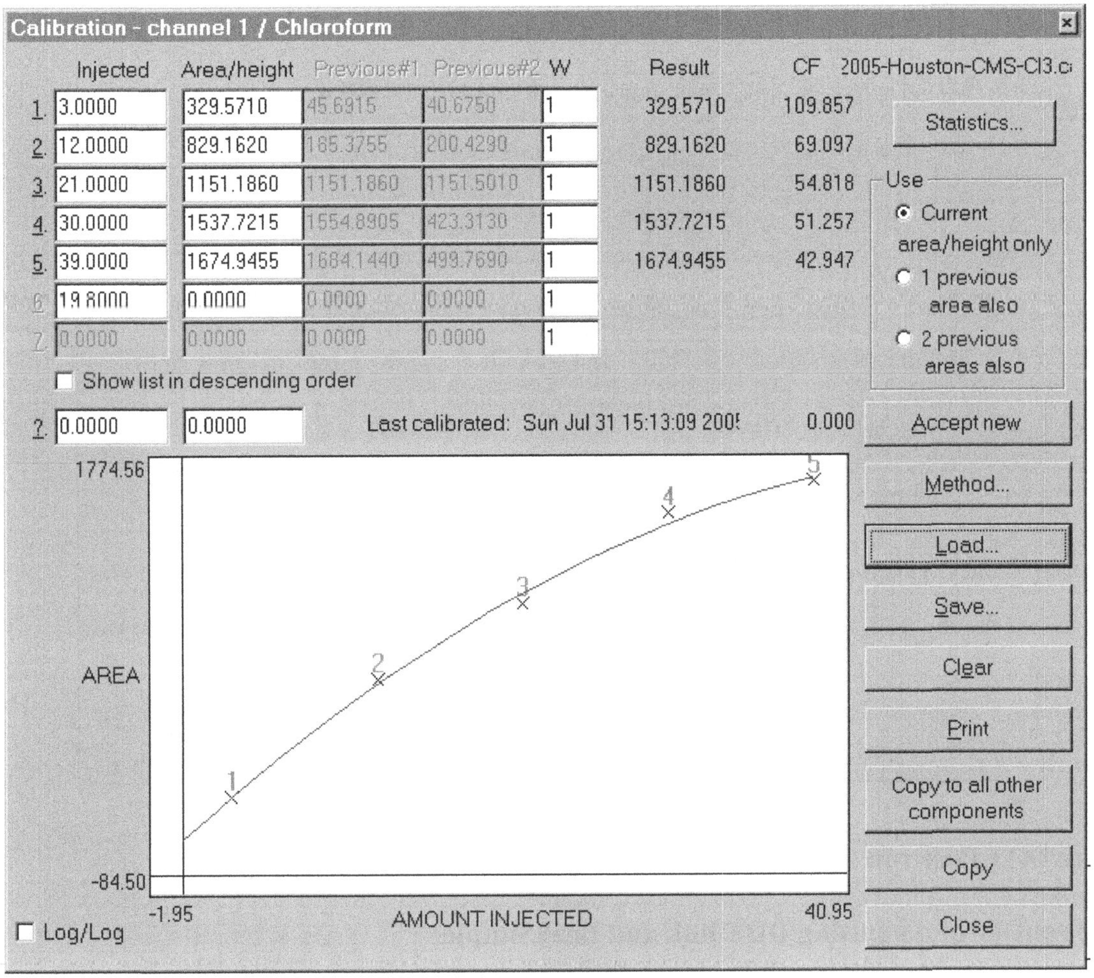

D.10 Calibration file for chloroform used during Houston study

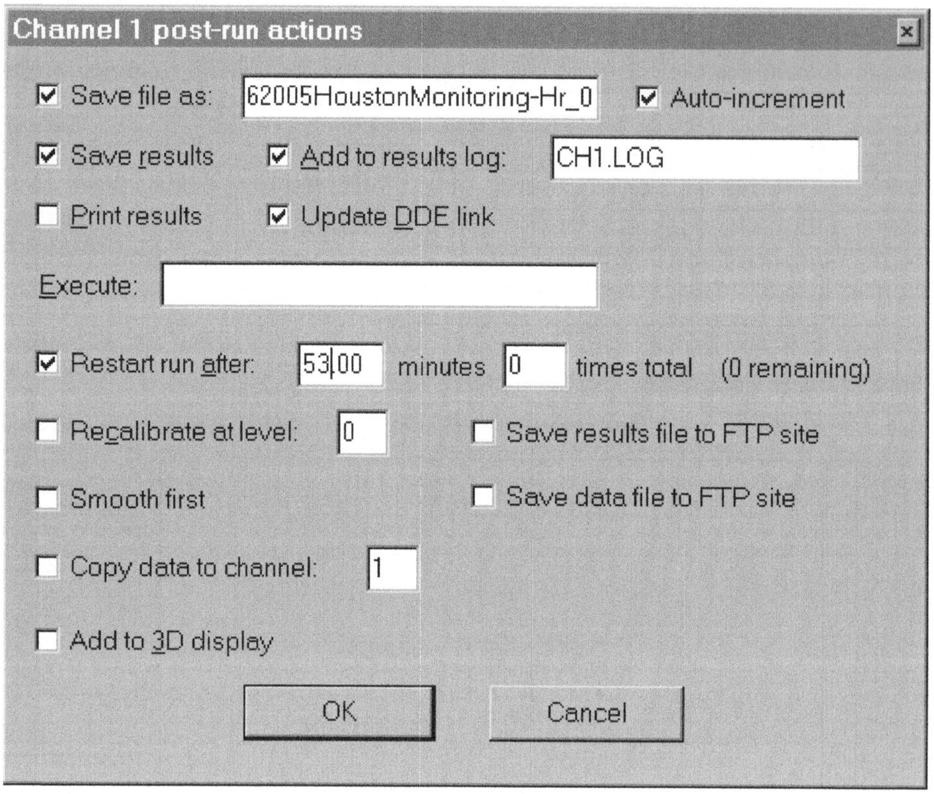

Figure D.11 Post-run actions for CMS-GC-ECD

D.3 Transfer of data using DDE link and PeakSimple

D.3.1 When all of the calibration standards are analyzed, the calibration plots can be inputted into the PeakSimple software as mentioned previously. Once this is done all data from future analyses will be entered immediately into the plots as peak area for each component and the corresponding concentration will be calculated.

D.3.2 These results are then added into a ***.log** file for storage or transfer. By using a DDE link to an EXCEL spreadsheet, the results can be put in a single desired row immediately after the analysis (Figure D.12). The file name, date, time, component, retention time, area, and concentration columns were labeled in an appropriately named EXCEL spreadsheet prior to beginning a study. To make a DDE link with the chosen spreadsheet a copy of the desired results log file must be made from PeakSimple. This copy is placed in the EXCEL worksheet by using the PasteSpecial (select link option) function in EXCEL. Place the copy in the first column underneath the previously mentioned headings. The results should appear in the order they were saved (chronological) and as they appear in the results log file. The spreadsheet must be open at all times to receive new results. Make sure that the update DDE link box in the PeakSimple post-run menu is selected.

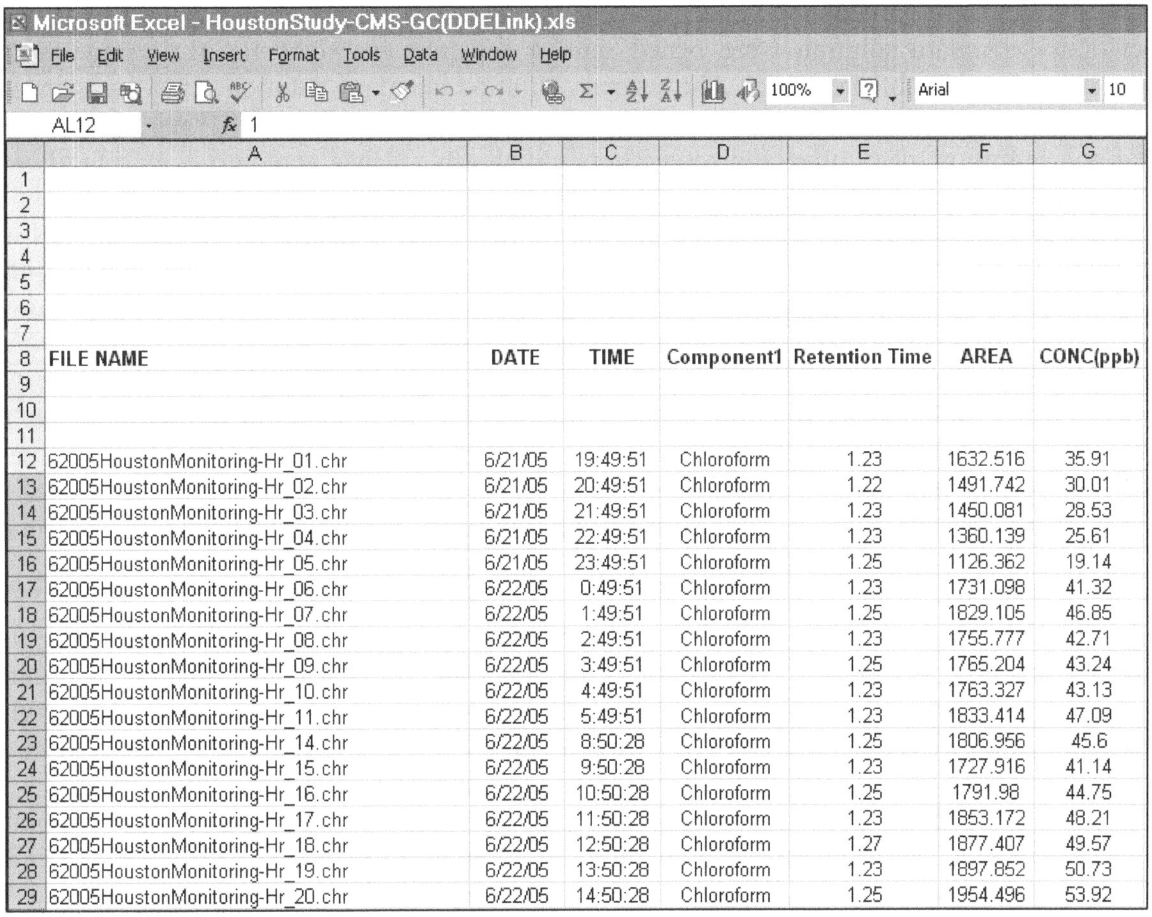

Microsoft Excel - HoustonStudy-CMS-GC(DDELink).xls

AL12 fx 1

FILE NAME	DATE	TIME	Component1	Retention Time	AREA	CONC(ppb)
62005HoustonMonitoring-Hr_01.chr	6/21/05	19:49:51	Chloroform	1.23	1632.516	35.91
62005HoustonMonitoring-Hr_02.chr	6/21/05	20:49:51	Chloroform	1.22	1491.742	30.01
62005HoustonMonitoring-Hr_03.chr	6/21/05	21:49:51	Chloroform	1.23	1450.081	28.53
62005HoustonMonitoring-Hr_04.chr	6/21/05	22:49:51	Chloroform	1.23	1360.139	25.61
62005HoustonMonitoring-Hr_05.chr	6/21/05	23:49:51	Chloroform	1.25	1126.362	19.14
62005HoustonMonitoring-Hr_06.chr	6/22/05	0:49:51	Chloroform	1.23	1731.098	41.32
62005HoustonMonitoring-Hr_07.chr	6/22/05	1:49:51	Chloroform	1.25	1829.105	46.85
62005HoustonMonitoring-Hr_08.chr	6/22/05	2:49:51	Chloroform	1.23	1755.777	42.71
62005HoustonMonitoring-Hr_09.chr	6/22/05	3:49:51	Chloroform	1.25	1765.204	43.24
62005HoustonMonitoring-Hr_10.chr	6/22/05	4:49:51	Chloroform	1.23	1763.327	43.13
62005HoustonMonitoring-Hr_11.chr	6/22/05	5:49:51	Chloroform	1.23	1833.414	47.09
62005HoustonMonitoring-Hr_14.chr	6/22/05	8:50:28	Chloroform	1.25	1806.956	45.6
62005HoustonMonitoring-Hr_15.chr	6/22/05	9:50:28	Chloroform	1.23	1727.916	41.14
62005HoustonMonitoring-Hr_16.chr	6/22/05	10:50:28	Chloroform	1.25	1791.98	44.75
62005HoustonMonitoring-Hr_17.chr	6/22/05	11:50:28	Chloroform	1.23	1853.172	48.21
62005HoustonMonitoring-Hr_18.chr	6/22/05	12:50:28	Chloroform	1.27	1877.407	49.57
62005HoustonMonitoring-Hr_19.chr	6/22/05	13:50:28	Chloroform	1.23	1897.852	50.73
62005HoustonMonitoring-Hr_20.chr	6/22/05	14:50:28	Chloroform	1.25	1954.496	53.92

Figure D.12 Screenshot of PeakSimple results transferred to EXCEL spreadsheet using a DDE link

D.4 Preparation of Stock Standards and Primary Dilution Standards for Analysis

D.4.1 All water used was distilled and then deionized by a Millipore Mill-Q Plus/Mill-RO® 10 Plus Water Purification System Glassware and sample containers were meticulously cleaned with concentrated detergent soaked for 24 hours and then rinsed with tap water three times followed by reagent grade water several times. All chemicals were HPLC reagent grade or better and were purchased from Fischer Scientific Inc. Stock standard solutions for the THM4 were purchased from Aldrich Chemical Company.

D.4.2 (Note all preparations should be carried out in hood) A separate disposable Pasteur pipet was used to transfer ~ 1 mL of each THM4 stock standard solution into separate 5 mL amber borasilicate vials with Teflon lined septum. Each vial was labeled accordingly. An additional four vials were each filled with methanol and labeled with each THM4 formula. These were used to wash the 10 μL gas tight syringes after transferring an aliquot from the stock standard solution.

D.4.3 A 10 mL grade (A) volumetric flask was filled with ~9.8 mL of methanol and allowed to sit until the alcohol-wetted surfaces had dried. Different aliquots (1.0 to 2.0 μL) of each stock standard solution were added with a separate gas tight 10 μL syringe without contacting the sides of the flasks. The primary dilution standard solution (PDS) was inverted gently three times to mix. The flask was labeled with the analytes included, their concentration, total concentration, date, time, and researchers initials (ex. THM4 standard, ~ 0.3 μg/μL each, 1.98 μg/μL total, 06/10/05, 11:00 am, MAB). This standard lasts ~3 to 4 weeks with minimal headspace. The solution was transferred into two 5 mL amber borasilicate vials. Both vials were labelled exactly as with the volumetric flask. The use of these vials limits the chance of spilling the solution as with round bottom flasks, reduces exposure inside the chemical refrigerator and reduces head space. The vials were kept chilled until needed.

D.4.4 This primary dilution standard solution was then added in desired amounts to a 100 mL grade (A) volumetric flask containing reagent grade water and inverted several times to produce different concentrations of the calibration standard solutions or check standard solutions. The standard lasts approximately one hour, but all standards were usually used in 15- 30 minutes

D.4.5 A prepared calibration standard solution can be analyzed by the CMS-GC-ECD only after the initial conditions for the instrument and software are reached. The peristaltic pump was attached downstream of the CMS at the sample outlet to avoid contamination. The sample inlet was placed in a clean 100 mL grade (A) volumetric flask containing reagent water and the peristaltic pump was turned on. The CMS was purged for at least 15 minutes with reagent water.

D.4.6 The sample inlet tubing was then placed in the first (low) level calibration standard solution. This standard was pumped for 15 minutes through the CMS. After 15 minutes, the channel 1 start run icon or the space bar was pressed to start the PeakSimple software. The injection valve rotated (with distinct noise) and the software started (turns from red to green in the CH 1 run icon) if working properly. The sample inlet was placed back in the reagent water after being washed off and wiped with a lint free cloth. This procedure was repeated for the remaining calibration standard solutions and the seven check standard solutions for a total time of ~ 12 hours for calibration. The results were then inputted into the calibration files as mentioned in section D.2.9. The CMS device can be disassembled for cleaning or replacement of the silicone membrane if necessary. High standards above 200 μg/L of total THM4 may require the CMS to be purged an additional 15 minutes if additional carry-over is evident.

APPENDIX E
CMS-FIA NCA-FL EXPERIMENTAL PROCEDURE

E.1 CMS-FIA Setup

E.1.1 Fluorescence detector must be turned on an hour before a study in order for the lamp to warm up.

E.1.2 Flow rates were checked and set according to maximum fluorescence intensity.

E.1.3 The heat bath was turned on and water was kept constantly at 97°C. The ice bath was prepared with fresh ice and changed regularly.

E.1.4 The solutions of 30% NCA and NaOH were prepared as described in section B.5.1 and B.5.2

E.1.5 Calibration Standards were prepared as described in Section B.5.3 and B.5.4.

E.1.6 Tubing connections were checked for leaks and calibration standards were analyzed.

E.2 Software Setup

E.2.1 The PeakSimple software (previously installed) was loaded. An initialization sequence took place between the software and the data acquisition board.

E.2.2 The Control File was loaded from the File pull-down menu as described in section D.2.2 (Control File should be named after the method used. For example, for this particular method control file should be named **CMS-FIA NCA-FL-control.con**.

E.2.3 The Channels menu was opened by clicking on a sub-menu '1234' (Figure D.3). Event File was opened from the Channels menu (Figure E.1). The Event File allowed controlling the ten-port injection valve and integrating HAA5 and THM4 peaks. Figure E.1 lists the final events that were added at each time depicted. At time 0.000 min, relay (C) and integration of baseline is turned off. At 8.000 min relay (C) is turned on to prepare for the injection valve actuation. At 8.010 min relay (C) is turned off, thus actuating the injection valve slightly past the beginning of an analysis. The same event takes place at 14.000 min and 14.010 min, respectively. Since the baseline tends to shift, an integration-based immediate function at 16.500, 20.000 and 26.500 min was applied to correct for the shifts in the baseline. The file was saved as **THM4_HAA5.evt**.

E.2.4 The Integration menu was opened (Figure D.7) and default settings were used throughout the study.

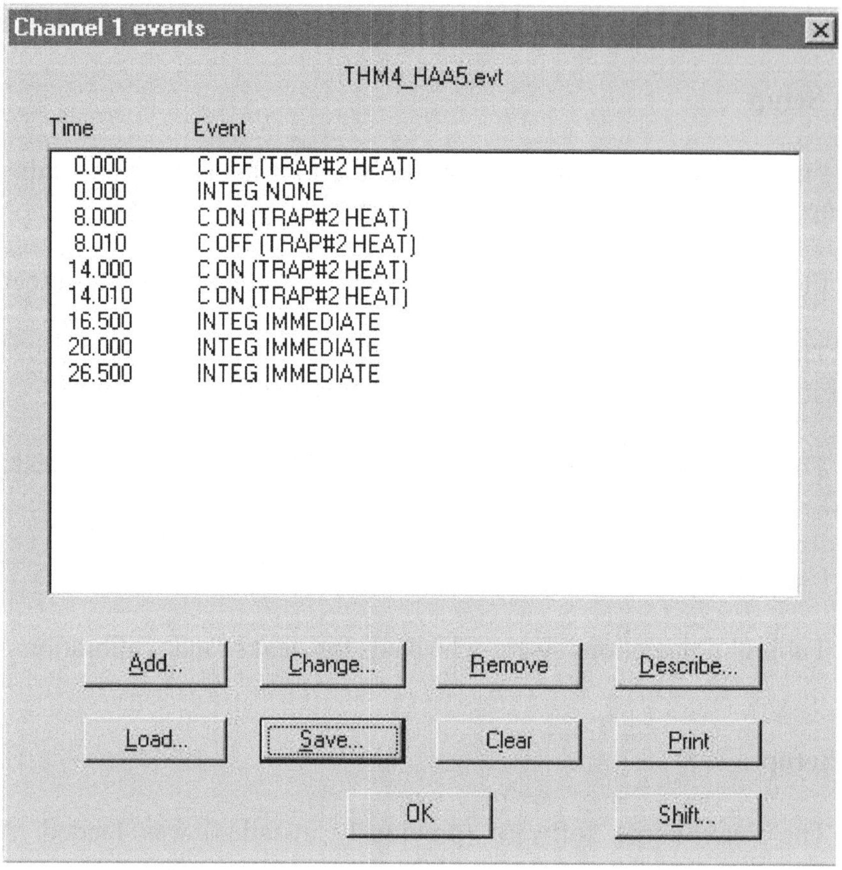

Figure E.1 Event file for CMS-FIA NCA-FL

E.2.5 The Component list was downloaded from the Channel menu (Figure E.2). The THM4 and HAA5 species were added according to a previously analyzed calibration standard where retention times were recorded. Addition of components was described in Section D.2.8. The component file was saved as **HAA5_THM4.cpt.**

E.2.6 After the Component list was created; a calibration file was prepared for THM4 and HAA5. The calibration file for THM4 was named THM4.cal and calibration file for HAA5 was named HAA5.cal. The single line (Ax+B) best fit was selected for the study since the THM4 and HAA5 species are linear over the concentration range of interest. The Injected column is the concentration of THM4 or HAA5 in the calibration standard. The area/height column is the corresponding height for the THM4 or HAA5 species in the calibration standard.

E.2.7 The Details box of the Channel menu was selected and all settings were left as default (Similar to Figure D.4). Changes were made for sampling time. Sampling rate of 5 Hz was chosen for the study. End of program was set at 45.0 minutes. Default display limits were set according to the fluorescence detector used.

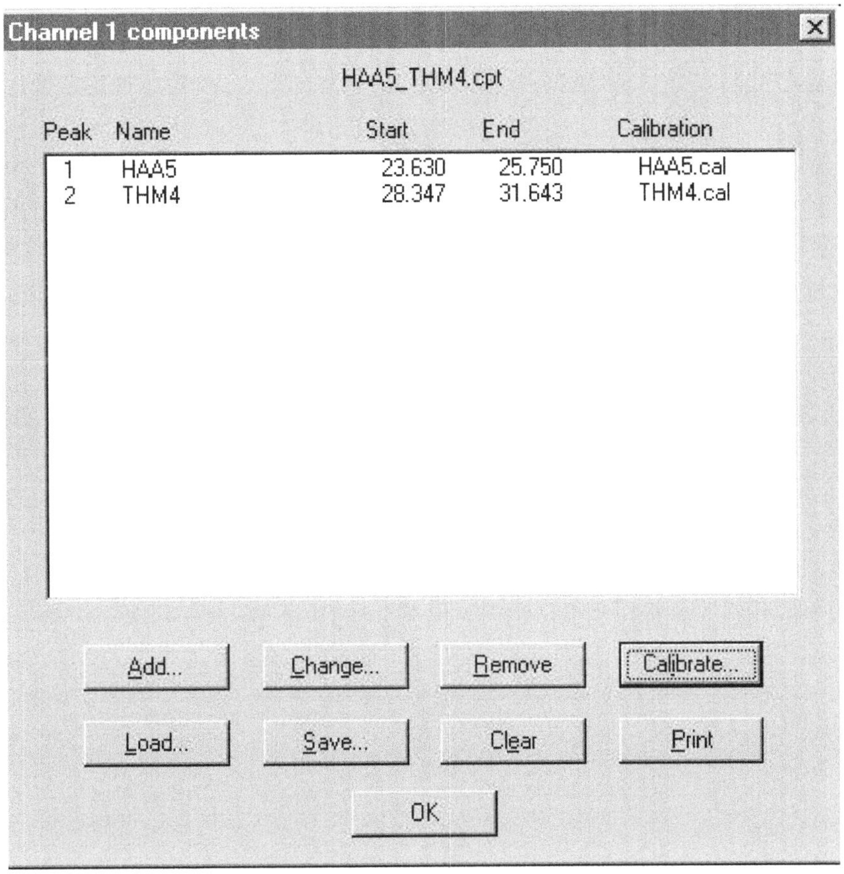

Figure E.2 Component list for CMS-FIA NCA-FL

E.2.8 The Post-run menu was opened and it was set that the event file would automatically restart after 15 minutes at the end of each run or until the operator stops it manually. During this procedure, the file names were inputted with _01.chr placed at the end for the first analysis. The auto increment box was selected to allow subsequent analysis to be named accordingly to a study performed (ex., Houston study for number of hours). All other settings were left as default.

APPENDIX F
PC-IC EXPERIMENTAL PROCEDURE

F.1 Start-up procedures

Note-1: Disconnect the detector from the Post-column system initially.
Note-2: Make sure PCIC.con, PCIC.tem, PCIC.cpt and PCIC.evt are in the root peaksimple directory (usually C:\Peak312\ -- if using PeakSimple version 3.12)
Note-3: Peristaltic pump tubing lasts between 12 to 15 hours.
Note-4: Check ice bath every 12 to 15 hours.

F.1.1 Fill the Heat bath, and turn it on and bring temperature to ~94 °C. Also, connect the luer adaptor to the inlet tubing on the detector, and using a 1mL syringe, flush with DI water. Then turn the detector on, and let it go through its startup procedure.

F.1.2 Fill the ice bucket with water and ice, place between detector and heat bath.

F.1.3 After the detector has booted, press the shift key, then the Home key. Press the "+/-" button to turn off "detector idle mode".

F.1.4 Set the gain to 25, the excitation wavelength to 365 nm, the emission wavelength to 455 nm, and the range to 10,000 EUFS. The detector is now ready for operation.

F.1.5 Start the PeakSimple software, then go to the "File" menu > open control file, then select and open the **PCIC.con** file (Figure F.1). This will load all the relevant settings used, and the event, component and temperature files. Also, open the relay window by going to the "View" menu > relay/pump window (Figure F.2)

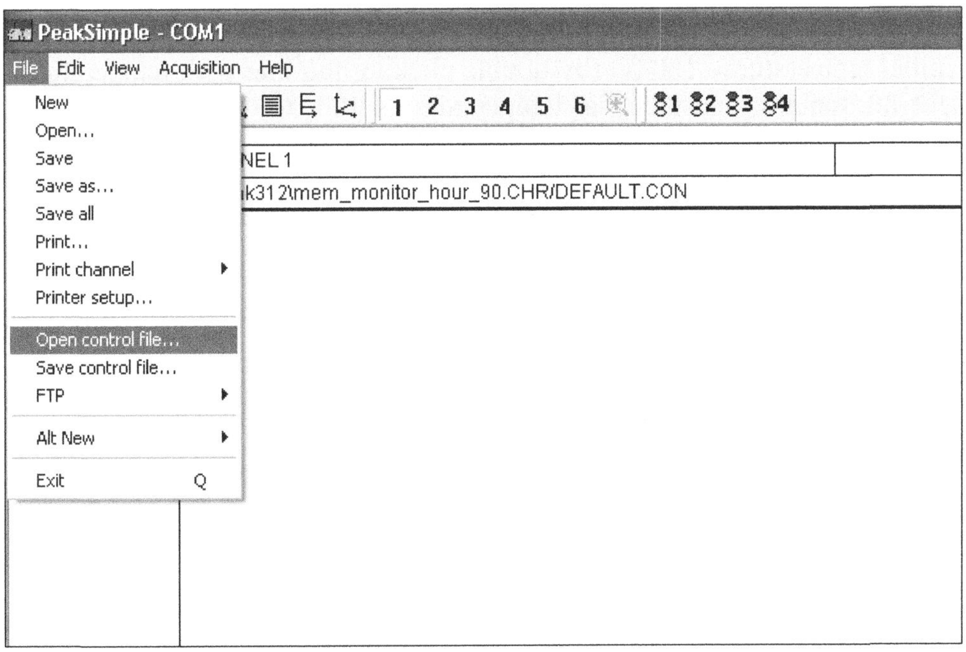

Figure F.1 Screenshot of opening the control file for PC-IC

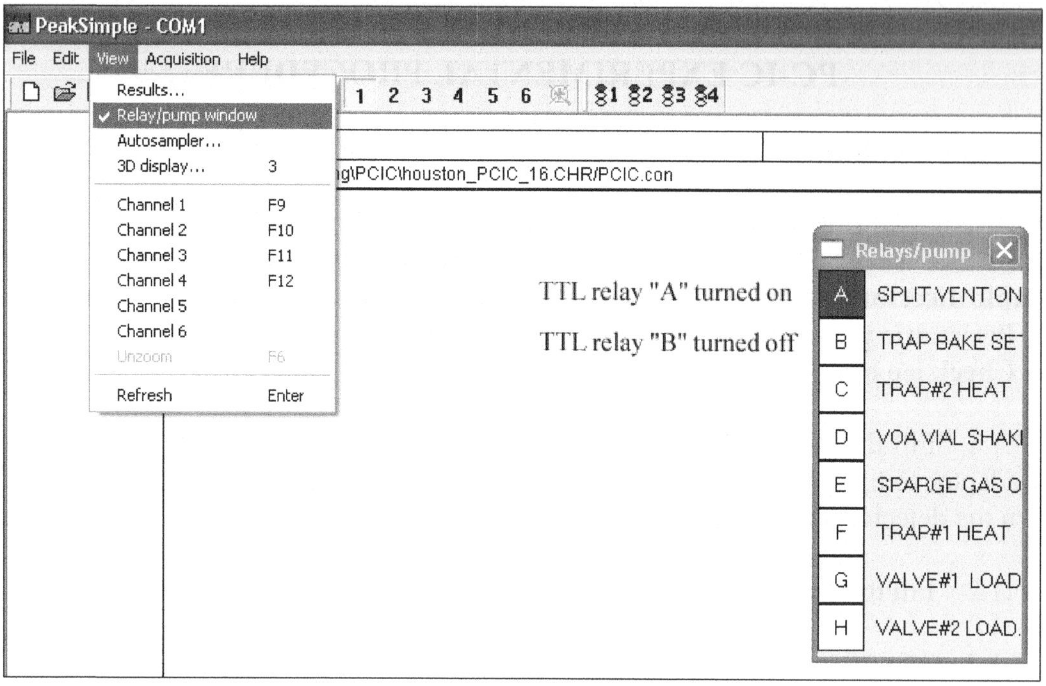

Figure F.2 Screenshot of opening the Relay/Pump window and illustrating an On and Off relay file for PC-IC

F.1.6 In the PeakSimple relay window, turn on TTL outs "A" and "B". Then turn them both off. Turn "A" and "B" back on. This resets the pump, and allows for automation by PeakSimple. Turning "A" on holds the gradient program from running. Turning "B" on turns the pump off, preventing it from pumping fluid. Later, "B" will be turned back on to do flow rate checks.

F.1.7 Proceed to the menu "Edit > Channels > Temperature on channel 1" (Figure F.3) Make the initial temperature ~100 °C (by double clicking the line with temperatures on it), the hold time for 60 minutes, and the final temperature as 100 °C (Figure F.4). Press OK three times, and return to the relay window. Turn relays "E" and "F" on then off again, this resets the peristaltic pump, readying it for use and it should be pumping at this point. Place the two PEEK reagent inlet sections into reagent water, while the flow rates are being adjusted.

F.1.8 Turn off the peristaltic pump by turning relays "E" and "F" on. Set the IC flow rate to 0.5 mL/min, and check using a 10mL graduated cylinder. Put the column out (after the analytical column) into the graduated cylinder and time it for 10 minutes. The flow should be approximately 0.5 mL/min. This should stay constant. Pour the column eluent into waste.

F.1.9 Connect the IC column out back into Mixing Tee (MT2), and turn the peristaltic pump back on, by turning off relays "E" and "F".

F.1.10 Disconnect the 2 m KOT from MT2, and measure the flow rate the same way the IC flow was measured. The total flow should be approximately 0.3 mL/min. Reconnect the KOT to MT2.

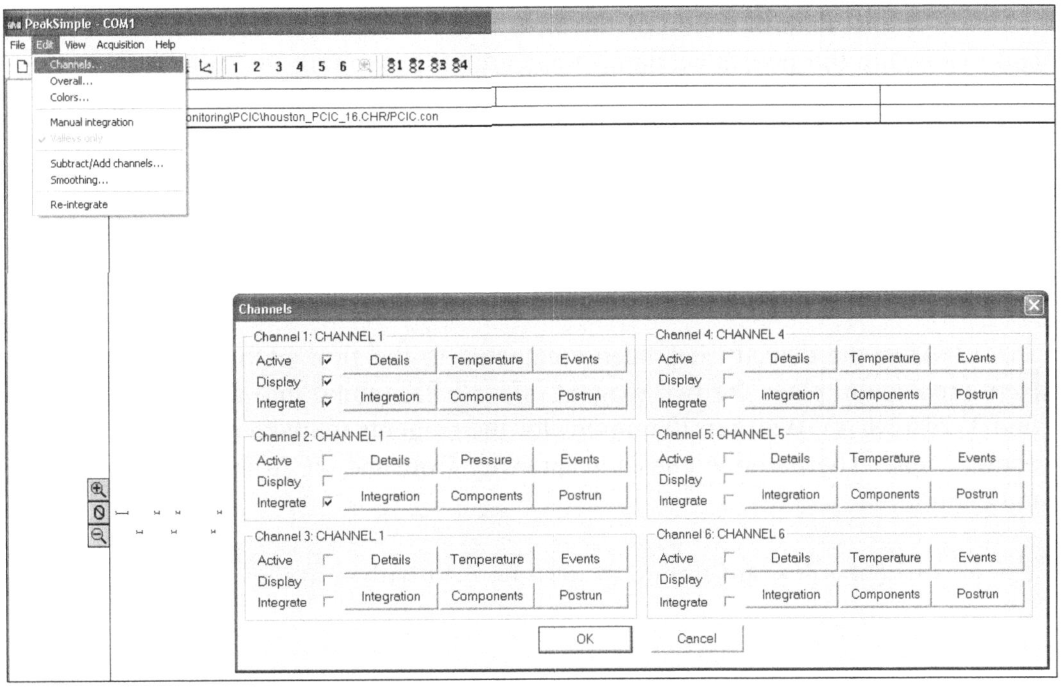

Figure F.3 Screenshot of opening Channels Area for PC-IC

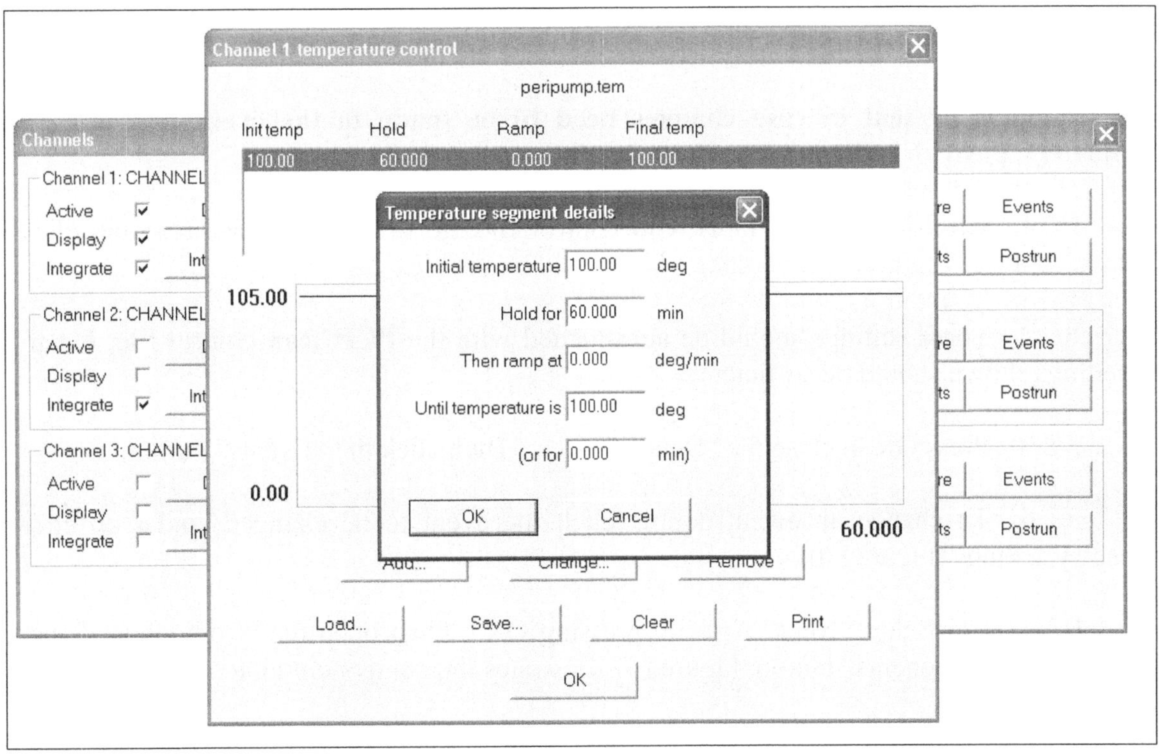

Figure F.4 Figure illustrating how to set temperature, which controls the peristaltic pump speed for PC-IC

F.1.11 Once the flow rates have been determined, connect the detector to the PC-IC system, and do a final determination of the flow rates. If the flow rate has slowed down, first check and make sure the peristaltic pump lines are still pumping by introducing an air bubble into the system, and making sure it flows more or less unimpeded. A small amount of backpressure is expected due to connecting the detector, the air bubble should bounce as it makes its way to MT2. If there are no problem with general flow, increase the temperature setting 10°C for every 0.05 mL/min the total flow is off, and measure the flow rate again, the total flow should be approximately 0.8 mL/min, where the total flow is the determined 0.5 mL/min (IC flow) + 0.3 mL/min (peristaltic pump flow).

F.1.12 After the detector has been connected to the flow path, and the flow rates have been determined again, place the two reagent inlet PEEK tubing into the separate reagents (2M NaOH and 37.5% NCA). Wait for 20 minutes for the reagents to make it through the heat and ice bath reaction traps. The baseline emission intensity from the NCA should be ~350 to 500 EU (Emission Units)

F.1.13 Place the 35m reaction trap into the heat bath, and the 2m cooling KOT into the ice bucket.

F.1.14 Inject standards using a female luer adapter connected to the sample inlet on the valve. Using a 1mL syringe, inject ~4 mL though the sample loop. Make sure PeakSimple is the active program on the computer, and hit space to start.

F.2 Other Software Settings

This section is present in case changes need to be made to the event file, integration parameters, postrun settings, or various details.

F.2.1 After loading the **PCIC.con** control file, go to the channels area, and click the "Details" button. Figure F.5 illustrates what should be visible.

F.2.2 These settings should be auto-loaded with the **PCIC.con** control file, but if not, the settings shown should be sufficient.

F.2.3 Press OK to close the "Details" Area. Then click the "Events" button.

F.2.4 To change an event, double click the event to be changed, and then alter the necessary settings (Figure F.6).

F.2.5 Close the "Event" Area by pressing OK. Then open the "Components" area by clicking the "Components" button. Figure F.7 illustrates the component table.

F.2.6 To change the time on a component, left clicking and holding the retention window bar, and moving is sufficient (Figure F.8).

F.2.7 To add a component, right click the around the time the component is too be added (Figure F.8).

F.2.8 After adding the component, right click the retention window bar, and left click, "Edit Component" (Figure F.9) In this dialog box, the peak number, name, and retention times can be edited, as well as other settings if needed (Figure F.10).

F.2.9 Click OK to close out to the "Component" area, and then press the "Integration" button.

F.2.10 Figure F.11 illustrates the "Integration" area; make sure the settings match those in the figure.

F.2.11 Close the "Integration" area by pressing OK, then open the "Postrun" area by clicking the "Postrun" button. Figure F.12 illustrates the "Postrun" area.

F.2.12 Be sure the postrun settings match the settings shown in Figure F.12. The only difference is the "Restart run after" check box. For monitoring purposes, the "Restart run after" check box needs to be checked. For calibration, turn this setting off by deselecting the box.

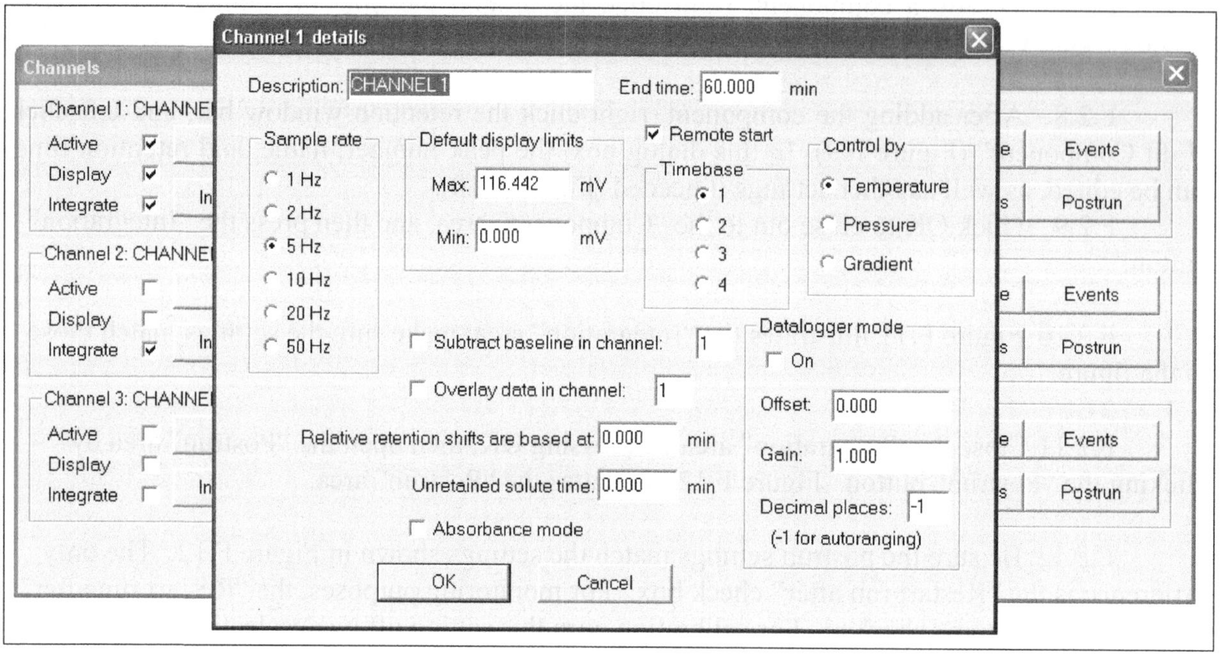

Figure F.5 Settings in the "Details" menu for PC-IC

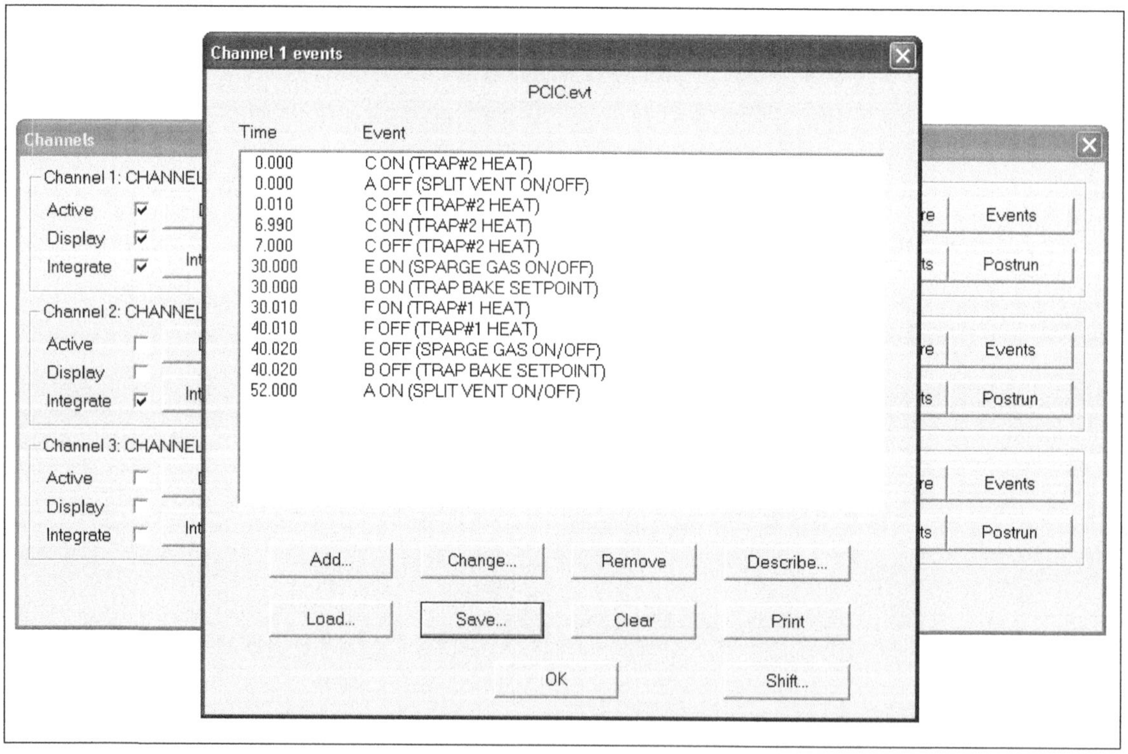

Figure F.6 Settings in the event file for PC-IC

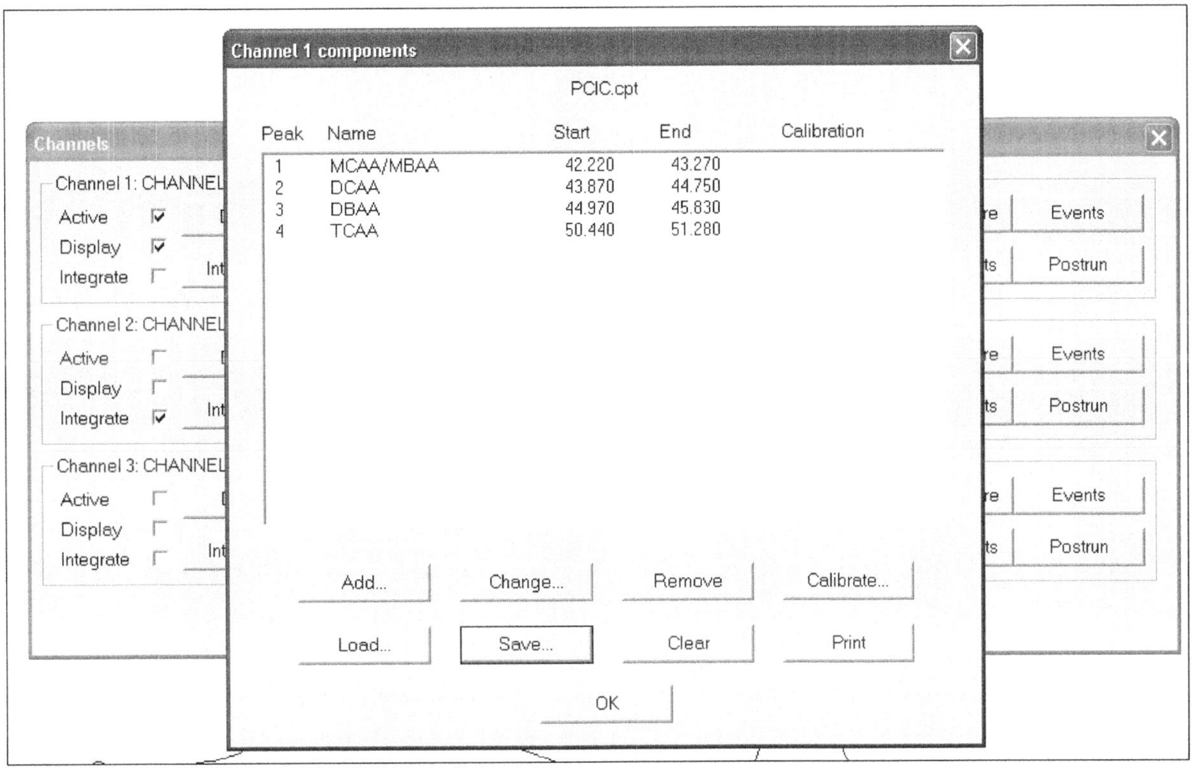

Figure F.7 List of components in the component table for PC-IC

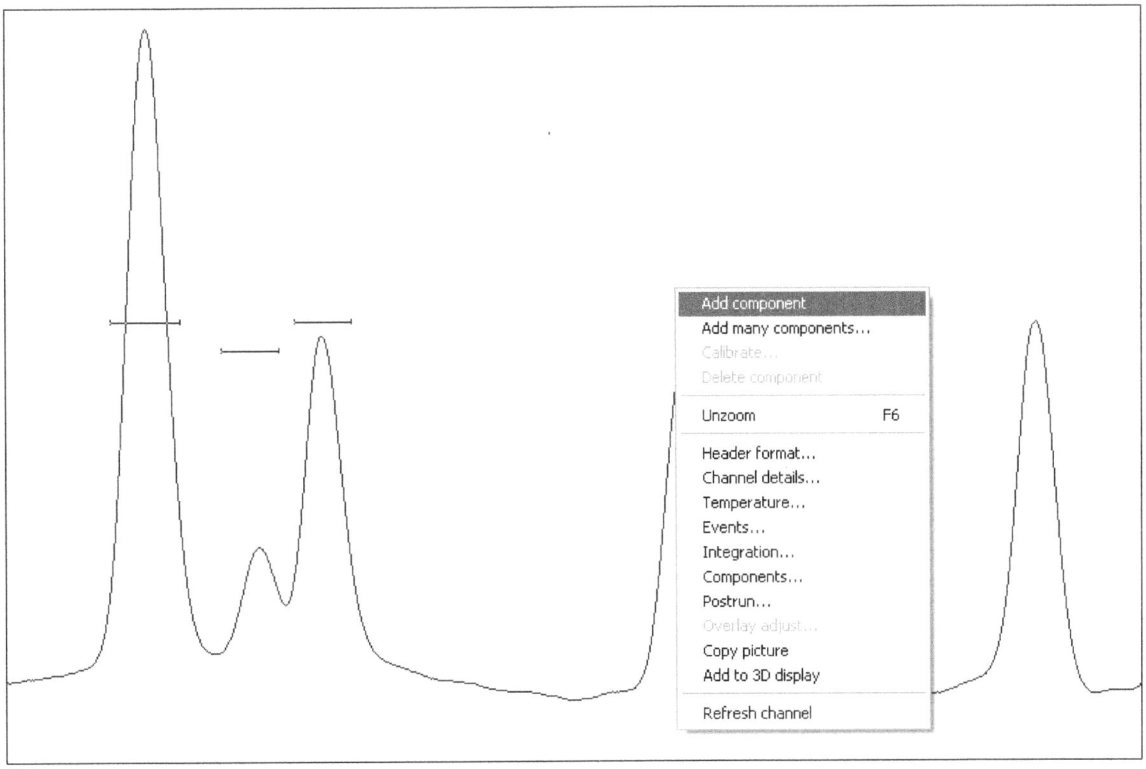

Figure F.8 Screenshot of adding a component for PC-IC

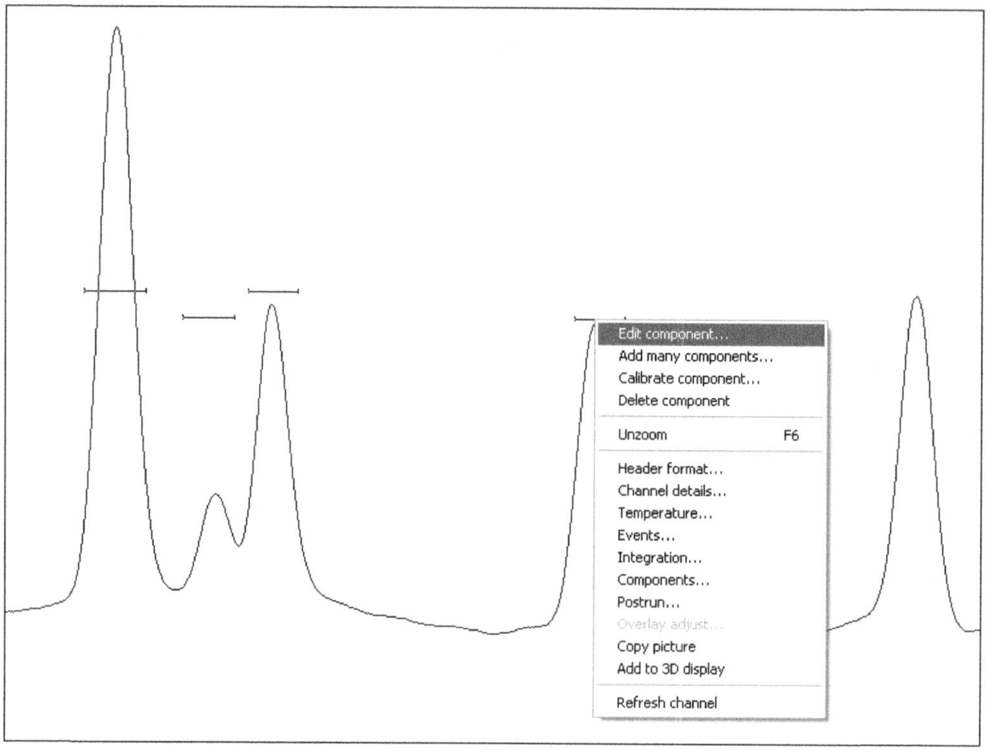

Figure F.9 Screenshot of editing a component for PC-IC

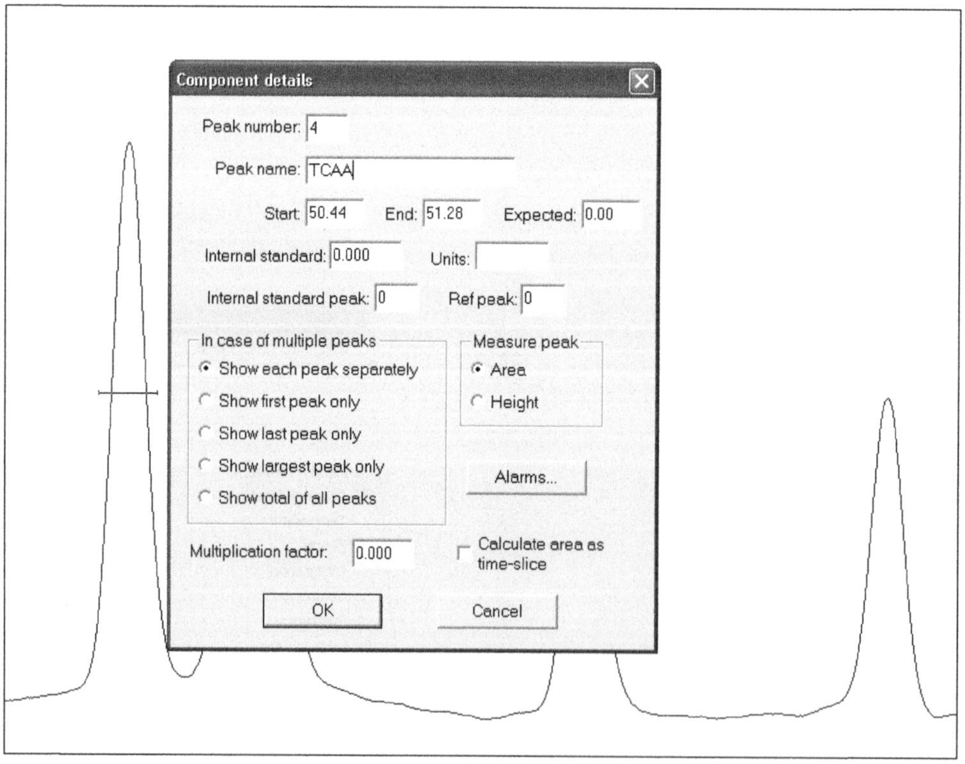

Figure F.10 Dialog box to add names, peak numbers, retention times and other settings for PC-IC

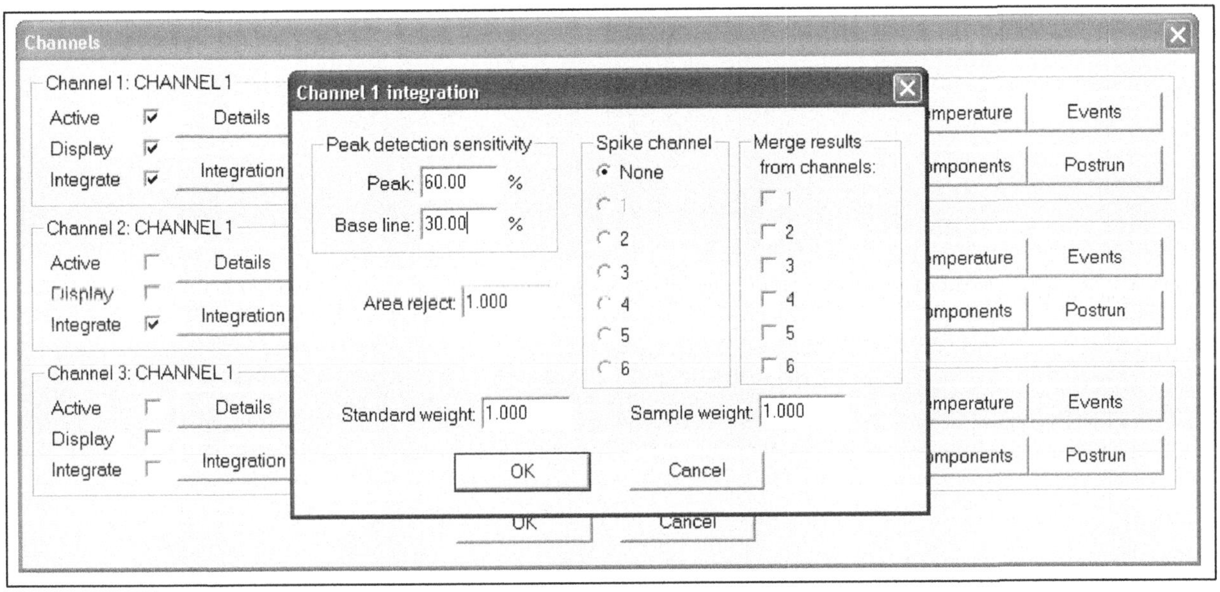

Figure F.11 Integration Settings for PC-IC

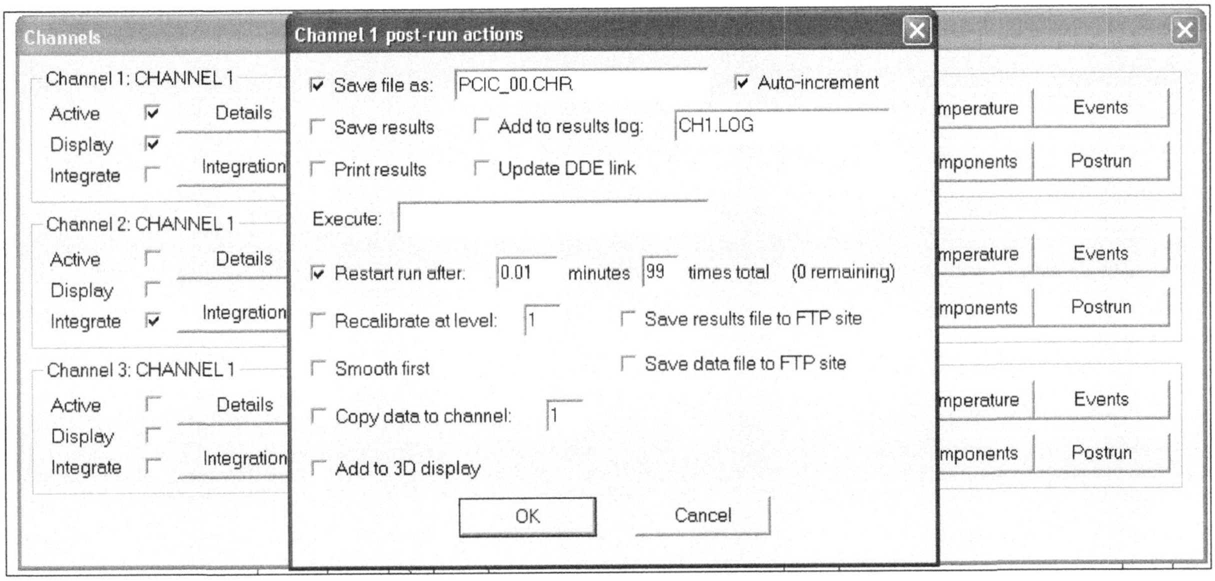

Figure F.12 Postrun settings for PC-IC

REFERENCES

Allonier, A.-S., M. Khalanski, A. Bermond and V. Camel. 2000. Determination of Trihalomethanes in Chlorinated Sea Water Samples Using a Purge-and-Trap System Coupled to Gas Chromatograph. *Talanta,* 51(3): 467-477.

Aoki, T. and K. Kawakami. 1989. Continuous Flow Method for Determination of Total Trihalomethanes in Drinking Water with Membrane Separation. *Wat. Res.,* 23(6): 739-742.

APHA, AWWA, and WEF. 1992. *Standard Methods for the Examination of Water and Wastewater.* 18th ed. Washington, D.C.: APHA.

Barron, L. and B. Paull. 2004a. Determination of Haloacetic Acids in Drinking Water Using Suppressed Micro-bore Ion Chromatography With Solid Phase Extraction. *Anal. Chim. Acta.* 522(2): 153-161

Barron, L. and B. Paull. 2004b. Direct Detection of Trace Haloacetates in Drinking Water using Microbore Ion Chromatography. Improved Detector Sensitivity using a Hydroxide Gradient and a Monolithic Ion-Exchage Type Suppressor. *J. Chromatogr. A.*, 1047(2): 205-212

Barron, L., P. Nesterenko and B. Paull. 2005. Use of Temperature Programming to Improve Resolution of Inorganic Anions, Haloacetic Acids and Oxyhalides in Drinking Water by Suppressed Ion Chromatography. *J. Chromatogr. A.*, 1072(2): 207-215

Blanchard, R. D. and J. K. Hardy. 1984. Use of a Permeation Sampler for Determination of Volatile Priority Pollutants. *Anal. Chem.,* 56(9): 1621-1624.

Brown, M. A., Emmert, G. L. 2006. On-line Monitoring of Trihalomethane Concentrations in Drinking Water Distribution Systems Using Capillary Membrane Sampling–Gas Chromatography. *Anal. Chim. Acta* 555, (2006), 75-83.

Cao, G. 2004. Further Explorations of Supported Capillary Membrane Sampling-Gas Chromatography. Master's Thesis. The University of Memphis, Memphis, TN.

Dean, J. A. 1985. *Lange's Handbook of Chemistry,* 13th ed. New York, NY: McGraw Hill.

Duty, C. N. 2000. Application of a Supported Capillary Membrane Sampler for the Analysis of Organic Compounds in Water. Master's Thesis. Arkansas State University, State University, Arkansas.

Emmert, G. L., G. Cao, C. Duty and W. Wolcott. 2004a. Measuring Trihalomethane Concentrations in Water Using Supported Capillary Membrane Sampling-Gas Chromatography. *Talanta,* 63(3): 675-682.

Emmert, G. L., G. Cao, G. Geme, N. Joshi, and M. Rahman. 2004b. *Methods for Real-Time Measurement of THMs and HAAs in Distribution Systems.* Denver, Col.: AwwaRF and AWWA.

Emmert, G. L. Brown, M. A., Liao, Z., Cao, G., Duty, C. 2006. Supported Capillary Membrane Sampling-Gas Chromatography on a Valve with a Pulsed Discharge Photoionization Detector. *Anal. Chim. Acta,* 560, 197-206.

Espenson, J.H. 1995. *Chemical Kinetics and Reaction Mechanisms.* 2nd ed. New York: McGraw-Hill.

Geme, G., M. A. Brown, P. S. Simone and G. L. Emmert. 2005. Measuring the Concentrations of Drinking Water Disinfection By-Products Using Capillary Membrane Sampling-Flow Injection Analysis. *Wat. Res.,* 39, 3827-3836.

Glaser, J. A. and D. L. Forest. 1981. Trace Analysis of Wastewaters. *Environ. Sci. Technol.,* 15(-): 1426-1435.

Global FIA INC.1999. Supported Capillary Membrane Sampler Operation Manual, Fox Harbor, WA.

Gordon, G., R.D. Gauw, G.L. Emmert, B.D. Walters, B. Bubnis. 2002. Treatment Technologies: Chemical Reduction Methods for Bromate ion Removal. *Jour. AWWA.*, 94(2): 91-98.

Gupta, K. S. and Y. K. Gupta. 1984. Hydrogen-Ion Dependence of Reaction Rates and Mechanism. J. Chem. Educ., 61 (11), 972-978.

Harris, D. C. 2001. *Quantitative Chemical Analysis.* 5th ed. New York, NY: W. H. Freeman and Company.

Kou, D., X. Wang, S. Mitra. 2004. Supported Liquid Membrane Microextraction With High-Performance Liquid Chromatography-UV Detection for Monitoring Trace Haloacetic Acids in Water. *J. Chromatogr. A.*, 1055(1-2): 63-69

Liao, Z. 2001. Development of a Supported Capillary Membrane Sampler-Gas Chromatograph for the Analysis of Volatile Organic Compounds in Water. Master's Thesis. Arkansas State University, State University, Arkansas.

Liu, Y. and S. Mou. 2003. Determination of Trace Levels of Haloacetic Acids and Perchlorate in Drinking Water by Ion Chromatography wit Direct Injection. *J. Chromatogr. A,* 997(1-2): 225-235.

Liu, Y., S. Mu, A. Lin, C. Jianhua, B. Du and J. Cui. 2003. Determination of Nine Haloacetic Acids and Six Anions in Drinking Water by Ion Chromatography with Large Volume Injection. *Chinese J. Chromatogr.,* 21(2): 181-183.

Lopez-Avila, V. 2000. *Development of an Experimental Procedure for Haloacetic Acids Using Ion Chromatography.* Denver, Col.: AwwaRF and AWWA.

Nicholson, B. C., B. P. Maguire and D. B. Bursill. 1984. Henry's Law Constants for the Trihalomethanes: Effect of Water Composition and Temperature. *Environ. Sci. Technol.*, 18(7): 518-521.

Nilsson, T., F. Pelusio, L. Montanarella, B. Larsen, S. Faccheti and J. Madsen. 1995. An Evaluation of Solid Phase Microextraction for Analysis of Volatile Organic Compounds in Drinking Water. *J. High Res. Chromatogr.*, 18(10): 617-624.

Pepich, B. V., M. M. Domino, T. A. Dattilio, P. S. Fair, D. J. Munch. 2004. Validating Sample Preservation Techniques and Holding Times for the Approved Compliance Monitoring Methods for Haloacetic Acids Under the US EPA's Stage 1 D/DBP Rule. *Wat. Res.*, 38: 895-902.

Reckhow, D. A. and P. D. Pierce. 1992. *A Simple Spectrometric Method for the Determination of THMs in Drinking Water.* Denver, Colo.: AwwaRF and AWWA.

Simone, P., S., G.T. Anderson, and G.L. Emmert. 2006. On-line Monitoring of μg/L Levels of Haloacetic Acids using Ion Chromatography with Post-Column Nicotinamide Reaction and Fluorescence Detection. *Anal. Chim. Acta.* 570, 259-266.

Skoog, D. A., F. J. Holler and T. A. Nieman. 1998. *Principles of Instrumental Analysis.* 5th ed. Orlando, FL: Saunders College Publishing.

SRI Instruments. 2002. *SRI Instruments Products Operation and Service Manual.*

Taylor, J. K. 1987. Quality Assurance of Chemical Measurements. Chelsea, MI. Lewis Publishers

USEPA. 1990. *Method 551: Determination of Chlorinated Disinfection By-products and Chlorinated Solvents in Drinking Water by Liquid-Liquid Extraction and Gas Chromatography with Electron Capture Detector,* Cincinnati, OH: Environmental Monitoring and System Laboratory.

USEPA. 1992. *Method 552.1: Determination of Haloacetic Acids and Dalapon in Drinking Water by Ion Exchange Liquid Solid Extraction, Derivatization and Gas Chromatography with Electron Capture Detection*, Cincinnati, OH.: Environmental Monitoring and System Laboratory.

USEPA. 1995a. *Method 524.2: Measurement of Purgeable Organic Compounds in Water by Capillary Gas Chromatography/Mass Spectrometry,* Cincinnati, OH: Environmental Monitoring and System Laboratory.

USEPA. 1995b. *Method 502.2: Volatile Organic Compounds in Water by Purge and Trap Capillary Gas Chromatography with a Photoionization and Electrolytic Conductivity Detector in Series.* Cincinnati, OH: Environmental Monitoring and System Laboratory.

USEPA. 1995c. *Method 552.2: Determination of Haloacetic Acids Dalapon in Drinking Water by Ion Exchange Liquid Solid Extraction, Derivatization and Gas Chromatography with Electron Capture Detector,* Cincinnati, OH: Environmental Monitoring and System Laboratory.

USEPA. 1996a. Appendix B, Part 136-Definition and Procedure for the Determination of the Method Detection Limit-Revision 1.11. *Fed. Reg*. 40 CFR 136,303-306.

USEPA. 1996b. *DBP/ICR Analytical Methods Manual*, USEPA, Office of Water. USEPA 814-B-96-002, Cincinnati, OH.

USEPA. 2003a. *Method 552.3: Determination of Haloacetic Acids and Dalapon in Drinking Water by Liquid-Liquid Microextraction, Derivatization and Gas Chromatography with Electron Capture Detector.* Cincinnati, OH: Environmental Monitoring and System Laboratory.

USEPA. 2003b. *Stage 2 Disinfectant-Disinfection By-Products Rule-Significant Excursion Guidance Manual [Draft] [Online]*. Available: <http://www.epa.gove/safewater/stage2/index.html>. [Cited July 30, 2003].

Walpole, R. E. and Myers, R. H. 1989. *Probability and Statistics for Engineers and Scientists,* 4th ed. New York, NY: Macmillan. Wang, X., C. Saridara, S. Mitra. 2005a. Microfluidic Supported Liquid Membrane Extraction. *Anal. Chim. Acta.*, 543(1-2): 92-98

Wang, H., M. S. Simmons and R. A. Deininger. 1995. Application of a Dynamic Headspace Analysis for Trihalomethanes in Flowing Water. *J. Chromatogr. Sci.,* 33(-): 109-115.

Wang, X., C. Saridara, S. Mitra. 2005a. Microfluidic Supported Liquid Membrane Extraction. *Anal. Chi.. Acta.*, 543(1-2): 92-98

Wang, X., D. Kou, S. Mitra. 2005b. Continuous, On-Line Monitoring of Haloacetic Acids via Membrane Extraction. *J. Chromatogr. A*,. 1089: 39-44.

Wingrove, A.S. and R.L. Caret. 1981. *Organic Chemistry*. New York, NY: Harper & Raw.

Zygmunt, B. 1996. Determination of Trihalomethanes in Aqueous Samples by Means of a Purge-and-Trap System with On-Sorbent Focusing Coupled to Gas Chromatography with Electron-Capture Detection. *J. Chromatogr. A,* 725(1): 157-163.

ABBREVIATIONS

/	per
>	greater than
<	less than
±	plus or minus
”	inch
~	approximately
a	coefficient of moles of reactant
A	known concentration of the check standard
A_0	initial concentration of reactant A
ACS	American Chemical Society
A/D	analog to digital
Ag	silver
Ag (I)	silver (I) ion
AGC	automatic gain control
amt.	amount
ANOVA	analysis of variance
AR	Arkansas
Avg.	average
b	coefficient of moles of reactant
B	measured concentration of the check standard
B_0	initial concentration of reactant B
Ba	barium
Ba (II)	barium (II) ion
BCAA	bromochloroacetic acid
BDCAA	bromodichloroacetic acid
BOR	borate
Br_3CCHO	tribromoacetaldehyde
BrO_2	bromine dioxide
°C	degress Celcius (centigrade)
CA	California
Ca (II)	calcium (II) ion
$CaCl_2$	calcium chloride
CAR	carbonate
CCl_4	carbon tetrachloride
Ch	channel

$CHBr_3$	bromoform
CH_2Br_2	dibromomethane
$C_3H_6Br_2$	1,2-dibromopropane
$CHBrCl_2$	bromodichloromethane
$CHBr_2Cl$	dibromochloromethane
$CHCl_3$	chloroform
C_6H_5F	fluorobenzene
Cl_3CCN	trichloroacetonitrile
$C_3H_4Cl_2O$	1,3-dichloroacetone
ClO_2	chlorine dioxide
cm	centimeter
°C/min	degrees Celsius per minute
CMS	capillary membrane sampling (sampler)
CMS-FIA	capillary membrane sampling flow injection analysis
CMS-FIA NCA-FL	capillary membrane sampling flow injection analysis with nicotinamide-fluorescence
CMS-GC	capillary membrane sampler-gas chromatograph(y)
CMS-GC1	portable CMS-GC (SRI 8610)
CMS-GC2	benchtop CMS-GC (Varian 3400)
CMS-GC-ECD	capillary membrane sampler-gas chromatograph with electron capture detection
Com	communications
Conc.	concentration
Corr.	corrected data
D	detector
$\overline{d}$	mean difference for an entire set of results for a particular analyte
DBAA	dibromoacetic acid
DBCAA	dibromochloroacetic acid
DBPs	disinfection by-products
DCAA	dichloroacetic acid
DELCD	dry electrolytic conductivity detector
Dev	deviation
d_i	bias of results for the two methods
DVI	digital valve interface
Δ	radiationless transition
Δ_0	difference in initial concentrations and reactants
E1	eluent one
E2	eluent two

ECD	electron capture detector
ELCD	electrolytic conductivity detector (Hall)
EPA	Environmental Protection Agency
eV	electron volt
EX	extended study
ext	external
ext/int	external to internal
F	F-test value to determine significance of linearity
FAC	free available chlorine
FEP	fluorinated ethylene propylene
FIA	flow injection analysis
FL	fluorescence
FL/FC	fast loop/flow cell
FS	full scan mode
f. t.	film thickness
FTP	file transfer protocol
g	gram
GC	gas chromatograph(y)
GC-ECD	gas chromatograph(y) with electron capture detection
GCOV	gas chromatography on a valve
GEC	gas extraction cell
GND	electrical ground
H	Houston sampling site
H^+	hydrogen ion
HAA	haloacetic acid
HAA5	the five regulated haloacetic acids
HAA9	the nine brominated and chlorinated haloacetic acids
HAA-NCA-FL	haloacetic acid-nicotinamide-fluorescent product
HAAs	haloacetic acids
Hall	electrolytic conductivity detector
H_3BO_3	boric acid
HBr	hydrobromic acid
HCl	hydrochloric acid
He	helium
He_2	diatomic helium
HID	helium ionization detector
HOCl	hypochlorous acid
HPLC	high performance liquid chromatography
hr	hour

HX	Houston extended study
IC	ion chromatography
IC-MSCD	ion chromatography with membrane suppressed conductivity detection
IC-UV	ion chromatography with ultraviolet detection
i.d.	internal diameter
int	internal
int/ext	internal to external
K	thousand
k_1	rate constant of pathway one
k_2	rate constant of pathway two
k_3	rate constant of pathway three
kg	kilogram
k_{mix}	mixed second order rate constant
$k_{mix\text{-}calc}$	calculated mixed second order rate constant
KOH	potassium hydroxide
KOT	knitted open tubular
L	liter
LA	Louisiana
LC	liquid chromatography
LC-MS	liquid chromatography with mass spectrometry
LOC	"lab-on-a-chip"
LOQ	limit of quantification
LVI	large volume injection
LVI-IC-MSCD	large volume injection-ion chromatography with membrane suppressed conductivity detection
$\lambda_{emission}$	fluorescence emission wavelength
λ_{excite}	fluorescence excitation wavelength
m	meter
M	moles per liter
M1	Memphis one sampling site
M2	Memphis two sampling site
MΩ	mega ohm
mA	milliampere
MBAA	monobromoacetic acid
MCAA	monochloroacetic acid
MCL	maximum contaminant level
MΩcm	mega ohm-centimeters (resistivity units)
MD	Maryland

MDL	method detection limit
meq	milliequivalents
mg	milligram
mg/L	milligrams per liter
min	minute(s)
MLGW	Memphis, Light, Gas, and Water
mL/min	milliliters per minute
mM	millimoles per liter
mm	millimeter
MS	mass spectrometer
MSCD	membrane suppressed conductivity detection
MT1	mixing tee one
MT2	mixing tee two
MTBE	methyl *tert*-butyl ether
mV	millivolt
mV*s	millivolt times seconds
MX	Memphis extended study (site one)
μ	micro
μg/L	micrograms per liter
μL	microliter
μm	micrometer
μS	micro Siemens
μS*s	micro Siemens times seconds
μV	microvolt
μV*s	microvolt times second
n, N	number of analyses or replicates
N_2	nitrogen
nA	nanoampere
NaOH	sodium hydroxide
$Na_2S_2O_3$	sodium thiosulfate
NCA	nicotinamide
NCA-FL	nicotinamide-fluorescence chemistry
n_{chk}	number of times a check standard was analyzed consecutively
$NHCl_2$	dichloroamine
NH_2Cl	monochloroamine
NH_4Cl	ammonium chloride
^{63}Ni	nickel-63
nm	nanometer
NOM	natural organic matter
n_{std}	number of calibration standards analyzed

OCl^-	hypochlorite ion
o.d.	outer diameter
OPTGC	on-line purge and trap gas chromatograph
OPTGC-DELCD	on-line purge and trap gas chromatograph with dry electrolytic conductivity detection
Org.	original (original data before cited corrections were made)
PAC	Project Advisory Committee
PC	personal computer
PC-IC	post column-ion chromatography
PC-IC NCA-FL	post column-ion chromatography with nicotinamide-fluorescence
PDECD	pulsed discharge photoionization detector operated in electron capture mode
PDHID	pulsed discharge in helium ionization mode
PDPID	pulsed discharge photoionization detector
PDS	primary dilution standard
PEEK	polyetheretherketone
PK	peak
PP	peristaltic pump
PP1	peristaltic pump one
PP2	peristaltic pump two
psi	pounds per square inch
pS*s	pico Siemens time seconds
PT	purge and trap
PTFE	polytetrafluoroethylene
PT-GC-ELCD/PID	purge and trap-gas chromatography with electrolytic conductivity detection in series with photoionization detection
PT-GC-MS	purge and trap-gas chromatography with mass spectrometry
PYD	pyridine
QA/QC	Quality Assurance/Quality Control
R1	reagent one
R2	reagent two
r^2	determination coefficient
RC	reaction coil
RC1	reaction coil one (hot)
RC2	reaction coil two (cold)
% REC	percent recovery calculated from check standard
% REC_{Bias}	percent recovery calculated from bias plot results from on-line monitoring studies
% $REC_{Monitor}$	percent recovery calculated from check standards run each shift during on-line monitoring studies

% RSD	percent relative standard deviation
[s]	concentration of internal standard
s	second
S	standard deviation of check standard analyses
s_b^2	variance of points about the function
sci.	science
SCMS	supported capillary membrane sampling probe
SCMS-GC	supported capillary membrane sampling-gas chromatograph(y)
SCMS-GC-ECD	supported capillary membrane sampling-gas chromatograph(y) with electron capture detection
SCMS-GC-MS	supported capillary membrane sampling-gas chromatograph(y) with mass spectrometry
SCMS-GCOV	supported capillary membrane sampling-gas chromatography on a valve
s.d.	standard deviation
s_d	standard deviation between two methods
SDS	secondary dilution standard
SFFS	stopped flow-fluorescence spectrophotometer
SIA	sequential injection analysis
SIM	selected ion monitoring mode
SPME	solid-phase microextraction
SS	stainless steel
SSS	stock standard solution
std.	standard
sub-μg/L	below microgram per liter
s_w^2	value of variance
t	time
$t_{(n-1,\ 1-alpha=0.98)}$	student's t-value for 98% confidence level with (n-1) degrees of freedom
T_1	sampling time one
T_2	sampling time two
TBAA	tribromoacetic acid
TCAA	trichloroacetic acid
$t_{calculated}$	value calculated by t-test
$t_{critical}$	value given from t-table for a two-tailed test
Temp.	temperature
THAA5	total haloacetic acid 5 concentration
THAA9	total haloacetic acid 9 concentration
THM	trihalomethane
THM4	the four regulated trihalomethanes
THMs	trihalomethanes
TN	Tennessee

TOC	total organic carbon
TP1	temperature program relay one
TP2	temperature program relay two
TTHM4	total trihalomethanes 4 concentration
TTL	transitor-transitor logic
U of M	University of Memphis
USEPA	United States Environmental Protection Agency
UV	ultraviolet
V	valve (in diagrams)
V	volt
Vis	Visible
VOCs	volatile organic carbon compounds
W	waste
WI	Wisconsin
[x]	concentration of analyte
XLVI	extra large volume injection
XLVI-IC-MSCD	extra large volume injection- ion chromatography with membrane suppressed conductivity detection
XLVI-IC-UV	extra large volume injection- ion chromatography with ultraviolet detection
XXLVI	double extra large volume injection
Y_0	initial instrument reading
Y_t	instrument reading at any time
Y_∞	final instrument reading

IWA Publishing's authorised EU representative for General Product Safety Regulations is Diane D'Arras, 15 rue Duret, 75116 Paris, France, e-mail: safety@iwap.co.uk.

Printed and bound by CPI Group (UK) Ltd, Croydon, CR0 4YY
17/09/2026
02218793-0010